COMPLEMENTARITY, DUALITY AND SYMMETRY IN NONLINEAR MECHANICS

Advances in Mechanics and Mathematics

Volume 6

COMPLEMENTARITY, DUALITY AND SYMMETRY IN NONLINEAR MECHANICS

Proceedings of the IUTAM Symposium

David Y. Gao
Department of Mathematics
Virginia Polytechnic Institute & State University
Blacksburg, VA 24061, U.S.A.
Email: gao@math.vt.edu

KLUWER ACADEMIC PUBLISHERS
Boston / Dordrecht / New York / London

Distributors for North, Central and South America:
Kluwer Academic Publishers
101 Philip Drive
Assinippi Park
Norwell, Massachusetts 02061 USA
Telephone (781) 871-6600
Fax (781) 871-6528
E-Mail <kluwer@wkap.com>

Distributors for all other countries:
Kluwer Academic Publishers Group
Post Office Box 322
3300 AH Dordrecht, THE NETHERLANDS
Telephone 31 78 6576 000
Fax 31 78 6576 474
E-Mail <orderdept@wkap.nl>

Electronic Services <http://www.wkap.nl>

Library of Congress Cataloging-in-Publication

Gao, David Y.
Complementarity, Duality and Symmetry in Nonlinear Mechanics: Proceedings of the IUTAM Symposium
ISBN HB 1-4020-7887-0
ISBN E-book 1-4020-7888-9

Printed on acid-free paper.

Printed in the United States of America

Contents

List of Figures

Preface

The concepts of complementarity and duality as used in our daily life mean the sort of harmony of two opposite parts through which they integrate into a whole. Symmetry and inner beauty in natural phenomena are bounded up with complementary duality, which has always been a rich source of inspiration in human understanding through the centuries, particularly in mathematics and science.

The study of complementarity, duality, and symmetry in mechanics has a long history since the well-known *Legendre transformation* was formally introduced in 1787. This elegant transformation plays a key role in complementary duality theory. In classical mechanical systems, each energy function defined in configuration space is linked via the Legendre transformation with a complementary energy in dual (source) space, and through which the Hamiltonian can be formulated. From geometrical point of view, Hamiltonian structure in convex systems shows a beautiful symmetry, which was widely studied by the founders of the subject. However, such a symmetry is broken in nonconvex systems. It turns out that in recent times, tremendous efforts and attentions have been focused on the role of symmetry and symmetry breaking in Hamilton mechanics in order to gain deep understanding into nonlinear and nonconvex phenomena.

In engineering mechanics, the complementary energy variational principle was first proposed by Hellinger in 1914. Since the boundary conditions were clarified by E. Reissner in 1953, complementary-dual variational principles and methods have been studied extensively for more than fifty years. The development of mathematical duality theory in convex variational analysis and optimization has a similar history since W. Fenchel proposed the well-known Fenchel transformation in 1949. After the revolutionary concepts of super-potential and subdifferential introduced by J.J. Moreau in the study of frictional mechanics, the modern mathematical theory of duality has been well-developed by celebrated mathematicians such as R. T. Rockafellar, F.H. Clarke, I. Ekeland and R. Temam. Mathematically speaking, in linear elasticity where the total potential energy is convex, the Hellinger-Reissner complementary variational principle in engineering mechanics is equivalent to a Fenchel-Moreau-Rockafellar type dual variational problem. The so-called generalized complementary variational principle is the well-known saddle Lagrangian duality theory, which serves the foundation for engineering finite element analysis and methods. During the last fifteen years, the so-called primal-dual interior point method has emerged as a revolutionary technic in mathematical programming and computational science.

Unfortunately, in nonlinear elasticity, where the total potential is usually nonconvex, the Fenchel transformation will produce a so-called duality gap between the potential energy and its dual function. This non zero duality gap shows that the Fenchel-Moreau-Rockafellar duality theory and method can be used only for convex systems. On the other side, the Hellinger-Reissner complementary energy principle holds for both convex and nonconvex problems. It is very interesting to note that in almost the same time period of Reissner, the generalized potential variational principle in finite deformation elastoplasticity was proposed independently by Hu Hai-chang (1955) and K. Washizu (1955). These two variational principles are perfectly dual to each other (with zero duality gap) and play important roles in large deformation mechanics and computational methods. The inner relations between the Hellinger-Reissner and Hu-Washizu principles were discovered by Wei-Zang Chien in 1964 when he proposed a systematic method to construct generalized variational principles in solid mechanics. Unfortunately, this important work was not able to be published due to the reason explained in the article by Gao-Lian Liu in this volume.

Mechanics and mathematics have been complementary partners since Newton's time and the history of science shows much evidence of the beneficial influence of these disciplines on each other. However, the independent developments on complementary-duality theory in mathematics and mechanics for more than a half century have generated a "duality gap" between the two partners. In modern analysis, the mathematical theory of duality was mainly based on the Fenchel transformation. During last three decades, many modified versions of the Fenchel-Moreau-Rockafellar duality have been proposed. One, the so-called relaxation method in nonconvex mechanics, can be used to solve the relaxed convex problems. However, due to the duality gap, these relaxed solutions are not equivalent to the real solutions to the nonconvex primal problems. Thus, tremendous efforts have been focused recently on finding the so-called perfect duality theory (i.e. without a duality gap) in global optimization. One the other hand, it seems that most engineers and scientists prefer the classical Legendre transformation. It turns out that their attentions have been mainly focused on how to use traditional Lagrange multiplier methods and complementary constitutive laws to formulate correctly complementary variational principles for numerical computation and application purposes. It is known the Hellinger-Reissner principle involves both the dual variable (the second Piola-Kirchhoff stress tensor) and the primal variable (displacement field), so it is not considered as a pure complementary (dual) energy principle. Though this principle has many important consequences in large deformation theory and computational mechanics, the extremality property of this well-known principles, as well as the Hu-Washizu principle, remained an open problem for more than forty years, and this raised many arguments in large deformation theory and nonconvex mechanics. Actually, this open problem has been completely solved recently, and it turns out that a potentially powerful canonical dual transformation and a so-called *triality theory* were proposed. This triality reveals a very interesting tri-duality pattern and inner beauty in nonconvex systems and natural phenomena, and will play important role in unified field theory and human understanding.

In view of the important roles of complementarity, duality, and symmetry played in nonlinear analysis and mechanics, and motivated by bridging the duality gap between mathematics and engineering science, the *IUTAM Symposium on Complementarity, Duality and Symmetry in Nonlinear Mechanics* was successfully held in Shanghai, China, during August 13-16, 2002. This IUTAM symposium brought together some of world's leading researchers in both mathematics and mechanics to provide an interdisciplinary but engineering flavored exploration of the field's foundation and state of the art developments. This volume contains 21 papers from selected lectures presented at the Symposium and a few invited papers by subject experts. These papers dealt with fundamental theory, methods, and applications of complementarity, duality and symmetry in multidisciplinary fields of nonlinear mechanics, including nonconvex/nonsmooth elasticity, dynamics, phase transitions, plastic limit and shakedown analysis of hardening materials and structures, bifurcation analysis, entropy optimization, free boundary value problems, minimax theory, fluid mechanics, periodic soliton resonance, finite element methods and computational mechanics. A special invited paper by Dr. Ken Chong, Director of Mechanics and Materials Program at the National Science Foundation, presented important research opportunities and challenges of the theoretical and applied mechanics as well as engineering materials in the exciting information age. It is expected that this book will provide a useful reference source not only for theoretical and applied mechanicians, but also for mathematicians, physicists, and engineering scientists who need to use the role of complementarity, duality, and symmetry in their work.

This book is dedicated to Professor Wei-Zang Chien, the President of Shanghai University, on the occasion of his 90th birthday. Professor Chien is one of founding masters of the singular perturbation theory as well as one of founders of Chinese modern mechanics and applied mathematics. During his exceptional academic career and social activities spanning over a half-century, Professor Chien has trained several generations of excellent scientists and educators, and has made profound influence on the development of certain fields in applied mathematics and mechanics, especially the field of generalized complementary variational methods.

Credit for this book publication is to be shared by all the authors who made their efforts for writing the high quality chapters. The success of this IUTAM symposium and the completion of this proceedings would not have been possible without the assistance from many colleagues and friends. As the symposium organizer and book editor, I wish to express my sincere appreciation to all those who helped. Special thanks go to Dr. Ken Chong at NSF, Professors Dick van Campen, Secretary-General of IUTAM, and Professor B. Freund at IUTAM Bureau for their guidance and generosity of financial support. Finally, I would like to thank John Martindale and his team, especially Miss Angela Quilici, at Kluwer Academic Publishers for their great enthusiasm and professional help in expediting the publication of this book.

DYG

This book is dedicated
to Professor Wei-Zang
Chien on the occasion
of his 90th birthday

Professor Wie-Zang Chien

Wei-Zang Chien and His Contributions to Applied Mathematics and Mechanics

Wei-Zang Chien is a world-renowned expert in applied mathematics, mechanics, education, Chinese information processing and social activities. He is one of the founders of modern mechanics in China as well as of the singular perturbation theory in the world. Though being 90 years old, he is still active both in scientific research and in educational and social affairs. He is now a professor and president of Shanghai University, director of Shanghai Institute of Appl. Maths. & Mechanics, honorary president of Jinan University and Nanjing University of Aeronautics & Astronautics, member of both the Chinese Academy of Sciences and the Polish Academy of Sciences, vice-chairman of the Chinese People's Political Consultative Conference and so on. In honor of his 90th anniversary a brief introduction of his contributions to science and education is given here for reference.

1. LIFE AND CAREER

W.Z. Chien was born on 9th October 1912, in the Village of Seven-Mansion Bridge, Wuxi, Jiangsu, China. His father, Shengyi Chien, was a primary school teacher and died when W.Z. Chien was only 16. His mother, Xiuzhen Wang, was a very kind and diligent woman. Apart from keeping the house, she raised silkworms, embroidered and made matchboxes to add to family income. Her diligence and the family's poverty taught W.Z. Chien to work hard and be self-reliant. His wife, Xiangying Kong, a fellow-student of his, graduated from the Department of Chinese Literature, National Tsinghua University, Beijing, and died only one year ago. Throughout their sixty years of married life, she had always been her husband's spiritual prop and supporter of his professional endeavors despite the difficulties they faced as a result of his being falsely labeled a political rightist in 1957. The couple had one son and two daughters. In September 1931, 19-year-old W.Z. Chien was admitted to Tsinghua University due to his excellent performance on literature and history in the entrance examination and was granted the Shanghai-based Yun-Chu Wu's Scholarship for Impoverished and Bright Students. Just 3 days after he entered the university, on 18th Sept. 1931, Japan's blitz aggressive occupation of Chinese Manchuria took place and motivated him to ask for transferring to the Department of Physics in order to make China stronger by arming her with modern science and technology. In that department he was greatly influenced by and benefited from some distinguished professors of physics, especially Youxun Wu and Qisun Ye. After graduation in 1935, he continued his study on the diffraction theory of X-ray in the graduate school of National Tsinghua University under the guidance of Prof. Youxun Wu for 2 years. He earned the Gao Men-dan Scholarship and became an intern researcher at the Institute of Physics of the Academia Sinica. In the period 1937-1939 he was giving lectures on thermodynamics in the National Southwestern Joint University and doing research on a unified intrinsic theory of elastic plates and shells.

In August 1939, WZ Chien, C.C.Lin and Y.H.Kuo passed an examination and were selected to study abroad with the support of Sino-British Boxer Indemnity. But when they got aboard the sea craft for Canada in Shanghai in January as required, they found there were Japanese visas on their passports. And they would pay a visit to Yokohama for three days. Chien and the others immediately decided to turn down the visas of the invaders and leave the craft, and they abandoned the chance of studying abroad. Till the early August of that year, they boarded a ship again and went to Canada. He entered Toronto University, studying elasticity under the guidance of Professor J.L. Synge, a well-known applied mathematician. C.C. Lin and Y.H. Kuo majored in fluid mechanics. During their first interview, Chien and Synge found they both were working on the unified intrinsic theory of elastic plates and shells. But there were differences: Synge was doing research on a macroscopic theory while Chien on microscopic theory. In 1942, Chien bridged the two theories and established the complete intrinsic theory of plates and shells, for which he got his PhD in Applied Mathematics. The theory has become one of the classic achievements in that field and had great and deep effects in scholastic circles. Furthermore, W.Z. Chien took part in a project of the National Research Council of Canada to study the radar antenna and proposed a method to calculate the resistance of radar wave-guides. He and Professor A. Weinstein jointly published a paper on the calculation of vibration frequency for clamped rectangular plates. Cooperating with Professor L. Infeld, he studied the problem of summing complicated trigonometric series.

At the end of 1942, Chien moved to California Institute of Technology, USA. There he, together with H.S. Tsien, C.C. Lin and Y.H. Kuo, worked in the Jet Propulsion Laboratory (JPL) under Prof. von Karman. Primarily, he was engaged in the aerodynamic design and calculation, and trajectory calculation of rockets, and the satellite orbit calculation, and so on. He also participated in some experimental work on the rocket launching spot. At the same time, he finished solving the torsion problem with variable twist and investigated the supersonic conical flow. That period, namely from 1942-1946, was his first prolific time of research.

1945 saw the victory of Chinese war against Japanese invasion. Missing his motherland, alma mater and relatives, Chien returned to China in May 1946. He accepted the appointment to be a professor of Mechanical Engineering in Tsinghua University, Beijing (Peking). He taught the engineering-major students applied mathematics, strength of materials, elasticity, theoretical mechanics, vibration theory, heat transfer, etc. From 1954, he also gave lectures on elasticity and applied mathematics to the teachers from universities, engineers and researchers in the city of Peking. In addition, he and Prof. Zhaolun Zeng, vice minister of the Higher Education Ministry, opened a "Graduate Class for Mechanics" sponsored by the Higher Education Ministry of China and Chinese Academy of Sciences. There he gave lectures on applied mathematics to the class. Simultaneously, he studied the large deflection theory and perturbation method for elastic plates, the rolling theory of metallic plates, hydrodynamic lubrication theory, snapping of shallow shells, etc. These problems are highly nonlinear, which made it extremely difficult to solve at that time when computers had just been invented and were still in their preliminary stage. In the corresponding papers, W.Z. Chien exhibited his outstanding creativity. We can say that that period was his second prolific time of research. Additionally, he presided over the Mechanics Division of the Institute of Mathematics of the Chinese Academy of Sciences from 1951 and took charge of setting up the Institute of Mechanics until the Institute was established in the early 1956. Then he was appointed deputy director of the Institute, of which H.S. Tsien was director. He was also vice-chairman of the board of directors of the Chinese Society of Theoretical & Applied Mechanics. In 1954, Chien was voted as an academician of Chinese Academy of Sciences. In 1955 he got the Second National Prize of Natural Science because of his contribution to the large deflection problem of circular thin plates. Later, in 1956, he became an academician of the Polish Academy of Sciences and vice president of Tsinghua University.

From 1957 when he was mistakenly labeled a rightist until 1976 when the Cultural Revolution ended, Chien was completely deprived of the rights to undertake normal scientific activities and publish papers, but he still stuck to academic research in such a difficult condition. His primary researches in this period included: the R&D of high energy zinc-air battery, the Lagrange multiplier method for generalized variational principle, the summing method for trigonometric series, armor penetration principle, the design of submarine keel and giant electro-machinery components and the stress analysis of chemical industrial tubes and plates, and so on.

The period, from 1977 up to now, was the most stable and happy one for W.Z. Chien both in work and in life, and is also his third academically prolific period. The main achievements included: generalized variational principle, finite element method, ring shell theory, the stress-strain analysis of corrugated tubes, Chinese information processing, the summing of trigonometric series, the non-Kirchhoff-Love theory of plates and shells, and so forth. During this period, he had published about 80 scientific papers and 15 monographs. For a person in his 70s to 90s, his vigorous energy, perseverance and creativity can be nothing but a miracle.

In addition, in 1982 his outstanding accomplishments on 'generalized variational principles' resulted in his getting the Second National Prize of Natural Science Award for the second time. In 1999, he obtained the HL & HL Science and Technology Achievement Prize for his preeminent contributions in academic research. From 1983, he became president of Shanghai University of Technology. During this period he founded the Shanghai Institute of Applied Mathematics & Mechanics and worked as director. In 1994, he became president of Shanghai University. Even as early as in 1978, he initiated and edited in chief the journal of *Applied Mathematics and Mechanics* (in both Chinese and English) and established the editor-recommending-examination system, which not only boosted the development of applied mathematics and mechanics in China, but also cultivated a lot of middle- and young-aged scientists.

So far Chien has published over 200 scientific papers (see the Appendix at the end of this paper), 18 monographs, and 500 other newspaper and magazine articles (mostly about scientific and educational problems). He served as editor-in-chief or editor of roughly 30 magazines, monographs or book series, dictionaries and encyclopedias. He is now editor-in-chief of *Applied Mathematics and Mechanics* (monthly, with English and Chinese editions), vice editor-in-chief of *Advances in Mechanics*, and editor or advisory editor of four international magazines

(*Thin-walled Structures, International Journal of Engineering Science, International Applied Mechanics, Finite Elements in Analysis & Design*) and many domestic magazines.

We will briefly go through the main achievements of W.Z. Chien hereinafter.

2. ACADEMIC CONTRIBUTIONS

2.1 Unified Intrinsic Theory Of Elastic Plates And Shells

(See Chien's publication list [A5, A6, A11, A12, A13, A33])[1]

Before 1941, the theory of elastic plates and shells had adopted various approximate simplifications in different situations so that their relationship to the three-dimensional elasticity theory was cut out. The situation was pretty disordered, so it was quite difficult to make a unified fair comparison and evaluation of their rationality and accuracy. For example, people treated plates quite differently from shells. They even handled the thin shells of different geometric shapes by different approximate methods. Generally, all methods began with the establishment of the equilibrium equations for the macroscopic internal force ingredients based on two-dimensional plate and shell elements. Then, the three assumptions of Kirchhoff-Love theory were used to determine the relationship between the internal force ingredients and the strains on the middle surface of the shell. In this way, three equilibrium differential equations that the three unknown displacements on the middle surface should satisfy were derived. Chien was long quite unsatisfied with these approximate theories of plates and shells and carried out an original study on this problem when he was still at home, in 1938-1939. He proposed to apply the microscopic stress-strain relation to rewrite the three-dimensional elasticity equilibrium equations in terms of the strain tensor. Simultaneously, he used the co-moving coordinates (x^0, x^1, x^2), namely the Lagrange coordinates (where x^1, $x^2$are the Gauss coordinates on the middle surface, and x^0 is perpendicular to them), and the tensor analysis to deduce the three equilibrium equations and three compatibility equations expressed by the extension tensor $p_{\alpha\beta}(\alpha,\beta=1, 2)$ and the curvature tensor $q_{\alpha\beta}$. Thus, an exact **microscopic** intrinsic theory of plates and shells was established.

In W.Z. Chien's first interview with his supervisor Synge in September 1940, he found they were studying the same problem. The only difference was that Synge considered the macroscopic intrinsic theory of plates and shells, in which he adopted the internal force ingredients (internal forces and internal moments) tensors of plates and shells to derive the (macroscopic) equilibrium equations. According to Synge's suggestion, Chien wrote one paper encompassing their two theories and had it published in a volume in honor of von Karman's 60th anniversary[A5]. What should be mentioned is that in that volume there were only 21 papers and the 26 authors except Chien, who was only 29, were all world leading scholars (including A. Einstein, von Neumann, von Mises, R. Courant, and S. Timoshenko and so on). This fact greatly stimulated Chien's courage and confidence to ascend the summit of science. That paper attracted considerable attention in the relevant academic circles. For example, Professor Rutten, a Holland dynamicist, sang high praise for it in his well-known book [1], believing that it inherited the work of Cauchy and Poisson of the early 19th century and added new vitality to the western literature on mechanics and it was the foundation of three-dimensional theory, because it used only the intrinsic variables (namely stresses and strains) characterizing the deformation state to rigorously derive from the 3-D theory the nonlinear equations that the thin shells of any shape should satisfy, and then transformed them to the two-dimensional approximate equations involving only 6 basic unknown variables ($p_{\alpha\beta}$ and $q_{\alpha\beta}$), and that were the most significant characteristics of Synge and Chien's work. In the 1950s and 1960s, this paper was referred to by scholars (such as E.L. Reiss, H.M. Mushturi, A.S. Wolmir) from many countries, including the United States and the Soviet Union. Later on, in 1942, Chien continued to deepen and extend his microscopic intrinsic theory in his doctoral dissertation (which was separately published in papers A11, A12, A13, A33). The main contributions were: (1) A systematic classification of plates and shells according to their thickness and curvature, which resulted in problems of 12 types for thin plates and 35 types for thin shells with different accuracy. Each problem was governed by six equations with $p_{\alpha\beta}$ and $q_{\alpha\beta}$ in them. They covered not only the conventional small displacement equations and some known large displacement equations, but also many unknown yet quite valuable equations, of which the flat shell equation was the most important. Thus, the latter equation, together with the circular cylindrical shell equation was widely named the "Chien's Equations" according to the suggestion in Ref. [2]. (2) Derivation of macroscopic equilibrium equations expressed in internal force ingredients by integrating the microscopic equilibrium equations along the thickness direction of plates and shells. This unified the Synge's microscopic theory with the Chien's macroscopic theory [A6,A33]. C. Truesdell also finished independently a similar work six years later, namely in 1948 and published it in the *Transactions of American Mathematical Society.* (3) In-depth analysis of the boundary effect in plates and shells. Later, in the 1960s, it had motivated a lot of study on the boundary effect in three-dimensional theories, including the work of A.E. Green, E. Reissner and P. Cicala.

In summary, Chien's intrinsic theory of plates and shells was the first rigorous, unified and exact theory. It pushed forward the theory of plates and shells to a new development stage. By 1989, it had been cited for over 100 times. Chien's corresponding papers were once the must-read materials for the postgraduates majoring in applied mechanics in the United States in the 1940s and 1950s and had greatly influenced the subsequent research work[3]. Even A.C. Eringen was inspired from Chien's intrinsic theory of plates and shells when he came to be engaged in founding rational mechanics.

2.2 Perturbation Theory[A29,A30,A32,A122]

In 1947, Chien in his paper [A30] initiated a perturbation method by using the central deflection w_m (scaled by the plate thickness h) as the perturbation parameter and successfully obtained for the first time the series solution to the nonlinear large-deflection differential equation von Karman deduced in 1910 for circular thin plates. This method has the following originalities.

(1) The perturbation parameter w_m in this method was unknown, contrary to the convention that the perturbation parameter should be a known value. The merit is that it improves a lot the convergence of the series. To elucidate this point, we analyze the series expression of the deflection w [A30,A41].

$$w = w_1(\eta) \cdot w_m + w_3(\eta) \cdot w_m^3 + w_5(\eta) \cdot w_m^5 + \cdots , \tag{2.1}$$

where $\eta = 1 - (r/a)^2$ and a is the outer radius of plates.

For the situation when the outer boundary of plates is clamped and simply supported, there are the boundary conditions at the central point $(\eta = 1)$ [A48],

$$w_1(1) = 1, w_{2i+1}(1) = 0, (\text{i=1, 2, 3, } \ldots\ldots) \tag{2.2a}$$

and simultaneously, from equation (2.1) we obtain

$$w(1) = w_m. \tag{2.2}$$

But at the outer radius $(\eta = 0)$, we have

$$w_{2i+1}(0) = 0, \quad (i = 1, 2, 3, \ldots\ldots). \tag{2.3}$$

From equations (2.2a), (2.2), and (2.3) we can see that for any finite value of w_m expansion Eq. (2.1) at the central point and the outer radius has only the first term(linear term) with all the higher order termsdisappeared. It shows that the selection of w_m as the perturbation parameter is very sensible and makes the perturbation solution converge most rapidly. But if we choose pressure load parameter $p[=\left(\frac{a}{h}\right)^4 \frac{q}{E}(1-\mu^2)]$ (where q is the pressure, μ the Poisson ratio, E the Young's modulus) as the perturbation parameter, as Vincent did in Ref. [4], the perturbation expansion of the deflection will be:

$$w = v_0(\eta) \cdot p + v_1(\eta) \cdot p^3 + v_2(\eta) \cdot p^5 + \cdots . \tag{2.4}$$

When the outer boundary is simply supported, we have at the central point

$$v_0(1) = \frac{3}{8}(0.5 + \beta),$$

$$v_1(1) = \frac{(1-\beta)}{184320\beta^2}(73 + 388\beta + 825\beta^2 + 840\beta^3 + 360\beta^4), \tag{2.5}$$

where $\beta = 2/(1+\mu)$.

From Eqs. (2.4) & (2.5), we can see that $v_1(1)$ doesn't equal zero. Ordinarily, $v_2(1)$, $v_3(1)$, etc do not equal zero, either. So, the relationship between w and p is nonlinear, and this just embodies the geometric nonlinearity of large-deflection problems. It's evident that the convergence of expansion Eq. (2.4) is worse than that of expansion Eq. (2.1). This Chien's method was widely used in the Soviet Union and called the Chien's Perturbation Method[5].

(2) The perturbation parameter w_m Chien chose does not appear at all in the von Karman's large deflection equation and in its boundary conditions. This turns out that it is quite similar to the method of artificial perturbation parameter. Ordinarily, the artificial parameter has no physical meaning[21], in contrast, the Chien's parameter w_m does have clear physical meaning so that it is very helpful to clarify and analyze the meaning involved in the solution results.

(3) Chien's result can also be obtained by the following renormalization method. First we get Eq. (2.4) according to Vincent by choosing p as the perturbation parameter. Then, substituting $\eta = 1$ into it, we get at the central point

$$w_m = w\,|_{\eta=1} = v_0(1) \cdot p + v_1(1) \cdot p^3 + v_2(1) \cdot p^5 + \cdots . \tag{2.6}$$

Then, inverting it, we obtain

$$p = \alpha_1 w_m + \alpha_2 w_m^3 + \alpha_3 w_m^5 + \cdots . \tag{2.7}$$

Substituting it into Eq. (2.4) to eliminate p will result in Chien's expansion Eq. (2.1).

(4) We can also treat Eqs. (2.4) and (2.6) as the consequence of the application of the strained parameter method. However, to get Chien's result, we have to obtain Eqs. (2.7) and (2.1) by inverting equation (2.6). Hence, we can regard Chien's perturbation method as the "inversion of the method of Strained Parameters".

The above (especially the points (1) and (2)) characteristics of Chien's perturbation method have greatly widened people's eyesight and techniques for choosing perturbation parameters and revealed his outstanding creativity and pushed forward a big stride the perturbation method research. The result agrees well with the experimental results of Mc Pherson, Rumberg and Levy [6].

When the deflection of plates and shells is very large, however, the above-mentioned method may still encounter convergence difficulty. In this situation, Chien studied in depth the problem of clamped circular thin plates under uniform pressure. He first found that there existed the boundary effect (similar to the boundary layer in fluid flow) that near the outer plates' boundary the variation rate of w is very large so that the conventional (regular) perturbation method did not work (the situation is just a typical one of the problems that later called the singular perturbation problems). To see clearly the variation there, he proposed using a "magnifier", that is, introducing a new magnified coordinate β instead of η and decomposed the solution into two parts. For example, write w in the following form:

$$w = F(\eta, \tau) + G(\beta, \tau), \tag{2.8}$$

where F is the outer deflection expressed by previous coordinate η and load parameter τ (proportional to$p^{1/3}$). G is the deflection correction term in the boundary layer expressed by the new amplified coordinate β and τ. Both F and G are expanded in series. First, using the boundary condition at $\eta = 1$ we get the outer perturbation solution F whose first term is just the Hencky's membrane solution. Then, we substitute Eq. (2.8) into the differential equation and utilize the boundary condition at $\eta = 0$ to get the inner corrected expansion in the boundary layer. The composite expansion Eq. (2.8) is the asymptotic solution that is valid for the whole plate area. This method is the method of composite expansion first suggested by Chien. It is obviously Chien's important creative work and is a breakthrough both conceptually and mathematically when compared with the Prandtl's method of matched asymptotic expansion. It is very effective and the numerical results agree well with the experiments of Ref. [6]. Similar methods [7-10] were published at least 8 years later than Chien's.

It is well known that the singular perturbation theory of boundary layer type stems from the Prandtl's method of matched asymptotic expansion (1905). The subsequent main advances are: Chien's composite expansion method (1948), Lighthill's strained coordinate method (or the PLK method) [11,12] (1949), Van Dyke's higher order matching law (1964). Obviously, Chien's method is the earliest one. Therefore, Chien is the first scientist who made splendid contributions to the development of singular perturbation theory. It should be mentioned that Lighthill himself advised that his method be used only for hyperbolic differential equations[13,14]. Chien's method, on the other hand, as proved in Refs. [A32, A122], is valid at least for elliptic equations (such as thin plate problems).

In addition, the method of boundary layer correction suggested by Lusternik et al. (1957) is widely used in Russia.[15] It is easy to see that its basic idea is quite similar to Chien's composite expansion method and it was published 9 years later than Chien's method.

In 1985, Chien and Chen improved the composite expansion method mainly in two aspects: (1) Choosing the central deflection w_m of thin plates as the perturbation parameter, which greatly increased the convergence speed. (2) All the boundary conditions were satisfied in all successive approximate solutions and the reliability of the results was improved.

In paper [A29] Chien used Karman-Moore's linear solution as the first order perturbation solution to the supersonic conical flow problem and got an asymptotic series solution by selecting conical angle ε and its following combinations as the perturbation parameters.

$$\varepsilon^2, \quad \varepsilon^4(\lg \tfrac{2}{\varepsilon}), \quad \varepsilon^4, \quad \varepsilon^6(\lg \tfrac{2}{\varepsilon})^2, \quad \varepsilon^6(\lg \tfrac{2}{\varepsilon}), \quad \varepsilon^6. \tag{2.9}$$

Generally speaking, it's rather hard to find the parameter combinations as (2.9), but Chien found them elegantly by expanding the Karman-Moore's linear solution in series. Chien assumed them to be six parameters with different scales (written as λ_1, λ_2, ..., λ_6 respectively) and developed the unknown solution φ into the following series with the multi-scales λ_i

$$\varphi = \sum_{i=1}^{6} \varphi_i(\theta)\lambda_i + \cdots .$$

It can be seen that the basic idea of this method is quite similar to the multi-scale expansion method first proposed in Ref. [16] in 1962.

But Chien used the multi-scale perturbation parameters instead of the multi-scale coordinates of Ref. [16]. So, we can call Chien's method "multi-scale parameter method".

From above it is obvious that Chien, through his papers [A29, A30, A32, A122], had made significant contributions, leading the world not only in the singular perturbation theory, but also in large deflection theory of plates.

2.3 Generalized Variational Principles and Finite Element Method[A55,A68,A108,A125,A127,A114,A117]

As early as in 1964, Chien put forward a systematic method to construct generalized variational principles[A55]. Starting from the potential energy principle and the complementary energy principle in elastostatics, he adjoined the variational constraints into the variational functional by Lagrange multipliers and then uniquely identified the multipliers by stationary conditions of the functional. So the generalized variational principle was rigorously deduced and the previous uniquely available try-and-error method for constructing generalized variational principles became outdated. This method is general in character, not limited to elasticity. Undoubtedly, it is an important contribution to the theory of generalized variational principles. But unfortunately, when paper [A55] was submitted to *Acta Mechanica Sinica*, it was rejected because of the referees' and editors' misunderstanding of the essence of the Lagrange multiplier method. Later, K. Washizu (1968, [17]) also proposed to use Lagrange multiplier method for deriving generalized variational principles, however, without going into details of some key points (such as how to identify multipliers). And only until 1977 did O.C. Zienkiewicz (in [18]) discuss it clearly and completely. But that was already 13 years later than Chien's work[A55]. After the 'reform and opening-up policy' of China issued in 1978, Chien was able to publish the paper [A68] which included and extended the main contents of [A55] supplemented by some new results on the finite element method. In 1982 Chien was awarded once again the Second National Prize of Natural Science of China.

In the further research on generalized variational principles, Chien found that sometimes the identified multiplier might be identically zero. For instance, using Lagrange multiplier method to derive the generalized variational principle from the Hellinger-Reissner variational principle, we encountered such a case that the stress-strain constraints could not be removed and hence the generalized variational principle could not be constructed. Chien called this phenomenon 'Variational Crisis'.

In Ref. [A108], Chien, extending the multiplier method, proposed the "High Order Lagrange Multiplier method" which can solve the problem of variational crisis completely. In this way, he successfully generalized the Hellinger-Reissner's variational principle and the Hu-Washizu's generalized variational principle to their corresponding more general forms. This is Chien's another breakthrough contribution to the generalized variational principles theory.

In 1987 and 1988, Chien in his papers [A125, A127] studied the variational principles of nonlinear (physically and/or geometrically) elasticity. What should be particularly mentioned are: (1) He established the stationary principle of complementary energy for large displacements in elasticity. It was admittedly recognized as an extremely difficult, unsolved problem[17]. (2) He successfully constructed the generalized variational principle families by using high order multiplier method.

Besides the variational principles in elasticity, Chien also made new achievements in the research on variational principles in fluid mechanics and magnetic field theory. For instance, he established the generalized variational principles and extremum principle of power dissipation for the Navier-Stokes equations of viscous flow by using weighted residual method and under proper simplification[A114]. He constructed all kinds of variational principles for the three-dimensional orthotropic nonlinear static magnetic field, especially the complementary energy principle and the derived therefrom generalized variational principle families that did not exist in the literature.

Chien not only laid a broad, rigorous and solid theoretical foundation for the finite element method via the above-mentioned research, but also added new creations to the finite element method itself. For instance, in Ref. [A96] he introduced the continuity condition at the interfaces of neighbor elements into the functional by using Lagrange multipliers. The multipliers identified by the stationary conditions of the functional were re-substituted into the functional to eliminate the multipliers. In this way, not only the degree of freedom is decreased, but also the stiffness matrix is simplified. It was a significant improvement to Pian & Tong's non-compatible hybrid element method and can be applied to non-elasticity problems. Moreover, in papers [A102-A104] on the finite element analysis of dynamic problems he put forward a new method to diagonalize the consistent mass matrix, which got rid of the shortcoming of lumped mass method due to insufficient theoretical justification and overcame the difficulty due to the non-diagonality of the consistent mass matrix. He proved that if the second order shape function is adopted the abovementioned difficulty can be avoided, and thus a very effective way to numerically solve the dynamics problems (such as impact, vibration problems etc.) is provided.

2.4 Ring Shell Theory And Its Application[A67,A78,A90]

The elastic analysis of ring shells has both theoretical and practical significance because the ring shell as a basic element of corrugated tubes is widely used in instruments and industrial piping systems. In his paper [A67] Chien derived in a unified manner the three known kinds of complex-variable equations (Tlke,1938, Clark, 1950 and Novozhilov, 1951) for axisymmetric circular ring shells. He also obtained the limit equation for thin ring shells (whose cross section radius is far less than the global ring radius) and the general solution to the homogeneous equation for the first time. Superposition of this solution with the Novozhilov's asymptotic solution to the non-homogeneous equation resulted in the general solution of thin ring shells. A seminar lecture based on Ref. [A67] was presented in Spring, 1984, at the Mechanical Engineering Department of California University, Berkley, which the attendants believed to be a significant contribution to the calculation of elastic components. The solution was put into use in the design of corrugated pipes and other instrument components and proved to be in good agreement with the existing experimental results[A78].

For axisymmetric ring shells (not limited to thin ring shells) there was no general solution for a long time until Chien et al. obtained in paper [A90] the general solution with great efforts. It was an important breakthrough in both ring shell theory and its application.

2.5 Torsion With Variable Twist[A27]

The variable-twist torsion problem of thin-walled cylinders with closed cross section (such as the case when the end is clamped so as to prevent the end surface from warping) was very difficult to solve without computers because in general the Saint-Venant solution for uniform-twist torsion cannot be used as the first approximate solution in the iterative solution procedure. But Chien noticed that in many practical problems, such as the torsion of wings with internal reinforcing ribs, the deformation of the cross section shape could be ignored while its warping is inevitable. This finding gained von Karman's complete affirmation. Based on this understanding, they let the axial stress and shearing stress satisfy the equilibrium equations and then successfully developed a practical theory for the variable-twist torsion problems[A27]. Von Karman was very satisfied with this paper, considering it to be full of classical character among all papers on elasticity that bore his name and that the solution procedure embodied the perfectness and conciseness of the classical applied mathematics.

2.6 Chinese Information Processing[A83,A107,A123]

Chien's in-depth academic background and creativity manifested not only in the field of applied mathematics and physics, but also in culture, literature and history. With the development and wide use of computers, he long keenly forecasted that in order to make full use of computers to the modernization of office work including writing, typing and communication in Chinese, it is necessary, first of all, to solve the most difficult and key problem of seeking a new advanced method for the input and output of Chinese characters. In 1980, he designed and manufactured a new Chinese character typewriter based on a strokes structure analysis of Chinese characters [A83,A107]. In 1986, he invented a new Chinese character coding method, namely the Chinese Character Macroscopic Stroke Coding method (abbreviated as Chien's Code). Before that, all coding methods of Chinese characters had the following two drawbacks: one is to try to avoid repeated codes, and the other is to input the Chinese single characters one by one, that is, one code per word. He first suggested adding a new "Choice Key" to overcome the repeated code difficulty. Then, he invented "Phrase Input Method" to simplify as well as accelerate the input of Chinese characters. Since then, these two innovations were widely adopted in most coding methods. Chien's Code was awarded the first grade prize in 1986 at a national evaluation meeting, and the first prize of Shanghai Science and Technology Progress Award.

In addition, Chien's research interests are very wide and concern many other academic fields, such as the summing problems of many kinds of trigonometric series[A66,A77]. He proposed a new method using Fourier transformation and compiled the table series "Summing of Fourier Series" consisting of over 10, 000 series. He studied also the bearing lubrication problem by perturbation method and deduced the Reynolds equation and higher order approximate equations from the Navier-Stokes equations and gave out their variational principles[A36]. A new conformal mapping method for determining the outflow angle of two-dimensional cascades was also proposed by Chien [A34]. He also developed a new rolling theory for thin plates [A37], in which he divided the contact area into 3 segments, and the solutions in each of them are treated separately by perturbation method and then are assembled by matching into a complete solution.

3. THOUGHTS ABOUT EDUCATION AND RESEARCH

The reason why Chien was able to make so great achievements is closely related to his thoughts and viewpoints about such problems as education, research, teaching, teaching and research as well as teaching and education. They were rather essential, original and complete, as can be seen hereinafter.

3.1 Thoughts on Education: The Dialectical Relationship Between Theory And Practice

From 1949 to 1957, while Chien was engaged in important teaching and administrative work (as vice-Dean of Studies, Dean of Studies, vice-president of Tsinghua University successively), he never stopped teaching personally all kinds of mechanics courses in science and engineering colleges. He even gave lectures on the theory of elasticity and applied mathematics to faculties in all the universities of Peking from 1954 to 1957. He taught again applied mathematics in the Graduate Class for Mechanics sponsored by the Chinese Academy of Sciences and Higher Education Ministry. The listeners amounted to 600.

After the end of the Cultural Revolution, from 1977 to 1987, Chien started lectures on variational methods and finite elements, tensor analysis, singular perturbation theory, armor penetrating mechanics, generalized variational principles, the application of Green function and variational method to the calculation of electromagnetic field and wave in Beijing, Shanghai, Lanzhou, Wuhan and so on. Some lectures had been repeated even for seven times. The total audience reached 3885 with good ovation. All these lectures reflected the latest and advanced research progresses of the world and Chien's own achievements, which later were put into books and published.

Through the teaching and lecture activities, especially the opening of the Graduate Class for Mechanics (two-year system), he had cultivated a whole generation of backbone college teachers and research staff in the discipline of mechanics. Among them, some have become academicians of Chinese Academy of Sciences or Chinese Academy of Engineering. This was really a strategic initiative. In 1996, he established the Department of Mechanics in Shanghai University, and opened the comprehensive class for applied physics, mathematics and mechanics. It was a new effort in this direction.

Chien's educational thoughts can best be summarized by his own saying 'to get rid of the four walls' (i.e. to promote the four combinations). They are as follows.

The wall between college and society. He insisted that the target of school education is to serve the country and society for the development and improvement of productivity and living. In other words, the teaching and research activities in school must be adapted to the demand of social economic construction and rapid scientific advancement. Theories must be combined with practice. He claimed that teaching and research must be geared to the demand of society, and win the understanding, trust and support of society. The students shall be acquainted with society through practice and social investigations, and by raising their consciousness and capability to serve the society.

The wall between specialties. The growth point of modern science all lies in its cross-disciplinarity. For the sake of this, people's standpoints must be changed, specialties developed and courses blended. As early as in the 1950s, he boldly advocated the combination of science and engineering colleges. The main purpose of this is to create favorable conditions for the mating of courses. Now, he further insists the cross development of science, engineering and humanities disciplines.

The wall between teaching and research. The basic target and task of universities is to educate talents of high quality. Teaching and research are two different but related ways to reach this target. Therefore, good universities should on the one hand be educational centers, on the other be research centers. But he objected to dividing the teachers into two parts, one for teaching and the other for research. Each teacher should be actively engaged in teaching and research simultaneously because research can boost teachers' teaching capability and inform them of new achievements through teaching materials, enhance students' interests and desire for creation and diligent study on their own. In other words, education can lay a solid foundation for basic theories and provide experimental techniques systematically for research. When a teacher is carrying on teaching work as well as research work, he/she is making contributions to our country in both educating talents and developing their own creativity.

The wall between teaching and learning. To get rid of this wall is to coordinate and improve the teachers' and students' two-way work. For this sake, the conventional educational thoughts and methodology, in which teachers instill knowledge to the students and students passively accept it, must be abrogated. Here the 'teaching' must be heuristic with the contents continuously renewed. Only in this way can students' initiative, enthusiasm for knowledge be cultivated and guided, and can the students really learn well. The initiative, as an internal factor, is the most important thing and differs from student to student. For the undergraduates, it is manifested in the fact that the students can learn on themselves and make notes. For the postgraduates, it symbolizes the ability by which they can search and read documents by themselves. For doctoral students, it means that they can raise new questions independently and find solutions on the basis of absorbing the state-of-the-art achievements at home and abroad. As far as students are concerned, the most critical thing of learning is exercising their initiative to think more (the genius come from diligence, while learning without thinking is useless.), grasp the method of independent thinking and learning, pay attention to moral qualities, and educate themselves to become morally, intellectually, physically and esthetically talented.

3.2 Thoughts on Academic Research

Chien believes that the purpose of research is to create something new. To tell a person if he/she is academically successful is to see whether he/she has creativity. Therefore he always requests the university to educate more talented persons with creative capabilities. He brought up an essential definition for "creation" to be used as a criterion for evaluating one's research achievements. Creation, he assumes, primarily shows in the three 'originalities': original problems, original ideas and original methods, of which original ideas are the most significant. In some work, new ideas and methods are put forward to solve new problems. It is the most excellent result. In another kind of work, new ideas are presented but the methods and problems are old. It is not too bad. And in still another kind of work, old ideas are used to solve new or old problems by new methods. This is not bad either. And still, there is a kind of work in which old ideas and methods are applied to solve new problems. It cannot be counted as too bad as well. The worst work is to solve old problems by using old ideas and old methods. We shall try our best to accomplish the first kind of research work.

For completing splendid creative work, Chien thinks the strategic and tactic respects should both be paid attention to. In strategy, the selection of new directions and new problems should first be made. To do this, one must have far-reaching eyesight, free and open brain, leading consciousness, overall and developing ideas. He/she has to face the world, face the future. In the tactic aspect, new ideas and methods should be proposed. To achieve this purpose, one ought to have highly scientific and dedicated spirit, work diligently and independently, seek truth from practices, and pay attention to methodology and never stop striving for the best. One must both be good at independent thinking and skillful learning and application, and communicate with others extensively. Chien himself is a good model in this aspect. As early as in

1938, he already foresaw that there would be more and more nonlinear problems and they would become increasingly important with the development of science and technology. They would, as he thought, become the central research task in mechanics and mathematics. Thus he chose the nonlinear intrinsic theory of plates and shells as his main research objectives. Later, he changed to study perturbation methods (in the period from the 1940s to 1960s), nonlinear elasticity variational principles and finite elements (in the period from the 1970s to 1990s), and many other fields for sixty years with great diligence and results.

Besides in the aspect of learning, Chien inherited the excellent tradition of Goettingen applied mathematics and mechanics school from Synge and von Karman, and developed it further. The Goettingen school believes that the task for applied mathematics is to solve practical problems and emphasizes that the understanding of the essence of physical processes is of primary importance. But it never hesitates to apply mathematic methods and always tries hard to use them in the most needed places. Necessarily, one even has to create new mathematic tools oneself. For the applied mathematicians and dynamicists, mathematics is nothing more than a tool for settling practical problems even though it is very important (see [23], pp. 589-590). Chien, attaching great importance to the coherent application of mathematics and physics in tackling practical problems, proposed the colligation of physics, mathematics and social science, as well as the colligation of antiquity and modernity. His invention of a new Chinese character input method (Chien's Code) is a good example for the first colligation. Later, he suggested using this colligation to investigate nonlinear social problems in finance and stock market. In addition, in 1983, he gave out the advice of 'removing sand by controlling water stream' method as used in ancient China (about 1,900 years ago) to solve the silting-up problem in Mawei harbor, Fujian province. It brought up unexpectedly good results and saved quite a lot of engineering cost. It was a model of the antiquity-modernity colligation.

4. CONCLUDING REMARKS

The above brief introduction adequately shows that Chien is academically versed, literarily talented, artistically versatile. He not only has got tactic eyesight and overrunning consciousness, but also worked diligently and practically, made great achievements, and turned out a good many publications. He well deserves to be one of the founders of Chinese modern mechanics and applied mathematics, and one of the founders of the singular perturbation theory in the world. His classic publications on nonlinear intrinsic theory of plates and shells have been one of the milestones in the history of this field. In the educational sphere of China, he has educated several generations of excellent scientific and teaching staff through his half-century effort, which has and will have produced profound stimulating effect in the development of science and technology in China. Though Chien has experienced many frustrations, he always remains calm, strives, and makes creations unremittingly and incessantly. This spirit fully shows that he is a patriotic scientist. At his 90th anniversary, we whole-heartedly wish he be well and long live and continue to lead us in gaining new accomplishments on the road to revitalize China through science and education.

Acknowledgment

I would like to express my warmest thanks to Prof. W.Z. Chien for his careful reading of this paper and for his very valuable suggestions and comments. I am also grateful to Prof. David Y. Gao (Virginia Tech., USA) for his invitation to write this paper. Thanks are also due to Prof. S.Z. Tang in the College of Foreign Languages, Shanghai University, for helping me check and improve the English version and to my graduate students: PhD candidate C. Chen and MS candidate Y. Tao for their great help in preparing the manuscript.

Gao-Lian Liu
Shanghai University
Shanghai Institute of Appl. Math. & Mechanics.

References

[1] Rutten H.S., The Theory and Design of Shells on the Basis of Asymptotic Analysis.1973.

[2] Fung Y.C. & Schler E.E., In Structure Mech. Symp. Proc. at Stanford Univ., Aug. 1958, edited by J.N.Goodier & N.J.Hoff.

[3] Gallagher R.H., Introductory Talk at the Plenary Session of the International Conf. on FEM, Aug. 2, 1982, Shanghai.

[4] Vincent J.J. Phil. Mag., vol.12 (1931) 185-196.

[5] Vol'mir A.S., Bending of Plates and Shells. Mir Press, Moscow, 1956.

[6] McPherson A.E., Ramburg W. & Levy S., NACA Report 744 (1942).

[7] Bromberg E., Commum. Pure & Appl. Math., vol.9 (1956)633-659.

[8] Srubshik L.S. & Ugowitz W.I., Trans. Academy of Sci., Soviet Union (DAN, USSR), Vol.139.No.2 1961.

[9] O'Malley R.E.Jr., SIAM Review, Vol.13 (1971)425-434.

[10] Latta G.E., Lectures on Advanced Ordinary Differential Equations, Stanford Univ. (1964).

[11] Lighthill M.J., Phil. Mag., Vol. 40 (1949) 1179-1201.

[12] Tsien, H.S., Advances Appl. Mech., Vol.4 (1956) 281-349.

[13] Van Dyke M., Perturbation Methods in Fluid Mechanics, Acad. Press, New York, 1964.

[14] Lin, Z.C. & Zhou, M.R., Perturbation Methods in Appl. Maths., Jiangsu Education Press, Nanjing, 1995.

[15] Vishik M.I. & Lusternik L.A., Advances in Mathematical Sci., Vol. 12, No.5 (1957) 3-122.

[16] Mahony J.J., J. Austral. Math. Soc. Vol. 2 (1962) 440-463.

[17] Washizu K., Variational Methods in Elasticity and Plasticity, Pergamon, 1968& 1975.

[18] Zienkiewicz O.C., The FEM, McGraw-Hill, 1977.

[19] von Karman Th, Bull. Amer. Math. Society, Vol. 46 (1940) 615-683

[20] Edson L., The wind and beyond, Th. von Karman, Pioneer in Aviation and Pathfinder in Space. Little Brown & Co., Boston, 1967.

[21] Liu G.L., Proceedings of 4th Intl. Conf. Nonlinear Mech., Shanghai 2002, edited by W.Z. Chien et al., Shanghai Univ. Press, pp. 829-835

[22] Chien W.Z., Selected Scientific Works of W.Z. Chien, Fujian People's Press, Fuzhou, 1989

[23] Chien W.Z., Selected Works of W.Z. Chien, Zhejiang S & T Publisher, Hangzhou, 1992

[24] Chien W.Z., Academic Works Secleted by the Author, Beijing Capital Normal University Press, 1994

[25] Chien W.Z., Thinking About Problems in Education and Teaching, Shanghai University Press, 2000

[26] Zhou W.B. & X.Y. Kung, Biography of W.Z. Chien. In [22], pp. 1-31

[27] Huang Q., & W.Z. Chien, – An appreciation. Applied Mathematics & Mechanics — Weizang Chien's 80th Anniversary Volume. Edited by Qian Huang & Lizhou Pan. Science Press, 1993, pp. 1-7

[28] Huang Q. & W.Z. Chien, Biography of Chinese Experts of Science and Technology: Volume: Engineering & Technology——Mechanics, Edited by Chinese Science & Technology Association. Chinese Science & Technology Press, 1993, pp. 166-195

[29] Dai S.Q., On W.Z. Chien's Study Idea And Academic Style. Shanghai University, October 2002.

Scientific Works of Wei-Zang Chien

1935

[1] Tsien, W.Z. and Ku, H.C., Atmospheric electricity in Peking, Annual Meeting of Chinese Society of Physics, Jun. 1935, Qingdao.

1937

[2] Tsien, W.Z., The Spectrum of Doubly Ionized Calcium (Ca III), *Chinese Journal of Physics*, 3 (1), 1-13, 1937.

1939

[3] Tsien, W.Z., Analysis of the Spectrum of Singly Ionized Cerim, *Chinese Journal of Physics*, 4 (1), 89-116, 1939.

[4] Tsien, W.Z., Highly Ionized Potasium and Calcium Spectra, *Chinese Journal of Physics*, 4 (1), 117-147, 1939.

1941

[5] Chien, W.Z. and Synge, J.L., The Intrisic Theory of Elastic Shells and Plates, Theodore von Karman Anniversary Volume, Applied Mechanics, 103-120, 1941.

1942

[6] Chien, W.Z., The intrinsic Theory of Elastic Shells and Plates, Ph.D Dissertation, University of Toronto, Canada, 1942.

[7] Chien, W.Z., Summery of Thesis, The Intrinsic Theory of Elastic Shells and Plates, Programme of the Final Oral Examination for PhD (Applied Mathematics), University of Toronto, Oct. 17, 1942, in the Senate Chamber.

1943

[8] Chien, W.Z., The Resistances of Antennae of Various Shapes and Positions in Rectangular and Circular Wave Guides, National Research Council of Canada, Special Committee on Applied Math., Radio Report No. 5, Feb., 1943. 13 pages.

[9] Chien, W.Z., The Reactance, Matching Conditions and Matching Resistance of a Circular Wave Guides in the Cases of a E Wave, National Research Council of Canada, Special Committee on Applied Math., Radio Report No. 6, Mar., 1943, 18 pages.

[10] Weinstein, A. and Chien, W.Z., On the Vibrations of a Clamped Plates under Tension, *Quarterly of Applied Mathematics*, 1 (1), 61-68, 1943.

1944

[11] Chien, W.Z., The Intrinsic Theory of Thin Shells and Plates, Part I: General Theory, *Quarterly of Applied Mathematics*, 1 (4), 297-327, 1944.

[12] Chien, W.Z., The Intrinsic Theory of Thin Shells and Plates, Part II: General Theory, *Quarterly of Applied Mathematics*, 2 (1), 43-59, 1944.

[13] Chien, W.Z., The Intrinsic Theory of Thin Shells and Plates, Part III: General Theory, *Quarterly of Applied Mathematics*, 2 (2), 120-135, 1944.

[14] Chien, W.Z., The Trajectoris of Missile XF10S1000, Progress Report, No. 4-1, Jet Propulsion Laboratory (JPL), Guggenheim Aeronautical Laboratory, California Instititute of Technology (CALCIT), Feb. 1944, Pasadena, California, USA.

[15] Chien, W.Z., The Differential Correction of Rocket Trajectories, Progress Report, No. 4-2, JPL, CALCIT, April, 1944, Pasadena, California, USA.

[16] Chien, W.Z., The Trajectories of Missile WacCorporal, Progress Report, No. 4-5, JPL, CALCIT, June, 1944, Pasadena, California, USA.

[17] Chien, W.Z., The testing report of missile WacCorporal, Progress Report, No. 4-8, JPL, CALCIT, Sept. 1944, Pasadena, California, USA.

1945

[18] Chien, W.Z., The Trajectories of Missile XF30L2000 (corporal), Progress Report, No. 4-7, JPL, CALCIT, Jan., 1945, Pasadena, California, USA.

[19] Chien, W.Z., The Dynamics of Parachute, Progress Report, No. 4-9, JPL, CALCIT, Feb., 1945, Pasadena, California, USA.

[20] Chien, W.Z., The Calculation of Satellite Orbites, Progress Report, No. 7-1, JPL, CALCIT, April, 1945, Pasadena, California, USA.

[21] Chien, W.Z., The Loss of Altitude per Revolution in Satellite Orbit, Progress Report, No. 7-3, JPL, CALCIT, June, 1945, Pasadena, California, USA.

[22] Chien, W.Z., The Trajectories of Missile Corporal E (11, 000 1b), Progress Report, No. 4-12, JPL, CALCIT, Oct., 1945, Pasadena, California, USA.

[23] Chien, W.Z., The Aerodynamic Coefficients of Missile WacCorpora, Progress Report, No. 4-15, JPL, CALCIT, Nov., 1945, Pasadena, California, USA.

[24] Chien, W.Z., Preliminary Report on the Snapping Pressure of a Thin Spherical Cap, 1945, unpublished. The main parts of this paper

are included in detail by H.C. Hu in his paper "The Snapping Problem of a Thin Spherical Cap." This paper appears in Chien, W.Z., Lin, H.S., Hu, H.C., Yeh, K.Y. "The Large Deflection Problems of Elastic Thin Spherical Cap", Science Press, Beijing, 1954, 76-98 (in Chinese).

[25] Chien, W.Z., The Design Formulae of Snapping Pressure of a Thin Spherical Cap, 1945 unpublished. The main parts of this paper are included in detail by H.C. Hu in his paper "The Snapping Problem of a Thin Spherical Cap." This paper appears in Chien, W.Z., Lin, H.S., Hu, H.C., Yeh, K.Y. "The Large Deflection Problems of Elastic Thin Spherical Cap", Science Press, Beijing, 1954, 76-98 (in Chinese).

1946

[26] Chien, W.Z., The Estimated Values of Aerodynamic Coefficients of the Corporal E, Progress Report, No. 4-20, JPL, CALCIT, Feb. 1946, Pasadena, California, USA.

[27] von Karman, Th. and Chien, W.Z., Torsion with Variable Twist, *Journal of Aeronautical Sciences*, 13 (10), 503-510, 1946.

1947

[28] Infeld, L., Smith, V.G., and Chien, W.Z., On Some Series of Bessel Functions, *Journal of Mathematical Physics*, 26 (1), 22-28, 1947.

[29] Chien, W.Z., Symmetrical Conical Flow at Supersonic Speed by Perturbation Method, The Engineering Reports of National Tsinghua University, 3 (1), 1-14, 1947.

[30] Chien, W.Z., Large Deflection of a Circular Clamped Plate under Uniform Pressure, *Chinese Journal of Physics*, 7 (2), 102-113, 1947.

1948

[31] Chien, W.Z., Ho, S.T., Asymptotic Method on the Problems of Thin Elastic Ring Shell with Rotational Symmetrical Load, The Engineering Reports of National Tsinghua University, 3 (2), 71-86, 1948.

[32] Chien, W.Z., Asymptotic Behavior of a Thin Clamped Circular Plate under Uniform Normal Pressure at Very Large Deflection, The Science Reports of National Tsinghua University, Series A, 5 (1), 71-94, 1948.

[33] Chien, W.Z., Derivation of the Equations of Equillibrium of an Elastic Shell from the General Theory of Elasticity, The Science Reports of National Tsinghua University, Series A, 5 (2), 240-251, 1948.

[34] Chien, W.Z., The True Leaving for Diaphragm and Bucket Wheel with Curved Guides at the Discharge End, The Engineering Reports of National Tsinghua University, 4 (1), 78-102, 1948.

1949

[35] Chien, W.Z., Infeld, L., Pounder, J.R., Stevenson, A.F. and Synge, J.L., Contributions to the Theory of Wave Guides, *Canadian Journal of Research*, (A), 27, 69-129, 1949.

[36] Chien, W.Z., Hydrodynamic Theory of Lubrication for Plane Sliders of Finite Width, *Chinese Journal of Physics*, 7 (4), 278-299, 1949.

1953

[37] Chien, W.Z. and Chen, C.T., Theory of rolling, *Chinese Journal of Physics*, 9, 2, 57-92 (1953)(in Chinese); Acta Scientia Sinica, 1, 2, 192-229 (1953)

[38] Chien, W.Z., Continuous beams with non-uniform stiffness, *Chinese Journal of Physics*, 9, 3,170-182 (1953) (in Chinese); *Acta Scientia Sinica*, 2, 2, 116-226 (1953).

[39] Chien, W.Z., Assumptions of Saint-Venant's solution for the torsion of elastic cylinder, *Chinese Journal of Physics*, 9, 4, 215-220 (1953) (in Chinese); *Acta Scientia Sinica*, 3, 2, 165-170 (1953)

1954

[40] Chien, W.Z., General theory of large symmetrical deflection of thin circular plates, in "The Large Deflection Problems of Elastic Circular Plates", Special Reports of Mechanics Section in Institute of Mathematics, *Academia Sinica, Series A*, No. 1, Problem of Mechanics Series, 1st Volume, May 1954, 1-22 (in Chinese); The Russian Translation in Moscow, TEO, 11-37 (1957)

[41] Chien, W.Z., The perturbation methods on the large deflection problems of circular thin plate,in "The Large Deflection Problems of Elastic Circular Plates", Special Reports of Mehancics Section in Institute of Mathematics, *Academia Sinica, Series A*, No. 1, Problems of Mechanics Series, 1st Volume, May, 1954, 37-55(in Chinese); *Acta Sinica*, 3, 4, 405-436(1954); Russian Translation in Moscow, TEO, 56-78 (1957).

[42] Chien, W.Z. and Yeh, K.Y., On large deflection of circular plates, *Chinese Journal of Physics*, 10, 3, 209-238(1954) (in Chinese); *Acta Sinica*, 3, 4, 405-436(1954); Russian Translation in Moscow, TEO, 178-207(1957).

[43] Chien, W.Z., Lin, H.S., Hu, H.C., Yeh, K.Y., Large deflection problems of thin elastic circular plates, Vol. 1, Problems in Mechanics. Mechanics Section, Institute of Math., *Academia Sinica*, 132 pages, Published by Academia Sinica, 1954 (in Chinese); Translated to Russian, A.C., 1957, All together 207 pages.

1955

[44] Chien, W.Z. and Yeh, K.Y., Design data on the large deflection of circular thin plates, *Chinese Journal of Mechanical Engineering*, 3, 1, 15-32 (1955) (in Chinese).

1956

[45] Chien, W.Z., Chinese classic architecture, *History of Science and Technology*, 1, 124-136 (1956) (in Chinese).

[46] Chien, W.Z., The large deflection problem of elastic circular thin plates, Inaugural Meetingof Members of Academia Sinica, (1956) (in Chinese).

[47] Chien, W.Z., The Large Deflection Problem of Elastic Circular Thin Plates (English Abstract), Abstract of the paper presented in the Inaugural Meeting of Members of Academia Sinica (1956).

[48] Chien, W.Z., Problem of Large Deflection of Circular Plate, *Nadbitkaz Archiwum Mechaniki Stosowanej*, Warszawa, 8 (1), 3-12, 1956.

[49] Chien, W.Z., Yeh, K.Y., Elasticity, Science Press, Beijing, 422 pages (1956) (in Chinese); Second Edition (1980).

[50] Chien, W.Z., Lin, H.S., Hu, H.C., Yeh, K.Y., Theory of torsion of elastic columns, Science Press, Beijing, 492 pages (1956) (in Chinese).

1957

[51] Chien, W.Z. and Hu, H.C., On the Snapping of a Thin Spherical Cap, Presented in IX International Congress of Applied Mechanics, Brussels, 1956, and Published in the Proceedings IX International Congress of Applied Mechanics, Brussels, 17 pages, 1957.

[52] Chien, W.Z. and Yeh, K.Y., On the Large Deflection of Rectangular Plate, Presented in IX International Congress of Applied Mechanics, Brussels, 1956, and Published in the Proceedings of IX International Congress of Applied Mechanics, Brussels, 403-412, 1957.

1961

[53] Chien, W.Z., Lecture Notes on Aeroelasticity, for Postgraduate Student in Tsinghua University, 13 chapters, 469, pages, 1961.

1963

[54] Chien, W.Z., On the Kirchhoff-Love assumptions in the approximation theory of classical small deflection shell problems, Published by Strenth of Material Laboratory, Tsinghua University, 1963 and in "Selected Workds of Prof. Weizang Chien", Fujian Education Press, 1390 pages (1989) (in Chinese).

1964

[55] Chien, W.Z., The generalized variational principles of elasticity and their application to the plate and shell problem (1964), Unpublished in 1960s and published in "Selected Works of Prof. Weizang Chien", 1390 pages, Fujian Education Press (1989) (in Chinese).

[56] Chien, W.Z., Discussion on the semi-infinite plane elasticity problems with a concentric forceacting on the groove bottom surface, *Acta Mechanica Sinica*, 16, 3, 251-259 (1964) (in Chinese)

1972

[57] Chien, W.Z., Internal resistance and current distribution of the electrode in zinc air cells, Report of Zinc Air Research Group, Tsinghua University (1972) (in Chinese).

[58] Chien, W.Z., The temperature rising computation of zinc cell and zinc air battery, the ninth part of "General Report of Zinc Air Battery Research Group", Tsinghua University (1972) (in Chinese).

1973

[59] Chien, W.Z., Song, J.Y., Meng, X.F., Liao, S.S. (published in the name of Zinc Air Cell Research Group), The study of manufacturing of zinc air (oxygen) battery, *Journal of Tsinghua University*, 1, 37-53 (1973) (in Chinese).

[60] Chien, W.Z., The temperature rising and heat radiation of electric cells and batteries, Report of Zinc Air Cell Group, Tsinghua University, 1973 (in Chinese).

[61] Chien, W.Z., Song, J.Y., Meng, X.F., Liao, S.S. (published in the name of Zinc Air Cell Research Group), Development and test of zinc air batteries for motor vehicles, *Journal of Tsinghua University*, 1973, 4, 1-9 (in Chinese).

[62] Chien, W.Z., Environment pollution and envirionment protection in foreign countries, Institute of Chinese Science and Technology Information, March, 1973, 22 pages (in Chinese).

1974

[63] Chien, W.Z., Song, J.Y., Meng, X.F., Liao, S.S. (published in the name of Zinc Air Cell Research Group), Zinc air battery for signal hand-lamp in railway services, *Journal of Science and Technology of Tshinghua University and Beijing University*, 1, 1, 167-178 (1974) (in Chinese).

[64] Chien, W.Z., Environment pollution in capitalistic countries, *Environment Protection*, 1, 1, 32-35 (1974) (in Chinese).

1975

[65] Chien, W.Z., (Translated and Edited in the name of Zinc Air Batteries Research Group), Advances in air (oxygen) zinc batteries, 409 pages, Science Press, Beijing (1975) (in Chinese).

1978

[66] Chien, W.Z., On the summation of sonic trigonometrical series, *Journal of Tsinghua University*, 18, 4, 53-78 (1978) (in Chinese); Papers in Applied Mathematics and Mechanics, Jiangsu Press in Science and Technology, 228-232 (1980) (in Chinese).

[67] Chien, W.Z., Zheng, S.L., Equations of symmetrical ring shells in complex quantities and their general solutions for slender ring shells, 6th Conference on Elastic Elements, Shanghai, Dec. 1978, (in Chinese); *Journal of Tsinghua University*, 19, 1, 27-47 (1979) (in Chinese); Papers in Applied Mathematics and Mechanics, Jiangsu Press in Science and Technology, 11-30 (1980) (in Chinese).

[68] Chien, W.Z., The studies of generalized variational principles in elasticity and their applications in finite element computation, Science Report of Tsinghua University No. TH78011, Nov. 1978, (in Chinese); National Finite Element Method Conference of Chinese Society of Mechanical Engineering, Society of Aeronautical Engineering Society of Naval Architecture Engineering at Bengbu, Nov. 1978 (in Chinese); *Mechanics and Practice*, 1, 1, 18-27 (1979) (in Chinese); *Chinese Journal of Mechanics Engineering*, 15, 2, 1-23 (1979) (in Chinese).

[69] Chien, W.Z., Manufacture, design, experiment and theory of corrugated tubes, Shanghai Conference of 6th National Elastic Element, Dec, 1978 (in Chinese); Applied Mathematics and Mechanics, Jiangsu Press in Science and Technology 68-83 (1980) (in Chinese).

[70] Chien, W.Z., Calculation of semi-circular arc corrugated tube – application of theory of slender ring shells, National 6th Conference of Elastic Elements in Shanghai, Dec. 1978 (in Chinese); *Journal of Tsinghua University*, 19, 1, 84-89 (1979) (in Chinese); Papers in Applied Mathematics and Mechanics, Jiangsu Press in Science and Technology, 94-109 (1980) (in Chinese).

1979

[71] Chien, W.Z., Studies on convergence problems of power series solution of ring shell equation and their related theorems on series convergence (in Chinese), *Journal of Lanzhou University, Special Issue in Mechanics*, 1-38 (1979); Papers in Applied Mathematics and Mechanics, Jiangsu Press in Science and Technology, 31-68 (1980).

[72] Chien, W.Z., Shieh, Z.C., Gu, Q.L., Yang, Z.F. and Zhou, C.T., The superposition of the finite element method on the singularity terms in determining the stress intensity factors, Wuchang Conference of Fracture Mechanics, Nov. 1979 (in Chinese); *Journal of Tsinghua University*, 20, 2, 15-24 (1980) (in Chinese); Applied Mathematics and Mechanics, Jiangsu Press in Science and Technology, 184-192, (1980) (in Chinese); *Engineering Fracture Mechanics*, 16, 1, 95-103 (1982).

[73] Chien, W.Z., Variatonal method and finite elements, 1st Volume, 604 pages, Science Press, Beijing, 1979 (in Chinese).

[74] Chien, W.Z., Papers in applied mathematics and mechanics, 256 pages, (in Chinese), Jiangsu Press in Science and Technology, Nanjing, 1979.

1980

[75] Chien, W.Z., The retional bases of small deflection theory of thin shells, Applied Mathematics and Mechanics, Jiangsu Press of Science and Technology, 1-10 (1982) (in Chinese).

[76] Chien, W.Z., Proof of two integration formulae, Applied Mathematics and Mechanics, Jiangsu Press of Science and Technology, 228-232 (1980) (in Chinese).

[77] Chien, W.Z., The numerical table of$\sum_{k=1}^{\infty}[k \pm s/m]^{-1}\cos kx$, $\sum_{k=1}^{\infty}[k \pm s/m]^{-1}\sin kx$, Applied Mathematics and Mechanics, Jiangsu Press in Science and Technology, 233-265 (1980) (in Chinese).

[78] Chien, W.Z., Non-homogeneous solutions of slender ring shell equations and their applications in instruments and meters design, *Chinese Journal of Scientific Instruments*, 1, 1, 89-112 (1980); Applied Mathematics and Mechanics, Jiangsu Press in Science and Technology, 69-83 (1980) (in Chinese).

[79] Chien, W.Z., Shieh, Z.C., Zheng, S.L. and Wang, R.W., Shape function of a compatible triangular finite element and its related stiffness matrix, *Journal of Mechanics Engineering*, 16, 4, 1-11 (1980); Applied Mathematics and Mechanics, Jiangsu Press in Science and Technology, 110-126 (1980) (in Chinese).

[80] Chien, W.Z., Finite element analysis of axisymmetric elastic body problems, Applied Mathematics and Mechanics, Jiangsu Press in Science and Technology, 167-177 (1980) (in Chinese); *Appl. Math. and Mech.*, 1, 1, 25-36 (Chinese Edition), 23-34 (English Edition) (1980).

[81] Chien, W.Z., The explicit form of field function in tetrahedron element with 16 and 20 degrees of freedom, Applied Mathematics and Mechanics, Jiangsu Press in Science and Technology, 178-183 (1980) (in Chinese); *Appl. Math. and Mech.*, 1, 2, 153-158 (Chinese Edition), 159-164 (English Edition) (1980).

[82] Chien, W.Z., Applied mathematics and mechancs, Jiangsu Press in Science and Technology (1980).

[83] Chien, W.Z., The stroke analysis of Chinese characters, and preliminary design for Chinese character type-writer, *Yuwen Xian Dai Hua (Modernization of Chinese Language)*, 2, 159-172 (1980) (in Chinese).

[84] Chien, W.Z., Foreword for applied mathematics and mechanics, *Appl. Math. and Mech.*, 1, 1, 1-2 (Chinese and English Edition), (1980).

[85] Chien, W.Z., Foreword for singular perturbation theory and its applications in mechanics, Singular Perturbation Theory and Its Applications in Mechanics, Science Press (1980) (in Chinese).

[86] Chien, W.Z., Foreword for the Chinese edition of continuum physics, Edited by A.C. Eringen, Jiangsu, Press in Science and Technology (1980) (in Chinese).

[87] Chien, W.Z., Variational method and finite elements, 1st Volume, Science Press (1980) (in Chinese).

[88] Chien, W.Z and Dai, F.L., Effective elastic constrants for thick perforated plates, Applied Mathematics and Mechanics, Jiangsu Press

in Science and Technology, 193-201 (1980) (in Chinese); *Acta Mechanica Sinica*, 4, 264-371 (1981) (in Chinese).

[89] Chien, W.Z., Recent advances in finite element method, Hangzhou Conference of Computational Structure Mechanics, Oct. 1980, Hangzhou; Mechanics and Practics, 4, 4-11 (1980) (in Chinese).

[90] Chien, W.Z. and Zheng, S.L., General solutions of axial symmetrical ring shells, *Appl. Math. and Mech.*, 1, 3, 287-300 (Chinese Edition), 305-318 (English Edition) (1980).

[91] Chien, W.Z. and Zheng, S.L., Calculation for semi-circular arc type corrugated tube – applications for general solutions of ring shell equation, Chongqing conference of Elasticity and Plasticity, April 1980, Chongqing (in Chinese); *Appl. Math. and Mech.*, 2, 1, 97-111 (Chinese Edition), 103-115 (English Edition) (1981).

[92] Chien, W.Z., Lecture note on Singular Perturbation Theory, Tsinghua University for Graduate Students (1980) (in Chinese).

[93] Chien, W.Z., Tensor calculus, Translated from the first volume of Modern Continuum Physics, Edited by Eringen, A.C., Jiangsu Press in Science and Technology (1980) (in Chinese).

[94] Chien, W.Z. and Wu, M.D., The non-linear characteristics of U-shaped bellows calculations by the method of perturbation, Shamen (Omen) 7th Conference of Elastic Element, Nov. 1981, Shamen (in Chinese); *Appl. Math. and Mech.*, 4, 5, 595-608 (Chinese Edition), 649-665(English Edition) (1983).

[95] Chien, W.Z., Incompatible plate elements based upon the generalized variational principles, International Symposium of Hybrid and Mixed Finite Element in Honor of Prof. T.H.H. Pian, Atlanta, Georgia, USA, April 8-10, 1981, Hybrid/Mixed Finite Element Method, Edited by Atluri, S.N., Gallagher, R.H. and Zienkiewicz, O.C., John Wiley, 381-404 (1983).

[96] Chien, W.Z., Incompatible elements and generalized variational principles, Proceedings of Symposium on Finite Element Method, May 20-24, 1981, Hefei, Anhui Province, Science Press, Beijing, Gordon and Breach, Science Publisher, New York, 252-329; Advances in Applied Mechanics, Vol. 24, Edited by Hutchson, J.W. and Theodore Y. Wu, Academic Press (1984), 93-153.

[97] Chien, W.Z., On Chinese information processing problems, Mr. Xue-In Gong, Xue In's Visiting Scholar Lecture in New Asia College, Hong Kong Chinese University, May 14, 1981 (in Chinese); *New Asia Life Monthly* (1981) (in Chinese).

[98] Chien, W.Z., Generalized variational principle, Science Lectures in Guizhou Province, Sept. 1981, 1-8 (in Chinese).

[99] Chien, W.Z., Non-linear finite element, Science Lectures in Guizhou Province, Sept. 1981 (in Chinese); Lecture Notes In Shanghai University of Technology, Nov. 3, 1981 (in Chinese).

[100] Chien, W.Z., Wang, Z.Z. and Wu, Y.G., The symmetrical deformation of circular membrane under the action of uniform distributes loads in its central portion, *App. Math. and Mech.*, 2, 6, 599-611(Chinese Edition), 653-668(English Edition) (1984).

1982

[101] Chien, W.Z., and Feng, S.C., References in corrugated tube, bordan tube, curved tube, expansion joint, ring shell, revolution shell, Section of Mechanics, Tsinghua University, Sept. 1982 (in Chinese).

[102] Chien, W.Z., Dynamic finite element with diagonalized consistent mass matrix and elastic-plastic impact calculation, *Appl. Math. and Mech.*, 3,3,281-296 (Chinese Edition), 319-334 (English Edition) (1982); Proceedings of International Conference of Finite Element Methods, Aug. 2-6,1982, Shanghai, Edited by He Guang Qian and Cheung, Y.K., Science Press, Gordon and Breach Science Publisher, New York, 47-58.

[103] Chien, W.Z., Diagonalized consistent mass matrix and the dynamic finite element analysis of elasto-plastic impact in axisymmetric problems, *Appl. Math. and Mech.*, 3,4,429-448 (Chinese Edition), 469-489 (English Edition) (1982); Dalian International Symposium Mixed/Hybrid Finite Element Method, 1982, Aug.11-29,1982, Dalian.

[104] Chien, W.Z.,Compatible dynamic finite element with diagonalized consistent mass matrix, *Appl. Math. and Mech.*, 3,5,565-567 (Chinese Edition), 609-622 (English Edition) (1982); Dalian International Symposium on Mixed/Hybrid Element Method, Aug. 11-29, 1982, Dalian.

[105] Chien, W.Z., The analytical solution of G.I. Taylor's theory of plastic deformation in impact of cylindrical projectiles and its improvement, *Appl. Math. and Mech.*, 3,6,743-755 (Chinese Edition), 801-815 (English Edition) (1982); Collected papers in Theoretical Physics and Mech., Edited by Wang, Z.X., etc., Science Press, 73-90 (1982).

[106] Chien, W.Z., History, present situation and problems to be solved of armour penetration mechanics, *Appl. Math. and Mech.* 1, 1, 1-15 (1982) (in Chinese).

[107] Chien, W.Z., Chinese language and Chinese typewriter, Report on National Committee of Chinese People's Political Consultative Conference, Jan. 1982, 1-19.

1983

[108] Chien, W.Z., Method of high-order Lagrange multipliers and generalized variational principles of elasticity with more general forms

of functionals, *Appl. Math. and Mech.*, 4, 2, 137-150 (Chinese Edition), 143-157 (English Edition) (1983).

[109] Chien, W.Z., Further study on generalized variational principles in elasticity – discussion with Mr. Hai-Chang Hu on the problem of equivalent theorem, *Acta Mechanica Sinica*, 4, 325-340 (1983) (in Chinese).

[110] Chien, W.Z., On nonlinear mechanics, *Advance in Mechanics*, 12, 2, 119 (1983) (in Chinese).

1984

[111] Chien, W.Z., Mechanics of armour penetration, National Defense Press (1984) (in Chinese).

[112] Chien, W.Z., Generalized variational principles, Knowledge Press (1984) (in Chinese).

[113] Chien, W.Z., Generalized variational principles in elasticity, Engineering Mechanics in Civil Engineering, Edited by Boral, A.P., Chong, K.P., Vol. 1, Academic Press (1984).

[114] Chien, W.Z., Variational principles and generalized principles in hydrodynamics of viscous fluids, *Appl. Math. and Mech.*, 5, 4, 305-322 (Chinese Edition), 1281-1295 (English Edition) (1984); Special Symposium Held by Holland Academy of Science at Endhoven (1984).

[115] Chien, W.Z., Further discussion on generalized variational principle and non-conditional variational principle – answer to Mr. Hai-Chang Hu concerning the above problems, *Acta Mechanica Sinica*, 3, 461-468 (1984) (in Chinese).

[116] Chien, W.Z., Classification of variational principles in elasticity, *Appl. Math. and Mech.*, 5, 6, 765-770, (Chinese Edition), 1737-1743 (English Edition).

[117] Chien, W.Z., Variational principles of magnetic energy for the problems of orthotropic nonlinear static magnetic field and their generalized variational principles, *Journal of Shanghai University of Technology*, 3, 1-14 (1984) (in Chinese).

[118] Chien, W.Z., Fan, D.J. and Hwang, C., Comparison of the calculations of three-convolutions cirlular arc corrugated diaphragm by toroidal shell theory and by orthogonal anisotropic plate theory, *Appl. Math. and Mech.*, 5, 1, 41-48 (Chinese Edition), 1019-1027 (English Edition) (1984).

[119] Chien, W.Z., Theory of circumerenciallly rib-reinforces monocoque cylinder with arbitrary cross-section, *Journal of Shanghai University of Technology*, 1, 1-30 (1984) (in Chinese).

[120] Chien, W.Z., The asymptotic solution of circumferencially rib-reinforced monocoque cylinder with arbitrary cross-section (especially

elliptical section) under uniformly distributed external pressure, *Journal of Shanghai University of Technology*, 2, 1-40 (1984) (in Chinese).

1985

[121] Chien, W.Z., Involutory transformations and variational principles with multi-variables in thin plate bending problems, *Appl. Math. and Mech.*, 6, 1, 15-40 (Chinese Edition), 25-39 (English Edition) (1985).

[122] Chien, W.Z. and Chen, S.L., The solution of large deflection problems of thin circular plate by the method of composite expansion. *Appl. Math. and Mech.*, 6, 2, 103-120 (Chinese Edition), 103-118 (English Edition) (1985).

1987

[123] Chien, W.Z., Cao, J.L., Feng, L.S. and Zhou, H., Macroscopic character code of Chinese character, Proc. Of Int. Conf. On Information Professing of Chinese Character, 1987, Beijing, 1, 24-31.

[124] Chien, W.Z., Further study of generalized variational principle in elasticity. The Advances of Applied Math. and Mech. in China, 1, 1-13, (1987).

[125] Chien, W.Z., Variational principles in elasticity with non-linear stress-strain relation, *Appl. Math. and Mech.*, 8, 7, 567-577 (Chinese Edition), 589-601 (English Edition) (1987).

[126] Chien, W.Z., Preface, *Advances in Applied Mathematics and Mechanics in China*, 1 (1987).

1988

[127] Chien, W.Z., Variational principles and generalized variational principles for non-linear elasticity with finite displacement. *Appl. Math. and Mech.*, 9, 1, 1-11 (Chinese Edition), 1-12 (English Edition) (1988).

[128] Chien, W.Z., On Lagrange multipliers and its uniqueness problem, *Acta Mechanica Sinica*, 20, 4, 311-324 (1988) (in Chinese).

1989

[129] Chien, W.Z., Scientific discoveries in Chinese history, Revised and Enlarged Edition, Chongqing Press, 172 pages (1989) (in Chinese).

[130] Chien, W.Z., Applications of Green functions and variational methods in electromagnetic field and wave computations, Shanghai Press of Science and Technology, 256 pages (1989). Revised edition: Shanghai University Press, 2000.

[131] Chien, W.Z., Selected works of Chien Weizang, Fujian Educational Press, 1390 pages (1989)(in Chinese, English, and Russian).

[132] Chien, W.Z., Summation of trigonometric series by Fourier transformations, *Appl. Math. and Mech.*, 10, 5, 371-384 (Chinese Edition), 385-397 (English Edition) (1989).

[133] Chien, W.Z. and Ning, J., The finite element method, combined with dynamic photoelastic analysis to determine stress intensity factors,

Appl. Math. and Mech., 10, 10, 865-870 (Chinese Edition), 909-914 (English Edition) (1989).

[134] Chien, W.Z. Wang, G., Rectangular plate elements based upon generalized variational principles, *Appl. Math. and Mech.*, 10, 11, 947-953 (Chinese Edition), 997-1003 (English Edition).

[135] Chien, W.Z., Loo, W.D. and Wang, S., Regional variational principles and general variational principles in elasto-dynamics, *Acta Mechanica Sinica*, 21, 3 (1989) (in Chinese).

1990

[136] Chien, W.Z., Wang, G., A new element for thin plate bending with curvilinear boundary quadrilateral element, *Appl. Math and Mech.*, 11, 4, 295-300 (Chinese Edition), 315-320 (English Edtion)(1990).

[137] Chien, W.Z., Torsional rigidity of shells of revolution, *Appl. Math. and Mech.*, 11, 7, 373-382 (Chinese Edition), 403-412 (English Edition) (1990).

[138] Chien, W.Z., Equation in complex variables of axisymmetrical deformation problems for general shells of revolution, *Appl. Math. and Mech.*, 12, 1, 565-579 (Chinese Edition), 605-620 (English Edition) (1990).

1991

[138] Huang, Q., Chien, W.Z. and Hoa S.V., Dual combined energy principles and finite elements with assumed partial stress and strain fields, Proceedings of First Canadian International Composites Conference and Exhibition, Montreal, 1E7 1-8 (1991).

[139] Feng, T., Chien, W.Z. and Sun, H.J., Numerical simulating of turbulent flow of Newtownian fluid and non-Newtonian fluid in a 180 square sectioned bend, *Appl. Math. and Mech.*, 12, 1, 1-14 (Chinese Edition), 1-14 (English Edition) (1991).

1992

[141] Chien, W.Z., Selected Readings of Weizang Chien, Jiangsu Press in Science and Technology (1992).

[142] Chien, W.Z., Applied Mathematics, Anhui Press of Science and Technology (1992).

[143] Chien, W.Z., Foundation of strength computation in electric machinery, Anhui Press of Science and Technology. (1992).

[144]Chien Wei-zang, Pan Lizhou Liu Xiao-ming, Large deflection problem of a clamped elliptical plate subjected to uniform pressure, *Journal of Shanghai University of Technology*, 1992, 13 (1): 1-26

[145]Chien Wei-zang, Pan Li-zhou, Liu Xiao-ming, Large deflection problem of a clamped elliptical plate subjected to uniform pressure, *Applied Mathematics and Mechanics*, 1992, 13 (10): 855-871

1994

[146]Chien Wei-zang, On the nonlinear sciences, *Science in the World*, 1994, (10): 18-19

[147]Chien Wei-zang, Ru Xueping, Preliminary report on the theory of elastic plates with no Kirchhoff-Love assumptions, *Journal of Shanghai University of Technology*, 1994, 15 (1): 1-26

[148]Chien Wei-zang, Huang Qian, Feng Wei, 3-D numerical study on the bending of symmetric composite laminates, *Applied Mathematics and Mechanics*, 1994, 15 (1): 1-6

[149]Chien Wei-zang, Huang Qian, Feng Wei, Three dimensional stress analysis of symmetric composite laminates under uniaxial extension and in-plane pure shear, *Applied Mathematics and Mechanics*, 1994, 15 (2): 95-103

1995

[150]Chien Wei-zang, Ru Xueping, A further study of the theory of elastic circular plates with non-Kirchhoff-Love assumptions, *Applied Mathematics and Mechanics*, 1995, 16 (2): 95-106

[151]Chien Wei-zang, Approximation theory of three dimensional elastic plates and its boundary conditions without using Kirchhoff-Love assumptions, *Applied Mathematics and Mechanics*, 1995, 16 (3): 189-209

[152]Chien Wei-zang, The second order approximation theory of three dimensional elastic plates and its boundary conditions without using Kirchhoff-Love assumptions, *Applied Mathematics and Mechanics*, 1995, 16 (5): 381-402

[153]Chien Wei-zang, About the nonlinear sciences, *Journal of Nature*, 1995,17 (1): 1-3

1997

[154]Chien Wei-zang, Ru Xueping, Preliminary report on the theory of elastic circular plate with no Kirchhoff-Love assumptions, *Journal of Shanghai University*, 1997, (1):1-14

[155]Xu Kaiyu, Zhou Zhewei, Chien Wei-zang, The subharmonic bifurcations of a laminated anisotropic circular cylindrical shell, *Journal of Shanghai University (Natural Science)*, 1997, 3 (1): 88-92

[156]Chien Wei-zang, The first order approximation of non-Kirchhoff-Love theory for elastic circular plate with fixed boundary under uniform surface loading (I), *Applied Mathematics and Mechanics*, 1997, 18 (1): 1-17

[157]Chien Wei-zang, The first order approximation of non-Kirchhoff-Love theory for elastic circular plate with fixed boundary under uniform surface loading (II), *Applied Mathematics and Mechanics*, 1997, 18 (2): 95-103

[158]Chien Wei-zang, Sheng Shangzhong, The first rrder approximation of non-Kirchhoff-Love theory for elastic circular plate with fixed

boundary under uniform surface loading (III)—-Numerical Results, *Applied Mathematics and Mechanics*, 1997, 18 (5): 385-393

1998

[159] Chien Wei-zang, Personal Statement at my 80th anniversary, Haitian Publishing House, 1998

2000

[160] Chien Wei-zang, Thinking about problems in education and teaching, Shanghai University Press, 2000

2002

[161]Chien Wei-zang, Second order approximation solution of nonlinear large deflection problems of YongJiang Railway Bridge in Ningbo, *Applied Mathematics and Mechanics*, 2002, 23 (5): 441-451

[162]Chien Wei-zang, Some methods for the positive integral solution of the equation:$x^2+y^2=z^2$, *Span the Century ¡kuayue shiji¿*, Shanghai: Shanghai University Press, 2002: 146-156

[163]Chien Wei-zang, General solution of the positive integral solution of the equation: $x^2+y^2=z^2$, *Span the Century ¡kuayue shiji¿*, Shanghai: Shanghai University Press, 2002: 157-162

[164]Chien Wei-zang, On the constructing features and non-uniqueness of Chinese Magic Cube, *Span the Century ¡kuayue shiji¿*, Shanghai: Shanghai University Press 2002: 43-94

Chapter 1

MECHANICS AND MATERIALS: RESEARCH AND CHALLENGES IN THE TWENTY-FIRST CENTURY

Ken P. Chong
Director of Mechanics and Materials Program, Directorate for Engineering
National Science Foundation, Arlington, VA 22230, U.S.A.

E-mail: *kchong@nsf.gov*

Abstract: Mechanics and materials, closely related and complementary to each other, are essential elements in all of the transcendent technologies in the twenty-first century and in the new economy. The transcendent technologies include nanotechnology, microelectronics, information technology and biotechnology. Research opportunities and challenges of the theoretical and applied mechanics as well as engineering materials in the exciting information age are presented and discussed.

Keywords: nanomechanics, transcendent technologies, micromechanics, multi-scales, model-based simulation.

1. Introduction

The National Science Foundation (NSF) has supported basic research in engineering and the sciences in the United States for a half century and it is expected to continue this mandate through the next century. As a consequence the United States is likely to continue to dominate vital markets because diligent funding of basic research does confer a preferential economic advantage [1]. Concurrently over this past half century, technologies have been the major drivers of the U. S. economy, and as well, NSF has been a major supporter of these technological developments. According to the former NSF Director for Engineering, Eugene Wong, there are three *transcendental* technologies [2]:

- Microelectronics – Moore's Law: doubling the capabilities every two years for the last 30 years; unlimited scalability; nanotechnology is essential to continue the miniaturization process.
- Information Technology [IT] – NSF and DARPA started the Internet revolution about three decades ago; confluence of computing and communications.
- Biotechnology – molecular secrets of life with advanced computational tools as well as advances in biological engineering, biology, chemistry, physics, engineering including mechanics and materials.

By promoting research and development at critical points where these technological areas intersect, NSF can foster major developments in engineering. The solid mechanics and materials engineering (M&M) communities will be well served if some specific linkages or alignments are made toward these technologies. Some thoughtful examples for the M&M communities are:

• Bio-mechanics/materials	• Simulations/modeling
• Thin-film mechanics/materials	• Micro-electro-mechanical systems (MEMS)
• Wave Propagation	• Smart materials/structures
• Nano-mechanics/materials	• Designer materials

Considerable NSF resources and funding will be available to support basic research related to these technologies. These opportunities will be available for the individual investigator, teams, small groups and larger interdisciplinary groups of investigators. Nevertheless, most of the funding at NSF will continue to support unsolicited individual investigator proposals on innovative "blue sky" ideas.

2. IT AND NANOTECHNOLOGY

In addition to NSF, there is also a sense that U. S. Federal agency research support is increasingly being driven by broad systemic initiatives. One of these initiative is *Information Technology* (IT), listed above as one of the *transcendental* technologies. The former President's Information Technology Advisory Committee (PITAC) advised that an "immediate and vigorous information technology research and development (R&D) effort in Information Technology be initiated. IT is essential for the United States to have economic growth and prosperity in the 21st Century." PITAC (www.ccic.gov) concluded that current U. S. Federal support for research in IT is inadequate and these current efforts also take "a short-term focus for immediate returns." PITAC recommends IT R&D with long term priorities focusing on "software development that is far more usable, reliable, and powerful, scalable information infrastructures that satisfy the demands of large numbers of users, high-end computing systems with both rapid calculation and rapid data movement, and IT education and training for the citizenry." Achieving these ends requires diversified modes of research support to foster projects of broader scope, longer duration and emphasis on projects involving multiple investigators over several years. Of most importance is using these new information technologies to advance critical application domains for the benefit our nation and the world.

Initiated by the author [KPC], with the organization and help of researchers from Brown [K. S. Kim, et al], Stanford, Princeton and other universities, a NSF Workshop on Nano- and Micro-Mechanics of Solids for Emerging Science and Technology was held at Stanford in October 1999. The following is extracted from the Workshop Executive Summary. Recent developments in science have advanced capabilities to fabricate and control material systems on the scale of nanometers, bringing problems of material behavior on the nanometer scale into the domain of engineering. Immediate applications of nanostructures and nano-devices include quantum electronic devices, bio-surgical instruments, micro-electrical sensors, functionally graded materials, and many others with great promise for commercialization. The branch of mechanics research in this emerging field can be termed nano- and micro-mechanics of materials. A particularly challenging aspect of fostering research in the nano- and micro-mechanics of materials is its highly cross-disciplinary character. Important studies of relevance to the area have been initiated in many different branches of science and engineering. A subset of these, which is both scientifically rich and technologically significant, has mechanics of solids as a distinct and unifying theme. The presentations at the workshop, and the open discussion precipitated by them, revealed the emergence of a range of interesting lines of investigation built around mechanics concepts which have potential relevance to microelectronics, information technology, bio-technology and other branches of nanotechnology. It was also revealed, however, that the study of complex behavior of materials on the nanometer scale is in its infancy. More basic research, which is well coordinated and which capitalizes on progress being made in other disciplines, is needed if this potential for impact is to be realized. In addition to the expected benefit to the target areas, such research invariably advances other technologies, conventional or emerging, through a spill-over effect; this serendipitous benefit can also be anticipated from focused mechanics research in nano technology.

Recognizing that this area of nanotechnology is in its infancy, substantial basic research must be done to establish an engineering science base; this link between the discoveries of basic science and the design of commercial devices must be completed to realize the potential of this area. Such a commitment to nano-and micro-mechanics will lead to a strong foundation of understanding and confidence underlying this technology based on capabilities in modeling and experiment embodying a high degree of rigor. The potential of various concepts in nanotechnology will be enhanced, in particular, by exploring the nano- and micro-mechanics of coupled phenomena and of multi-scale phenomena. Examples of coupled phenomena discussed in this workshop include modification of quantum states of materials caused by mechanical strains, ferroelectric transformations induced by electric field and mechanical stresses, chemical reaction processes biased by mechanical stresses, and change of bio-molecular conformality of proteins caused by environmental mechanical strain rates. Multi-scale phenomena arise in situations where properties of materials to be exploited in applications at a certain size scale are controlled by physical processes occurring on a size scale which are orders of magnitude smaller. Important problems of this kind arise, for example, in thermo-mechanical behavior of thin-film nano-structures, evolution of surface as well as bulk nano-structures caused by various material defects, nano-indentation, nano-tribological response of solids, and failure processes of MEMS structures. Details of this workshop report can be found in:

[http://en732c.engin.brown.edu/nsfreport.html].

The following figure [Fig. 1] depicts the instruments and techniques that are available for experimental micro and nano mechanics, courtesy of K. S. Kim of Brown University.

Field of View (Gage Length in m)

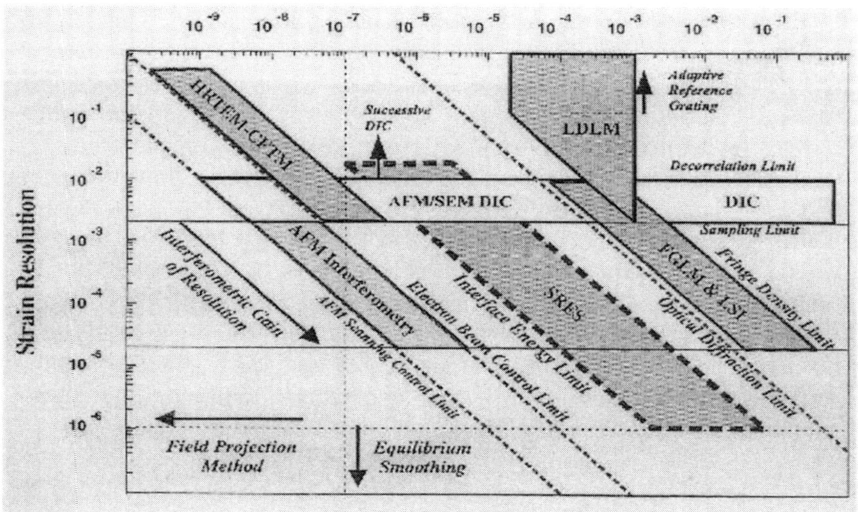

Fig. 1. Instruments and techniques for experimental micro and nano mechanics [courtesy of K. S. Kim of Brown University] where,

HRTEM High Resolution Transmission Electron Microscopy
LDLM Large Deformation Laser Moire
CFTM Computational Fourier Transform Moire
FGLM Fine Grating Laser Moire
AFM Atomic Force Microscopy
LSI Laser Speckle Interferometry
SEM Scanning Electron Microscopy
DIC Digital Image Correlation
SRES Surface Roughness Evolution Spectroscopy.

Coordinated by M. Roco, NSF recently announced a program [NSF 02-148; see: www.nsf.gov] on collaborative research in the area of nanoscale science and engineering. The goal of this program is to catalyze synergistic science and engineering research in emerging areas of nanoscale science and technology, including: biosystems at nanoscale; nanoscale structures, novel phenomena, and quantum control; device and system architecture; design tools and nanosystems specific software; nanoscale processes in the environment; multi-scale, multi-phenomena modeling and simulation at the nanoscale; and studies on societal implications of nanoscale science and engineering. This program or initiative provides support for: Nanoscale Interdisciplinary Research Teams (NIRT), Nanoscale Science and Engineering Centers (NSEC), and Nanoscale Exploratory Research (NER). Key research areas have been identified in advanced materials, nanobiotechnology (e.g. nano-photosynthesis), nanoelectronics, advanced healthcare, environmental improvement, efficient energy conversion and storage, space exploration, economical transportation, and bionanosensors. The National Nanotechnology Initiative (NNI - published on February 7, 2000 and is available on www.nano.gov) will ensure that investments in this area are made in a coordinated and timely manner (including participating federal agencies – NSF, DOD, DOE, EPA and others), and will accelerate the pace of revolutionary discoveries now occurring . In addition, individual investigator research in nanoscale science and engineering will continue to be supported in the relevant Programs and Divisions outside of this initiative.

3. DISCUSSION

Consistent with this theme, the National Science Foundation Directorate for Engineering in collaboration with other NSF Directorates and Federal agencies recently announced several IT-related initiatives. The NSF Directorate for Computer and Information Science and Engineering (CISE) and the Directorate for Engineering in cooperation with Division of International Programs jointly announced the *Wireless Information Technology and Networks* initiative (Program Announcement NSF 99-68) in early 1999. "The great demand for Internet services, wireless cable television distribution, and information technology, makes the development of broadband wireless mobile communication systems a national imperative in the 21st Century." Researchers face many technical challenges, but "data rates of tens of megabits per second (enabling broadband Internet access, for example) are apparently realizable in the near future."

The Engineering Directorate research initiative on *Engineering Microsystems: "XYZ on a Chip"* (Program Announcement NSF 99-31) focuses on non-electronic applications of microelectronic technologies and exploration of non-electrical processes at the micro-scale. Here *XYZ* refers to any non-electrical phenomena such as "biology, genomics, chemistry, optics, mechanics, materials, sensors, actuators, and software." Examples of solid mechanics and materials engineering *XYZ* related phenomena could include MEMS, nanomechanics, microactuation, etc.

The initiative *Engineering Sciences for Modeling and Simulation-Based Life-Cycle Engineering* (Program Announcement NSF 99-56) is a three-year collaborative research program by NSF and the Sandia National Laboratories (Sandia) focusing on advancing the fundamental knowledge needed to support advanced computer simulations. This collaborative initiative capitalizes on the missions of both organizations. NSF's mission is to advance the fundamental science and engineering base of the United States. Sandia has the responsibility to provide solutions to a wide range of engineering problems pertinent to national security and other national issues. It is moving toward engineering processes in which decisions are based heavily on computational simulations [see e.g. Ref. 3]; thus, capitalizing on the available high performance computing platforms. This initiative has sought modeling and simulation advances in key engineering focus areas such as thermal sciences, mechanics and design. About 24 awards were made totaling $7m.

The NSF-wide research initiative, *Biocomplexity: Research on the Functional Interrelationships between Microorganisms and Biological, Chemical, Geological, Physical and Social Systems* [NSF 00-22], is a long-term effort to support "integrated research on the functional interrelationships between microorganisms and the biological, chemical, geological, physical, and/or social systems that jointly comprise complex systems." Understanding biological complexity requires sophisticated approaches of integrating scientists from a range of disciplines. These include: "biology, physics, chemistry, geology, hydrology, social sciences, statistics, mathematics, computer science and engineering (including mechanics and materials)." These collaborations cannot be constrained by institutional, departmental or disciplinary boundaries.

The NSF Civil and Mechanical Systems (CMS) Division developed an initiative on *Model-based Simulation* (MBS), see (NSF 00-26). Model-based simulation is a process that integrates physical test equipment with system simulation software in a virtual test environment aimed at dramatically reducing product development time and cost. This initiative will impact many civil/mechanical areas: "structural, geotechnical, materials, mechanics, surface science, and natural hazards (e.g., earthquake, wind, tsunami, flooding and landslides)." MBS would involve "combining numerical methods such as finite element and finite difference methods, together with statistical methods and reliability, heuristics, stochastic processes, etc., all combined using super-computer systems to enable simulations, visualizations, and virtual testing." Expected results could be fewer physical testing, or at best, better strategically planned physical testing in the conduct of R&D. Examples of the use of MBS in research, design and development exist in the atmospheric sciences, biological sciences, and the aerospace, automotive and defense industries. The manufacturing of the prototype Boeing 777 aircraft, for example, was based on computer-aided design and simulation.

In the future one should expect the continued introduction of bold innovative research initiatives related to important national agenda such as the environment, civil and mechanical infrastructure, the service industry and the business enterprises.

4. CHALLENGES

The challenge to the mechanics and materials research communities is: How can we contribute to these broad-base and diverse research agenda? Although the mainstay of research funding will support the traditional programs for the foreseeable future, considerable research funding will be directed toward addressing these research initiatives of national focuses. At the request of the author [KPC] a NSF research workshop has been organized by F. Moon of Cornell University to look into the research needs and challenges facing the mechanics communities. Findings of the workshop are to be published and/or posted in the web.

Mechanics and materials engineering are really two sides of a coin, closely integrated and related. For the last decade this cooperative effort of the M&M Program has resulted in better understanding and designed of materials and structures across all physical scales, even though the seamless and realistic modeling of different scales from nano-level to system integration-level (Fig. 2) is not yet attainable. In the past, engineers and material scientists have been involved extensively with the characterization of given materials. With the availability of advanced computing, and new developments in material sciences, researchers can now characterize processes, design and manufacture materials with desirable performance and properties. One of the challenges is to model short-term micro-scale material behavior, through meso-scale and macro-scale behavior into long-term structural systems performance. Accelerated tests to simulate various environmental forces and impacts are needed [NSF 98-42]. Supercomputers and/or workstations used in parallel are useful tools to solve this scaling problem by taking into account the large number of variables and unknowns to project micro-behavior into infrastructure systems performance, and to model or extrapolate short term test results into long term life-cycle behavior [4, 5, 7].

MATERIALS		**STRUCTURES**		**INFRASTRUCTURE**
nano-level (10^{-9})	**micro-level (10^{-6})**	**meso-level (10^{-3})**	**macro-level (10^{+0})**	**systems-level (10^{+3}) m**
Molecular Scale	***Microns***		***Meters***	***Up to Km Scale***
***nano-mechanics**	***micro-mechanics**	***meso-mechanics**	***beams**	***bridge systems**
***self-assembly**	***micro-structures**	***interfacial-structures**	***columns**	*** lifelines**
***nano-fabrication**	***smart materials**	***composites**	***plates**	***airplanes**

Fig. 2. Physical scales in materials and structural systems [6]

An interesting biomechanics research project involves the multiscale mechanics of bone by Iwona M. Jasiuk of the Georgia Tech , Atlanta, GA. This NSF research award investigates the mechanics of bone, treating bone as a multiscale material. A major goal of the research is to be able to pass from the molecular level comprising thousands of mineral fibers and collagen matrix (micrometer scale) to an effec tive continuum level (scale of mm) – see Fig. 3. Local fields and constitutive laws of bone for elastic, inelastic, and fracture/damage regimes, on those larger scales will be sought. The project will involve numerical and analytical investigation but clo se contact with the experiments being concurrently conducted at Emory University will be maintained. These experiments, along with already published results, will provide a basis for verification of the computational micromechanics modeling.

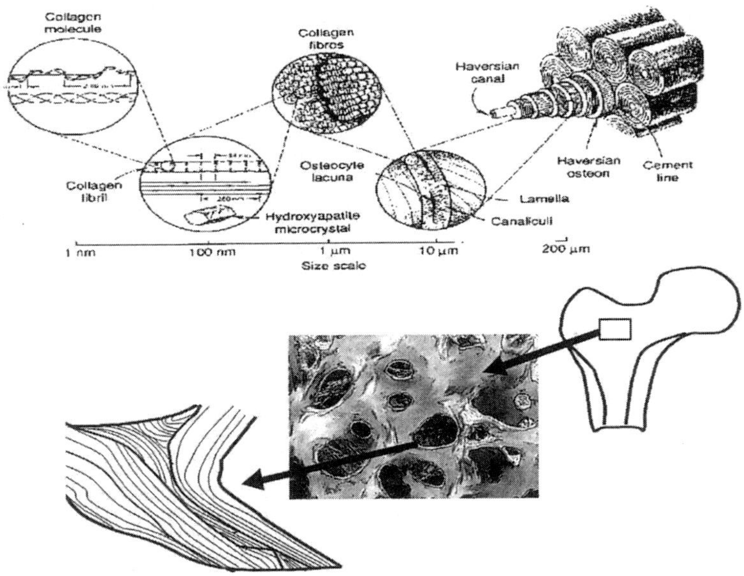

Fig. 3. Different scales in the bone structure [Courtesy of Iwona M. Jasiuk, Georgia Tech]

5. Acknowledgments and Feedback

The authors would like to thank their colleagues and many members of the research communities for their comments and in puts during the writing of this opinion paper. We would appreciate further feedback from the research communities at large. Information on NSF initiatives, announcements and awards can be found in the NSF website:[www.nsf.gov].

6. Disclosure

The opinions expressed in this article are the authors only, not necessarily those of the National Science Foundation nor representative of their positions at NSF. This paper was written by a U.S. Government employee.

REFERENCES

[1] Wong, E. "An Economic Case for Basic Research", *Nature*, Vol. 381, pp.187-188, May. (1996)

[2] Wong, E. "Engineering the Service Industry," www.eng.nsf.gov/engnews/1999, 5 pp. (1999).

[3] Boresi, A. P., Chong, K. P. and Saigal, S., *Approximate Solution Methods in Engineering Mechanics*, John Wiley, New York (2003).

[4] NSF "Long Term Durability of Materials and Structures: Modeling and Accelerated Techniques", NSF 98-42, National Science Foundation, Arlington, VA. (1998)

[5] Chong, K. P. "Smart Structures Research in the U.S.", *Keynote paper, Proc. NATO Adv. Res. Workshop on Smart Structures*, held in Pultusk, Poland, 6/98, *Smart Structures,* Kluwer Publ. (1999) pp. 37-44.

[6] Boresi, A. P. and Chong, K. P. *Elasticity in Engineering Mechanics*, John Wiley, New York (2000).

[7] Chong, K. P., Saigal, S., Thynell, S. and Morgan, H. (eds.), *Modeling and Simulation-based Life Cycle Engineering*, Spon, London and New York (2002).

Chapter 2

NON-CONVEX DUALITY

Ivar Ekeland
The Pacific Institute for the Mathematical Sciences
1933 West Mall, University of British Columbia Vancouver, BC V6T 1Z2
ekeland@pims.math.ca

Abstract Many problems of classical mechanics are variational in nature, but not convex. This paper shows how the duality theory of convex optimization can be extended to classical mechanics. It is shown in particular that there is a duality theory for functions of square matrices which factor through the determinant.

Keywords: Non-convex duality, variational principles, quasi-convex functionals

1. The least action principle.

I shall introduce the least action principle in the Lagrangian setting. Everything I have to say carries over to the Hamiltonian setting, where very striking results have been obtained. I refer to [2] for a full exposition of the theory, proofs and references, and to [3] for a history of the subject.

Consider the problem of seeking T-periodic solutions to a motion driven by a potential in R^n:

$$\frac{d^2x}{dt^2} + V'(x) = 0, \ x(t+T) = x(t)$$

Assume $V(x) \to \infty$ when $x \to \infty$ (nonlinear spring). It is variational problem, and its solutions are the critical points of the functional:

$$\Phi(x) = \int_0^T \left[-\frac{1}{2} \left\| \frac{dx}{dt} \right\|^2 + V(x) \right] dt$$

on the space of $x(t)$ satisfying $x(0) = x(T)$. This is the *least action principle.*

Even though the least action principle has been known since the 18th century, surprisingly little has been done with it. In fact, the only place wher it is actually used in the classical literature is to prove Noether's theorem. This is because, from the mathematical point of view, the least action principle is very impractical. One would want to solve $\Phi'(x) = 0$, and the natural way to do so would be to maximize or minimize the function Φ. But this function turns out to be unbounded from above and from below:

- $\inf \Phi(x) = -\infty$ (just consider $x_n(t) = \xi \sin(2i\pi nt/T)$,and let $n \to \infty$)
- $\sup \Phi(x) = +\infty$ (just consider $x_n(t) = n\xi$ and let $n \to \infty$)

The analysis can be refined to show that Φ cannot have local maxima or local minima either. Until very recently, the mathematical tools to find critical points which were neither local maxima not local minima simply were not available, except in the exceptional cases were everything can be computed explicitly. Certainly the classical calculus of variations tells us nothing about such situations. The situation has now changed, and I am going to show one of the ways around this problem.

2. Clarke's duality principle.

Let X a Banach space. Define a function $\Phi : X \to \mathbf{R} \cup \{+\infty\}$ by

$$\Phi(x) = \frac{1}{2} < Ax, x > +F(x)$$

where:

- $A : X \to X^*$ a closed linear operator with $A^* = A$
- $F : X \to \mathbf{R} \cup \{+\infty\}$ a convex l.s.c function

Note that the quadratic form $< Ax, x >$ is not positive, so Φ is not a convex function of x, and there is no sense in trying to minimize it. We shall say that x is a *critical point* of Φ if

$$0 \in Ax + \partial F(x)$$

where $\partial F(x)$ denotes the subgradient of F in the sense of convex analysis. We shall also use the notation F^* to denote the Fenchel transform of F, defined by:

$$F^*(x^*) = \sup\{< x^*, x > -F(x) \mid x \in X\}$$

Introduce the function $\Psi : X \to \mathbf{R} \cup \{+\infty\}$ defined by:

$$\Psi(x) = \frac{1}{2} < Ax, x > + F^*(-Ax)$$

where F^* is the Legendre-Fenchel transform of F. We shall assume a constraint qualification condition, namely that the set

$$\{x^* - Ax \mid F(x^*) < \infty,\ x \in X\} \subset X^*$$

contains 0 in its interior. Then:

Theorem. x *is a critical point of* Φ *if and only all the* $y \in x + \ker A$ *are critical points of* Ψ.

This is a *dual action principle.* It was first given by Clarke in the case of convex Hamiltonian systems. The general formulation I have given here is due to Ekeland and Lasry. Its main impact is to change the problem of finding critical points of Φ to the problem of finding critical points of Ψ, and it turns out that the function Ψ is much better behaved than the function Φ.

3. Finding critical points

Let us go back to our original problem:

$$\frac{d^2x}{dt^2} + V'(x) = 0,\ x(t+T) = x(t) \tag{3.1}$$

This amounts to finding critical points of Φ on X, where:

$$\begin{aligned}\Phi(x) &= \int_0^T \left[-\frac{1}{2}\left\|\frac{dx}{dt}\right\|^2 + V(x)\right] dt \\ X &= \left\{x \in H^1([0,T]; R^n) \mid x(0) = x(T)\right\}\end{aligned}$$

If V is convex, we have $\Phi(x) = \frac{1}{2} < Ax, x > + F(x)$ with $A = d^2/dt^2$ and $F(x) = \int V(x(t))\, dt$. The dual then is:

$$\Psi(x) = \int_0^T \left[-\frac{1}{2}\left\|\frac{dx}{dt}\right\|^2 + V^*\left(\frac{d^2x}{dt^2}\right)\right]$$

Ψ is much better behaved than Φ, because the higher-order terms are contained in V^*, but not in V. We can take advantage of this to prove that the boundaty-value problem (2.1) we started form has at least one non-constant solution (existence results), or even several (multiplicity results), and we can even use the dual variational principle to find qualitative properties of the solutions found by this method, for instance their stability and their minimal period. Here we give just a sample of such results (see [2] for more).

3.1 Existence theory.

If $V(x)$ grows like $\frac{1}{p}\|x\|^p$, then $V^*(y)$ behaves like $\frac{1}{q}\|y\|^q$, with $\frac{1}{p}+\frac{1}{q}=1$. Clearly then:

$$\Phi(x) \text{ behaves like } \int_0^T \left[-\frac{1}{2}\left\|\frac{dx}{dt}\right\|^2 + \frac{1}{p}\|x\|^p\right] dt$$

$$\Psi(x) \text{ behaves like } \int_0^T \left[-\frac{1}{2}\left\|\frac{dx}{dt}\right\|^2 + \frac{1}{q}\left\|\frac{d^2x}{dt^2}\right\|^q\right]$$

If $p<2$, so $q>2$, then $\Psi(x)\to\infty$ when $\|x\|\to\infty$ in X. As we noted above, this is not the case for Φ, which is unbounded from above and from below. So Ψ has a minimizer, and this gives a critical point of Φ, and hence a solution of problem (2.1). Further analysis shows that it is non-constant.

If $p>2$, so $q<2$ then Ψ is unbounded from above and from below, but a non-trivial critical point of Ψ is found by the mountain-pass procedure.

3.2 Qualitative information

Let x be a critical point of Ψ. Then the Hessian $\Psi''(x)$ has only finitely many negative eigenvalues, while the Hessian $\Phi''(x)$ has infinitely many positive and negative eigenvalues.

$$(\Psi''(x)\,y,y) = \int_0^T \left[-\left\|\frac{dy}{dt}\right\|^2 + \left(V''(x(t))\,\frac{d^2x}{dt^2},\frac{d^2x}{dt^2}\right)\right] dt$$

By definition, the *Morse index* of x is the number of negative eigenvalues of $\Psi''(x)$. This is a non-negative integer, and the procedure used to find critical points of Ψ will give us information on that number. For instance, in the case $p<2$ which we just discussed, a critical point is found by minimizing Ψ, and therefore its Morse index must be zero.

We will translate this information in more geometrical terms. The critical point x we are investigating is also a solution of the boundary-value problem (2.1). Define a time s to be *conjugate to* 0 if the boundary-value problem

$$\frac{d^2y}{dt^2} + V''(x(t))\,y = 0,\ y(0) = y(s)$$

has a non-zero solution. The multiplicity of s is the number of linearly independent solutions to that problem. Define the *Jacobi index* of x to the number (counted with multiplicities) of times $s \in (0, T]$ which are conjugate to 0.

It can be shown that the Morse index is equal to the Jacobi index. This resutl then is used to prove that a critical point gotten by minimizing Ψ or applying to it the mountain-pass procedure has minimal period T.

4. Open problems.

4.1 The nonlinear wave equation.

The nonlinear wave equation can be cast into that mold in two ways at least. Consider the equation:

$$u_{tt} - u_{xx} + p(u) = 0 \tag{4.1}$$

which is the Euler equation of:

$$\Phi(u) = \int \left[\frac{1}{2} u_x^2 - \frac{1}{2} u_t^2 + P(u) \right] dxdt$$

If P is convex, this can be written as $\Phi(u) = \frac{1}{2} < Au, u > +F(u)$ in two different ways, yielding two different dual functionals Ψ_1 and Ψ_2.

$$A = \left(\frac{\partial^2}{\partial t^2} - \frac{\partial^2}{\partial x^2} \right), \; F(u) = \int P(u) \, dxdt$$

$$A = \frac{\partial^2}{\partial t^2}, F(u) = \int \left[\frac{1}{2} u_x^2 + P(u) \right] dxdt$$

The first way has been used by Brezis, Coron and Nirenberg [1] to give another proof of a result by Rabinowitz, according to which the boudary-value problem:

$$\begin{aligned} && u_{tt} - u_{xx} + p(u) = 0 \\ && 0 \le t \le T, \; 0 \le x \le 1 \\ u(0, x) &=& u(T.x), \; u(t, 0) = u(t, 1) = 0 \end{aligned}$$

has a non-trivial solution, provided T is a rational number. This is clearly a very particular result, and I think this problem should be revisited, perhaps using the second formulation of the dual variational principle.

4.2 Nonlinear elasticity

The Legendre transform F^L of a function F is defined by:

$$F^L(y) = \{ xy - F(x) \mid F'(x) = y \}$$

Unfortunately, it may be multivalued. In the convex case, it is not, so F^L is well-defined, and coincides with the Fenchel transform F^*. But there are other, non-convex cases, where F^L is well-defined, although it does not coincide with the Fenchel transform. Here is a striking example. Let X be the space of $n \times n$ matrices, and the duality pairing:

$$< x, y >= \sum_{i,j} x^{ij} y_{ij}$$

Proposition. If $F(x) = \frac{1}{p}(\text{Det}[x])^p$, *then* $F^L(y) = \frac{1}{q}(\text{Det}[y])^q$ *with* $\frac{1}{p} + \frac{1}{q} = n$.

The duality theorem extends to that case, with F^L replacing F^*, and we thus get pairings of variational principles in nonlinear elasticity. For instance, consider the functionals:

$$\begin{aligned}
\Phi(u) &= \int_\Omega \left[\frac{1}{2}\|\nabla u\|^2 + \frac{1}{p}(\text{Det}[\nabla u])^p\right] d\omega \\
\Psi(X) &= \int_\Omega \left[\frac{1}{2}\left\|\nabla\Delta^{-1}\text{div}X\right\|^2 + \frac{1}{q}(\text{Det}[X])^q\right] d\omega \\
&= \int_\Omega \left[\frac{1}{2}\left(\Delta^{-1}\text{div}X, \text{div}X\right) + \frac{1}{q}(\text{Det}[X])^q\right] d\omega
\end{aligned}$$

where $\Omega \subset R^n$ is open and bounded, u is a function on Ω, and $X = \left(x^{ij}\right)$ is a field of $n \times n$ matrices. Here $(\text{div}X)^i = \sum \partial x^{ij}/\partial\omega^i$. Then critical points of Φ, in the class of maps $u \in H^1(\Omega)$ satisfying given boundary conditions, are related to critical points of Ψ, in the class of vector fields $X \in L^2(\Omega, R^n)$ with corresponding boundary conditions, by the equation:

$$\text{div}X = \Delta u$$

I hope such ideas could be exploited to show existence theorems and qualitative results, as in the theory of Hamiltonian systems.

The proposition is a particular case of the following more general theorem:

Proposition Let $f : \mathbf{R} \to \mathbf{R}$ *be such that the map* $t^{n-1} f'(t)^n$ *is invertible, and denote by* g *its inverse, so that* $s = t^{n-1} f'(t)^n$ if and only if $t = g(s)$. *Then the map* $F : R^{n\times n} \to R$ *defined by*

$$F(x) = f(\text{Det}[x])$$

has a Legendre transform F^L *given by:*

$$F^L(y) = n\, g((\text{Det}[y]))\; f' \circ g(\text{Det}[y]) - f \circ g(\text{Det}[y])$$

Proof. By definition:

$$\begin{aligned}\Phi^L(y) &= \left\{\sum y_{ij}x^{ij} - \Phi(x) \,\middle|\, y_{ij} = \frac{\partial\Phi}{\partial x^{ij}}\right\} \\ &= \left\{\sum y_{ij}x^{ij} - f(\mathrm{Det}[x]) \,\middle|\, y_{ij} = \frac{\partial\Phi}{\partial x^{ij}}\right\}\end{aligned}$$

We have:

$$\frac{\partial\Phi}{\partial x^{ij}} = f'(\mathrm{Det}[x])\frac{\partial \mathrm{Det}[x]}{\partial x^{ij}}$$

The preceding formula then becomes:

$$\begin{aligned}\Phi^L(y) &= f'(\mathrm{Det}[x])\sum\frac{\partial \mathrm{Det}[x]}{\partial x^{ij}}x_{ij} - f(\mathrm{Det}[x]) \\ &= n\, f'(\mathrm{Det}[x])\,\mathrm{Det}[x] - f(\mathrm{Det}[x]) \qquad (4.2)\end{aligned}$$

Set $X_{ij} = \mathrm{Cofactor}[x^{ij}]$. It is known that $[\mathrm{Det}[x]]^{n-1} = \mathrm{Det}[X]$. Since:

$$y_{ij} = \frac{\partial\Phi}{\partial x^{ij}} = f'(\mathrm{Det}[x])\,X_{ij}$$

we have

$$\mathrm{Det}[y] = f'(\mathrm{Det}[x])^n\,[\mathrm{Det}[x]]^{n-1}$$

so that $\mathrm{Det}[x] = g(\mathrm{Det}[y])$, and the result follows by substitution this into formula (4.2) ∎

References

[1] Haim Brezis, Jean-Michel Coron, Louis Nirenberg, "Free vibrations for a nonlinear wave equation and a theorem of P. Rabinowitz", *Comm. Pure Applied Math.* 33 (1980), 667-684.

[2] Ivar Ekeland, "Convexity methods in Hamiltonian mechanics", Springer-Verlag.

[3] Ivar Ekeland, "Le meilleur des mondes possibles", Seuil; English translation, *"The best of all possible worlds"* to appear at Chicago University Press.

Chapter 3

DUALITY, COMPLEMENTARITY, AND POLARITY IN NONSMOOTH/NONCONVEX DYNAMICS

David Y. Gao
Department of Mathematics
Virginia Tech, Blacksburg, VA 24061, USA
gao@math.vt.edu

Abstract This paper presents a brief survey and a unified theory in nonsmooth and nonconvex dynamical systems. The canonical dual/polar transformation methods and the associated bi-duality and triality theories proposed recently in nonconvex variational problems are generalized into fully nonlinear dissipative dynamical systems governed by nonsmooth constitutive laws and boundary conditions. It is shown that by this method, nonsmooth and nonconvex Hamilton systems can be reformulated into certain smooth dual-complementary variational problems. An bi-polarity variational principle is established for 3-D nonsmooth elastodynamical systems, and a potentially powerful complementary variational principle can be used for solving unilateral variational inequality problems governed by nonsmooth boundary conditions. Applications are illustrated by dynamically post-buckling analysis of extended beam model. A very interesting new phenomenon: meta-buckling transition period in chaotic vibration is revealed and a bifurcation criterion for dissipative Duffing system is obtained by the dual analysis. It is shown that the chaotical solutions form an invariant set in canonical dual phase space. This invariant set and the bifurcation criterion play key roles in feedback controlling against chaos.

Keywords: Nonsmooth elastodynamics, Nonconvexity, Nonconservativity, Chaotic Dynamics, Nonconvex dissipative Hamilton systems, Complementarity, Duality, Polarity, Triality, Critical point theory, contact mechanics, Chaotical vibrations, Meta-chaos, Variational method

1. Nonsmooth and Nonconvex Problems

The problem in nonsmooth, nonconvex, and nonconservative systems is of significant practical importance and has attracted considerable attention during the last years. As it was shown in the author's recent work (see Gao, 1998-2001) that by introducing certain suitable bilinear forms, many nonlinear, nonconvex systems, either conservative or dissipative, can be written in the following stationary variational problem

$$(\mathcal{P}): \quad \Pi(u) = K(\partial_t u) - P(u) \quad \to \text{sta} \;\; \forall u \in \mathcal{U}_k, \tag{1.1}$$

where the feasible space $\mathcal{U}_k$ is a convex, non-empty subset of a vector space $\mathcal{U}$ over an open space-time domain $\Omega_t = \Omega \times (0, t_c) \subset \mathbb{R}^n \times \mathbb{R}^+$, in which, the essential boundary-initial conditions and certain constraints are prescribed; $\partial_t : \mathcal{U}_k \to \mathcal{V}$ is a time-differential operator from $\mathcal{U}_k$ to the space of velocity $\mathcal{V}$; $K(v)$ stands for the total kinetic energy of the system and $P : \mathcal{U}_k \to \mathbb{R}$ is the total potential of the system; the notation $\Pi(u) \to \text{sta} \;\forall u \in \mathcal{U}_k$ stands for finding the stationary (or critical) points of Π over the feasible space $\mathcal{U}_k$.

The general variational form of the problem $(\mathcal{P})$ covers a great variety of situations. In Newtonian mechanics, the operator ∂_t is simply the time derivative $\partial/\partial t$, the kinetic energy $K(v)$ is usually a convex (quadratic) differentiable functional, and the velocity-momentum relation $p = DK(v)$ is invertible. The complementary kinetic energy $K^*(p)$ can be obtained easily by the classical Legendre transformation. If the total potential $P(u)$ is convex, then the total action $\Pi(u)$ of the system is a so-called *d.c. functional*, i.e., the difference of convex functionals. In this case, the classical Hamiltonian $H(u,p) = K^*(p) + P(u)$ is a convex functional and the stationary condition of (1.1) leads to a particularly symmetrical structure in classical conservative dynamical systems (cf. e.g. Marsden and Ratiu, 1995)

$$u_{,t} = D_p H(u,p), \;\; p_{,t} = -D_u H(u,p).$$

However, this beautiful symmetry is broken in nonconservative finite deformation systems, where, the total potential P is usually nonconvex, or even nonsmooth, which leads to many substantial difficulties in solving the problem (1.1).

Nonsmooth, nonconvex and nonconservative phenomena arise naturally from real-life systems. Many problems in modern sciences (such as hysteresis and phase transitions, super-conductivity, cosmology, mathematical economics, composite and smart materials, frictional contact problems and damage/fracture mechanics, nonlinear bifurcation and post-buckling analysis of large deformed structures, etc), require the consideration of non-differentiability and nonconvexity for their accurate mathematical modeling (see, for examples, Moreau *et al* 1968, 1988; Panagiotopoulos, 1985; Pfeiffer and Glocker, 1996; Dem'yanov *et al*, 1996; Brokate and Sprekels, 1996; Kalies and Holmes, 1996; Kibble, 1997; Mistakidis and Stavroulakis, 1998; Motreanu and Panagiotopoulos, 1999; Gao, 1999; Gao *et al*, 2000; Stewart, 2000 and much more).

Generally speaking, due to the non-differentiability of the nonsmooth total potential, traditional direct methods for solving nonsmooth and nonconvex variational problems are usually very difficult, or even impossible. Duality theory, however, plays a potentially powerful role in nonsmooth and nonconvex analysis.

For example, let us consider a mass-spring system with linear damping such that the total kinetic energy is a weighted quadratic form $K(\partial_t u) = \int_I \frac{1}{2} e^{\nu t} \rho u_{,t}^2 \, dt$ over the time domain $I = (0, t_c)$, where $e^{\nu t}$ is a weight function (see, Gao, 1999b). For spring with jumping effect (see Fig. 3.2a), the total potential $P(u)$

$$P(u) = \begin{cases} \int_I e^{\nu t} \frac{1}{2} a u^2 \, dt & \text{if } u \le u_a, \\ \\ \int_I e^{\nu t} [\frac{1}{2} a u_a^2 + \frac{1}{2} b (u - u_a)^2 + u_b^* (u - u_a)] \, dt & \text{if } u > u_a, \end{cases} \tag{1.2}$$

is a nonsmooth functional (see Fig. 3.1(a)), where a, b, u_a and u_b^* are elastic constants. At the point $u = u_a$, the *sub-differential* of P is a convex set in $\mathcal{U}^*$

$$u^* \in \partial P(u) = \begin{cases} a u & \text{if } u \le u_a, \\ (u_a^*, u_b^*) & \text{if } u = u_a, \\ b(u - u_a) + u_b^* & \text{if } u \ge u_a, \end{cases} \tag{1.3}$$

which is discontinuous at the point $u = u_a$ (see Fig. 3.2a). Thus, the stationary condition of the variational problem (1.1) leads to a discontinuous second order differential inclusion

$$-\rho u_{,tt} - \nu \rho u_{,t} \in \partial P(u) \;\; \forall t \in I.$$

By the duality in classical Legendre transformation (cf. e.g. Gao, 1999b), the conjugate P^* of the nonsmooth P is, however, a smooth function (see Fig. 3.1(b)):

$$P^*(u^*) = \begin{cases} \int_I e^{\nu t} \frac{1}{2a} u^{*2} \, dt & \text{if } u^* \le u_a^*, \\ \\ \int_I e^{\nu t} [\frac{1}{2a} u_a^{*2} + u_a (u^* - u_a^*)] \, dt & \text{if } u_a^* < u^* \le u_b^*, \\ \\ \int_I e^{\nu t} [\frac{1}{2a} u^{*2} + u_a (u^* - u_a^*) + \frac{1}{2b} (u^* - u_b^*)^2] \, dt & \text{if } u^* > u_b^*. \end{cases} \tag{1.4}$$

Its Gâteaux derivative is a continuous function (see Fig. 3.2b)

$$u = DP^*(u^*) = \begin{cases} \frac{1}{a}u^* & \text{if } u^* \le u_a^*, \\ u_a & \text{if } u_a^* < u^* < u_b^*, \\ u_a + \frac{1}{b}(u^* - u_b^*) & \text{if } u^* \ge u_b^*. \end{cases} \quad (1.5)$$

Thus, the dual problem will be much easier than the nonsmooth primal problem.

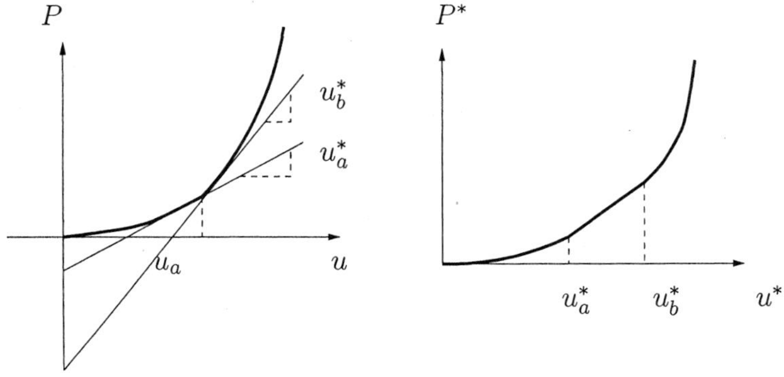

(a) Graph of $P(u)$. (b) Graph of $P^*(u^*)$.

Figure 3.1. Nonsmooth function and its smooth Legendre conjugate

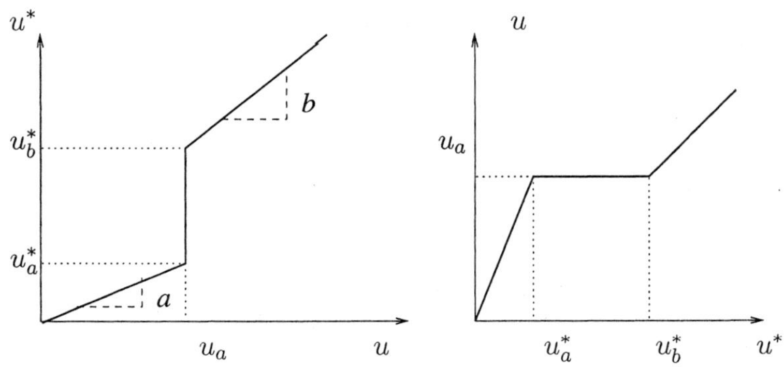

(a) $u^* \in \partial P(u)$ (b) $u = DP^*(u^*)$

Figure 3.2. Discontinuous constitutive law and continuous inverse form

In finite deformation theory, the total potential of systems is generally nonconvex. Actually, the concept of non-convexity is synonymous with the notion of multi-equilibrium states of systems. In post-buckling analysis of large deformation structures, the total potential has more than one potential well. The multiple local minimizers of a given variational problem lead to possible post-buckling states of the deformed structures. To see this, let us consider the modified version of the large deformation extended beam model proposed by Gao (1997,2000a)

$$\begin{aligned} & u_{,xx} + \beta u_{,yy} = 0 \;\; \forall (x,y,t) \in \Omega_t = (0,\ell) \times (-h,h) \times (0,t_c), \\ & \rho w_{,tt} + \rho \nu w_{,t} = \left(\tfrac{3\alpha^2}{2} w_{,x}^2 + \beta - \lambda\alpha \right) w_{,xx} + \tfrac{\beta}{2h} [u_{,x}]_{y=\pm h} + f, \\ & u_{,y}(x,\pm h,t) + w_{,x}(x,t) = \pm \mu^{\pm}(x,t), \;\; \forall (x,t) \in (0,\ell) \times (0,t_c), \;\; y = \pm h, \end{aligned} \tag{1.6}$$

where $u(x,y,t)$ measures the shear deformation of the beam at the point (x,y), while $w(x,t)$ is the deflection of the beam; the parameters $\alpha, \beta > 0$ are material constants; $[u_{,x}]_{y=\pm h} = u_{,x}(x,h,t) - u_{,x}(x,-h,t)$; $\lambda > 0$ represents the axial load and $\mu^{\pm}(x,t)$ the shear load on the upper and lower beam surfaces ($y = \pm h$). The total potential associated with this model is a nonconvex functional

$$P(u,w) = \int_{\Omega_t} e^{\nu t} \frac{1}{2} [(u_{,x}^2 + \frac{1}{2} \alpha w_{,x}^2 - \lambda)^2 + \beta (u_{,y} + w_{,x})^2] \, \mathrm{d}\Omega_t - F(u,w), \tag{1.7}$$

where $F(u,w)$ is the external energy. If shear load $\mu^{\pm}(x,t)$ is given at $y = \pm h$, the external energy F can be simply defined by

$$F(u,w) = \int_0^{t_c} \int_0^{\ell} e^{\nu t} [\mu^+ u(x,h,t) + \mu^- u(x,-h,t) + f w] \, \mathrm{d}x \, \mathrm{d}t. \tag{1.8}$$

In static problem, if the beam is subjected only to a compressive load λ at $x = \ell$ and the transverse load $f(x)$ such that the shear deformation u can be ignored, then the total potential P depends on w according to

$$P_\mu(w) = W(w_{,x}) - F(w) = \int_0^{\ell} \frac{1}{2} (\frac{1}{2} \alpha w_{,x}^2 - \lambda)^2 \, \mathrm{d}x - \int_0^{\ell} f w \, \mathrm{d}x,$$

where the stored energy $W(\epsilon) = \int_0^{\ell} \frac{1}{2} (\frac{1}{2} \alpha \epsilon^2 - \lambda)^2 \, \mathrm{d}x$ is the well-known van der Waals double-well function (see Fig. 3.3a) of the linear strain $\epsilon = w_{,x}$. For certain given loads $\lambda > 0$ and f, the total potential P_μ has at most three critical points: two local minimizers, corresponding to two possible stable buckled states, and one local maximizer, corresponding to an unstable buckled state. The global minimizer of P_μ depends on the lateral load f (see Fig. 3.3b).

For general statically contact problems, if the beam is supported on a strictly concave obstacle $\psi(x)$ (see Fig. 3.4), the external energy $F(u,w)$

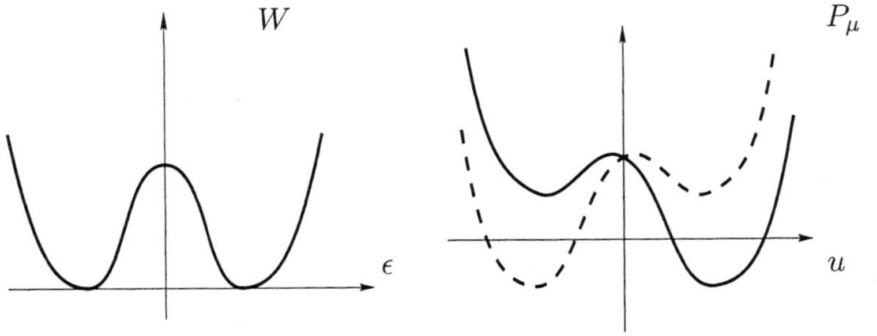

(a) Graph of $W(\epsilon)$ (b) Graphs of $P_\mu(u)$ ($f > 0$ solid, $f < 0$ dashed)

Figure 3.3. Double-well energy and nonconvex potential

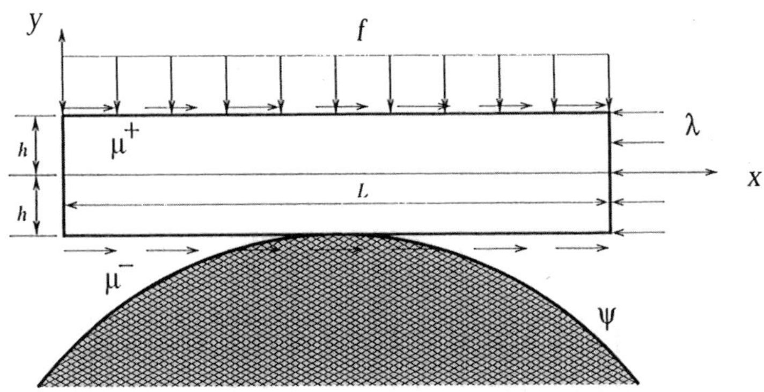

Figure 3.4. Unilateral buckling beam with concave obstacle $\psi(x)$.

is usually a nonsmooth functional

$$F(u, w) = \int_0^\ell wf \, dx + j(u),$$

where $j(u)$ is the so-called *super-potential*. For Coulomb's law of dry friction, this super-potential is a convex, lower semicontinuous functional

$$j(u(x, -h)) = \sup_{|\mu^-| \leq -\nu f} \int_0^\ell -\mu^-(x) u(x, -h) \, dx,$$

where ν is a coefficient of friction. The frictional law

$$\mu^-(x) \in \partial^- j(u(x, -h)), \;\; \forall x \in (0, \ell).$$

on the contact boundary $y = -h$ can be expressed as

$$|\mu^-(x)| < -\nu f \;\;\Rightarrow\;\; u(x,-h) = 0 \;\; \forall x \in I_c,$$
$$|\mu^-(x)| = -\nu f \;\;\Rightarrow\;\; u(x,-h) = -\lambda^* \mu^-(x) \;\; \forall x \in I_c \;\text{ for some } \lambda^* \geq 0.$$

Here, $\lambda^* \geq 0$ is a physical parameter and $I_c \subset (0, \ell)$ is the contact region, defined by

$$I_c := \{x \in (0,\ell) |\;\; w(x) = \psi(x) \;\; \forall x \in (0,\ell)\},$$

and its boundary are unknown until the problem is solved.

For a lubricated contact problem, the lubrication implies that $\mu^-(x) = 0$. If the beam is clamped at $x = 0$, simply supported at $x = \ell$ and is subjected to the static loads $f(x) < 0$ and $\lambda > 0$, the kinematically admissible space $\mathcal{U}_k$ is a convex set with an inequality constraint

$$\mathcal{U}_k = \left\{ \begin{array}{ll} w \in C^2(0,\ell) | & w(0) = w_{,x}(0) = w(\ell) = w_{,xx}(\ell) = 0, \\ & w(x) \geq \psi(x) \;\; \forall x \in (0,\ell) \end{array} \right\}.$$

In this case, the solution $\bar{w}(x)$ of the post-buckling problem is usually a local minimizer of the total potential P_μ, satisfies the complementarity condition

$$\bar{w}(x) - \psi(x) \geq 0, \;\; \bar{f}(x) \leq 0, \;\;\; (\bar{w}(x) - \psi(x))\bar{f}(x) = 0 \;\; \forall x \in (0,\ell),$$

where

$$\bar{f}(x) = \left(\frac{3\alpha^2}{2} \bar{w}_{,x}^2 - \lambda\alpha \right) \bar{w}_{,xx} + f(x).$$

In elastoplastic beam contact problems, the plastic yield condition will lead to the so-called bi-complementarity problems. The traditional direct methods for solving these kinds of unilateral variational inequality and nonlinear complementarity problems are proved to be very difficult. However, the analytic solutions can be obtained easily by using canonical dual transformation method (see, Gao, 1998a, 1999b).

If the beam is subjected to a periodically dynamical load $f(x,t)$, the two local minimizers of P_μ become extremely unstable. For the given post-buckling load $\lambda = 1.075\lambda_c$, ($\lambda_c > 0$ is the Euler buckling load) and $f = f_0 \cos(3.76t)$, Fig. 3.5(a) shows that the beam is in chaotic post-buckling state for the amplitude $f_0 \in (0.8, 2.5)$. To see the intrinsic structure of this chaotic vibration, for each given f_0, the trajectories in the phase space are shown in Fig. 3.5(b), where we can see clearly that the so-called chaos is actually the random vibration of the system around either potential well. Mathematically speaking, the chaotic phenomenon of a nonlinear dynamical system is mainly due to the nonconvexity of the total potential of the system. Very small perturbations

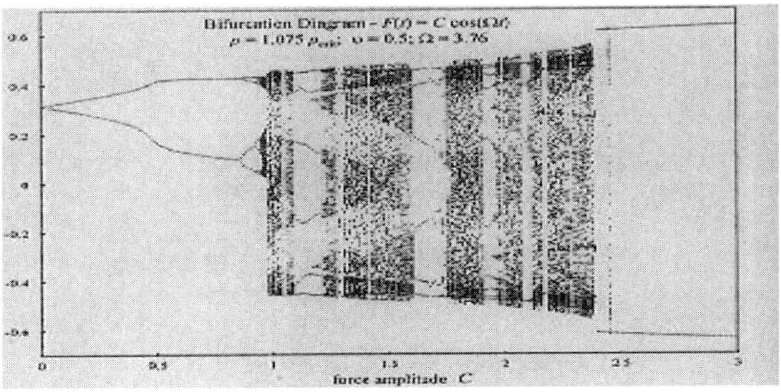

(a) Chaotic bifurcation diagram: displacement w vis force amplitude f_0

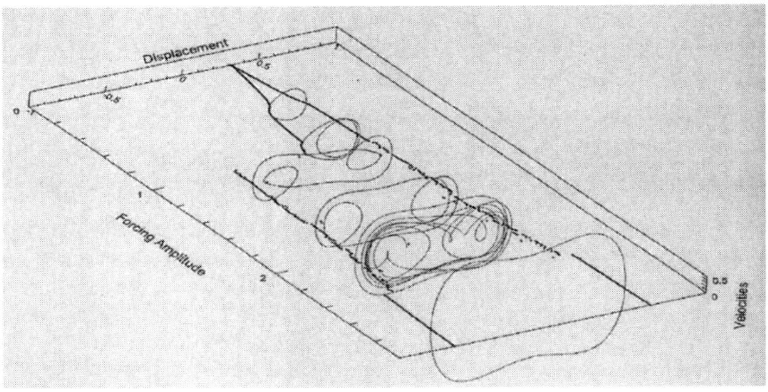

(b) Trajectories of chaotic beam vibration in force-phase space $(w, w_{,t}, f_0)$

Figure 3.5. Chaotic bifurcation for pre-bucked extended beam model

of the system's initial conditions and parameters may lead the system to different potential wells with significantly different performance characteristics. The numerical results vary with the methods used (cf. e.g., Gao, 1999b). This is the one of main reasons why the traditional perturbation analysis and the direct approaches cannot successfully be applied to chaotic systems. Based upon these observations and in order to handle the nonconvex dynamical problems, many new methods and techniques have been developed (see, e.g., Fowler, 1989; Ott *et al*, 1990; Chen and Dong, 1993; Ogorzalek, 1993; Antoniou *et al*, 1996; Ghezzi and Piccardi, 1997; Arada and Raymond, 1999; Koumboulis and Mertzios, 2000).

Duality theory plays fundamental roles in natural phenomena. In engineering mechanics, the study on the complementary-dual variational principles has a long history (cf. e.g., Hellinger, 1914; Reissner, 1955; Oden and Reddy, 1983; Tabarrok and Rimrott, 1994). During the last decade, the so-called *primal-dual interior point method* has emerged as the most important and efficient revolutionary technique in mathematical programming (cf. e.g., Wright, 1998). The advantage of the primal-dual approaches relying on a common mathematical structure that underlies many physical theories. A self-contained comprehensive presentation of the mathematical duality theory in general nonconvex, nonsmooth systems was given recently by Gao (1999b). The aim of this article is to generalize the author's previous results on nonconvex static problems into nonsmooth dissipative dynamical systems. The rest of this paper is divided into six main sections. The next section set up the notation used in the paper and describes the problems. A general framework and the canonical dual transformation method in fully nonlinear, nonsmooth, nonconservative dynamical systems are discussed. Section 3 presents an extended Lagrangian duality theory in general nonconvex Hamilton systems, by which, the dual action can be obtained. The critical points in fully nonlinear systems are classified. In section 4, the interesting triality theory is generalized into canonical dynamical systems with general geometrically nonlinear operator Λ. Section 5 and 6 are devoted mainly to the super-Lagrange duality theory in geometrically linear systems. The nice bi-polarity theory is proposed for nonsmooth elastodynamics and a complementary action is presented for solving nonconservative dynamical system governed by nonsmooth boundary conditions. Section 7 devotes applications to post-buckling analysis of dynamical beam model, and chaotic vibration of dissipative Duffing system. An interesting new phenomenon is discovered in chaotic vibration and a bifurcation criterion is proposed and verified both theoretically and numerically.

2. Framework in Nonconvex, Nonsmooth and Nonconservative Dynamical Systems

Let $\mathcal{U}$ and $\mathcal{U}^*$ be two real linear spaces, finite- or infinite dimensional, placed in duality by a bilinear form $\langle \cdot , \cdot \rangle : \mathcal{U} \times \mathcal{U}^* \to \mathbb{R}$. For a given extended real-valued function $P : \mathcal{U} \to \bar{\mathbb{R}} = \mathbb{R} \cup \{-\infty\} \cup \{+\infty\}$, the sub- and super-differentials of P at $\bar{u} \in \mathcal{U}$ are defined by

$$\partial^- P(\bar{u}) = \{\bar{u}^* \in \mathcal{U}^* | P(u) - P(\bar{u}) \geq \langle \bar{u}^*, u - \bar{u} \rangle \ \forall u \in \mathcal{U}\},$$

$$\partial^+ P(\bar{u}) = \{\bar{u}^* \in \mathcal{U}^* | P(u) - P(\bar{u}) \leq \langle \bar{u}^*, u - \bar{u} \rangle \ \forall u \in \mathcal{U}\},$$

respectively. Clearly, we always have $\partial^+ P = -\partial^-(-P)$. In convex analysis, it is convention that ∂^- is simply written as ∂. In this paper, ∂ stands for either ∂^- or ∂^+, i.e.,

$$\partial = \{\partial^-,\, \partial^+\}.$$

If P is finite and Gâteaux-differentiable on a subspace $\mathcal{U}_s \subset \mathcal{U}$, then

$$\partial P(\bar{u}) = \partial^- P(\bar{u}) = \partial^+ P(\bar{u}) = \{DP(\bar{u})\},$$

where $DP : \mathcal{U}_s \to \mathcal{U}^*$ denotes the Gâteaux derivative of P at $\bar{u}$. The following notations and definitions, used in Gao (1999b), will be of convenience in nonconvex, nonsmooth variational analysis.

Definition 1 The set of functions $P : \mathcal{U} \to \bar{\mathbb{R}}$ which are either convex or concave is denoted by $\Gamma(\mathcal{U})$. In particular, let $\check{\Gamma}(\mathcal{U})$ denote the subset of functions $P \in \Gamma(\mathcal{U})$ which are convex and $\hat{\Gamma}(\mathcal{U})$ the subset of $P \in \Gamma(\mathcal{U})$ which are concave.

(D1) The *canonical function space* $\Gamma_G(\mathcal{U}_s)$ is a subset of functions $P \in \Gamma(\mathcal{U}_s)$ which are Gâteaux differentiable on $\mathcal{U}_s \subset \mathcal{U}$.

(D2) The *extended canonical function space* $\Gamma_0(\mathcal{U})$ is a subset of functions $P \in \Gamma(\mathcal{U})$ which are either convex, lower semicontinuous or concave, upper semicontinuous, and if P takes the values $\pm\infty$, then P is identically equal to $\pm\infty$. ♢

By the *Fenchel transformation*, the *super-conjugate function* of an extended function $P : \mathcal{U} \to \bar{\mathbb{R}}$ is defined by

$$P^\sharp(u^*) = \sup_{u\in\mathcal{U}}\{\langle u, u^*\rangle - P(u)\}.$$

From the theory of convex analysis we know that $P^\sharp : \mathcal{U}^* \to \bar{\mathbb{R}}$ is always convex and lower semicontinuous, i.e., $P^\sharp \in \check{\Gamma}_0(\mathcal{U}^*)$. Dually, the *sub-conjugate function* of P, defined by

$$P^\flat(u^*) = \inf_{u\in\mathcal{U}}\{\langle u, u^*\rangle - P(u)\},$$

is always concave and upper semicontinuous, i.e., $P^\flat \in \hat{\Gamma}_0(\mathcal{U}^*)$, and $P^\flat = -P^\sharp$. Both the super- and sub-conjugates are called Fenchel conjugate functions and we write $P^* = \{P^\flat, P^\sharp\}$. Thus the extended Fenchel transformation can be written as

$$P^*(u^*) = \text{ext}\{\langle u, u^*\rangle - P(u) |\;\; u \in \mathcal{U}\}. \tag{2.1}$$

Clearly, if $P \in \Gamma_0(\mathcal{U})$, we have the *Fenchel duality relations*, namely,

$$u^* \in \partial P(u) \;\;\Leftrightarrow\;\; u \in \partial P^*(u^*) \;\;\Leftrightarrow\;\; P(u) + P^*(u^*) = \langle u\,,\, u^*\rangle. \tag{2.2}$$

The pair (u, u^*) is called the *Fenchel duality pair* on $\mathcal{U} \times \mathcal{U}^*$ if and only if equation (2.2) holds on $\mathcal{U} \times \mathcal{U}^*$.

The conjugate pair (u, u^*) is called the *Legendre duality pair* on $\mathcal{U}_s \times \mathcal{U}_s^* \subset \mathcal{U} \times \mathcal{U}^*$ if and only if the equivalent relations

$$u^* = DP(u) \;\Leftrightarrow\; u = DP^*(u^*) \;\Leftrightarrow\; P(u) + P^*(u^*) = \langle u, u^* \rangle. \quad (2.3)$$

hold on $\mathcal{U}_s \times \mathcal{U}_s^*$.

In nonconvex Hamilton systems, the total potential $P : \mathcal{U} \to \bar{\mathbb{R}}$ is usually nonconvex. In order to study duality theory, we introduce a pair of the so-called *intermediate variable spaces* $(\mathcal{E}, \mathcal{E}^*)$, placed in duality by the second bilinear form $\langle \cdot\, ; \,\cdot \rangle : \mathcal{E} \times \mathcal{E}^* \to \mathbb{R}$. We assume that there exists a Gâteaux differentiable operator $\Lambda : \mathcal{U} \to \mathcal{E}$ and canonical functionals $F \in \Gamma_0(\mathcal{U})$, $E \in \Gamma_0(\mathcal{E})$ such that

$$P(u) = E(\Lambda(u)) - F(u). \quad (2.4)$$

In finite deformation theory, Λ is usually a gradient-like operator and $E \in \Gamma_0(\mathcal{E})$ is called the stored potential energy. The element $\xi = \Lambda(u) \in \mathcal{E}$ is called canonical strain measure (see Gao, 1999b) and the duality relation $\xi^* \in \partial_\xi J_\xi(u, \xi) = \partial E(\xi)$ is called the *constitutive law*. The functional $F \in \Gamma_0(\mathcal{U})$ stands for the external energy, and the duality relation $u^* \in \partial F(u)$ leads to the boundary conditions (see, Gao 1999b). Let $\mathcal{U}_a \subset \mathcal{U}$ and $\mathcal{E}_a \subset \mathcal{E}$ be two subsets on which the extended canonical functionals $F \in \Gamma_G(\mathcal{U}_a)$, $E \in \Gamma_G(\mathcal{E}_a)$ are Gâteaux differentiable. Let $\mathcal{U}_a^* \subset \mathcal{U}^*$ and $\mathcal{E}_a^* \subset \mathcal{E}^*$ be the ranges of the duality mappings $DF : \mathcal{U}_a \to \mathcal{U}^*$ and $DE : \mathcal{E}_a \to \mathcal{E}^*$, respectively. Then on $(\mathcal{U}_a, \mathcal{U}_a^*)$ and $(\mathcal{E}_a, \mathcal{E}_a^*)$, we have the Legendre duality relations

$$u^* = DF(u) \;\Leftrightarrow\; u = DF^*(u^*) \;\Leftrightarrow\; F(u) + F^*(u) = \langle u \,,\, u^* \rangle,$$

$$\xi^* = DE(\xi) \;\Leftrightarrow\; \xi = DE^*(\xi^*) \;\Leftrightarrow\; E(\xi) + E^*(\xi^*) = \langle \xi \,;\, \xi^* \rangle.$$

The directional derivative of ξ at $\bar{u}$ in the direction $u \in \mathcal{U}$ is defined by

$$\delta\xi(\bar{u}; u) := \lim_{\theta \to 0^+} \frac{\xi(\bar{u} + \theta u) - \xi(\bar{u})}{\theta} = \Lambda_t(\bar{u})u, \quad (2.5)$$

where $\Lambda_t(\bar{u}) = D\Lambda(\bar{u})$ denotes the Gâteaux derivative of the operator Λ at $\bar{u}$. For a given $\xi^* \in \mathcal{E}^*$, $G_{\xi^*}(u) = \langle \Lambda(u) \,;\, \xi^* \rangle$ is a real-valued function of u on $\mathcal{U}$. Its Gâteaux derivative at $\bar{u} \in \mathcal{U}_a$ in the direction $u \in \mathcal{U}$ reads

$$\delta G_{\xi^*}(\bar{u}; u) = \langle \Lambda_t(\bar{u})u \,;\, \xi^* \rangle = \langle u \,,\, \Lambda_t^*(\bar{u})\xi^* \rangle,$$

where $\Lambda_t^*(\bar{u}) : \mathcal{E}^* \to \mathcal{U}^*$ is the adjoint operator of Λ_t associated with the two bilinear forms. The *complementary operator* $\Lambda_c(u)$ is defined by the geometrical operator decomposition (Gao and Strang, 1989):

$$\Lambda(u) = \Lambda_t(u)u + \Lambda_c(u).$$

We have

$$G_{\xi^*}(u) = \langle \Lambda(u) \; ; \; \xi^* \rangle = \langle u \; , \; \Lambda_t^*(u)\xi^* \rangle - G(u, \xi^*), \tag{2.6}$$

where $G(u, \xi^*) = \langle -\Lambda_c(u) \; ; \; \xi^* \rangle$ is the so-called *complementary gap functional*, introduced by Gao and Strang (1989). This gap function plays an essential role in nonconvex variational analysis.

For dynamical problems, we let $\mathcal{V}$ and $\mathcal{V}^*$ be the velocity and momentum space, respectively, placed in duality by the third bilinear form $\langle * \; , \; * \rangle : \mathcal{V} \times \mathcal{V}^* \to \mathbb{R}$. Let $\mathcal{V}_a \subset \mathcal{V}$ be a subspace in which certain essential time initial/boundary conditions are given. Dually, let $\mathcal{V}_a^*$ be a subset of $\mathcal{V}^*$, in which the complementary (natural) conditions are given. Thus, for a given time-differential operator $\partial_t : \mathcal{U}_a \to \mathcal{V}_a$, the adjoint $\partial_t^* : \mathcal{V}_a^* \to \mathcal{U}^*$ can be defined by

$$\langle \partial_t u \; , \; p \rangle = \langle u \; , \; \partial_t^* p \rangle \;\; \forall u \in \mathcal{U}_a, \;\; \forall p \in \mathcal{V}_a^*$$

For Newtonian systems, the kinetic energy $K : \mathcal{V} \to \mathbb{R}$ and its Legendre conjugate $K^* : \mathcal{V}_a^* \to \mathbb{R}$ are usually quadratic functionals:

$$K(v) = \frac{1}{2}\langle v \; , \; \rho v \rangle, \quad K^*(p) = \frac{1}{2}\langle \rho^{-1} p \; , \; p \rangle.$$

Thus on $\mathcal{V}_a \times \mathcal{V}_a^*$, the canonical physical relations between the duality pair (v, p) are linear, i.e.,

$$p = DK(v) = \rho v \;\; \Leftrightarrow \;\; v = DK^*(p) = \rho^{-1} p.$$

Let $\Phi(u, v, \xi) = K(v) - E(\xi) + F(u) : \mathcal{U} \times \mathcal{V} \times \mathcal{E} \to \bar{\mathbb{R}}$ be the extended canonical functional with effective domain dom $\Phi = \mathcal{U}_a \times \mathcal{V}_a \times \mathcal{E}_a$ such that $\Pi(u) = \Phi(u, \partial_t u, \Lambda(u))$. Then the *kinematically admissible space* $\mathcal{U}_k$ can be written in a standard canonical form

$$\mathcal{U}_k = \{ u \in \mathcal{U}_a | \;\; \Lambda(u) \in \mathcal{E}_a, \;\; \partial_t u \in \mathcal{V}_a \}. \tag{2.7}$$

In terms of the canonical variables (u, v, ξ), the primal problem $(\mathcal{P})$ can be reformulated in the canonical form:

Problem 1 (Canonical Primal Problem) Suppose that $\Phi : \mathcal{U}_a \times \mathcal{V}_a \times \mathcal{E}_a \to \mathbb{R}$ is an canonical functional such that $\mathcal{U}_k$ is not empty and $\Pi(u) = \Phi(u, \partial_t u, \Lambda(u))$ for each $u \in \mathcal{U}_k$. Find the critical point of Π such that

$$(\mathcal{P}_c) : \quad \Pi(u) = \Phi(u, \partial_t u, \Lambda(u)) \;\; \to \text{sta} \;\; \forall u \in \mathcal{U}_k. \tag{2.8}$$

The following classification for dynamical systems was originally introduced in nonlinear variational/boundary value problems by Gao (1998, 1999b).

Definition 2 Suppose that for a given problem $(\mathcal{P})$, the total action $\Pi(u)$ is well-defined on its domain $\mathcal{U}_k \subset \mathcal{U}$. If the generalized geometrical operator $\Lambda : \mathcal{U} \to \mathcal{E}$ can be chosen in such a way that $\Pi(u) = \Phi(u, \partial_t u, \Lambda(u))$, and $\Phi \in \Gamma_G(\mathcal{U}_a) \times \Gamma_G(\mathcal{V}_a) \times \Gamma_G(\mathcal{E}_a)$, then

(1) the transformation $\{\Pi; \mathcal{U}_k\} \to \{\Phi; \mathcal{U}_a \times \mathcal{V}_a \times \mathcal{E}_a\}$ is called the *canonical transformation*, and $\Phi : \mathcal{U}_a \times \mathcal{V}_a \times \mathcal{E}_a \to \mathbb{R}$ is called the *canonical action associated with* Λ;

(2) the problem $(\mathcal{P})$ is called *geometrically nonlinear (resp. linear)* if $\Lambda : \mathcal{U} \to \mathcal{E}$ is nonlinear (resp. linear); it is called *physically nonlinear* (resp. linear) if the duality mapping $D\Phi : \mathcal{U}_a \times \mathcal{V}_a \times \mathcal{E}_a \to \mathcal{U}_a^* \times \mathcal{V}_a^* \times \mathcal{E}_a^*$ is nonlinear (resp. linear); it is called *fully nonlinear* if it is both geometrically and physically nonlinear. ♢

The canonical transformation plays a fundamental role in duality theory of nonconvex systems. Clearly, if $\Phi \in \Gamma_G(\mathcal{U}_a) \times \Gamma_G(\mathcal{V}_a) \times \Gamma_G(\mathcal{E}_a)$ is a canonical functional, the Gâteaux derivative $D\Phi : \mathcal{U}_a \times \mathcal{V}_a \times \mathcal{E}_a \to \mathcal{U}_a^* \times \mathcal{V}_a^* \times \mathcal{E}_a \subset \mathcal{U}^* \times \mathcal{V}^* \times \mathcal{E}^*$ is a monotone mapping, i.e., the duality relations

$$u^* = D_u\Phi(u, v, \xi), \quad p = D_v\Phi(u, v, \xi), \quad \xi^* = D_\xi\Phi(u, v, \xi) \tag{2.9}$$

are reversible on $\mathcal{U}_a \times \mathcal{V}_a \times \mathcal{E}_a$. Thus, on $\mathcal{U}_k$, the criticality condition

$$\begin{aligned} \delta\Pi(\bar{u}; u) &= \langle u \, , \, DF(\bar{u})\rangle + \langle \partial_t u \, , \, DK(\partial_t \bar{u})\rangle - \langle \Lambda_t(\bar{u})u \, ; \, DE(\Lambda(\bar{u}))\rangle \\ &= \langle u \, , \, \bar{u}^*\rangle + \langle u \, , \, \partial_t^* \bar{p}\rangle - \langle u \, ; \, \Lambda_t^*(\bar{u})\bar{\xi}^*\rangle = 0 \quad \forall u \in \mathcal{U}_k \end{aligned}$$

leads to the *fundamental inclusion*

$$0 \in \partial_t^* \partial K(\partial_t u) - \Lambda_t^*(u)\partial E(\Lambda(u)) + \partial F(u). \tag{2.10}$$

In terms of canonical variables, this fundamental inclusion can be written in the *trio-canonical forms*, namely,

$$\begin{array}{ll} \text{(1) geometrical equations:} & v = \partial_t u, \quad \xi = \Lambda(u), \\ \text{(2) physical relations:} & p = DK(v), \quad \xi^* \in \partial E(\xi), \quad u^* \in \partial F(u), \\ \text{(3) balance equation:} & u^* = \Lambda_t^*(u)\xi^* - \partial_t^* p. \end{array} \tag{2.11}$$

The framework for the fully nonlinear system is shown in Fig. 3.6. Extensive illustrations of the canonical transformation and the trio-canonical forms in mathematical physics and variational analysis were given in the monograph by Gao (1999b).

$$
\begin{array}{ccccc}
v \in \mathcal{V}_a \subset \mathcal{V} & \longleftarrow & \langle v \,,\, p \rangle & \longrightarrow & \mathcal{V}^* \supset \mathcal{V}^*_a \ni p \\
\partial_t \Big\uparrow & & & & \Big\downarrow \partial_t^* \\
u \in \mathcal{U}_a \subset \mathcal{U} & \longleftarrow & \langle u \,,\, u^* \rangle & \longrightarrow & \mathcal{U}^* \supset \mathcal{U}^*_a \ni u^* \\
\Lambda_t + \Lambda_c = \Lambda \Big\downarrow & & & & \Big\uparrow \Lambda_t^* = (\Lambda - \Lambda_c)^* \\
\xi \in \mathcal{E}_a \subset \mathcal{E} & \longleftarrow & \langle \xi \,;\, \xi^* \rangle & \longrightarrow & \mathcal{E}^* \supset \mathcal{E}^*_a \ni \xi^*
\end{array}
$$

Figure 3.6. Framework in fully nonlinear Newtonian systems

3. Canonical Hamiltonian, Total Complementary Energy and Dual Action

In this section, we shall study extended Lagrange forms and the associated dual actions in fully nonlinear, nonsmooth Hamilton systems.

Definition 3 Let $\mathcal{Z} = \mathcal{U} \times \mathcal{V}^* \times \mathcal{E}^*$ be the so-called *extended canonical phase space.*

(D1) The functional $\Theta : \mathcal{Z} \to \bar{\mathbb{R}}$ defined by

$$\Theta(u, p, \xi^*) = K^*(p) - E^*(\xi^*) - F(u) \in \Gamma(\mathcal{U}) \times \Gamma(\mathcal{V}^*) \times \Gamma(\mathcal{E}^*) \tag{3.1}$$

is called the *extended canonical Hamiltonian* associated with Π;

(D2) The functional $\Xi : \mathcal{Z} \to \bar{\mathbb{R}}$ defined by

$$\begin{aligned} \Xi(u, p, \xi^*) &= \langle \partial_t u \,,\, p \rangle - \langle \Lambda(u) \,;\, \xi^* \rangle - \Theta(u, p, \xi^*) \\ &= \langle \partial_t u \,,\, p \rangle - \langle \Lambda(u) \,;\, \xi^* \rangle - K^*(p) + E^*(\xi^*) + F(u) \end{aligned} \tag{3.2}$$

is called the *total complementary energy* of $(\mathcal{P})$ associated with Λ. $\diamondsuit$

The total complementary energy Ξ is also called the *extended Lagrangian* in [19]. Since $K^* \in \Gamma_0(\mathcal{V}^*)$, $E^* \in \Gamma_0(\mathcal{E}^*)$ are canonical functionals, it is easy to show that the fundamental form (2.10) is equivalent to the *extended canonical Hamilton forms*

$$\begin{aligned} \partial_t u &\in \partial_p \Theta(u, p, \xi^*), \\ \partial_t^* p - \Lambda_t^*(u)\xi^* &\in \partial_u \Theta(u, p, \xi^*), \\ -\Lambda(u) &\in \partial_{\xi^*} \Theta(u, p, \xi^*). \end{aligned} \tag{3.3}$$

In the case that $K \in \Gamma_G(\mathcal{V}_a)$, $F \in \Gamma_G(\mathcal{U}_a)$ and $E \in \Gamma_G(\mathcal{E}_a)$ are Gâteaux differentiable canonical functionals, then, the extended Lagrangian Ξ is Gâteaux-differentiable on $\mathcal{Z}_a = \mathcal{U}_a \times \mathcal{V}_a^* \times \mathcal{E}_a^*$,, and the criticality condition $D\Xi(\bar{u}, \bar{p}, \bar{\xi}^*) = 0$ leads to the following *extended Lagrange equations*

$$\begin{array}{l} \partial_t u = DK^*(p), \\ \partial_t^* p = \Lambda_t^*(u)\xi^* - DF(u), \\ \Lambda(u) = DE^*(\xi^*). \end{array} \tag{3.4}$$

Since K, E and F are canonical functionals, we know that, by the Legendre duality theory, any critical point of Ξ solves the variational problem $(\mathcal{P})$. Thus, the canonical primal problem $(\mathcal{P}_c)$ is equivalent to the general variational problem on $\mathcal{Z}_a$:

$$(\Xi): \quad \Xi(u, p, \xi^*) \to \text{sta} \ \ \forall (u, p, \xi^*) \in \mathcal{Z}_a. \tag{3.5}$$

By the fact that $E(\xi) = \text{ext}\{\langle \xi \, ; \, \xi^* \rangle - E^*(\xi^*) | \ \ \forall \xi^* \in \mathcal{E}^*\}$, then for any fixed $(u, p) \in \mathcal{U} \times \mathcal{V}$, we have

$$L(u, p) = \text{ext}\{\Xi(u, p, \xi^*) | \ \ \forall \xi^* \in \mathcal{E}^*\} = \langle \partial_t u \ , \ p \rangle - K^*(p) - P(u),$$

which is the well-known classical Lagrangian.

In general, for a given $u \in \mathcal{U}$, the extended Lagrangian $\Xi : \mathcal{V}^* \times \mathcal{E}^* \to \bar{\mathbb{R}}$ is an extended canonical functional, and

$$\Pi(u) = \text{sta}\{\Xi(u, p, \xi^*) | \ \ \forall (p, \xi^*) \in \mathcal{V}^* \times \mathcal{E}^*\}. \tag{3.6}$$

Particularly, if $E \in \check{\Gamma}_G(\mathcal{E}_a)$, $K \in \check{\Gamma}_G(\mathcal{V}_a)$, then $\Xi : \mathcal{V}_a^* \times \mathcal{E}_a^* \to \mathbb{R}$ is a saddle functional for any given $u \in \mathcal{U}_a$. In this case, the stationary problem (3.6) is actually the saddle minimax problem

$$\Pi(u) = \inf_{\xi^* \in \mathcal{E}_a^*} \sup_{p \in \mathcal{V}_a^*} \Xi(u, p, \xi^*) = \sup_{p \in \mathcal{V}_a^*} \inf_{\xi^* \in \mathcal{E}_a^*} \Xi(u, p, \xi^*) \ \ \forall u \in \mathcal{U}_a. \tag{3.7}$$

On the other hand, for any given $(p, \xi^*) \in \mathcal{V}_a^* \times \mathcal{E}_a^*$, the convexity of $\Xi(u, \cdot, \cdot) \to \mathbb{R}$ depends on the operator Λ and $\xi^* \in \mathcal{E}^*$. The dual action $\Pi^d : \mathcal{V}_a^* \times \mathcal{E}_a^* \to \mathbb{R}$ can be defined by

$$\begin{aligned} \Pi^d(p, \xi^*) &= \text{sta}\{\Xi(u, p, \xi^*) | \ \ \forall u \in \mathcal{U}_a\} \\ &= E^*(\xi^*) - K^*(p) - F^\Lambda(p, \xi^*) \ \ \forall (p, \xi^*) \in \mathcal{V}_a^* \times \mathcal{E}_a^*, \end{aligned} \tag{3.8}$$

where $F^\Lambda(p, \xi^*)$ is the so-called Λ-dual functional of $F(u)$ defined by the *canonical dual transformation*

$$F^\Lambda(p, \xi^*) = \text{sta}\{\langle \Lambda(u) \, ; \, \xi^* \rangle - \langle \partial_t u \ , \ p \rangle - F(u) | \ \ \forall u \in \mathcal{U}\}. \tag{3.9}$$

For any given $(p, \xi^*) \in \mathcal{V}^* \times \mathcal{E}^*$, the solution $\bar{u}$ of this stationary variational problem satisfying the balance equation:

$$\Lambda_t^*(\bar{u})\xi^* - \partial_t^* p \in \partial F(\bar{u}). \tag{3.10}$$

Let $\mathcal{W}_s^* \subset \mathcal{V}_a^* \times \mathcal{E}_a^*$ be the so-called *dual feasible space* such that for any given $(p, \xi^*) \in \mathcal{W}_s^*$, the critical point $\bar{u}$ of the stationary problem (3.9) can be determined as $\bar{u} = \bar{u}(p, \xi^*)$, the problem that is dual to the canonical primal problem $(\mathcal{P})$ can be proposed as the following.

Problem 2 (Canonical Dual Variational Problem) Suppose that the dual feasible space $\mathcal{W}_s^*$ is not empty. Find the critical point of Π^d such that

$$(\mathcal{P}^d): \quad \Pi^d(p, \xi^*) \quad \to \text{sta } \forall (p, \xi^*) \in \mathcal{W}_s^*. \tag{3.11}$$

The following lemma plays a key role in duality theory for nonlinear dynamical systems.

Lemma 1 *Let $\Xi(u, p, \xi^*)$ be a given extended Lagrangian associated with $(\mathcal{P}_c)$ and $\Pi^d(p, \xi^*)$ the dual action defined by (3.8). Suppose that $\mathcal{Z}_r = \mathcal{U}_r \times \mathcal{V}_r^* \times \mathcal{E}_r^*$ is an open subset of $\mathcal{Z}_a$ and $(\bar{u}, \bar{p}, \bar{\xi}^*) \in \mathcal{Z}_r$ is a critical point of Ξ on $\mathcal{Z}_r$, Π is Gâteaux differentiable at $\bar{u}$, and Π^d is Gâteaux differentiable at $(\bar{p}, \bar{\xi}^*)$. Then $D\Pi(\bar{u}) = 0$, $D\Pi^d(\bar{p}, \bar{\xi}^*) = 0$, and*

$$\Pi(\bar{u}) = \Xi(\bar{u}, \bar{p}, \bar{\xi}^*) = \Pi^d(\bar{p}, \bar{\xi}^*). \tag{3.12}$$

This Lemma can be proved by using some fundamental results in calculus of variations (cf. e.g., Gao, 1999b). Lemma 1 shows that the critical points of the extended Lagrangian are also the critical points for both the primal and dual variational problems. The extremality of the primal and dual actions will be discussed in the next section.

In the case that $F(u)$ is a linear functional, then by the Riesz representation theory we know that there exists an element $\bar{u}^* \in \mathcal{U}^*$ such that $F(u) = \langle u \ , \ \bar{u}^* \rangle$. Thus, $DF(\bar{u}) = \bar{u}^*$ and the critical point $\bar{u}$ of the problem (3.9) satisfies the balance equation

$$\partial_t^* p - \Lambda_t^*(\bar{u})\xi^* + \bar{u}^* = 0. \tag{3.13}$$

If for any given $(p, \xi^*) \in \mathcal{W}_s^*$, the critical point $\bar{u} = \bar{u}(p, \xi^*)$ can be well determined by this balance equation, then by the canonical operator decomposition $\Lambda = \Lambda_t + \Lambda_c$, we have

$$F^\Lambda(p, \xi^*) = -G^d(p, \xi^*), \ \ s.t. \ (p, \xi^*) \in \mathcal{W}_s^*, \tag{3.14}$$

where $G^d(p, \xi^*) = \langle -\Lambda_c(\bar{u}(p, \xi^*)) \ ; \ \xi^* \rangle = G(\bar{u}, \xi^*)$ is the *pure complementary gap function* (see, Gao, 1999b). In this case, the dual action reads

$$\Pi^d(p, \xi^*) = E^*(\xi^*) + G^d(p, \xi^*) - K^*(p). \tag{3.15}$$

Clearly, $G^d(p,\xi^*) = 0$ if Λ is a linear operator. In this case, $\Xi(u,p,\xi^*)$ is a canonical functional and Π^d can be considered as the Fenchel-Rockafellar dual action. Lemma 1 shows that in nonconvex systems, the duality gap existing in the Fenchel-Rockafellar duality theory is recovered by the complementary gap functional.

4. Triality Theory in Fully Nonlinear Dynamical Systems

In order to clarify the extremality conditions of the extended Lagrangian, in this section we shall assume that $E \in \check{\Gamma}_G(\mathcal{E}_a)$, $K \in \check{\Gamma}_G(\mathcal{V}_a)$ are convex canonical functionals and that $F(u) = \langle u\ ,\ \bar{u}^* \rangle$ is a linear functional on $\mathcal{U}_a$. Let $\mathcal{Z}_c \subset \mathcal{Z}_a = \mathcal{U}_a \times \mathcal{V}_a^* \times \mathcal{E}_a^*$ be a critical point set of Ξ, i.e.,

$$\mathcal{Z}_c = \{(\bar{u},\bar{p},\bar{\xi}^*) \in \mathcal{Z}_a |\ \ \delta\Xi(\bar{u},\bar{p},\bar{\xi}^*; u,p,\xi^*) = 0\ \ \forall (u,p,\xi^*) \in \mathcal{Z}_a\}.$$

For any given critical point $(\bar{u},\bar{p},\bar{\xi}^*) \in \mathcal{Z}_c$, we let $\mathcal{Z}_r = \mathcal{U}_r \times \mathcal{V}_r^* \times \mathcal{E}_r^* \subset \mathcal{Z}_a$ be its *neighborhood* such that on $\mathcal{Z}_r$, $(\bar{u},\bar{p},\bar{\xi}^*)$ is the only critical point of Ξ. The following extremum results are of fundamental importance in the stability analysis and critical point theory of nonlinear dynamical systems.

Theorem 1 (Triality Theorem) *Suppose that $(\bar{u},\bar{p},\bar{\xi}^*) \in \mathcal{Z}_c$, and $\mathcal{Z}_r$ is a neighborhood of $(\bar{u},\bar{p},\bar{\xi}^*)$. If the functional $G_{\xi^*}(u) = \langle \Lambda(u)\ ;\ \bar{\xi}^* \rangle$ is concave on $\mathcal{U}_r$, then on $\mathcal{Z}_r$,*

$$\begin{aligned}\Xi(\bar{u},\bar{p},\bar{\xi}^*) &= \min_u \max_p \min_{\xi^*} \Xi(u,p,\xi^*) = \max_p \min_u \min_{\xi^*} \Xi(u,p,\xi^*) \\ &= \min_{u,\xi^*} \max_p \Xi(u,p,\xi^*) = \max_p \min_{u,\xi^*} \Xi(u,p,\xi^*). \qquad (4.1)\end{aligned}$$

However, if $G_{\xi^*}(u)$ is convex on $\mathcal{U}_r$, then on $\mathcal{Z}_r$ we have either

$$\begin{aligned}\Xi(\bar{u},\bar{p},\bar{\xi}^*) &= \min_u \max_p \min_{\xi^*} \Xi(u,p,\xi^*) = \min_p \max_u \min_{\xi^*} \Xi(u,p,\xi^*) \\ &= \min_{\xi^*,u} \max_p \Xi(u,p,\xi^*) = \min_{p,\xi^*} \max_u \Xi(u,p,\xi^*). \qquad (4.2)\end{aligned}$$

or

$$\begin{aligned}\Xi(\bar{u},\bar{p},\bar{\xi}^*) &= \max_u \min_{\xi^*} \max_p \Xi(u,p,\xi^*) = \max_p \min_{\xi^*} \max_u \Xi(u,p,\xi^*) \\ &= \min_{\xi^*} \max_{u,p} \Xi(u,p,\xi^*) = \max_{u,p} \min_{\xi^*} \Xi(u,p,\xi^*). \qquad (4.3)\end{aligned}$$

Proof. Since $E^* \in \check{\Gamma}(\mathcal{E}_a^*)$, $K^* \in \check{\Gamma}(\mathcal{V}_a^*)$, if $G_{\xi^*}(u) = \langle \Lambda(u)\ ;\ \bar{\xi}^* \rangle$ is concave on $\mathcal{U}_r$, then for the given $\bar{\xi}^*$, $\Xi \in \check{\Gamma}(\mathcal{U}_r) \times \hat{\Gamma}(\mathcal{V}_a^*)$ is a saddle functional.

Thus the equality (4.1) follows from the saddle-Lagrangian duality theorem (cf. e.g., Gao, 1999b). However, if $\langle \Lambda(u) \; ; \; \bar{\xi}^* \rangle$ is convex on $\mathcal{U}_r$, then for any given $\xi^* \in \mathcal{E}_r^*$, the extended Lagrangian $\Xi \in \hat{\Gamma}(\mathcal{U}_r) \times \hat{\Gamma}(\mathcal{V}_a^*)$ is a *super-critical functional* (see Gao, 1999b). By the *super-Lagrangian duality theorem* proved in Gao (1999b), we have either (4.2) or (4.3). □

Theorem 2 (Tri-Duality Theorem) *Suppose that* $(\bar{u}, \bar{p}, \bar{\xi}^*) \in \mathcal{Z}_c$ *is a critical point of* Ξ *and* $\mathcal{Z}_r = \mathcal{U}_r \times \mathcal{V}_r^* \times \mathcal{E}_r^*$ *is a neighborhood of* $(\bar{u}, \bar{p}, \bar{\xi}^*)$ *such that* $\mathcal{V}_r^* \times \mathcal{E}_r^* \subset \mathcal{W}_s^*$. *If* $\langle \Lambda(u) \; ; \; \bar{\xi}^* \rangle$ *is concave on* $\mathcal{U}_r$, *then*

$$\Pi(\bar{u}) = \min_{u \in \mathcal{U}_r} \Pi(u) \quad \textit{iff} \quad \Pi^d(\bar{p}, \bar{\xi}^*) = \max_{p \in \mathcal{V}_r^*} \min_{\xi^* \in \mathcal{E}_r^*} \Pi^d(p, \xi^*). \tag{4.4}$$

However, if $\langle \Lambda(u) \; ; \; \bar{\xi}^* \rangle$ *is convex on* $\mathcal{U}_r$, *then*

$$\Pi(\bar{u}) = \min_{u \in \mathcal{U}_r} \Pi(u) \quad \textit{iff} \quad \Pi^d(\bar{p}, \bar{\xi}^*) = \min_{(p, \xi^*) \in \mathcal{W}_s^*} \Pi^d(p, \xi^*); \tag{4.5}$$

$$\Pi(\bar{u}) = \max_{u \in \mathcal{U}_r} \Pi(u) \quad \textit{iff} \quad \Pi^d(\bar{p}, \bar{\xi}^*) = \max_{p \in \mathcal{V}_r^*} \min_{\xi^* \in \mathcal{E}_r^*} \Pi^d(p, \xi^*). \tag{4.6}$$

Proof. This theorem can be proved by combining Lemma 1 and the triality theorem. □

As we have seen that there are two variational arguments p and ξ^* in the canonical dual action problem, subjected to the constraint $(p, \xi^*) \in \mathcal{W}_s^*$. The so-called *polar variational method* proposed by Gao (1999b) in static systems can be used to relax this constraint.

5. Bi-Polarity Principle in Nonsmooth Elastodynamics

The goal of this section is to discuss the duality principles in geometrically linear nonconservative Hamilton systems, i.e., for a given total action $\Pi(u)$, there exists a linear operator $\Lambda : \mathcal{U} \to \mathcal{E}$ and canonical functionals $F \in \Gamma_G(\mathcal{U}_a)$, $E \in \Gamma_G(\mathcal{E}_a)$ such that $\Pi(u) = K(\partial_t u) - E(\Lambda u) + F(u)$. General framework was established recently by Gao (2000d). Here we shall study the particularly important problems and concrete applications in boundary/initial value problems of nonsmooth elastodynamics with dissipation.

Let a deformable body in an undeformed state occupy an open, simply connected, bounded domain $\Omega_t = \Omega \times I \subset \mathbb{R}^3 \times \mathbb{R}^+$ with boundary $\Gamma = \partial\Omega_t = \partial\Omega \times \partial I$. Let the general admissible deformation space $\mathcal{U}$ be a subset of Lebesgue integrable space $\mathcal{L}^\alpha(\Omega_t; \mathbb{R}^3)$ $(1 < \alpha < \infty)$, in which certain differentiable conditions are given[1]. The dual space $\mathcal{U}^* \subset \mathcal{L}^{\alpha^*}(\Omega_t; \mathbb{R}^3)$ of $\mathcal{U}$ is the admissible force space, α^* is the dual number of α $(1/\alpha + 1/\alpha^* = 1)$. The element $\mathbf{u}^* \in \mathcal{U}^*$ is specified as the body force in the domain Ω_t, and the surface traction on the boundary Γ. The bilinear form $\langle *, * \rangle : \mathcal{U} \times \mathcal{U}^* \to \mathbb{R}$ is defined by

$$\langle \mathbf{u} \, , \, \mathbf{u}^* \rangle = \int_{\Omega_t} e^{\omega(t,\mathbf{x})} \mathbf{u} \cdot \mathbf{u}^* \mathrm{d}\Omega_t + \int_\Gamma e^{\omega(t,\mathbf{x})} \mathbf{u} \cdot \mathbf{u}^* \, \mathrm{d}\Gamma,$$

where $\omega(\mathbf{x}, t) : \Omega_t \to \mathbb{R}$ is a given weight function.

In mixed boundary/initial value problems, the boundary Γ can be split into two disjoint parts, i.e., $\Gamma = \Gamma_f \cup \Gamma_u$, where $\Gamma_f = \partial\Omega_f \cup \partial I_f$, and $\Gamma_u = \partial\Omega_u \cup \partial I_u$ satisfying $\Gamma_f \cap \Gamma_u = \emptyset$. On Γ_f, the surface traction $\bar{\mathbf{t}}$ is given; while on the remaining part Γ_u, the displacement is prescribed. For problems with homogeneous boundary condition, the admissible deformation space $\mathcal{U}_a \subset \mathcal{U}$ can be defined by

$$\mathcal{U}_a = \{ \mathbf{u} \in \mathcal{U} | \ \mathbf{u}(\mathbf{x}, t) = 0 \ \ \forall (\mathbf{x}, t) \in \Gamma_u \}. \tag{5.1}$$

If the external force is given as $\bar{\mathbf{u}}^*(\mathbf{x}, t) = \{ \bar{\mathbf{b}}(\mathbf{x}, t) \ \ (\text{in } \Omega_t); \ \bar{\mathbf{t}}(\mathbf{x}, t) \ \ (\text{in } \Gamma_f) \}$, the external potential $F : \mathcal{U} \to \bar{\mathbb{R}}$ can be defined as

$$F(\mathbf{u}) = \langle \mathbf{u}, \bar{\mathbf{u}}^* \rangle = \int_{\Omega_t} e^{\omega(\mathbf{x},t)} \mathbf{u} \cdot \bar{\mathbf{b}} \, \mathrm{d}\Omega_t + \int_{\Gamma_f} e^{\omega(\mathbf{x},t)} \mathbf{u} \cdot \bar{\mathbf{t}} \, \mathrm{d}\Gamma, \tag{5.2}$$

which is a linear functional defined on a convex domain $\mathcal{U}_a$. Thus, the duality relation

$$\mathbf{u}^* = DF(\mathbf{u}) = \begin{cases} \bar{\mathbf{b}} & \text{in } \Omega_t, \\ \bar{\mathbf{t}} & \text{on } \Gamma_f \end{cases} \quad \forall \mathbf{u} \in \mathcal{U}_a \tag{5.3}$$

gives the natural boundary condition.

In Newtonian systems, the kinetic energy is a quadratic functional on the Hilbert space $\mathcal{V}_a = \mathcal{V} = \mathcal{H}(\Omega_t; \mathbb{R}^3)$

$$K(\mathbf{v}) = \frac{1}{2} \langle \mathbf{v} \, , \, \rho \mathbf{v} \rangle = \int_{\Omega_t} \frac{1}{2} e^{\omega(\mathbf{x},t)} \rho \mathbf{v} \cdot \mathbf{v} \, \mathrm{d}\Omega_t.$$

For the linear operator $\partial_t = \partial / \partial t$, its adjoint $\partial_t^* : \mathcal{V}^* \to \mathcal{U}^*$ can be obtained by using integration by parts

$$\begin{aligned}
\langle \partial_t \mathbf{u} \, , \, \mathbf{p} \rangle &= \int_{\Omega_t} e^{\omega(\mathbf{x},t)} \frac{\partial \mathbf{u}}{\partial t} \cdot \mathbf{p} \, \mathrm{d}\Omega_t \\
&= \int_{\Omega_t} [e^{\omega(\mathbf{x},t)} \mathbf{u} \cdot \mathbf{p}]_{,t} \, \mathrm{d}\Omega_t - \int_{\Omega_t} \mathbf{u} \cdot (e^{\omega(\mathbf{x},t)} \mathbf{p})_{,t} \, \mathrm{d}\Omega_t \\
&= \int_\Omega e^{\omega(\mathbf{x},t)} n \mathbf{u} \cdot \mathbf{p} |_{\partial I_f} \, \mathrm{d}\Omega - \int_{\Omega_t} e^{\omega(\mathbf{x},t)} \mathbf{u} \cdot (\mathbf{p}_{,t} + \omega_{,t} \mathbf{p}) \, \mathrm{d}\Omega_t \\
&= \langle \mathbf{u} \, , \, \partial_t^* \mathbf{p} \rangle.
\end{aligned}$$

Thus, the adjoint operator ∂_t^* is defined by

$$\partial_t^* = \begin{cases} -\frac{\partial}{\partial t} - \omega_{,t} & \text{in } \Omega_t \\ n & \text{on } \partial I_f, \end{cases}$$

where $n = \pm 1$ is a unit vector normal to the time boundary ∂I. For linear time-dissipative systems, the weight function is usually a linear function of $t \in I$, say, $\omega = \nu t$, where $\nu > 0$ is a given constant (see, Gao, 1999b).

For an infinitesmall deformation system, the canonical strain measure $\xi = \boldsymbol{\epsilon}$ is a symmetrical tensor field

$$\boldsymbol{\epsilon} = \Lambda \mathbf{u} = \frac{1}{2}(\nabla \mathbf{u} + (\nabla \mathbf{u})^T). \tag{5.4}$$

Thus, the geometrical operator $\Lambda = \nabla_s = \frac{1}{2}(\nabla + \nabla^T)$ is a linear differential mapping, which assigns to each displacement vector $\mathbf{u} \in \mathcal{U}$ a symmetrical tensor in $\mathbb{R}^{3\times 3}(\Omega_t)$. Let

$$\mathcal{E} = \{\boldsymbol{\epsilon} \in \mathcal{L}^\beta(\Omega_t; \mathbb{R}^{3\times 3})| \ \boldsymbol{\epsilon} = \boldsymbol{\epsilon}^T \ \text{ in } \Omega_t\}$$

be a space of all second-order, symmetrical, Lebesgue integrable tensor functions with domain in Ω_t and range in $\mathbb{R}^{3\times 3}$. For each given $\boldsymbol{\epsilon} \in \mathcal{E}$, the conjugate stress measure $\boldsymbol{\sigma} = \xi^*$ is also a second-order symmetrical tensor field, defined by the general nonsmooth constitutive law

$$\boldsymbol{\sigma} \in \partial U(\boldsymbol{\epsilon}) : \mathcal{E} \to \mathcal{E}^* \tag{5.5}$$

where $U : \mathcal{E} \to \bar{\mathbb{R}}$ is the so-called stored strain energy density, and

$$\mathcal{E}^* = \{\boldsymbol{\sigma} \in \mathcal{L}^{\beta^*}(\Omega_t; \mathbb{R}^{3\times 3})| \ \boldsymbol{\sigma} = \boldsymbol{\sigma}^T \ \text{ in } \Omega_t\}$$

is the dual space of $\mathcal{E}$ with $1/\beta + 1/\beta^* = 1$. The bilinear form between $\mathcal{E}$ and $\mathcal{E}^*$ can be defined by

$$\langle \boldsymbol{\epsilon} \, ; \boldsymbol{\sigma} \rangle = \int_{\Omega_t} e^{\omega(t,\mathbf{x})} \boldsymbol{\epsilon} : \boldsymbol{\sigma} \, d\Omega_t,$$

where $\boldsymbol{\epsilon} : \boldsymbol{\sigma} = \text{tr}(\boldsymbol{\epsilon} \cdot \boldsymbol{\sigma})$. By using the Gauss-Green formula, we have

$$\begin{aligned} \langle \Lambda \mathbf{u} \, ; \boldsymbol{\sigma} \rangle &= \int_{\Omega_t} e^{\omega(t,\mathbf{x})} (\nabla_s \mathbf{u}) : \boldsymbol{\sigma} \, d\Omega_t \\ &= \int_{\Omega} [\nabla \cdot (e^{\omega(t,\mathbf{x})} \mathbf{u} \cdot \boldsymbol{\sigma}) - \mathbf{u} \cdot (\nabla \cdot (e^{\omega(t,\mathbf{x})} \boldsymbol{\sigma}))] \, d\Omega_t \\ &= \int_I \oint_{\partial\Omega} e^{\omega(\mathbf{x},t)} \mathbf{u} \cdot \boldsymbol{\sigma} \cdot \mathbf{n} d\partial\Omega \, dt \\ &\quad - \int_{\Omega_t} e^{\omega(\mathbf{x},t)} \mathbf{u} \cdot [\nabla \cdot \boldsymbol{\sigma} + (\nabla \omega(\mathbf{x}, t)) \cdot \boldsymbol{\sigma}] \, d\Omega_t \\ &= \langle \mathbf{u}, \Lambda^* \boldsymbol{\sigma} \rangle, \end{aligned}$$

where $\mathbf{n} \in \mathbb{R}^3$ is a unit vector normal to $\partial\Omega$. Thus, the adjoint operator $\Lambda^* : \mathcal{E}^* \to \mathcal{U}^*$ associated with these bilinear forms is defined by

$$\Lambda^* \boldsymbol{\sigma} = \begin{cases} -\nabla \cdot \boldsymbol{\sigma} - (\nabla \omega(\mathbf{x}, t)) \cdot \boldsymbol{\sigma} & \text{in } \Omega_t, \\ \mathbf{n} \cdot \boldsymbol{\sigma} & \text{on } \partial\Omega. \end{cases}$$

Let the admissible strain space $\mathcal{E}_a$ be a subset of $\mathcal{E}$, on which, the stored strain energy density $U(\boldsymbol{\epsilon})$ is finite and convex; $\mathcal{E}_a^* \subset \mathcal{E}^*$ be the range of the constitutive mapping $\partial U : \mathcal{E}_a \to \mathcal{E}^*$ such that the Fenchel duality relations

$$\boldsymbol{\sigma} \in \partial U(\boldsymbol{\epsilon}) \quad \Leftrightarrow \quad \boldsymbol{\epsilon} \in \partial U^*(\boldsymbol{\sigma}) \quad \Leftrightarrow \quad U(\boldsymbol{\epsilon}) + U^*(\boldsymbol{\sigma}) = \mathrm{tr}(\boldsymbol{\epsilon} \cdot \boldsymbol{\sigma})$$

hold on $\mathcal{E}_a \times \mathcal{E}_a^*$. Thus, on the kinematically admissible space $\mathcal{U}_k = \{\mathbf{u} \in \mathcal{U}_a | \ \partial_t \mathbf{u} \in \mathcal{V}_a, \ \ \nabla_s \mathbf{u} \in \mathcal{E}_a\}$, the primal problem

$$\begin{aligned} \Pi(\mathbf{u}) &= K(\partial_t \mathbf{u}) - E(\Lambda \mathbf{u}) + F(\mathbf{u}) \\ &= \int_{\Omega_t} e^{\omega(t,\mathbf{x})} \left(\frac{1}{2} \rho \mathbf{u}_{,t} \cdot \mathbf{u}_{,t} - U(\nabla_s \mathbf{u}) \right) \mathrm{d}\Omega_t + F(\mathbf{u}) \to \mathrm{sta} \ \ \forall \mathbf{u} \in \mathcal{U}_k \end{aligned}$$

is a geometrically linear variational problem. The criticality condition of Π leads to

the standard geometrically linear trio-canonical forms in the domain Ω_t

$$\begin{cases} \mathbf{v} = \partial_t \mathbf{u} = \mathbf{u}_{,t}, \ \ \boldsymbol{\epsilon} = \Lambda \mathbf{u} = \nabla_s \mathbf{u}, \\ \mathbf{p} = DK(\mathbf{v}) = \rho \mathbf{v}, \ \ \boldsymbol{\sigma} \in \partial U(\boldsymbol{\epsilon}), \ \ \mathbf{u}^* \in \partial F(\mathbf{u}), \\ \mathbf{u}^* = -\partial_t^* \mathbf{p} + \Lambda^* \boldsymbol{\sigma} = \mathbf{p}_{,t} + \omega_{,t} \mathbf{p} - \nabla \cdot \boldsymbol{\sigma} - (\nabla \omega) \cdot \boldsymbol{\sigma}, \end{cases} \tag{5.6}$$

and the natural boundary condition $\mathbf{u}^* = DF(\mathbf{u}) = \bar{\mathbf{t}}$ on Γ_f.

To see a unified beauty of duality theory in geometrically linear systems, we introduce two product spaces $\mathcal{W} = \mathcal{V} \times \mathcal{E}$ and $\mathcal{W}^* = \mathcal{V}^* \times \mathcal{E}^*$, placed in duality by the bilinear form $\langle * {:} * \rangle : \mathcal{W} \times \mathcal{W}^* \to \mathbb{R}$. For each given $\boldsymbol{\varpi} = (\mathbf{v}, -\boldsymbol{\epsilon})^T \in \mathcal{W}$, the canonical function $W : \mathcal{W} \to \bar{\mathbb{R}}$, defined by

$$\begin{aligned} W(\boldsymbol{\varpi}) &= W(\mathbf{v}, \boldsymbol{\epsilon}) = K(\mathbf{v}) - E(\boldsymbol{\epsilon}) \\ &= \int_{\Omega_t} e^{\omega(\mathbf{x},t)} \left[\frac{1}{2} \rho \mathbf{v} \cdot \mathbf{v} - U(\boldsymbol{\epsilon}) \right] \mathrm{d}\Omega_t, \end{aligned} \tag{5.7}$$

is called the *stored action* of the system. For each given $\boldsymbol{\varpi}^* = (\mathbf{p}, \boldsymbol{\sigma})^T \in \mathcal{W}^*$, the conjugate functional of W is simply defined by

$$\begin{aligned} W^*(\boldsymbol{\varpi}^*) &= \mathrm{ext}\{\langle \boldsymbol{\varpi} {:} \boldsymbol{\varpi}^* \rangle - W(\boldsymbol{\varpi}) | \ \forall \boldsymbol{\varpi} \in \mathcal{W}\} \\ &= \mathrm{ext}\{\langle \mathbf{v} \ , \ \mathbf{p} \rangle - K(\mathbf{v}) - \langle \boldsymbol{\epsilon} \ ; \ \boldsymbol{\epsilon}^* \rangle + E(\boldsymbol{\epsilon}) | \ \forall \mathbf{v} \in \mathcal{V}, \ \ \boldsymbol{\epsilon} \in \mathcal{E}\} \\ &= \int_{\Omega_t} e^{\omega(\mathbf{x},t)} \left[\frac{1}{2} \rho^{-1} \mathbf{p} \cdot \mathbf{p} - U^*(\boldsymbol{\sigma}) \right] \mathrm{d}\Omega_t = K^*(\mathbf{p}) - E^*(\boldsymbol{\epsilon}^*). \end{aligned}$$

Thus, the Fenchel duality relations

$$\varpi^* \in \partial W(\varpi) \;\Leftrightarrow\; \varpi \in \partial W^*(\varpi^*) \;\Leftrightarrow\; W(\varpi) + W^*(\varpi) = \langle \varpi : \varpi^* \rangle$$

holds on $\mathcal{W} \times \mathcal{W}^*$.

Let $\mathcal{W}_a = \mathcal{V}_a \times \mathcal{E}_a$. By introducing a space-time operator $\Upsilon = (\partial_t, -\Lambda)^T : \mathcal{U}_a \to \mathcal{W}_a$, the geometrical equations in (5.6) can be written in the vector form

$$\varpi(\mathbf{u}) = \begin{pmatrix} \mathbf{v}(\mathbf{u}) \\ -\boldsymbol{\epsilon}(\mathbf{u}) \end{pmatrix} = \Upsilon \mathbf{u} = \begin{pmatrix} \partial_t \mathbf{u} \\ -\Lambda \mathbf{u} \end{pmatrix} \in \mathcal{W}_a, \tag{5.8}$$

and for any given $\mathbf{u} \in \mathcal{U}_a$ the balance operator $\Upsilon^* : \mathcal{W}_a^* = \mathcal{V}_a^* \times \mathcal{E}_a^* \to \mathcal{U}^*$ can be defined by

$$\langle \mathbf{u} \,,\, \Upsilon^* \varpi^* \rangle = \langle \Upsilon \mathbf{u} : \varpi^* \rangle \;\Rightarrow\; \Upsilon^* = (\partial_t^*, \, -\Lambda^*). \tag{5.9}$$

Thus, the Lagrangian $\Xi : \mathcal{Z} = \mathcal{U} \times \mathcal{W}^* \to \bar{\mathbb{R}}$ in geometrically linear dynamical systems has the standard form

$$\begin{aligned} \Xi(\mathbf{u}, \varpi^*) &= \langle \Upsilon \mathbf{u} : \varpi^* \rangle - W^*(\varpi^*) + F(\mathbf{u}) \\ &= \int_{\Omega_t} e^{\omega(\mathbf{x},t)} \left[\mathbf{u}_{,t} \cdot \mathbf{p} - (\nabla_s \mathbf{u}) : \boldsymbol{\sigma} - \frac{1}{2} \rho^{-1} \mathbf{p} \cdot \mathbf{p} + U^*(\boldsymbol{\sigma}) \right] d\Omega_t + F(\mathbf{u}), \end{aligned}$$

which is a canonical functional on $\mathcal{U}_a \times \mathcal{Z}_a$. For any given $\varpi^* \in \mathcal{W}_a^*$, the dual action can be obtained by the standard canonical Lagrange dual transformation

$$\Pi^d(\varpi^*) = \text{sta}\{\Xi(\mathbf{u}, \varpi^*) | \; \forall \varpi^* \in \mathcal{W}_a^*\} = F^*(\Upsilon^* \varpi^*) - W^*(\varpi^*), \tag{5.10}$$

where $F^*(\mathbf{u}^*)$ is defined by

$$F^*(\Upsilon^* \varpi^*) = \text{sta}\{\langle \Upsilon \mathbf{u} : \varpi^* \rangle + F(\mathbf{u}) | \; \forall \mathbf{u} \in \mathcal{U}_a\}. \tag{5.11}$$

The criticality condition of this problem leads to the balance equation

$$0 \in \Upsilon^* \varpi^* + \partial F(\bar{\mathbf{u}}). \tag{5.12}$$

For each critical point $\bar{\mathbf{u}} \in \mathcal{U}_a$, the general solution for this linear, non-homogeneous equation can be written as

$$\varpi^* = \varpi_o^* + \varpi_p^*, \quad \Upsilon^* \varpi_o^* = 0, \quad 0 \in \Upsilon^* \varpi_p^* + \partial F(\bar{\mathbf{u}}),$$

where $\varpi_p^* \in \mathcal{W}_a^*$ is a particular solution and ϖ_o^* is a homogeneous solution.

For mixed boundary-value problem such that $F(\mathbf{u}) = \langle \mathbf{u} \ , \ \bar{\mathbf{u}}^* \rangle$ is a linear functional, the dual feasible space is a hyper-plane in the product space $\mathcal{W}_a^*$

$$\mathcal{W}_s^* = \{\boldsymbol{\varpi}^* \in \mathcal{W}_a^* | \ \langle \mathbf{u}, \Upsilon^* \boldsymbol{\varpi}^* + \bar{\mathbf{u}}^* \rangle = 0 \ \ \forall \mathbf{u} \in \mathcal{U}_a\}.$$

Thus, on $\mathcal{W}_s^*$, we have $F^*(\Upsilon^* \varpi^*) = 0$, and the dual variational problem takes a very simple form

$$\begin{aligned} \Pi^d(\boldsymbol{\varpi}^*) &= -W^*(\boldsymbol{\varpi}^*) \\ &= \int_{\Omega_t} e^{\omega(\mathbf{x},t)} \left[U^*(\boldsymbol{\sigma}) - \frac{1}{2}\rho^{-1}\mathbf{p}\cdot\mathbf{p} \right] d\Omega_t \ \to \text{sta} \ \forall \boldsymbol{\varpi}^* \in \mathcal{W}_s^*. \end{aligned}$$

In order to relax the balance constraint in $\mathcal{W}_s^*$, we let $\boldsymbol{\varpi}^* = \boldsymbol{\varpi}_o^* + \boldsymbol{\varpi}_p^*$. For the given source $\bar{\mathbf{u}}^*$, the particular solution $\boldsymbol{\varpi}_p^*$ can by determined by solving the linear equation $\Upsilon^* \boldsymbol{\varpi}_p^* + \bar{\mathbf{u}}^* = 0$. Let $\mathcal{U}^o$ be the so-called *polar configuration space* (see Gao, 1999b), placed in duality with $\mathcal{U}^{o*}$ by the bilinear form $\langle *,* \rangle : \mathcal{U}^o \times \mathcal{U}^{o*} \to \mathbb{R}$. For the given geometrically linear system and $\Upsilon : \mathcal{U} \to \mathcal{W}$, a linear mapping $\Upsilon^o : \mathcal{U}^o \to \mathcal{W}^*$ is called a *null-source polar operator* of Υ if

$$\langle \Upsilon \mathbf{u} : \Upsilon^o \mathbf{u}^o \rangle = 0. \tag{5.13}$$

Clearly, if there exists a subset $\mathcal{U}_a^o \subset \mathcal{U}^o$ such that (5.13) holds for any given $\mathbf{u}^o \in \mathcal{U}_a^o \subset \mathcal{U}^o$, then

$$\Upsilon^{o*} \Upsilon \mathbf{u} = 0 \in \mathcal{U}^{o*}, \tag{5.14}$$

which is the so-called *compatibility condition.* On the other hand, if (5.13) holds for any $\mathbf{u} \in \mathcal{U}_a$, then the equation

$$\Upsilon^* \Upsilon^o \mathbf{u}^o = 0 \in \mathcal{U}^* \tag{5.15}$$

is called *polar compatibility condition* (Gao, 1999b). For a given operator $\Upsilon : \mathcal{U}_a \to \mathcal{W}$, the *null-source admissible polar configuration space* $\mathcal{U}_a^o \subset \mathcal{U}^o$ is defined by

$$\mathcal{U}_a^o = \{\mathbf{u}^o \in \mathcal{U}^o | \ \langle \Upsilon \mathbf{u} : \Upsilon^o \mathbf{u}^o \rangle = 0 \ \ \forall \mathbf{u} \in \mathcal{U}_a\}. \tag{5.16}$$

Thus, for any given $\mathbf{u}^o \in \mathcal{U}_a^o$, the balance condition $\boldsymbol{\varpi}^* = \boldsymbol{\varpi}_o^* + \boldsymbol{\varpi}_p^* \in \mathcal{W}_s^*$ is satisfied and $\boldsymbol{\varpi}_o^* = \Upsilon^o \mathbf{u}^o$ is in the null space of the balance operator Υ^*. A diagrammatic representation for the geometrically linear polar system is shown in Fig.3.7.

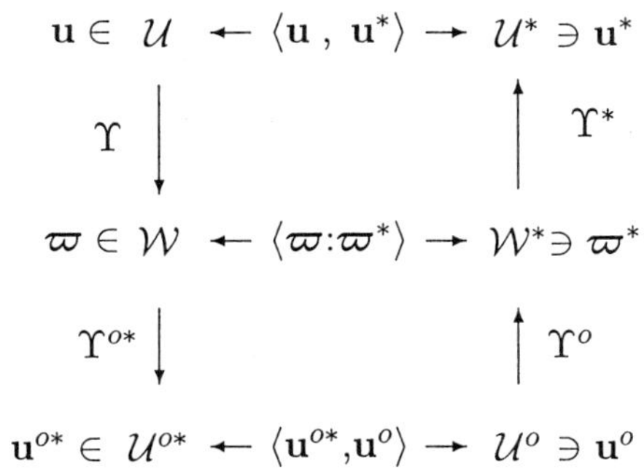

Figure 3.7. Structure of geometrically linear system and its polar

There are many choices for the null-source polar operator. In static systems, the polar configuration variable $\mathbf{u}^o$ possesses certain beautiful geometrical meanings (see Gao, 1999b). For dynamical systems such that the geometrical operator $\Upsilon = (\partial_t, \Lambda)^T : \mathcal{U} \to \mathcal{W} = \mathcal{V} \times \mathcal{E}$ is linear, if $\partial_t \Lambda = \Lambda \partial_t$, we can simply let $\Upsilon^o = (\Lambda^*, \ \ \partial_t^*)^T$ (see Gao, 2000d). Thus,

$$\langle \Upsilon \mathbf{u} ; \Upsilon^* \mathbf{u}^o \rangle = \langle \partial_t \mathbf{u} \ , \ \Lambda^* \mathbf{u}^o \rangle - \langle \Lambda \mathbf{u} \ ; \ \partial_t^* \mathbf{u}^o \rangle = \langle \Lambda \partial_t \mathbf{u} - \partial_t \Lambda \mathbf{u} , \mathbf{u}^o \rangle = 0.$$

Let $\varpi_p^* \in \mathcal{W}_s^*$ be a particular solution of the linear balance equation $\Upsilon^* \varpi^* + \bar{\mathbf{u}}^* = 0$. Replacing ϖ^* by $\varpi^* = \Upsilon^o \mathbf{u}^o + \varpi_p^* \in \mathcal{W}_s^*$, and letting $W^o(\mathbf{u}^o) = W^*(\Upsilon^o \mathbf{u}^o + \varpi_p^*)$, the canonical Hamiltonian Θ can be written as

$$\Theta^o(\mathbf{u}, \mathbf{u}^o) = W^o(\Upsilon^o \mathbf{u}^o) + \langle \mathbf{u} \ , \ \bar{\mathbf{u}}^* \rangle : \mathcal{U} \times \mathcal{U}^o \to \bar{\mathbb{R}}, \tag{5.17}$$

which we call the *polar Hamiltonian*. The associated *polar Lagrangian* $\Xi^o : \mathcal{U} \times \mathcal{U}^o \to \bar{\mathbb{R}}$ is defined by

$$\Xi^o(\mathbf{u}, \mathbf{u}^o) = \langle \Upsilon \mathbf{u} ; \Upsilon^o \mathbf{u}^o \rangle - W^o(\Upsilon^o \mathbf{u}^o). \tag{5.18}$$

Clearly, $\Xi^o \in \Gamma_0(\mathcal{U}) \times \Gamma_0(\mathcal{U}^o)$ is an extended canonical functional. On $\mathcal{U}_a \times \mathcal{U}_a^o$, the criticality condition leads to the polar governing equations

$$D\Xi^o(\mathbf{u}, \mathbf{u}^o) = 0 \quad \Leftrightarrow \quad \left\{ \begin{array}{l} \Upsilon^{o*} D W^o(\Upsilon^o \mathbf{u}^o) = \Upsilon^{o*} \Upsilon \mathbf{u}, \\ \Upsilon^* \Upsilon^o \mathbf{u}^o = 0. \end{array} \right. \tag{5.19}$$

By the fact that $W \in \Gamma_G(\mathcal{W}_a)$, we have $(W^*(\varpi^*))^* = W(\varpi)$. Thus, for any given $\mathbf{u} \in \mathcal{U}_k$,

$$\Pi(\mathbf{u}) = \text{ext}\{\Xi^o(\mathbf{u}, \mathbf{u}^o) | \ \ \forall \mathbf{u}^o \in \mathcal{U}^o\}.$$

On the other hand, by introducing the *polar feasible space*

$$\mathcal{U}_k^o = \{\mathbf{u}^o \in \mathcal{U}_a^o | \ \Upsilon^o \mathbf{u}^o \in \mathcal{W}_a^*\}, \tag{5.20}$$

the *polar action* can be obtained by

$$\Pi^o(\mathbf{u}^o) = \text{sta}\{\Xi^o(\mathbf{u}, \mathbf{u}^o) | \ \forall \mathbf{u} \in \mathcal{U}\} = -W^o(\Upsilon^o \mathbf{u}^o). \tag{5.21}$$

Then the *polar variational problem* $(\mathcal{P}^o)$ can be proposed as

$$(\mathcal{P}^o): \qquad \Pi^o(\mathbf{u}^o) \to \text{sta} \ \forall \mathbf{u}^o \in \mathcal{U}_k^o. \tag{5.22}$$

The critical condition $D\Pi^o(\mathbf{u}^o) = 0$ leads to the polar governing equation:

$$\Upsilon^{o*} DW^o(\Upsilon^o \mathbf{u}^o) = \Upsilon^{o*} \Upsilon \mathbf{u}, \tag{5.23}$$

where $\mathbf{u} \in \mathcal{U}_a$ is the Lagrange multiplier for the equilibrium condition in $\mathcal{U}_a^o$.

Theorem 3 (Bi-Polarity Theorem) *Suppose that for a given total action $\Pi(\mathbf{u}) = K(\partial_t \mathbf{u}) - E(\Lambda \mathbf{u}) + \langle \mathbf{u}, \bar{\mathbf{u}}^* \rangle$ such that $K \in \check{\Gamma}(\mathcal{V}_a)$ and $E \in \check{\Gamma}(\mathcal{E}_a)$ are convex, and $\Pi^o : \mathcal{U}^o{}_k \to \mathbb{R}$ is the associated polar action. If $(\bar{\mathbf{u}}, \bar{\mathbf{u}}^o)$ is a critical point of Ξ^o, then $\Pi(\bar{\mathbf{u}}) = \Xi^o(\bar{\mathbf{u}}, \bar{\mathbf{u}}^o) = \Pi^o(\bar{\mathbf{u}}^o)$ and*

$$\Pi(\bar{\mathbf{u}}) = \inf_{\mathbf{u} \in \mathcal{U}_k} \Pi(\mathbf{u}) \quad \Leftrightarrow \quad \inf_{\mathbf{u}^o \in \mathcal{U}_k^o} \Pi^o(\mathbf{u}^o) = \Pi^o(\bar{\mathbf{u}}^o), \tag{5.24}$$

$$\Pi(\bar{\mathbf{u}}) = \sup_{\mathbf{u} \in \mathcal{U}_k} \Pi(\mathbf{u}) \quad \Leftrightarrow \quad \sup_{\mathbf{u}^o \in \mathcal{U}_k^o} \Pi^o(\mathbf{u}^o) = \Pi^o(\bar{\mathbf{u}}^o). \tag{5.25}$$

Proof. For linear operator $\Upsilon = (\partial_t, \ \Lambda)^T : \mathcal{U} \to \mathcal{W}$, the extended Lagrangian $\Xi : \mathcal{U}_a \times \mathcal{W}_a \to \mathbb{R}$ associated with Π is

$$\Xi(\mathbf{u}, \mathbf{p}, \boldsymbol{\xi}^*) = \langle \partial_t \mathbf{u} \ , \ \mathbf{p} \rangle - K^*(\mathbf{p}) - \langle \Lambda \mathbf{u} \ ; \ \boldsymbol{\xi}^* \rangle + E^*(\boldsymbol{\xi}^*) + \langle \mathbf{u}, \bar{\mathbf{u}}^* \rangle. \tag{5.26}$$

By Lemma 1, we know that if $(\bar{\mathbf{u}}, \bar{\mathbf{u}}^o, \bar{\boldsymbol{\epsilon}}^*)$ is a critical point of Ξ, then $\Pi(\bar{\mathbf{u}}) = \Pi^d(\bar{\mathbf{p}}, \bar{\boldsymbol{\epsilon}}^*)$. Replacing $\bar{\boldsymbol{\varpi}}^* = (\bar{\mathbf{p}}, \bar{\boldsymbol{\epsilon}}^*)$ by $\Upsilon^o \bar{\mathbf{u}}^o + \boldsymbol{\varpi}_p^*$ leads to $\Pi(\bar{\mathbf{u}}) = \Xi^o(\bar{\mathbf{u}}, \bar{\mathbf{u}}^o) = \Pi^o(\bar{\mathbf{u}}^o)$.

Since $E^*(\boldsymbol{\xi}^*)$ and $K^*(\mathbf{p})$ are convex, for any given $\mathbf{u} \in \mathcal{U}_a$, $\Xi \in \hat{\Gamma}(\mathcal{V}_a^*) \times \check{\Gamma}_G(\mathcal{E}_a^*)$ is a left-saddle functional, hence

$$\sup_{\mathbf{p} \in \mathcal{V}_a^*} \inf_{\boldsymbol{\xi}^* \in \mathcal{E}_a^*} \Xi(\mathbf{u}, \mathbf{p}, \boldsymbol{\xi}^*) = \inf_{\boldsymbol{\xi}^* \in \mathcal{E}_a^*} \sup_{\mathbf{p} \in \mathcal{V}_a^*} \Xi(\mathbf{u}, \mathbf{p}, \boldsymbol{\xi}^*) = \Pi(\mathbf{u}) \ \ \forall \mathbf{u} \in \mathcal{U}_k.$$

On the other hand, for a given $(\mathbf{u}, \mathbf{p}) \in \mathcal{U}_a \times \mathcal{V}_a^*$, $\inf_{\boldsymbol{\xi}^* \in \mathcal{E}^*} \Xi(\mathbf{u}, \mathbf{p}, \boldsymbol{\xi}^*)$ leads to the classical Lagrangian

$$L(\mathbf{u}, \mathbf{p}) = \inf_{\boldsymbol{\xi}^* \in \mathcal{E}^*} \Xi(\mathbf{u}, \mathbf{p}, \boldsymbol{\xi}^*) = \langle \partial_t \mathbf{u} \ , \ \mathbf{p} \rangle - K^*(\mathbf{p}) - E(\Lambda \mathbf{u}) + \langle \mathbf{u}, \bar{\mathbf{u}}^* \rangle,$$

which is a super-critical functional on $\mathcal{U}_a \times \mathcal{V}_a^*$. By the super-Lagrange duality theory (cf. Gao, 1999b), we know that $\inf_{\mathbf{u}} \sup_{\mathbf{p}} L(\mathbf{u},\mathbf{p}) = \inf_{\mathbf{p}} \sup_{\mathbf{u}} L(\mathbf{u},\mathbf{p})$, and $\sup_{\mathbf{u}} \sup_{\mathbf{p}} L(\mathbf{u},\mathbf{p}) = \sup_{\mathbf{p}} \sup_{\mathbf{u}} L(\mathbf{u},\mathbf{p})$. Thus,

$$\begin{aligned}\inf_{\mathbf{u}} \Pi(\mathbf{u}) &= \inf_{\mathbf{u}} \sup_{\mathbf{p}} \inf_{\boldsymbol{\xi}^*} \Xi(\mathbf{u},\mathbf{p},\boldsymbol{\xi}^*) = \inf_{\mathbf{p}} \sup_{\mathbf{u}} \inf_{\boldsymbol{\xi}^*} \Xi(\mathbf{u},\mathbf{p},\boldsymbol{\xi}^*) \\ &= \inf_{\mathbf{p}} \inf_{\boldsymbol{\xi}^*} \sup_{\mathbf{u}} \Xi(\mathbf{u},\mathbf{p},\boldsymbol{\xi}^*) = \inf_{\mathbf{p},\boldsymbol{\xi}^*} \Pi^d(\mathbf{p},\boldsymbol{\xi}^*).\end{aligned}$$

Replacing $(\mathbf{p},\boldsymbol{\xi}^*)$ by $\boldsymbol{\varpi}^* = \Upsilon^o \mathbf{u}^o + \boldsymbol{\varpi}_p^*$, we have $\sup_{\mathbf{u}} \Xi(\mathbf{u},\mathbf{p}(\mathbf{u}^o),\boldsymbol{\xi}^*(\mathbf{u}^o)) = \Pi^o(\mathbf{u}^o)$, and this prove the statement (5.24)

Similarly, the statement (5.25) can be proved by the bi-duality theorem. □

For conservative systems with standard bilinear forms ($\omega = 0$), the geometrically linear operator Υ and its polar operator can be chosen as

$$\Upsilon = \begin{pmatrix} \partial/\partial t \\ -\nabla_s \end{pmatrix}, \quad \Upsilon^o = \begin{pmatrix} \text{div} \\ \partial/\partial t \end{pmatrix}. \tag{5.27}$$

Using Gauss-Green theorem and integration by parts, we have

$$\begin{aligned}\langle \Upsilon \mathbf{u} ; \Upsilon^o \mathbf{u}^o \rangle &= \int_{\Omega_t} (\mathbf{u}_{,t} \text{div}\mathbf{u}^o - (\nabla_s \mathbf{u}) \cdot \mathbf{u}^o{}_{,t}) d\Omega_t \\ &= \int_{\Omega_t} \mathbf{u}(-\frac{\partial}{\partial t}(\text{div}\mathbf{u}^o) + \text{div}\mathbf{u}^o{}_{,t}) d\Omega_t + \int_{\Omega} \mathbf{u} \cdot (\text{div}\mathbf{u}^o)|_{\partial I} \, d\Omega \\ &\quad - \int_I \oint_{\partial\Omega} \mathbf{u} \cdot \mathbf{u}^o{}_{,t} \cdot \mathbf{n} d\partial\Omega \, dt.\end{aligned}$$

Thus, for a given admissible space $\mathcal{U}_a$, the condition

$$\langle \Upsilon \mathbf{u} ; \Upsilon^o \mathbf{u}^o \rangle = \langle \mathbf{u} \, , \, \Upsilon^* \Upsilon^o \mathbf{u}^o \rangle = 0 \;\; \forall \mathbf{u} \in \mathcal{U}_a$$

leads to the polar equilibrium condition in $\mathcal{U}_a^o \subset \mathcal{U}^o$. For example, if we let $\mathcal{U}_a = \{\mathbf{u} \in \mathcal{U} | \;\; \mathbf{u}(t,x) = 0 \;\; \forall t \in \partial I\}$, then $\mathcal{U}_a^o = \{\mathbf{u}^o \in \mathcal{U}^o | \;\; \mathbf{n} \cdot \mathbf{u}^o{}_{,t}(t,x) = 0 \;\; \forall \mathbf{x} \in \partial\Omega\}$. Thus, on $\mathcal{U}_a$ and $\mathcal{U}_a^o$, we have $\partial_t^* = (\partial/\partial t)^* = -\partial/\partial t$, $\Lambda^* = \text{grad}^* = -\text{div}$, i.e.,

$$\Upsilon^* = (-\partial/\partial t, \text{div}), \quad \Upsilon^{o*} = (-\text{grad}, \; -\partial/\partial t), \tag{5.28}$$

and Υ^o is a null-source polar operator. For any given $\mathbf{u}^o \in \mathcal{U}_a^o$, $\Upsilon^o \mathbf{u}^o$ is an homogeneous solution of the balance equation, i.e.,

$$\Upsilon^* \Upsilon^o \mathbf{u}^o = \partial_t^* \text{grad}^* \mathbf{u}^o - \text{grad}^* \partial_t^* \mathbf{u}^o = \frac{\partial}{\partial t}(\text{div}\mathbf{u}^o) - \text{div}\left(\frac{\partial}{\partial t}\mathbf{u}^o\right) = 0 \quad \text{in } \Omega_t.$$

The particular solution of the balance equation $\Upsilon^* \varpi_p^* = DF(\mathbf{u})$ can be defined by $\varpi_p^* = (\bar{\mathbf{p}}, 0)^T$, where $\bar{\mathbf{p}}$ is a particular solution of the non-homogeneous equation $\partial_t^* \bar{\mathbf{p}} = -DF(\mathbf{u})$. Thus, replacing $\varpi^* = (\mathbf{p}, \boldsymbol{\sigma})^T$ by $\Upsilon^o \mathbf{u}^o + \varpi_p^* = (\mathrm{div}\mathbf{u}^o + \bar{\mathbf{p}}, \mathbf{u}^o_{,t})^T$, the polar variational problem for this geometrically linear dynamical system has a very simple form

$$\Pi^o(\mathbf{u}^o) = \int_{\Omega_t} \left[U^*(\mathbf{u}^o_{,t}) - \frac{1}{2}\rho^{-1} |\nabla \cdot \mathbf{u}^o + \bar{\mathbf{p}}|^2 \right] \mathrm{d}\Omega_t \quad \to \mathrm{sta} \ \forall \mathbf{u}^o \in \mathcal{U}_k^o,$$

where $|*|$ stands for the Euclidean norm in $\mathbb{R}^3$. In linear elastodynamics such that the stored energy density $U(\boldsymbol{\epsilon}) = \frac{1}{2}\boldsymbol{\epsilon} : \mathbf{H} : \boldsymbol{\epsilon}$ is a quadratic function, where $\mathbf{H} = \{H_{ijkl}\}$ is a fourth-order symmetrical elastic tensor, satisfies $H_{,ijkl} = H_{jikl} = H_{klij}$, the complementary energy density is simply defined as $U^*(\boldsymbol{\sigma}) = \frac{1}{2}\boldsymbol{\sigma} : \mathbf{H}^{-1} : \boldsymbol{\sigma}$. In this case, the polar action Π^o is similar to Tabarrok's complementary action (Tabarrok, 1987) and Bui's dual formulation (Bui, 1992).

6. Complementary Variational Principle for Contact Mechanics

We now assume that the stored action $W : \mathcal{W} \to \mathbb{R}$ is a quadratic functional:

$$W(\varpi) = \frac{1}{2}\langle \varpi : \mathbf{C}\varpi \rangle = \frac{1}{2}\langle \mathbf{v} \ , \ \rho \mathbf{v} \rangle - \frac{1}{2}\langle \boldsymbol{\epsilon} \ ; \ \mathbf{H} : \boldsymbol{\epsilon} \rangle, \quad \mathbf{C} = \begin{pmatrix} \rho & 0 \\ 0 & -\mathbf{H} \end{pmatrix},$$

where $\mathbf{C} : \mathcal{W} \to \mathcal{W}^*$ is a symmetrical operator. However, the external energy $F : \mathcal{U} \to \bar{\mathbb{R}}$ is a nonsmooth extended canonical functional

$$F(\mathbf{u}) = \langle \mathbf{u} \ , \ \bar{\mathbf{u}}^* \rangle + J(\mathbf{u}), \tag{6.1}$$

where $J(\mathbf{u}) \in \Gamma_0(\mathcal{U})$ is a nonsmooth super-potential.

For example, for frictional contact problems, the boundary is split into three parts: $\Gamma = \Gamma_f \cup \Gamma_u \cup \Gamma_c$, where Γ_c is the contact surface. The external $F(\mathbf{u})$ can be written as

$$F(\mathbf{u}) = \int_{\Omega_t} e^{\omega(\mathbf{x},t)} \mathbf{u} \cdot \bar{\mathbf{b}} \, \mathrm{d}\Omega_t + \int_{\Gamma_f} e^{\omega(\mathbf{x},t)} \mathbf{u} \cdot \bar{\mathbf{t}} \, \mathrm{d}\Gamma + \int_{\Gamma_c} j(\mathbf{u}) \, \mathrm{d}\Gamma, \tag{6.2}$$

where $j(\mathbf{u})$ is the frictional super-potential as discussed in section one. The duality relation $\mathbf{u}^* \in \partial F(\mathbf{u})$ gives the balance equation $\mathbf{u}^* = \bar{\mathbf{b}}$ in the domain Ω_t, the natural boundary condition on Γ_f and the contact condition on Γ_c:

$$\mathbf{u}^* \in \partial F(\mathbf{u}) = \begin{cases} \bar{\mathbf{t}} & \text{on } \Gamma_f, \\ \partial j(\mathbf{u}) & \text{on } \Gamma_c. \end{cases}$$

We further assume that the admissible displacement space $\mathcal{U}_a$ is defined by (5.1). Thus, the primal problem takes the form of

$$\begin{aligned}\Pi(\mathbf{u}) &= \frac{1}{2}\langle \Upsilon\mathbf{u} ; \mathbf{C}\Upsilon\mathbf{u}\rangle + F(\mathbf{u}) \\ &= \frac{1}{2}\langle \mathbf{u} \,,\, A\mathbf{u}\rangle + \langle \mathbf{u} \,,\, \bar{\mathbf{u}}^*\rangle + J(\mathbf{u}) \quad \to \text{sta } \forall \mathbf{u} \in \mathcal{U}_k,\end{aligned}$$

where $A = \Upsilon^* C \Upsilon = \partial_t^* \rho \partial_t - \Lambda^* \mathbf{H} \Lambda$ is a self-adjoint operator. Since $A : \mathcal{U} \to \mathcal{U}^*$ is not monotone, $\Pi(\mathbf{u})$ may have many critical points. The primal feasible space in this case can be defined by $\mathcal{U}_k = \{\mathbf{u} \in \mathcal{U}_a | \ \mathbf{u} \in \text{dom } A \cap \text{dom } J\}$. A critical point of Π is defined to be a point $\bar{\mathbf{u}} \in \mathcal{U}_k$ satisfying

$$0 \in A\bar{\mathbf{u}} + \bar{\mathbf{u}}^* + \partial J(\bar{\mathbf{u}}). \tag{6.3}$$

The polar Lagrangian $\Xi^o : \mathcal{U}_a \times \mathcal{U}_a^o \to \mathbb{R}$ associated with this problem is

$$\begin{aligned}\Xi^o(\mathbf{u}, \mathbf{u}^o) &= \langle \Upsilon\mathbf{u} ; \Upsilon^o \mathbf{u}^o\rangle + J(\mathbf{u}) \\ &\quad -\frac{1}{2}\langle \mathbf{C}^{-1}(\Upsilon^o\mathbf{u}^o + \varpi_p^*) ; (\Upsilon^o\mathbf{u}^o + \varpi_p^*)\rangle,\end{aligned}$$

where $\varpi_p^* \in \mathcal{W}^*$ is a particular solution of the balance equation $\Upsilon^* \varpi_p^* + \bar{\mathbf{u}}^* = 0$. By definition, the polar action Π^o is obtained as

$$\begin{aligned}\Pi^o(\mathbf{u}^o) &= \text{ext}\{\Xi^o(\mathbf{u}, \mathbf{u}^o) | \ \forall \mathbf{u} \in \mathcal{U}_a\} \\ &= -\frac{1}{2}\langle \mathbf{C}^{-1}(\Upsilon^o\mathbf{u}^o + \varpi_p^*) ; (\Upsilon^o\mathbf{u}^o + \varpi_p^*)\rangle - J^o(-\Upsilon^*\Upsilon^o\mathbf{u}^o),\end{aligned}$$

where $J^o : \mathcal{U}^o \to \bar{\mathbb{R}}$ is defined by

$$J^o(-\Upsilon^*\Upsilon^o\mathbf{u}^o) = \text{ext}\{\langle \mathbf{u} \,,\, -\Upsilon^*\Upsilon^o\mathbf{u}^o\rangle - J(\mathbf{u}) | \ \forall \mathbf{u} \in \mathcal{U}_a\}.$$

Thus, on the polar feasible space $\mathcal{U}_k^o$, the bi-polarity theorem holds. By the fact that J^o is a smooth functional over its effective domain, this polar variational problem should be easier than the original nonsmooth primal problem.

Since $A = \Upsilon^* \mathbf{C} \Upsilon : \mathcal{U}_k \to \mathcal{U}^*$ is a self-adjoint operator, if $\mathcal{U}_k$ is a reflexive Banach space, we can let $\mathcal{U}^o = \mathcal{U}$ and choose the polar operator $\Upsilon^o = \mathbf{C}\Upsilon$. Thus, for a given particular solution $\mathbf{u}_p$ satisfying $A\mathbf{u}_p + \bar{\mathbf{u}}^* = 0$, replacing $\mathbf{u}^o$ by $\mathbf{u} - \mathbf{u}_p$, and let $\varpi_p^* = \mathbf{C}\Upsilon\mathbf{u}_p$, then Π^o can be written in the form

$$\Pi^o(\mathbf{u} - \mathbf{u}_p) = -\frac{1}{2}\langle \mathbf{u} \,,\, A\mathbf{u}\rangle - J^o(-A(\mathbf{u} - \mathbf{u}_p)).$$

Let $\mathcal{U}_k^c = \{\mathbf{u} \in \mathcal{U}_a | \ A\mathbf{u} + \bar{\mathbf{u}}^* \in \text{dom } J^o\}$ be the so-called *complementary feasible space*, and on which, we introduce the so-called *complementary action* $\Pi^c : \mathcal{U} \to \bar{\mathbb{R}}$

$$\Pi^c(\mathbf{u}) = -\Pi^o(\mathbf{u} - \mathbf{u}_p) = \frac{1}{2}\langle \mathbf{u} \ , \ A\mathbf{u}\rangle + J^o(-A\mathbf{u} - \bar{\mathbf{u}}^*). \tag{6.4}$$

Then, the complementary action principle can be proposed as

$$(\Pi^c): \quad \Pi^c(\mathbf{u}^c) = \text{sta } \Pi^c(\mathbf{u}) \ \ \forall \mathbf{u} \in \mathcal{U}_k^c. \tag{6.5}$$

Since $J^o \in \Gamma_0(\mathcal{U}^*)$ is an extended canonical functional, the criticality condition of Π^c leads to the following complementary Euler-Lagrange inclusion

$$0 \in A\mathbf{u}^c + \partial[J^o \circ (-A)](\mathbf{u}^c - \mathbf{u}_p). \tag{6.6}$$

By the theory of convex analysis (cf. e.g. Gao, 1999b), we know that $\partial(J^o \circ A)(\mathbf{u}) = A^*\partial J^o(A\mathbf{u})$. Since $A = A^*$ is a self-adjoint operator, the complementary Euler-Lagrange inclusion can be written in the form

$$\mathbf{u}^c \in \partial J^o(-A\mathbf{u}^c - \bar{\mathbf{u}}^*). \tag{6.7}$$

Theorem 4 (Complementary Theorem)
Let $\mathcal{N}(A) = \{\mathbf{u}_o \in \mathcal{U}_k | \ A\mathbf{u}_o = 0\}$ be the null-space of the operator A.

(T1) If $\bar{\mathbf{u}}$ is a critical point of Π, then for any given $\mathbf{u}_o \in \mathcal{N}(A)$, $\mathbf{u}^c = \mathbf{u}_o + \bar{\mathbf{u}}$ is a critical point of Π^c.

(T2) Conversely, if $\mathcal{U}_k^c \neq \emptyset$ and if $\mathbf{u}^c$ is a critical point of Π^c, then $\bar{\mathbf{u}} = \mathbf{u}_o + \mathbf{u}^c$ is a critical point of Π for some given $\mathbf{u}_o \in \mathcal{N}(A)$.

(T3) If $\mathcal{U}_k^c \neq \emptyset$, $\bar{\mathbf{u}}$ and $\mathbf{u}^c$ are critical points of Π and Π^c, respectively, then the complementary conditions

$$\begin{array}{l}\Pi(\bar{\mathbf{u}}) + \Pi^c(\bar{\mathbf{u}} + \mathbf{u}_o) = 0 \\ \Pi(\mathbf{u}_o + \mathbf{u}^c) + \Pi^c(\mathbf{u}^c) = 0\end{array} \quad \forall \mathbf{u}_o \in \mathcal{N}(A) \tag{6.8}$$

hold.

Proof. (T1) If $\bar{\mathbf{u}}$ is a critical point of Π, then $0 \in A\bar{\mathbf{u}} + \bar{\mathbf{u}}^* + \partial J(\bar{\mathbf{u}})$. By the Fenchel duality formula, this is equivalent to $\bar{\mathbf{u}} \in \partial J^o(-A\bar{\mathbf{u}} - \bar{\mathbf{u}}^*)$. This means that $\bar{\mathbf{u}}$ is a critical point of Π^c. Since $\Pi^c(\mathbf{u})$ is invariant under translation in $\mathcal{N}(A)$, i.e., for any given $\mathbf{u}_o \in \mathcal{N}(A)$, $\Pi^c(\mathbf{u} + \mathbf{u}_o) = \Pi^c(\mathbf{u}) \ \forall \mathbf{u} \in \mathcal{U}_k^c$, then all the points in $\bar{\mathbf{u}} + \mathcal{N}(A)$ are also critical points of Π^c.

(T2) Conversely, let $\mathbf{u}^c$ be a critical point of Π^c. Since $\mathcal{U}_k^c$ is not empty, then for any given $\mathbf{u}_o \in \mathcal{N}(A)$ we have $0 \in A(\mathbf{u}^c + \mathbf{u}_o) + \partial[J^o \circ (-A)](\mathbf{u}^c + \mathbf{u}_o - \mathbf{u}_p)$, i.e., $\mathbf{u}^c + \mathbf{u}_o \in \partial J^o(-A(\mathbf{u}^c + \mathbf{u}_o) - \bar{\mathbf{u}}^*)$. By the Fenchel duality we know that this is equivalent to $-A(\mathbf{u}^c + \mathbf{u}_o) - \bar{\mathbf{u}}^* \in \partial J(\mathbf{u}^c + \mathbf{u}_o)$, or,

$0 \in A(\mathbf{u}^c + \mathbf{u}_o) + \bar{\mathbf{u}}^* + \partial J(\mathbf{u}^c + \mathbf{u}_o)$ since $A\mathbf{u}_p + \bar{\mathbf{u}}^* = 0$. This shows that $\bar{\mathbf{u}} = \mathbf{u}^c + \mathbf{u}_o$ is indeed a critical point of Π for any given $\mathbf{u}_o \in \mathcal{N}(A)$.

(T3) This statement follows from the Bi-Polarity Theorem. □

In the case that $F(\mathbf{u}) = \langle \mathbf{u}, \bar{\mathbf{u}}^* \rangle + J(\mathbf{u})$ is a nonsmooth convex functional, defined by (6.2), then for homogeneous boundary conditions, the total action is a nonsmooth d.c. functional

$$\begin{aligned}\Pi(\mathbf{u}) &= \frac{1}{2}\langle \mathbf{u} \, , \, A\mathbf{u}) + \langle \mathbf{u} \, , \, \bar{\mathbf{u}}^* \rangle + J(\mathbf{u}) \\ &= \int_{\Omega_t} \frac{1}{2}[\rho \mathbf{u}_{,t} \cdot \mathbf{u}_{,t} - (\nabla \mathbf{u})^T : \mathbf{H} : (\nabla \mathbf{u}) + \mathbf{u} \cdot \bar{\mathbf{b}}] \, d\Omega_t \\ &\quad + \int_{\Gamma_f} \mathbf{u} \cdot \bar{\mathbf{t}} \, d\Gamma + \int_{\Gamma_c} j(\mathbf{u}) \, d\Gamma,\end{aligned}$$

where $A : \mathcal{U}_k \to \mathcal{U}^*$ is a self-adjoint wave operator

$$A\mathbf{u} = \begin{cases} -\frac{\partial}{\partial t}\left(\rho \frac{\partial}{\partial t}\mathbf{u}\right) + \operatorname{div} \mathbf{H} \operatorname{grad}\mathbf{u} & \text{in } \Omega_t, \\ \mathbf{n} \cdot (\mathbf{H} : (\nabla \mathbf{u})) & \text{on } \partial\Omega. \end{cases}$$

On the complementary feasible space $\mathcal{U}_k^c$, the complementary action has a very simple form

$$\begin{aligned}\Pi^c(\mathbf{u}) &= \frac{1}{2}\langle \mathbf{u} \, , \, A\mathbf{u} \rangle + J^o(-A\mathbf{u} - \bar{\mathbf{u}}^*) \\ &= \int_{\Omega_t} \frac{1}{2}\left[\rho \mathbf{u}_{,t} \cdot \mathbf{u}_{,t} - (\nabla \mathbf{u})^T : \mathbf{H} : (\nabla \mathbf{u})\right] \, d\Omega_t \\ &\quad + \int_{\Gamma_c} j^o(-\mathbf{n} \cdot (\mathbf{H} : \nabla \mathbf{u})) \, d\Gamma. \qquad (6.9)\end{aligned}$$

As we discussed in Section 1 that if the super-potential $j(\mathbf{u})$ is nonsmooth on $\mathcal{U}_k$, its conjugate $j^o(\mathbf{u}^*)$, however, could be a smooth functional. Then, the complementary variational problem will be much easier than the primal problem.

7. Applications in Nonconvex Dynamical Structures

We now turn our attention to the geometrically nonlinear systems by considering the dynamically post-buckling problem of the beam model proposed in (1.6)

$$\Pi(u, w) = K(w_{,t}) - P(u, w) \to \text{sta} \;\; \forall (u, w) \in \mathcal{U}_k, \qquad (7.1)$$

where the kinetic energy is a quadratic form of $w(x,t)$

$$K(w_{,t}) = \frac{1}{2}\int_0^{t_c}\int_0^{\ell} e^{\nu t}\rho w_{,t}^2 \, dx \, dt,$$

and the momentum $p(x,t)$ depends linearly on $w_{,t}$, i.e., $p = DK(w_{,t}) = \rho w_{,t}(x,t)$. However, the total potential $P(u,w)$ is a nonconvex functional

$$P(u,w) = \int_{\Omega_t} \frac{1}{2} e^{\nu t}[(u_{,x} + \frac{1}{2}\alpha w_{,x}^2 - \lambda)^2 + \beta(u_{,y} + w_{,x})^2] \, d\Omega_t - F(u,w). \tag{7.2}$$

The displacement $\mathbf{u} = (u,w)^T \in \mathcal{U} = C^1(\Omega_t;\mathbb{R}^2)$ is a continuous, differentiable vector in $\mathbb{R}^2$. If the beam is clamped at $x = 0$ and simply supported at $x = \ell$, the admissible space $\mathcal{U}_a \subset \mathcal{U}$ can be defined as

$$\mathcal{U}_a := \left\{ \mathbf{u} = \begin{pmatrix} u \\ w \end{pmatrix} \in \mathcal{U} \;\middle|\; \begin{array}{l} w(0,t) = w_{,x}(0,t) = w(\ell,t) = 0, \\ w(x,0) = w(x,t_c) = 0, \\ u(0,y,t) = u_{,x}(\ell,y,t) = 0 \;\; \forall y \in [-h,h] \end{array} \right\}.$$

For the given external force vector $\bar{\mathbf{u}}^* = (\mu^{\pm}(x,\pm h,t), \;\; f(x,t))^T \in \mathbb{R}^2$, the external potential $F(\mathbf{u}) = \langle \mathbf{u}, \bar{\mathbf{u}}^* \rangle$ is a linear functional defined by (1.8).

Since for each given u, the total potential $P(w,u)$ is a double-well functional of the lateral displacement w, the primal variational problem is very sensitive to the external load f and $\lambda > 0$. As it was discussed in section 1 (Fig. 3.5), for a fixed post-buckling load $\lambda > \lambda_c$, the beam will go through a chaotic vibration transition period when the transverse force amplitude is increasing. On the other hand, if we fix the transverse load and increase gradually the axial load λ, a very interesting phenomenon, which could be called the meta-buckling transition period, can be observed in this nonconvex dynamical system by the numerical experiments. From Fig. 3.8 we can see that as the axial load λ increases, the amplitude of oscillation of the beam grows, i.e., the frequency of the vibration decreases. But at a certain *meta-buckling value* $\lambda = \lambda_m < \lambda_c$, the amplitude of the vibration starts to decrease rapidly until λ reaches the buckling load $\lambda = \lambda_c$. After the chaotic transition region when λ is bigger than the so-called *post-chaos load* λ_{pc}, the beam moves again periodically around one of the stable potential wells (Fig. 3.8(a)). For the given transverse load $f = \cos(3.76t)$, a closed view of this *meta-buckling/chaotic transition* is shown in Fig. 3.8(b), where we can see that the meta-buckling load $\lambda_m \approx .94\lambda_c$ while the post-chaos load $\lambda_{pc} \approx 1.1\lambda_c$. The author believes that this phenomenon exists also in many other nonconvex dissipative Hamiltonian systems, and deserves

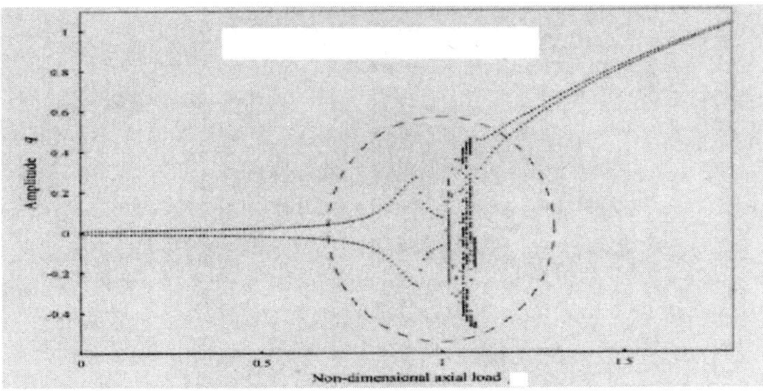

(a) Pre and post-chaotic vibration: Amplitude q vis the axial load λ

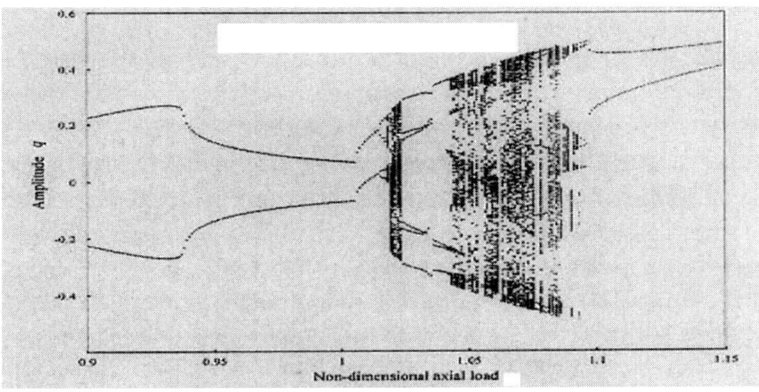

(b) Meta-buckling transition regions: $\lambda_m \leq \lambda \leq \lambda_{pc}$

Figure 3.8. Chaos vase: A new phenomenon in chaotic vibration of the dissipative extended beam model a detailed study. Here we will use the canonical dual transformation method to determine these parameters.

In order to develop the duality approach for this nonconvex dynamical system, we let $\boldsymbol{\xi} = (\xi, \gamma)^T$ be the generalized strain vector, defined by the geometrical equation

$$\begin{aligned} \boldsymbol{\xi} &= \begin{pmatrix} \xi(\mathbf{u}) \\ \gamma(\mathbf{u}) \end{pmatrix} = \Lambda(\mathbf{u}) \\ &= \begin{pmatrix} \frac{\partial}{\partial x} & \frac{1}{2}\alpha w_{,x}\frac{\partial}{\partial x} \\ \frac{\partial}{\partial y} & \frac{\partial}{\partial x} \end{pmatrix} \begin{pmatrix} u \\ w \end{pmatrix} = \begin{pmatrix} u_{,x} + \frac{1}{2}\alpha w_{,x}^2 \\ u_{,y} + w_{,x} \end{pmatrix}. \end{aligned}$$

Thus, by using the quadratic operator $\Lambda(\mathbf{u})$, we have the canonical transformation $P(w, u) = E(\Lambda(w, u)) - F(w, u)$, where the stored energy E is the strictly convex (quadratic) functional of the canonical strain $\boldsymbol{\xi}$

$$E(\boldsymbol{\xi}) = \int_{\Omega_t} \frac{1}{2} e^{\nu t}[(\xi - \lambda)^2 + \beta\gamma^2]\, \mathrm{d}\Omega_t.$$

The conjugate stress vector $\boldsymbol{\xi}^*$ depends linearly on $\boldsymbol{\xi}$

$$\boldsymbol{\xi}^* = \begin{pmatrix} \varsigma \\ \tau \end{pmatrix} = DE(\boldsymbol{\xi}) = \begin{pmatrix} 1 & 0 \\ 0 & \beta \end{pmatrix} \begin{pmatrix} \xi - \lambda \\ \gamma \end{pmatrix}. \tag{7.3}$$

Thus, the primal problem (7.1) is a geometrically nonlinear problem. From the Legendre-Fenchel transformation, the complementary energy is uniquely defined by

$$E^*(\boldsymbol{\xi}^*) = \sup_{\boldsymbol{\xi}}\{\langle \boldsymbol{\xi}; \boldsymbol{\xi}^* \rangle - E(\boldsymbol{\xi})\} = \int_{\Omega_t} e^{\nu t}[\frac{1}{2}\varsigma^2 + \lambda\varsigma + \frac{1}{2\beta}\tau^2]\, \mathrm{d}\Omega_t.$$

The canonical decomposition of the geometrical operator $\Lambda(\mathbf{u}) = \Lambda_t(\mathbf{u})\mathbf{u} + \Lambda_c(\mathbf{u})$ in this example is

$$\Lambda_t(\mathbf{u})\mathbf{u} = \begin{pmatrix} \frac{\partial}{\partial x} & \alpha w_{,x}\frac{\partial}{\partial x} \\ \frac{\partial}{\partial y} & \frac{\partial}{\partial x} \end{pmatrix} \begin{pmatrix} u \\ w \end{pmatrix} = \begin{pmatrix} u_{,x} + \alpha w_{,x}^2 \\ u_{,y} + w_{,x} \end{pmatrix},$$

$$\Lambda_c(\mathbf{u}) = \Lambda_c(w) = \begin{pmatrix} -\frac{1}{2}\alpha w_{,x}^2 \\ 0 \end{pmatrix}.$$

With the weighted bilinear forms, the canonical Lagrangian associated with the potential variational problem is

$$\begin{aligned}\Xi(\mathbf{u}, p, \boldsymbol{\xi}^*) &= \langle w_{,t}\ , \ p\rangle - K^*(p) + F(\mathbf{u}) - \langle \Lambda(\mathbf{u}); \boldsymbol{\xi}^*\rangle + E^*(\boldsymbol{\xi}^*) \\ = & \int_0^{t_c}\int_0^{\ell} e^{\nu t}[fw + \mu^+ u(x,h,t) + \mu^- u(x,-h,t)]\, \mathrm{d}x\, \mathrm{d}t \\ - & \int_{\Omega_t} e^{\nu t}[\varsigma(\frac{1}{2}aw_{,x}^2 + u_{,x}) + \tau(w_{,x} + u_{,y}) - (\frac{1}{2}\varsigma^2 + \lambda\varsigma + \frac{1}{2\beta}\tau^2)]\, \mathrm{d}\Omega_t \\ + & \int_0^{t_c}\int_0^{\ell} e^{\nu t}[w_{,t}p - \frac{1}{2}\rho^{-1}p^2]\, \mathrm{d}x\, \mathrm{d}t. \end{aligned} \tag{7.4}$$

The criticality condition leads to the abstrac canonical Euler-Lagrange equation

$$\Lambda_t^*(\mathbf{u})\boldsymbol{\xi}^* - \partial_t^* p = \bar{\mathbf{u}}^*,$$

or, in explicit forms

$$\begin{aligned} & -(\varsigma_{,x} + \tau_{,y}) = 0 \quad \forall (x,y,t) \in \Omega_t, \\ & p_{,t} + \nu p - \int_{-h}^{h} (\tau + \alpha w_{,x}\varsigma)_{,x}\, \mathrm{d}y = f \quad \forall (x,t) \in (0,\ell) \times (0,t_c), \\ & \int_{-h}^{h} (\tau + \alpha w_{,x}\varsigma)\, \mathrm{d}y = 0 \quad \text{at } x = \ell, \\ & \tau(x, \pm h, t) = \mu^{\pm}(x,t) \quad \forall (x,t) \in (0,\ell) \times (0,t_c). \end{aligned}$$

Let $\mathcal{W}_s^* \subset \mathcal{W}^*$ be the dynamically admissible space

$$\mathcal{W}_s^* = \left\{ \left(\begin{array}{c} p \\ \boldsymbol{\xi}^* \end{array} \right) \left| \begin{array}{l} \varsigma_{,x} + \tau_{,y} = 0 \ \text{in}\Omega_t, \ \ \int_{-h}^{h} \varsigma\, \mathrm{d}y \neq 0 \ \ \forall x \in (0,\ell), \\ \tau(x,\pm h, t) = \mu^{\pm}, \ \ \int_{-h}^{h}\tau\, \mathrm{d}y = \int_{-h}^{h} \varsigma\, \mathrm{d}y = 0 \ \ \text{at } x = \ell \end{array} \right. \right\}.$$

For any given $\boldsymbol{\varpi}^* \in \mathcal{W}_s^*$, the solution of the stationary problem

$$F^{\Lambda}(\boldsymbol{\varpi}^*) = \mathrm{sta}\,\{\langle \Lambda(\mathbf{u}); \boldsymbol{\xi}^*\rangle - \langle w_{,t}\ , \ p\rangle - F(\mathbf{u}) |\ \ \forall \mathbf{u} \in \mathcal{U}_a\}$$

is unique and given by

$$\begin{aligned} w_{,x} &= \frac{\int_0^x [p_{,t}(s,t) + \nu p(s,t) - f(s,t)]\, \mathrm{d}s - \int_{-h}^{h} \tau\, \mathrm{d}y}{\alpha \int_{-h}^{h} \varsigma\, \mathrm{d}y} \\ &= \frac{g(x,t) - \bar{\tau}(x,t)}{\alpha\bar{\varsigma}(x,t)}, \end{aligned} \tag{7.5}$$

where

$$g(x,t) = \int_0^x [p_{,t}(s,t) + \nu p(s,t) - f(s,t)]\, \mathrm{d}s,$$

$$\bar{\tau}(x,t) = \int_{-h}^{h} \tau \, dy, \quad \bar{\varsigma}(x,t) = \int_{-h}^{h} \varsigma \, dy.$$

On substituting of (7.5) into F^Λ, the pure complementary energy functional Π^d for this nonlinear beam is then obtained as

$$\begin{aligned} \Pi^d(p, \boldsymbol{\xi}^*) &= \operatorname{sta}\{L(\mathbf{u}, p, \boldsymbol{\xi}^*) | \ \forall \mathbf{u} \in \mathcal{U}_a\} \\ &= E^*(\boldsymbol{\xi}^*) - K^*(p) - F^\Lambda(\boldsymbol{\varpi}^*) \\ &= \int_{\Omega_t} e^{\nu t}[\frac{1}{2}\varsigma^2 + \lambda\varsigma + \frac{1}{2\beta}\tau^2] \, d\Omega_t - \int_0^{t_c}\int_0^\ell \frac{1}{2} e^{\nu t}\rho^{-1}p^2 \, dx \, dt \\ &\quad + \int_0^{t_c}\int_0^\ell e^{\nu t}\frac{[g(x,t) - \bar{\tau}(x,t)]^2}{2\alpha\bar{\varsigma}(x,t)} \, dx \, dt. \end{aligned}$$

For the sake of simplicity, if the shear deformation can be ignored and the dynamical deflection $w(x,t)$ can be separated variables as $w(x,t) = q(t)X(x)$, then the primal problem of this nonlinear post-buckling beam model is equivalent to the following simple nonconvex variational problem

$$\Pi(q) = \int_I e^{\nu t}[\rho q'^2 - \frac{1}{2}a(\frac{1}{2}q^2 - \lambda)^2 + fq] \, dt \quad \to \operatorname{sta} \ \forall q \in \mathcal{U}_k. \tag{7.6}$$

Over the kinematically admissible space $\mathcal{U}_k$, defined by

$$\mathcal{U}_k = \{q \in \mathcal{L}^4(0,t_c) | \ q' \in \mathcal{L}^2(0,t_c), \ q(0) = q_0, \ q'(0) = v_0\},$$

the stationary condition of Π leads to the well-known dissipative Duffing equation

$$\rho(q'' + \nu q') - aq(\lambda - \frac{1}{2}q^2) = f(t), \ \ \forall t \in I = (0, t_c). \tag{7.7}$$

In terms of the nonlinear canonical measure $\xi = \Lambda(q) = \frac{1}{2}q^2$, the energy density $U(\xi)$ and its conjugate $U^*(\varsigma)$ are convex functions:

$$U(\xi) = \frac{1}{2}a(\xi - \lambda)^2, \quad U^*(\varsigma) = \frac{1}{2a}\varsigma^2 + \lambda\varsigma.$$

The extended Lagrangian for this nonconvex system is

$$\Xi(q,p,\varsigma) = \int_I e^{\nu t}\left(pq' - \varsigma(\frac{1}{2}q^2 - \lambda) - \frac{1}{2\rho}p^2 + \frac{1}{2a}\varsigma^2\right) dt + \int_I e^{\nu t} fq \, dt. \tag{7.8}$$

The criticality condition $D_q\Xi(\bar{q}, p, \varsigma) = 0$ leads to the equilibrium equation

$$p' + \nu p + \bar{q}\varsigma = f \ \ \forall t \in I.$$

Clearly, the critical point $\bar{q} = (f - p' - \nu p)/\varsigma$ is well-defined for any nonzero ς. Thus, the dual feasible space can be defined as

$$\mathcal{W}_s^* = \left\{ (p, \varsigma) \in \mathcal{C}^1(I) \;\middle|\; \begin{array}{l} p(0) = \rho v_0, \;\; -\lambda a \leq \varsigma(t) < +\infty, \\ \varsigma(t) \neq 0 \;\; \forall t \in I, \;\; \varsigma(0) = a(\frac{1}{2}q_0^2 - \lambda) \end{array} \right\}.$$

Substituting $\bar{q} = (f - p' - \nu p)/\varsigma$ into the Λ-dual functional (3.9), we have

$$F^\Lambda(p, \varsigma) = e^{\nu t_c} p(t_c) q(t_c) - \rho v_0 q_0 + \int_I e^{\nu t} \frac{(p' + \nu p - f)^2}{2\varsigma} \, \mathrm{d}t.$$

Thus, the dual action for this problem is obtained as

$$\begin{aligned} \Pi^d(p, \varsigma) &= e^{\nu t_c} p(t_c) q(t_c) - \rho v_0 q_0 \\ &\quad + \int_I e^{\nu t} [\frac{1}{2a}\varsigma^2 + \lambda\varsigma + \frac{(p' + \nu p - f)^2}{2\varsigma} - \frac{1}{2\rho} p^2] \, \mathrm{d}t, \end{aligned}$$

which is well defined on $\mathcal{W}_s^*$. The criticality condition for Π^d leads to the *dual Duffing system* in the time domain $I \subset \mathbb{R}$

$$\left(\frac{1}{\varsigma}(p' + \nu p - f) \right)' + \frac{1}{\rho} p = 0, \tag{7.9}$$

$$\varsigma^2 (\frac{1}{a}\varsigma + \lambda) = \frac{1}{2}(f - p' - \nu p)^2. \tag{7.10}$$

This system consists of the so-called *differential-algebraic equations* (DAE's), which arise naturally in many applications (cf., e.g., Brenan *et al*, 1996). Clearly, the algebraic equation (7.10) has zero solution $\varsigma = 0$ if and only if $g = (f - p' - \nu p) = 0$. Otherwise, for any nonzero $g(t) = f(t) - p'(t) - \nu p$, the algebraic equation (7.10) has at most three real roots $\varsigma_i(t)$ $(i = 1, 2, 3)$, each of them leads to the state solution $q_i(t) = (f(t) - p_i'(t) - \nu p_i(t))/\varsigma_i(t)$. By the definition introduced by Gao (2000d), the index $i_d \in \mathbb{R}$ of the Duffing system is the number of negative roots $\bar{\varsigma}$ of the dual Euler-Lagrange equation (7.10).

Theorem 5 (Stability and Bifurcation Criteria) *For a given parameter $\lambda > 0$, initial data (q_0, v_0) and the external load $f(t)$, if at a certain time period $I_s \subset I = (0, t_c)$,*

$$\lambda_p(t) = \frac{3}{2} \left(\frac{f(t) - p'(t) - \nu p(t)}{a} \right)^{2/3} > \lambda, \;\; t \in I_s \tag{7.11}$$

then $i_d = 0$ and the Duffing system possesses only one stable solution set $(\bar{q}(t), \bar{p}(t), \bar{\varsigma}(t))$ satisfying $\bar{\varsigma}(t) > 0$ $\forall t \in I_s$, and over the period I_s,

$$\Pi(\bar{q}) = \min \Pi(q) \quad \textit{iff} \quad \Pi^d(\bar{p}, \bar{\varsigma}) = \min \Pi^d(p, \varsigma), \tag{7.12}$$

$$\Pi(\bar{q}) = \max \Pi(q) \quad \textit{iff} \quad \Pi^d(\bar{p}, \bar{\varsigma}) = \max_p \min_\varsigma \Pi^d(p, \varsigma). \tag{7.13}$$

However, if at a certain time period $I_b \subset I = (0, t_c)$ we have $\lambda_p(t) < \lambda$, then, $i_d = 2$ and the system possesses three sets of different solutions $(\bar{q}_i, \bar{p}_i(t), \bar{\varsigma}_i(t))$, $i = 1, 2, 3$. In the case that the three solutions $\varsigma_i(t)$ are in the following ordering

$$-a\lambda \le \bar{\varsigma}_3(t) \le \bar{\varsigma}_2(t) \le 0 \le \bar{\varsigma}_1(t) \;\; \forall t \in I_b, \tag{7.14}$$

then over the period I_b, the solution set $(\bar{q}_1(t), \bar{p}_1(t), \bar{\varsigma}_1(t))$ satisfies either (7.12) or (7.13); while the solution sets $(\bar{q}_i(t), \bar{p}_i(t), \bar{\varsigma}_i(t))$ $(i = 2, 3)$ satisfy

$$\Pi(\bar{q}_i) = \min_q \Pi(q) = \max_p \min_\varsigma \Pi^d(p, \varsigma) = \Pi^d(\bar{p}_i, \bar{\varsigma}_i) \quad i = 2, 3. \tag{7.15}$$

Proof. The dual Euler-Lagrange equation (7.10) is the so-called singular algebraic curve in (ς, g)-space (see Fig. 3.9). For a given $f(t)$, if $\lambda_p(t) > \lambda$, the cubic algebraic equation (7.10) possesses only one real positive solution $\bar{\varsigma}(t)$ and thus, $i_d = 0$. In the case when $\lambda_p(t) < \lambda$, the cubic algebraic equation (7.10) possesses three different real solutions $\bar{\varsigma}_i(t)$ and two of them are negative, then we have $i_d = 2$. By Theorem 3.4.4 proved by the author (Gao, 1999b), if $\bar{\varsigma}_i(t)$ $(i = 1, 2, 3)$ are the three solutions of the dual Euler-Lagrange equation (7.10) in the order of (7.14), then the associated $\bar{q}_1(t)$ is a global minimizer of the total potential

$$P(q) = \int_I e^{\nu t} \frac{1}{2} a (\frac{1}{2} q^2 - \lambda)^2 \, \mathrm{d}t - \int_I e^{\nu t} g(t) q \, \mathrm{d}t;$$

while $\bar{q}_2(t)$ is a local minimizer of P and $\bar{q}_3(t)$ is a local maximizer of P. Thus the theorem follows by the triality theorem. $\Box$.

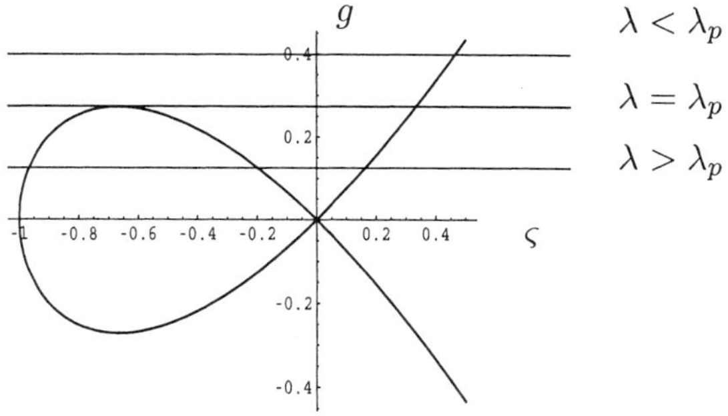

Figure 3.9. Dual solution set and bifurcation criteria

Fig. 3.9 also shows that if the continuous function $g(t) = f(t) - p'(t) - \nu p(t)$ is one-signed on certain time interval $I_b \subset I = (0, t_c)$, each root $\bar{\varsigma}(t)$ of (7.10) is also one-signed on I_i. Actually, the two negative solutions $\bar{\varsigma}(t)$ are the sources that lead to the chaotic motion of the system. Theoretically speaking, for the same initial conditions, the Duffing equation (7.7) and its dual system (7.9-7.10) should have the same solution set. Numerically, both the primal and dual solutions vary with the methods used. For the given data $a = 1, \lambda = 1.0, q_0 = 0, v_0 = .2$, $\nu = 0$ and $f = 0$, Figure 3.10 shows the numerical primal (solid line) and dual (dashed line) solutions. From the dual trajectories in the dual phase space ς-p-$p_{,t}$ (Fig. 3.10(2)) we can see that at the point $\varsigma_3(t) = -a\lambda$, if the function $g(t) = f(t) - p_{,t}(t) - \nu p(t)$ changes its sign, the state $q(t)$ crosses the origin and goes to another potential well in the phase space $\mathcal{U} \times \mathcal{V}^*$. The bifurcation is then occurred. Thus, based on this observation and the bifurcation criterion proposed in this paper, a dual feedback control method can be suggested against the chaotic vibration of the Duffing system.

For the forced vibration with linear damping, the numerical results are extremely sensitive to the parameters. Fig. 3.11 and 3.12 show that the trajectories are completely chaotic in phase spaces q-p (Fig.3.11-3.12 (b)) and ς-p-g(Fig.3.11-3.12 (d)). However, the trajectory in the dual phase space ς-g is an invariant (see Fig.3.11-3.12 (c)), which depends only on the parameters λ, a and the amplitude of the driving force $f(t)$. The stability and bifurcation criteria are verified numerically by Fig.3.11-3.12 (a), where we can see clearly that bifurcations occur (i.e., the beam jumps from one potential well to another) only during time intervals $I_b = [t_a, t_b]$ for which the characteristic load $\lambda_p(t) \leq \lambda \ \forall t \in I_b$ and there are two intermediate points $t_1, t_2 (t_a < t_1 < t_2 < t_b)$ such that $\lambda_p(t)$ is decreasing on $[t_a, t_1]$ from $\lambda_p(t_a) = \lambda$ to $\lambda_p(t_1) = 0$, and increasing in $[t_1, t_2]$ from 0 to λ, and then decreasing again on $[t_2, t_b]$ (see Fig.3.11-3.12 (a)).

As is known, nonconvex dynamical systems are very sensitive to both the parameters and numerical methods used. For the given periodic driving force $f(t) = 1.5\cos(2.75t)$ and $\nu = 0.1$, Fig. 3.12 shows that different numerical methods produce very different "chaotic results". However, solutions in dual phase space ς-g form an invariant set (Fig.3.12 (c)). This important fact and the triality theorem will play fundamental roles in stability and bifurcation analysis of chaotic systems.

The concept of duality is one of the most successful ideas in modern mathematics and science. The inner beauty of duality theory owes much to the fact that many different natural phenomena can be put in a unified trio-canonical framework (cf. e.g., Gao, 1999b). By the fact that the

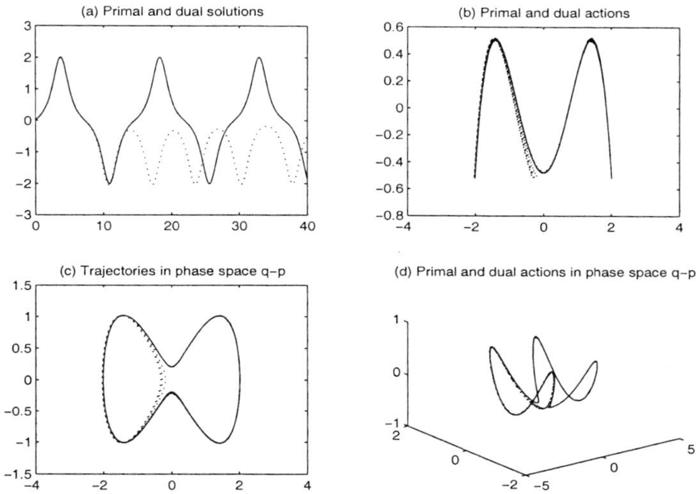

(1) Primal and dual solutions in primal phase space

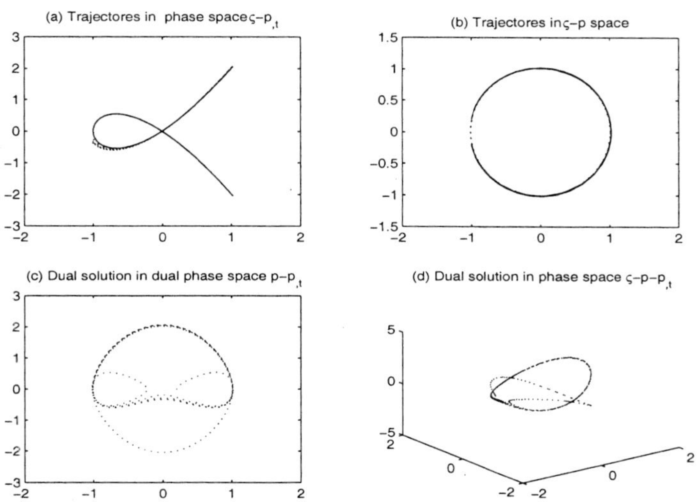

(2) Duffing solutions in dual phase spaces

Figure 3.10. Primal and dual solutions for conservative Duffing system

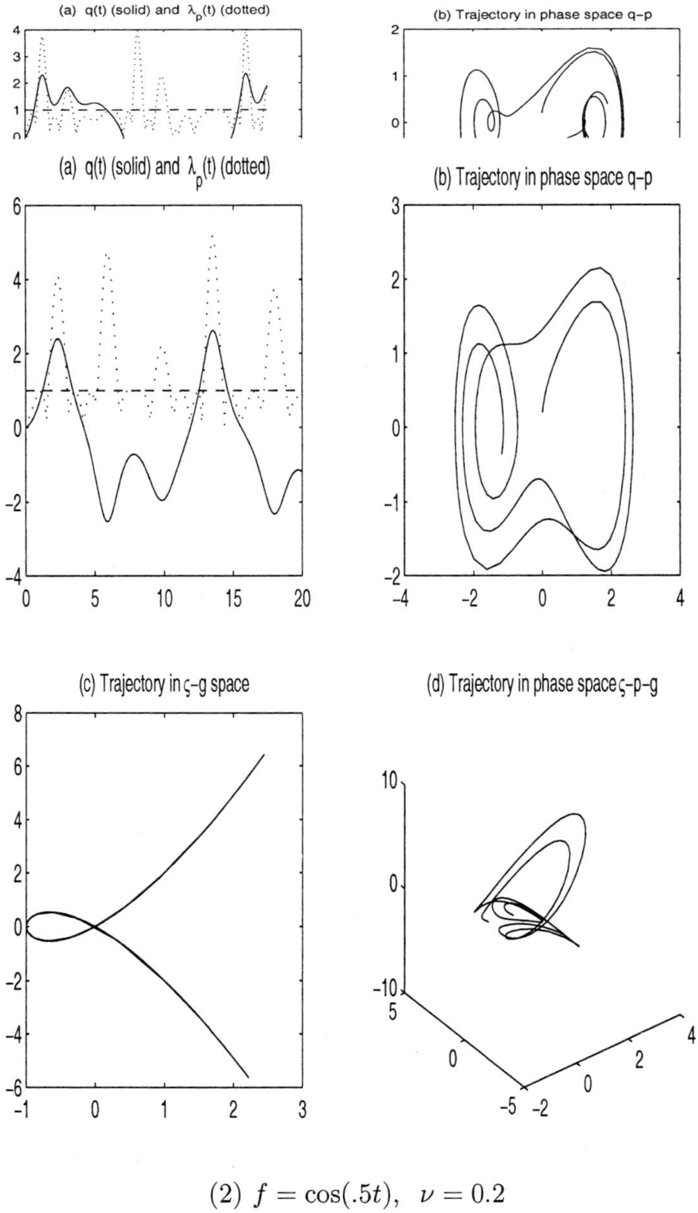

$$(2)\ f = \cos(.5t), \quad \nu = 0.2$$

Figure 3.11. Chaotic solutions, invariant sets, and bifurcation criterion in dissipative Duffing system

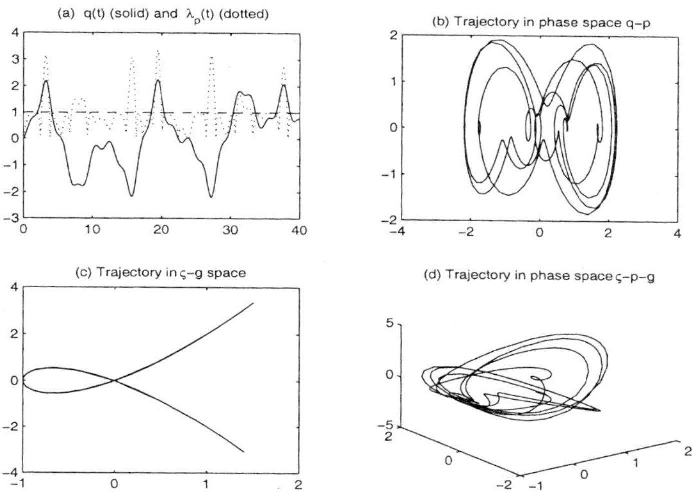

(1) Numerical results computed by "ode23"

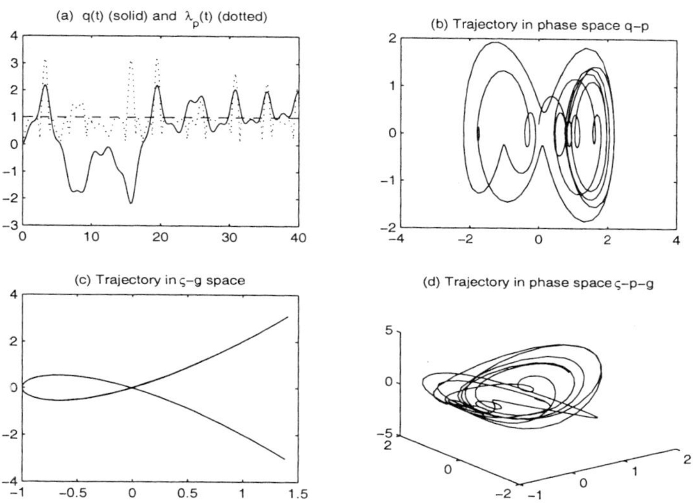

(2) Numerical results computed by "ode15s"

Figure 3.12. Chaos: numerical results by two differential numerical methods in MATLAB canonical physical variables appear always in pairs, the canonical dual transformation method can be used to solve many problems in natural systems. The associated extended Lagrange duality and triality theories may possess profound computational impacts. For any given nonconvex Newtonian dynamical system, as long as there exists a geometrical operator Λ such that the tri-canonical forms can be characterized correctly, the canonical dual transformation method and the associated triality principles can be used to establish nice theories and to develop efficient alternative algorithms for robust computations. By using the so-called sequential canonical dual transformation method (see Gao, 2000b), the results presented in this paper can be generalized into non-Newtonian systems, where the time operator ∂_t is nonlinear, as well as nonconvex systems with multiple potential wells.

Acknowledgements

The author is grateful to Dr. L.M. Moreschi for the important help in some of numerical computations. The referee's important comments and suggestions are gratefully acknowledged.

Notes

1. Since this paper does not concern the existence of stationary points, these differentiable conditions are not important for our current discussion on extremality conditions of the nonconvex variational problems.

References

[1] Antoniou, I., Basios, V. and Bosco, F. (1996). Probabilistic control of chaos: The β-adic Renyi map under control, *Int. J. Bifurcation Chaos,* **6**, 1563-1573.

[2] Arada, N. and Raymond, J.-P. (1999). Minimax control of parabolic systems with state constraints, *SIAM J. Control Optim.* **38**, 254-271.

[3] Auchmuty, G (1983), Duality for non-convex variational principles, *J. Diff. Equations*, **vol. 50**, pp 80-145

[4] Barron, E.N. (1990). The pontryagin maximum principle for minimax problems of optimal control, *Nonlinear Analysis*, **15**, 1155-1165.

[5] Brenan, K.E., Campbell, S.L. and Petzold, L.R. (1996). *Numerical solution of Initial-Value Problems in Differential-Algebraic Equations*, SIAM, Philadelphia.

[6] Brokate, M. and Sprekels, J. (1996). *Hysteresis and Phase Transitions*, Springer-Verlag, New York.

[7] Bui, H.D. (1992). On the variational boundary integral equations in elastodynamics with the use of conjugate functions, *J. Elasticity*, **28**, 247-256.

[8] Chen, G. and Dong, X. (1993). From chaos to order-perspectives and methodologies in controlling chaotic nonlinear dynamic systems, *Int. J. Bifurcation Chaos*, **3**, 1363-1409.

[9] Clarke, F.H. (1985). The dual action, optimal control, and generalized gradients, *Mathematical Control Theory*, Banach Center Publ., 14, PWN, Warsaw, pp. 109-119.

[10] Dem'yanov, V.F., Stavroulakis, G.E., Polyakova, L.N. and Panagiotopoulos, P.D., (1996). *Quasidifferentiability and Nonsmooth Modelling in Mechanics, Engineering and Economics.* Kluwer Academic Publishers: Dordrecht.

[11] Ekeland, I (1990), *Convexity Methods in Hamiltonian Mechanics*, Springer-Verlag, 247pp.

[12] Ekeland, E. and Temam, R. (1976). *Convex Analysis and Variational Problems.* Dunod: Paris.

[13] Ericksen, J.L., Equilibrium of bars, *J. Elasticity* **5** (1975), 191-202.

[14] Fowler, T.B. (1989). Application of stochastic control techniques to chaotic nonlinear systems, *IEEE Trans. Autom. Contr.*, **34**, 201-205.

[15] Gao, D.Y. (1997). Dual extremum principles in finite deformation theory with applications in post-buckling analysis of nonlinear beam model, *Appl. Mech. Reviews, ASME* **50**, S64-S71.

[16] Gao, D.Y.(1998a). Bi-complementarity and duality: A framework in nonlinear equilibria with applications to contact problem of elastoplastic beam theory. *J. Math. Anal. Appl.*, **221**, 672-697.

[17] Gao, D.Y.(1998b). Duality, triality and Complementary extremum principles in nonconvex parametric variational problems with applications, *IMA J. Appl. Math.*, **61**, 199-235.

[18] Gao, D. Y. (1999a) General Analytic Solutions and Complementary Variational Principles for Large Deformation Nonsmooth Mechanics *Meccanica*, **34**, 169-198.

[19] Gao, D.Y. (1999b). *Duality Principles in Nonconvex Systems: Theory, Methods and Applications*, Kluwer Academic Publishers.

[20] Gao, D.Y. (2000a). Finite deformation beam models and triality theory in dynamical post-buckling analysis. *Int. J. Non-Linear Mechanics*, **35**, 103-131.

[21] Gao, D.Y. (2000b). Analytic solutions and triality theory for nonconvex and nonsmooth variational problems with applications, *Nonlinear Analysis*, **42**, 1161-1193.

[22] Gao, D.Y. (2000c). Canonical dual transformation method and generalized triality theory in nonsmooth global optimization, *J. Global Optimization*, to appear.

[23] Gao, D. Y. (2000d). Nonsmooth and nonconvex dynamics: Duality, polarity and complementary extrema, in *Nonsmooth/Nonconvex Mechanics,* D.Y. Gao, R.W. Ogden and G. E. Stavroulakis (eds), Kluwer Academic Publishers, 2000, 95-140.

[24] Gao, D.Y. and Strang, G. (1989). Geometric nonlinearity: Potential energy, complementary energy, and the gap function, *Quart. Appl. Math.* **47**, 487-504.

[25] Ghezzi, L.L. and Piccardi, C. (1997). PID control of a chaotic system: an application to an epidemiological model, *Automatica*, **33**, 181-191.

[26] Hellinger, E. (1914). Die allgemeine Ansätze der Mechanik der Kontinua. Encyklopädie der Mathematischen Wissenschaften IV, 4, pp. 602-94.

[27] Kalies, W.D. and Holmes, P.J. (1996). On a dynamical model for phase transformation in nonlinear elasticity, *Fields Institute Communications*, **Vol. 5**, 255-269.

[28] Kibble, T.W.B. (1997). Phase transitions and topological defects in the early universe, *Aust. J. Phys.* **50**, 697-722.

[29] Koumboulis, F.N. and Mertzios, B.G. (2000). Feedback controlling against chaos, *Chaos, Solitons and Fractals*, **11**, 351-358.

[30] Marsden, J. and Ratiu, T. (1995). *Introduction to Mechanics and Symmetry,* Springer.

[31] Mistakidis, E.S. and Stavroulakis, G.E. (1998), *Nonconvex Optimization in Mechanics, Algorithms, Heuristics and Engineering Applications by the F.E.M.*, Kluwer Academic Publishers, Dordrecht /Boston / London, 285pp.

[32] Moreau, J.J. (1968). La notion de sur-potentiel et les liaisons unilatérales en élastostatique, *C.R. Acad. Sc. Paris,* 267 A, 954-957.

[33] Moreau, JJ, Panagiotopoulos, PD and Strang, G (1988), Topics in nonsmooth mechanics. Birkhuser Verlag, Basel-Boston, MA, 329pp.

[34] Motreanu, D. and Panagiotopoulos, P. D. (1999). *Minimax Theorems and Qualitative Properties of the Solutions of Hemivariational Inequalities.* Kluwer Academic Publishers, Dordrecht.

[35] Oden, J.T. and Reddy, J.N. (1983). *Variational Methods in Theoretical Mechanics.* Springer-Verlag.

[36] Ogden, R.W., (1984). *Non-linear elastic deformations*, Ellis Horwood: Chichester.

[37] Ogorzalek, M.J. (1993). Taming chaos - Part II: control, *IEEE Trans. Circuits Syst.*, **40**, 700-706.

[38] Ott, E., Grebogi, C. and Yorke, J.A. (1990). Controlling chaos, *Phys. Rev. Lett.*, **64**, 1196-1199.

[39] Panagiotopoulos, P.D., (1985). *Inequality Problems in Mechanics and Applications.* Birkhäuser: Boston.

[40] Pfeiffer, F. and Glocker, C. (1996). *Multibody Dynamics with Unilateral Contacts*, John Wiley, New York.

[41] Reissner, E (1953), On a variational theorem for finite elastic deformations, *J. Math. & Phys.* **32**, 129-153.

[42] Reissner, E. (1996). *Selected works in applied mechanics and mathematics*, Jones and Bartlett Publishers: Boston.

[43] Rockafellar, R.T. and Wets, R.J.B. (1997). *Variational analysis*, Springer: Berlin, New York.

[44] Stewart, D.E. (2000). Rigid-body dynamics with friction and impact, *SIAM Review*, **42**, 3-39.

[45] Strang, G (1986), *Introduction to Applied Mathematics*, Wellesley-Cambridge Press, 758pp.

[46] Tabarrok, B. (1984). Complementary variational principles in elastodynamics, *Computers & Structures*, **19**, (1-2), 239-246.

[47] Tabarrok, B. and Rimrott, F.P.J. (1994). *Variational methods and complementary formulations in dynamics.* Kluwer Academic Publishers: Dordrecht.

[48] Tonti, E. (1972), A mathematical model for physical theories, *Fisica Matematica, Serie VIII, Vol. LII.*

[49] Tuy, H (1995), D.C. optimization: theory, methods and algorithms, in *Handbook of Global Optimization*, eds. R. Horst and P. Pardalos, Kluwer Academic Publishers, 149-216.

[50] Wright, M. H. (1998). The interior-point revolution in constrained optimization, in *High-Performance Algorithms and Software in Nonlinear Optimization* (R. DeLeone, A. Murli, P. M. Pardalos, and G. Toraldo, eds.) 359–381, Kluwer Academic Publishers, Dordrecht.

Chapter 4

TRI-DUALITY THEORY IN PHASE TRANSFORMATIONS OF FERROELECTRIC CRYSTALS WITH RANDOM DEFECTS

David Y. Gao

Department of Mathematics

Virginia Tech, Blacksburg, VA, 24061, USA. E-mail: gao@vt.edu

Jie-Fang Li & D. Viehland

Department of Materials Science and Engineering

Virginia Tech, Blacksburg, VA 24061, USA

Abstract This paper presents a general solution theory to the Landau-Ginzburg (LG) equation in the presence of a high concentration of point defects of which a small concentration is charged, and subsequently applied to disordered ferroelectric crystals. Beginning with a canonical dual transformation method, a triality theorem is developed for solving the non-convex variational problem associated with LG equation. The approach allows for an exact solution of the LG equation in the case of strong Ginzburg contributions to the total energy, avoiding conventional linearization of the order parameter into Fourier components. It is demonstrated that the Ginzburg and random-field terms have similar effects, both acting as a driving force. The imperfect contribution from high concentrations of uncharged point defects acts to decrease the stability of the double potential well, favoring a single well. It is shown that the order parameter is unstable with increasing magnitude of the Ginzburg/random-field and imperfect contributions. Both theoretical and experimental results demonstrate with increasing polarization nonuniformity, that ferroelectric domains breakup into subdomains, resulting in pseudo-cubic symmetry.

Keywords: Landau-Ginzburg equation, nonconvex analysis, duality, triality, Imperfect Ferroelectrics

1. Primal Problem and Motivations

The Landau-Ginzburg (LG) equation has been applied to many problems in materials science and physics including: ferroelectricity, ferromagnetism, ferroelasticity, and superconductivity (cf. e.g., [1-5]). All approaches are founded upon the existence of an order parameter within a region of space. Landau terms describe the energy of the limiting case of uniform order within that space, whereas Ginzburg terms describe the gradients of the order parameter contained there within. Solutions to the LG equation have focused on approximations of the order parameter using a plane wave (see [1-4]), which is essentially the classical Fourier approach. In conjunction with the Liftshitz in-variant condition, these approaches have been successfully applied to incommensurate modulations of order parameters [2], modulated or tweed domains [3,4], and various fluctuation effects [1-4]. However, this type of approach inherently involves linearization, which limits its applicability to the case where the Ginzburg term only acts to perturbate the homogenous Landau solution. However, real materials are not confined to only this limiting case. Strongly inhomogeneous materials exist that have order parameters within spatially limited regions, where the Ginzburg contributions to the total energy are significant. To date, understanding of LG theory in the case of finite non-uniformity has not developed. The LG equation has proven difficult to solve, due to fact that the total potential energy associated with this equation is a nonconvex (double well) functional. Traditional direct analysis and related numerical methods for solving this nonconvex variational problem have proven to date unsuccessful.

The concept of *duality* is one of the most successful ideas in modern mathematics and science. Duality in physics, roughly speaking, is a fundamental concept that underlies many aspects of conservation laws and extremum principles in natural systems. Eigenvectors, Hamiltonian canonical forms and equilibrium equations in many field theories are all critical points of certain energy functionals on some appropriate constrained sets or manifolds. Considerable attention has been attracted on this fascinating research subject during the last years. A comprehensive study of duality in general nonconvex systems was given in [10]. Based on the general duality theory developed recently in [8-12], we present, in this paper, a unique approach to the solution of the LG equation, which begins with a canonical dual transformation method, from which a triality theorem is developed for solving the non-convex variational problem associated with the LG equation. Our approach allows for both direct and numerical analysis of the LG equation in the case of both Ginzburg and random-field contributions. In ferroelectric materials with spontaneous polarization, both theoretical and experimental results demonstrate with increasing nonuniformity, that domains breakup into subdomains, resulting in pseudo-cubic symmetry.

2. Primal Problem

Let the region of space $\Omega \subset \mathbb{R}^3$ occupied by the material be a smooth, bounded simply-connected domain with boundary $\partial\Omega$. The spontaneous polarization (i.e., dielectric displacement) $\eta(\mathbf{x})$ is a real-valued function (i.e. the so-called order-parameter). Mathematically, $\eta : \Omega \to \mathbb{R}$ is mapped into space as a real number; Physically, $\eta(\mathbf{x})$ is used to denote a field whose values describe the phase of the system under consideration. On the boundary of Ω, η is prescribed. Then, the 1-D Landau-Ginzburg energy G_0 for a ferroelectric can be written as

$$G_0(\eta) = \int_\Omega \frac{D}{2}|\nabla\eta|^2 \,\mathrm{d}\Omega + \int_\Omega Q_r\eta \,\mathrm{d}\Omega + \int_\Omega \phi(\eta) \,\mathrm{d}\Omega. \tag{2.1}$$

The first term is the Ginzburg energy, the 2nd is the random-field term where Q_r is the defect-induced field conjugate to the order parameter, and the 3rd term is Landau's "coarse-grain" free energy [1], where

$$\phi(\eta) = \phi_0 + \frac{A}{2}\eta^2 + \frac{B}{4}\eta^4 + \frac{\delta A}{2}\eta^2 + \frac{\delta B}{4}\eta^4, \tag{2.2}$$

in which, the last two terms represents the defect or imperfect contribution. Equation 2.2 applies to second order ferroelectric transformations. The approach equally applies to the case of first order transitions, where the 6th power expansion in η is included. For an isolated point defect located at randomly fixed $\mathbf{x} = \mathbf{x}_i$ (i.e., ones of the random-field type [6]), the coefficient δA can be represented as a series (cf. e.g. [1]):

$$\begin{aligned}
\delta A(\mathbf{x} - \mathbf{x}_i) &= A_1\delta(\mathbf{x} - \mathbf{x}_i) + A_2\delta"(\mathbf{x} - \mathbf{x}_i) + ..., \\
\delta B(\mathbf{x} - \mathbf{x}_i) &= B_1\delta(\mathbf{x} - \mathbf{x}_i) + B_2\delta"(\mathbf{x} - \mathbf{x}_i) + ...
\end{aligned}$$

The derivatives for δA and δB are of even orders only, as the local symmetry is conserved. This term describes changes due to the presence of point defects, however it does not describe the effect of any charges they might have. The perturbation of the local field (Q_r) due to charges on these point defects located at randomly fixed $\mathbf{x} = \mathbf{x}_i$ (i.e., ones of the random-field type [6]) can be represented as a series (cf. e.g.[1]).

$$Q_r = Q_1\delta(\mathbf{x} - \mathbf{x}_i) + Q_2\delta'(\mathbf{x} - \mathbf{x}_i) + Q_3\delta''(\mathbf{x} - \mathbf{x}_i) + Q_4\delta'''(\mathbf{x} - \mathbf{x}_i) + ...$$

The derivatives for Q_r are of all orders, as the defect-induced conjugate can break local symmetry. Our approach allows for high concentration of point defects, however the random-field term we use in equation (1) considers only low concentration of charges. To consider higher concentrations of charged defects, the defect conjugate coupling would need to extend to higher powers of η. Thus, in the approach presented here, not all point defects are charged. Only even orders of Q_i are considered, thus we restrict the model to the case where the local symmetry is conserved.

By the duality nature of the *canonical constitutive laws* for general materials (see [10]), we believe that there exists certain material parameters $\alpha > 0$ and $\beta \in \mathbb{R}$ such that the Landau potential energy density can be written in the following canonical form:

$$\phi(\eta) = \frac{\alpha}{2}(\frac{1}{2}\eta^2 - \beta)^2. \tag{2.3}$$

Thus, we have

$$\phi_0 = \frac{\alpha}{2}\beta^2 > 0, \quad \alpha = 2(B + \delta B), \quad \beta = -\frac{A + \delta A}{2(B + \delta B)}, \quad A + \delta A = -\alpha\beta.$$

This canonical form (2.3) for the Landau "coarse grain" total energy is our "new" physical assumption. It is based on the existence of $\phi_0 = \frac{1}{2}\alpha\beta^2$. This follows directly from the traditional polynomial form (2.2). Clearly, if $\beta \leq 0$, the Landau potential ϕ is a strictly convex function of η. While if $\beta > 0$, the Landau potential $\phi(\eta)$ is a *double-well* function, whose wells define the phase of materials. In this case, $A + \delta A$ must be negative if $\alpha > 0$.

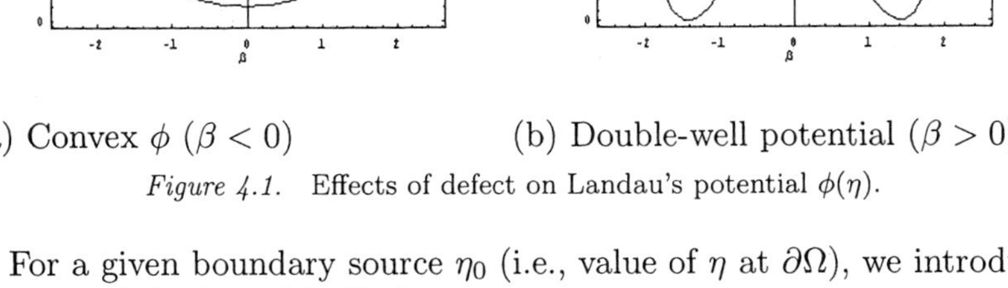

(a) Convex ϕ ($\beta < 0$) (b) Double-well potential ($\beta > 0$).

Figure 4.1. Effects of defect on Landau's potential $\phi(\eta)$.

For a given boundary source η_0 (i.e., value of η at $\partial\Omega$), we introduce the so-called admissible displacement space

$$\mathcal{U}_k = \{\eta \in \mathcal{L}^4(\Omega; \mathbb{R}) | \ \nabla\eta \in \mathcal{L}^2(\Omega, \mathbb{R}^3), \ \eta(\mathbf{x}) = \eta_0 \ \forall \mathbf{x} \in \partial\Omega\},$$

where $\mathcal{L}(\Omega;\mathbb{R})$ is the standard notation for the Lebesgue integrable space with domain in Ω and range in $\mathbb{R}$ ($\eta \in \mathcal{L}(\Omega;\mathbb{R})$ means that $\eta : \Omega \to \mathbb{R}$ is Lebesgue integrable on the domain Ω). Thus, the primal variational problem (($\mathcal{P}_{\min}$) in short) for the Landau-Ginzburg equation is to find a global minimizer of G for all $\eta \in \mathcal{U}_k$, i.e.

$$(\mathcal{P}_{\min}): \quad G(\eta) \quad \to \min \ \ \forall \eta \in \mathcal{U}_k. \tag{2.4}$$

The criticality condition of the total potential $G(\eta)$ leads to the boundary-value problem (($\mathcal{P}_{bv}$) in short) of the Landau-Ginzburg equation

$$D\Delta\eta(\mathbf{x}) - Q_r(\mathbf{x}) + \alpha\eta(\mathbf{x})(\beta - \frac{1}{2}\eta(\mathbf{x})^2) = 0 \ \ \forall \mathbf{x} \in \Omega, \tag{2.5}$$

$$\eta(\mathbf{x}) = \eta_0(\mathbf{x}) \ \ \forall \mathbf{x} \in \partial\Omega. \tag{2.6}$$

From the point view of convex analysis we know that if $\beta \leq 0$, the L-G energy G is convex on $\mathcal{U}_k$. In this case, the primal variational problem ($\mathcal{P}_{\min}$) and the BVP ($\mathcal{P}_{\text{bv}}$) are equivalent in the sense that they have same solution set, i.e. the solutions of ($\mathcal{P}_{\text{bv}}$) are global minimizers of $G(\eta)$. However, if $\beta > 0$, the problem ($\mathcal{P}_{\text{bv}}$) for the Landau-Ginzburg equation may have more than one solution over the domain Ω, and all these solutions are critical points of $G(\eta)$ over the feasible space $\mathcal{U}_k$, i.e. each solution of ($\mathcal{P}_{\text{bv}}$) could be either local (global) minimizer or maximizer. In this case, the two problems ($\mathcal{P}_{\min}$) and ($\mathcal{P}_{\text{bv}}$) are not equivalent.

Due to the nonconvexiy of G , the traditional analytic methods and associated algorithms for solving phase transition problem are very difficult. The numerical results of most direct approaches depend on the initial iteration point chosen. One way to solve this nonconvex variational problem and to clarify the phase states (i.e. the minimizers of $G(\eta)$) is to formulate it as a canonical dual variational problem.

3. Canonical Dual Transformation

The key idea of the canonical dual transformation method introduced in [10] is to choose certain geometrical measure $\boldsymbol{\xi}$ such that the stored energy $\bar{V}$ of the system is a strictly convex function of this measure. The convexity of $\bar{V}(\boldsymbol{\xi})$ leads to a one-to-one physical relation between ξ and its dual, i.e. the canonical constitutive equation: $\boldsymbol{\xi}^* = D\bar{V}(\boldsymbol{\xi})$. For the LG system considered in this paper, this geometrical measure $\boldsymbol{\xi}$ can be defined by

$$\boldsymbol{\xi} = \Lambda(\eta) = (\text{grad}\,\eta\,,\ \frac{1}{2}\eta^2 - \beta)^T = (\epsilon,\ \xi)^T, \tag{3.1}$$

where $\boldsymbol{\epsilon} = \text{grad}\ \eta$ is an infinite-small change in η, and $\xi = \frac{1}{2}\eta^2 - \beta$ is a so-called *canonical strain measure* (see[8-12]). Thus, in terms of $\boldsymbol{\xi} = (\boldsymbol{\epsilon},\ \xi)^T$, the Landau-Ginzburg potential can be re-written as $G_0(\eta) = \int_\Omega [\bar{V}(\Lambda(\eta)) + Q_r \eta]\, d\Omega$, where $\bar{V}$ is the quadratic form

$$\bar{V}(\boldsymbol{\xi}) = \frac{D}{2}\boldsymbol{\epsilon} \cdot \boldsymbol{\epsilon} + \frac{\alpha}{2}\xi^2 = \frac{1}{2}\boldsymbol{\xi}^T C \boldsymbol{\xi},$$

in which C is a positive-definite matrix defined by

$$C = \begin{pmatrix} D & 0 \\ 0 & \alpha \end{pmatrix}.$$

The dual variable $\boldsymbol{\varsigma} = (\boldsymbol{\sigma},\ \varsigma)^T = \boldsymbol{\xi}^*$ of the general strain measure $\boldsymbol{\xi} = (\boldsymbol{\epsilon}, \xi)^T$ is defined by the canonical constitutive law

$$\boldsymbol{\varsigma} = (\boldsymbol{\sigma},\ \varsigma)^T = D\bar{V}(\boldsymbol{\xi}) = (D\boldsymbol{\epsilon}, \alpha\xi)^T.$$

Since Λ is a quadratic operator, $\boldsymbol{\xi}$ can be considered as a Cauchy-Green type strain vector in finite deformation theory, or a Riemann metric tensor in differential geometry, and its dual variable $\boldsymbol{\varsigma}$ is a second Piola-Kirchhoff type stress (see [10]). The directional derivative of $\Lambda(\eta)$ at $\bar{\eta}$ in the direction η (i.e. the perturbation of $\bar{\eta}$) is given simply as

$$\delta\boldsymbol{\xi}(\bar{\eta}; \eta) = (\text{grad}\ \eta\ , \bar{\eta}\eta)^T = \Lambda_t(\bar{\eta})\eta, \tag{3.2}$$

where $\Lambda_t(\bar{\eta}) = (\text{grad}\ , \bar{\eta})^T$ is the Gâteaux derivative of Λ at $\bar{\eta}$. Let $\mathcal{E} = \mathcal{L}^2(\Omega; \mathbb{R}^3)$ and $\mathcal{S} = \mathcal{E}^*$ be the paired vector spaces in duality by the standard bilinear form $\langle *; * \rangle : \mathcal{E} \times \mathcal{S} \to \mathbb{R}$ defined by

$$\langle \boldsymbol{\xi}; \boldsymbol{\varsigma} \rangle = \int_\Omega \boldsymbol{\xi} \cdot \boldsymbol{\varsigma}\, d\Omega = \int_\Omega (\epsilon_1\sigma_1 + \epsilon_2\sigma_2 + \xi\varsigma)\, d\Omega.$$

From the Gauss formula,

$$\begin{aligned} \langle \Lambda_t(\bar{\eta})\eta; \boldsymbol{\varsigma} \rangle &= \int_\Omega [(\text{grad}\eta) \cdot \boldsymbol{\sigma} + \bar{\eta}\eta\varsigma]\, d\Omega \\ &= \int_\Omega \eta(-\text{div}\boldsymbol{\sigma} + \varsigma\bar{\eta})\, d\Omega + \int_{\partial\Omega} \eta(\mathbf{n} \cdot \boldsymbol{\sigma})\, d\Gamma \\ &= \langle \eta, \Lambda_t^*(\bar{\eta})\boldsymbol{\varsigma} \rangle \quad \forall \eta \in \mathcal{U}_k, \end{aligned}$$

the adjoint operator Λ_t^* of Λ_t associated with the standard bilinear form $\langle , \rangle : \mathcal{U} \times \mathcal{U}^* \to \mathbb{R}$ can be defined by

$$\Lambda_t^*(\bar{\eta})\boldsymbol{\varsigma} = \begin{cases} -\text{div}\boldsymbol{\sigma} + \bar{\eta}\varsigma & \text{in } \Omega, \\ \mathbf{n} \cdot \boldsymbol{\sigma} & \text{on } \partial\Omega. \end{cases}$$

Thus, in terms of the canonical dual pair $(\boldsymbol{\xi}, \varsigma)$, the Ginzburg-Landau equation can be written in the so-called trio-canonical forms:

$$\left.\begin{array}{lll} (1) & \text{Geometrical equation:} & \boldsymbol{\xi} = \Lambda(\eta); \\ (2) & \text{Constitutive equation:} & \varsigma = C\boldsymbol{\xi}; \\ (3) & \text{Balance equation:} & \Lambda_t^*(\eta)\varsigma = -Q_r \end{array}\right\} \quad \Lambda_t^*(\bar{\eta})C\Lambda(\bar{\eta}) = -Q_r. \tag{3.3}$$

By the definition introduced in Gao [10], the Landau-Ginzburg equation belongs to a geometrically nonlinear canonical system, i.e. the geometrical mapping $\Lambda(\eta)$ is nonlinear, while the constitutive equation is linear. This classification is fundamentally important in understanding the bifurcation in static system. and chaotic phenomena in nonlinear dynamic systems. The *canonical primal variational problem* ((Π) in short) associated with this canonical boundary value problem is to find all *stationary points* $\bar{\eta}$ of $\Pi(\eta)$ on $\mathcal{U}_k$ such that

$$(\Pi): \quad \Pi(\bar{\eta}) = \text{sta} \int_\Omega [\bar{V}(\Lambda(\eta)) + Q_r\eta] \, d\Omega \quad \forall \eta \in \mathcal{U}_k. \tag{3.4}$$

Clearly, this canonical variational problem is equivalent to the boundary value problem $(\mathcal{P}_{\text{bv}})$ in the sense that they have same solution set. Based on this canonical variational form, we are able to develop its dual form and associated triality theorem for solving LG equation.

4. Triality Theory

Since the canonical stored energy $\bar{V}$ is a quadratic function of the canonical strain $\boldsymbol{\xi}$, the canonical constitutive relation $\varsigma = C\boldsymbol{\xi}$ is reversible, i.e. $\boldsymbol{\xi} = C^{-1}\varsigma$. Thus, the complementary stored energy $\bar{V}^c$ can be simply obtained by the traditional Legendre transformation

$$\bar{V}^c(\varsigma) = \boldsymbol{\xi}(\varsigma) \cdot \varsigma - \bar{V}(\boldsymbol{\xi}(\varsigma)) = \frac{1}{2}\varsigma^T C^{-1}\varsigma = \frac{1}{2}D^{-1}\boldsymbol{\sigma}\cdot\boldsymbol{\sigma} + \frac{1}{2}\alpha^{-1}\varsigma^2. \tag{4.1}$$

By introducing an admissible configuration space $\mathcal{U}_a$ defined by

$$\mathcal{U}_a = \{\eta \in \mathcal{L}^2(\Omega;\mathbb{R}) | \ \nabla\eta \in \mathcal{L}^2(\Omega;\mathbb{R}^3), \ \eta(\mathbf{x}) = \eta_0 \ \forall \mathbf{x} \in \partial\Omega\}, \tag{4.2}$$

the so-called extended Lagrangian $L : \mathcal{U}_a \times \mathcal{S} \to \mathbb{R}$ associated with the canonical primal problem is [8-12]

$$\begin{aligned} L(\eta, \varsigma) &= \int_\Omega [\Lambda(\eta)\cdot\varsigma - \bar{V}^c(\varsigma) + Q_r\eta] \, d\Omega \\ &= \int_\Omega [(\frac{1}{2}\eta^2 - \beta)\varsigma + (\nabla\eta)\cdot\boldsymbol{\sigma} - \frac{1}{2}D^{-1}\boldsymbol{\sigma}\cdot\boldsymbol{\sigma} - \frac{1}{2}\alpha^{-1}\varsigma^2 + Q_r\eta] \, d\Omega \end{aligned}$$

It is easy to see that for each given $\eta \in \mathcal{U}_a$, $L : \mathcal{S} \to \mathbb{R}$ is a strictly concave. However, the convexity of $L : \mathcal{U}_a \to \mathbb{R}$ depends on the sign of ς. The critical condition $\delta L(\eta, \boldsymbol{\varsigma}) = 0$ leads to the mixed canonical equations

$$\Lambda_t^*(\eta)\boldsymbol{\varsigma} = -\mathrm{div}\boldsymbol{\sigma} + \eta\varsigma = -Q_r, \tag{4.3}$$

$$\Lambda(\eta) - D\bar{V}^c(\boldsymbol{\varsigma}) = \begin{cases} \frac{1}{2}\eta^2 - \beta - \alpha^{-1}\varsigma = 0 \\ \nabla\eta - D^{-1}\boldsymbol{\sigma} = 0 \end{cases} \tag{4.4}$$

Let $\mathcal{S}_a \subset \mathcal{S}$ be the so-called *statically admissible canonical stress field*, defined by

$$\mathcal{S}_a = \{(\boldsymbol{\sigma}, \varsigma) \in \mathcal{S} | \ \varsigma(\mathbf{x}) \neq 0 \ \ \forall \mathbf{x} \in \Omega\} \tag{4.5}$$

For each given $\boldsymbol{\varsigma} = (\boldsymbol{\sigma}, \varsigma) \in \mathcal{S}_a$, the state $\eta(\mathbf{x})$ can be well-determined by the equilibrium equation (4.3) as $\eta(\mathbf{x}) = (\mathrm{div}\boldsymbol{\sigma}(\mathbf{x}) - Q_r(\mathbf{x}))/\varsigma(\mathbf{x})$. Thus, the canonical dual functional $\Pi^d : \mathcal{S}_a \to \mathbb{R}$ can be obtained as

$$\begin{aligned} \Pi^d(\boldsymbol{\varsigma}) &= \mathrm{sta}_{\eta \in \mathcal{U}_a} L(\eta, \boldsymbol{\varsigma}) \\ = & - \int_\Omega [\frac{1}{2\varsigma}(\mathrm{div}\boldsymbol{\sigma} - Q_r)^2 + \frac{1}{2D}\boldsymbol{\sigma}\cdot\boldsymbol{\sigma} + \frac{1}{2}\alpha^{-1}\varsigma^2 + \beta\varsigma] \, \mathrm{d}\Omega \\ & + \int_{\partial\Omega} \eta_0 \mathbf{n}\cdot\boldsymbol{\sigma} \, \mathrm{d}\Gamma \end{aligned} \tag{4.6}$$

Clearly, $\Pi^d : \mathcal{S}_a \to \mathbb{R}$ is a nonconvex functional. The canonical dual problem ((Π^d) in short) associated with the LG equation can be proposed as: finding the critical point $\bar{\boldsymbol{\varsigma}} \in \mathcal{S}_a$ of Π^d over the canonical dual space $\mathcal{S}_a$, i.e.

$$(\Pi^d): \quad \Pi^d(\bar{\boldsymbol{\varsigma}}) = \mathrm{sta}\, \Pi^d(\boldsymbol{\varsigma}) \ \ \forall \boldsymbol{\varsigma} \in \mathcal{S}_a. \tag{4.7}$$

The criticality condition of (Π^d) leads to the following canonical dual system

$$2\varsigma^2(\alpha^{-1}\varsigma + \beta) = (\mathrm{div}\boldsymbol{\sigma} - Q_r)^2, \quad \text{in } \Omega \tag{4.8}$$

$$\mathrm{grad}[\varsigma^{-1}(\mathrm{div}\boldsymbol{\sigma} - Q_r)] - D^{-1}\boldsymbol{\sigma} = 0 \quad \text{in } \Omega \tag{4.9}$$

in the domain Ω and the boundary condition

$$\varsigma^{-1}(\mathrm{div}\boldsymbol{\sigma} - Q_r) = \eta_0(\mathbf{x}) \quad \text{on } \partial\Omega. \tag{4.10}$$

Clearly, for each given stress field $\boldsymbol{\sigma}(\mathbf{x})$ and defect field $Q_r(\mathbf{x})$, if $f = \mathrm{div}\boldsymbol{\sigma} - Q_r = 0$, the algebraic equation (4.8) has only one solution set $\varsigma = \{0, -\alpha\beta\}$. Otherwise, for any nonzero $f(\mathbf{x}) = \mathrm{div}\boldsymbol{\sigma}(\mathbf{x}) - Q_r$, the algebraic equation (4.8) has at most three real roots $\varsigma_i(\mathbf{x})$ $(i = 1, 2, 3)$

(see Fig. 4.2) satisfying $\varsigma_3(\mathbf{x}) < \varsigma_2(\mathbf{x}) < 0 < \varsigma_1(\mathbf{x})$, and each of them leads to the corresponding state solution

$$\eta_i(\mathbf{x}) = \frac{\mathrm{div}\boldsymbol{\sigma}(\mathbf{x}) - Q_r(\mathbf{x})}{\varsigma_i(\mathbf{x})}, \quad i = 1, 2, 3. \tag{4.11}$$

We now present three theorems based on the solution to the primal problem. The proof of these theorems can be given by the general duality theory developed by Gao, 2000.

Theorem 4.1 (Complementary-Duality Theorem) *The canonical primal problem* (Π) *and its dual problem* (Π^d) *are equivalent in the sense that they have same solution set. If* $(\bar{\eta}, \bar{\varsigma})$ *is a critical point of* L*, then* $(\bar{\eta}, \bar{\varsigma})$ *solves the boundary value problem* $(\mathcal{P}_{\mathrm{bv}})$ *and*

$$\Pi(\bar{\eta}) = L(\bar{\eta}, \bar{\varsigma}) = \Pi^d(\bar{\varsigma}). \tag{4.12}$$

This theorem proves that the solutions of our dual problem solve the canonical primal one and the boundary value problem of the LG equation in the sense that there is no duality gap. Further more, by the general triality theory developed in [8-12] for general geometrically nonlinear systems, we have the following even more interesting results.

Theorem 4.2 (Tri-Duality Theorem) *Suppose that* $(\bar{\eta}, \bar{\varsigma})$ *is a critical point of* L*. If on a subdomain* $\Omega^+ \subset \Omega$ *such that* $\bar{\varsigma}(\mathbf{x}) > 0$ *for* $\mathbf{x} \in \Omega^+$*, then* $(\bar{\eta}, \bar{\varsigma})$ *is a saddle-critical point of* L *in the domain* Ω^+ *and*

$$\min \Pi(\eta) = \min_{\eta \in \mathcal{U}_a} \max_{\varsigma \in \mathcal{S}_a} L(\eta, \varsigma) = L(\bar{\eta}, \bar{\varsigma}) = \max_{\varsigma \in \mathcal{S}_a} \min_{\eta \in \mathcal{U}_a} L(\eta, \varsigma) = \max \Pi^d(\varsigma); \tag{4.13}$$

However, if on a subdomain $\Omega^- \subset \Omega$ *such that* $\bar{\varsigma}(\mathbf{x}) < 0 \ \forall \mathbf{x} \in \Omega^-$*, then* $(\bar{\eta}, \bar{\varsigma})$ *is a super-critical point of* L *in the domain* Ω^- *and on the neighborhood of* $(\bar{\eta}, \bar{\varsigma})$ *we have either*

$$\min \Pi(\eta) = \min_{\eta} \max_{\varsigma} L(\eta, \varsigma) = L(\bar{\eta}, \bar{\varsigma}) = \min_{\varsigma} \max_{\eta} L(\eta, \varsigma) = \min \Pi^d(\varsigma), \tag{4.14}$$

or

$$\max \Pi(\eta) = \max_{\eta} \max_{\varsigma} L(\eta, \varsigma) = L(\bar{\eta}, \bar{\varsigma}) = \max_{\varsigma} \max_{\eta} L(\eta, \varsigma) = \max \Pi^d(\varsigma). \tag{4.15}$$

This theorem clarifies the extremality conditions of the critical points in our solutions of the canonical variational problems to the LG equation. This theorem has important physical meaning, which will be discussed in the next section.

5. Application to Imperfect Ferroelectric Crystals

Based on the dual Landau-Ginzburg system and triality theory estabilished above, we are able to discuss the effects of both Ginzburg term and random-fields on the trio-solutions to the LG equation.

5.1 Ginzburg and Random-field Effects

The effect of the Ginzburg and random-field terms were found to be identical. This can readily be seen in equation (19), as div$\boldsymbol{\sigma}$ and Q_r are linearly combinatorial, both contributions act as a driving force in the equation. For the given positive parameters $\alpha, \beta > 0$, the graph of $\psi(\varsigma) = \pm\varsigma\sqrt{2(\varsigma/\alpha + \beta)}$, which is the left hand-side of equation 4.8, is the so-called singular elliptic curve (see Figure 4.2). The singularity means that the curve passes through the origin. It has previously been shown in [8-12] that for fixed parameters $\alpha, \beta > 0$, there exists a critical load $f_c = \alpha(2\beta/3)^{3/2}$; such that if $f < f_c$, the level set $\psi(\varsigma) = f$ has three cross points (see Fig. 4.2). Thus, the algebrac equation (4.8) has three real roots $\varsigma_i (i = 1, 2, 3)$, satisfying $\varsigma_3 < \varsigma_2 < 0 < \varsigma_1$. Based on this simple algebraic property, we have the following theorem.

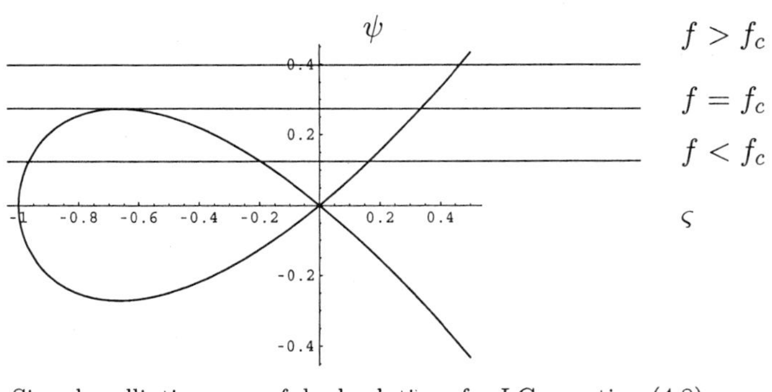

Figure 4.2. Singular elliptic curve of dual solutions for LG equation (4.8)

Theorem 4.3 (Clarification for Ginzburg/Random-field term) *For the given distributed parameters $\alpha, \beta > 0$ and the boundary input η_0, let $(\bar{\eta}, \bar{\boldsymbol{\sigma}}, \bar{\varsigma})$ be a critical point of L, and $f(\mathbf{x}) = div\bar{\boldsymbol{\sigma}}(\mathbf{x}) - Q_r(\mathbf{x})$. If $|f(\mathbf{x})| > f_c = \alpha(2\beta/3)^{3/2}$ $\forall \mathbf{x} \in \Omega$, then the critical point $(\bar{\eta}, \bar{\boldsymbol{\sigma}}, \bar{\varsigma})$ is the only solution of the Landau-Ginzburg equation on Ω, and the $\bar{\eta}$ is a global minimizer of G. Otherwise, there exists a subdomain $\Omega_m \subset \Omega$ such that $|f(\mathbf{x})| < f_c$ $\forall \mathbf{x} \in \Omega_m$, then the Landau-Ginzburg equation may have three sets of solutions $(\bar{\eta}_i, \bar{\boldsymbol{\sigma}}_i, \bar{\varsigma}_i)$ (i=1,2,3) in Ω_m satisfying*

$\bar{\varsigma}_3 < \bar{\varsigma}_2 < 0 < \bar{\varsigma}_1$. *In this case,* $\bar{\eta}_1$ *is a global minimizer of* G, $\bar{\eta}_2$ *is a local minimizer of* G *and* $\bar{\eta}_3$ *is local maximizer of* G.

Graphic illustration of this clarification theorem is given in Figure 4.2, which shows solutions to equation (4.8) as a function of $\psi(\varsigma) = \pm\varsigma\sqrt{2(\varsigma/\alpha+\beta)}$, at a constant point defect concentration. Lines are dashed in for various values of $\psi = f$. Each crossing of the dashed line and the elliptic curve represents a solution to the LG equation. Neither the Ginzburg or random-field contributions altered the shape of the singular elliptic curve, rather it only affected the cross point of the curve with the dashed lines. For $f < f_c$, the dashed lines crossed the curve at three points yielding the three solutions $\varsigma_3 < \varsigma_2 < 0 < \varsigma_1$, whereas for $f > f_c$ only one crossing was found.

To determine the influence of the Ginzburg term on the potential energy surface, the total potential as a functional of η was determined using

$$G_0(\eta) = \int_\Omega [\frac{\alpha}{2}(\frac{1}{2}\eta^2 - \beta)^2 - \eta f(\eta)]\,\mathrm{d}\Omega + \int_\Gamma \eta\mathbf{n}\cdot\boldsymbol{\sigma}\,\mathrm{d}\Gamma, \qquad (5.1)$$

where $f = \operatorname{div}\boldsymbol{\sigma} - Q_r$ is the force of the Ginzburg/random-field, $\boldsymbol{\sigma} = D\,\operatorname{grad}\eta$ and $\beta = -\frac{1}{2}(A+\delta A)/(B+\delta B)$. Figure 4.3 illustrates the effect of f on the potential diagrams at a constant degree of imperfection for: (a) $f > f_c$, (b) $f = f_c$, and (c) $f < f_c$. Both the stability (Figure 2) and potential (Fig. 4.3) curves were calculated at iso-Ginzburg/random-field contributions. Increasing f resulted in changes in the relative depths of the two minimizers and in the height of the local maximizer. The value η at which the minimizer(s) occurred shifted slightly with f. For $f > f_c$ the results demonstrate that the potential energy surface has a single potential well which is a global minimizer, whereas for $f < f_c$ it has a double potential well which has a local and a global minimizer with a local maximizer in-between.

The important effect of the Ginzburg/random-field contributions is to change the symmetry of the potential well. For $f = 0$, a double well potential was found, where the depths of the potential wells were equivalent. However, for $f > f_c$, an anharmonic single well was found, which had its global minimum at a non-zero value of η. The value of η was found to increase slightly with increasing value of f, as can be seen by comparisons of Figures 4.3(a) and (c). The anharmonic single well will result in a significantly higher dielectric compliance, due to an enhanced flatness of the well. Thus, increasing Ginzburg/random-field contributions make ferroelectric materials "softer". This means for non-stationary solutions there will be: (i) a lowering of the coercive field

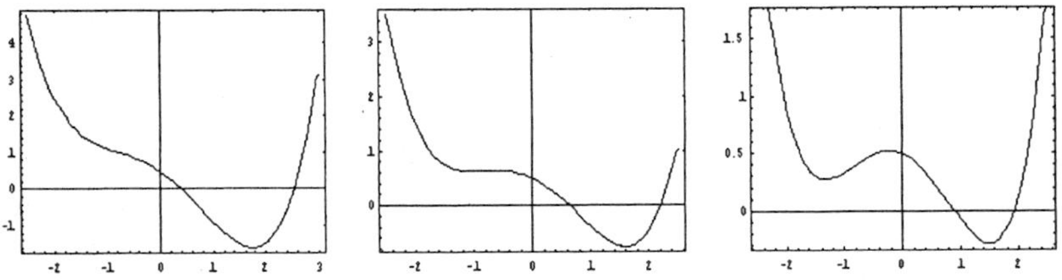

(a) $f > f_c$ (b) $f = f_c$ (c) $f < f_c$

Figure 4.3. Effect of Ginzburg contribution on potential diagrams at constant random-field contribution.

for switching between local and global minimizers, and (ii) enhanced dielectric and piezoelectric constants. "Extrinsic" contributions are often referred to in the experimental literature, they can be accounted for in this LG approach, simply by changes in the symmetry of the potential diagrams and stability solutions.

The solutions and potential diagrams presented in this section hold for all functional forms of $\eta(x)$ and all space-inhomogeneous distributions thereof. It is a complete solution for the nonlinear LG equation containing a space inhomogeneous order parameter in the presence of random-fields.

5.2 Effects of High Concentration of Imperfections

The previous section did not discuss the effect of high concentration of imperfections. We shall now include these effects, but are limited to the constraint that only a subset of these defects is charged. The dual solutions ς_i to equation (4.8) are graphically illustrated in Figure 4.4 for various values of the imperfection term $\beta = -\frac{1}{2}(A + \delta A)/(B + \delta B)$, at a constant value of the Ginzburg/random-field terms $f = \text{div}\boldsymbol{\sigma} - Q_r$. ¿From these algebraic equations, we can develop the following interesting theorem concerning the effect of the imperfection contributions.

Theorem 4.4 (Clarification Theorem for imperfection term)

Let $(\bar{\eta}, \bar{\boldsymbol{\sigma}}, \bar{\varsigma})$ *be a critical point of the Lagrangian* L, *and* $f(\mathbf{x}) = \text{div}\bar{\boldsymbol{\sigma}}(\mathbf{x}) - Q_r$. *If* $|f(\mathbf{x})|$ *has extremum values* $f_{\min}$ *and* $f_{\max}$ *on* Ω *such that* $f^2_{\min} \leq f^2(\mathbf{x}) \leq f^2_{\max}$, $\forall \mathbf{x} \in \Omega$, *then there exist two constants*

$$\beta_{\min} = 1.5(\alpha^{-1} f_{\min})^{2/3}, \quad \beta_{\max} = 1.5(\alpha^{-1} f_{\max})^{2/3}. \tag{5.2}$$

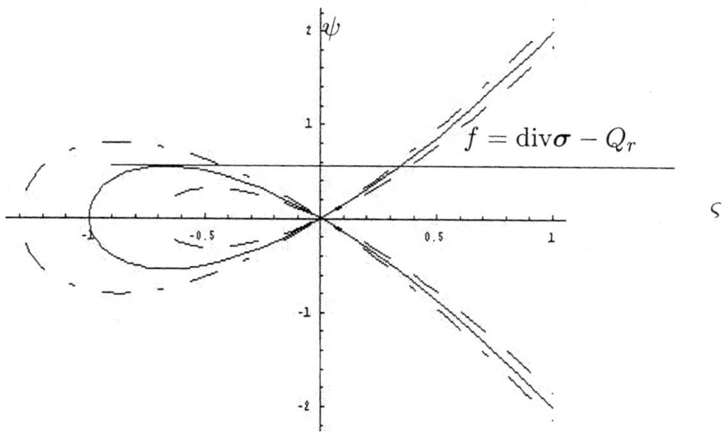

Figure 4.4. Illustration of effect of imperfection contribution on stability of solution at constant Ginzburg/random-field contribution.

If the parameter $\beta \in (-\infty, \beta_{\min})$, then the critical point $(\bar{\eta}, \bar{\epsilon}^*, \bar{\varsigma})$ is the only solution of the Ginzburg-Landau equation on Ω, and $\bar{\eta}$ is a global minimizer of P.

If the parameter $\beta \in (\beta_{\min}, \beta_{\max})$, then there exists a subdomain $\Omega_m \subset \Omega$ such that on Ω_m, the Landau-Ginzburg equation may have three solutions.

If $\beta \in (\beta_{\max}, +\infty)$, the Landau-Ginzburg equation may have three solutions on the whole domain Ω.

The effects of imperfections were found to be different than that of the Ginzburg/random-field contributions. The imperfection term only affected the homogeneous or Landau portion of the LG equation, resulting in a renormalization of the temperature dependent 2nd order Landau coefficient. Graphic illustration of this clarification theorem is given in Figure 4.4, which shows solutions to equation (4.8) as a function of $f = \text{div}\boldsymbol{\sigma} - Q_r$, for various values of $\beta = -\frac{1}{2}(A + \delta A)/(B + \delta B)$. The dashed line represents a constant value of f, which can serve as a reference to observe changes caused by β. As can be seen, the imperfection contribution altered the shape of the singular elliptic curve. Decreasing β resulted in a decrease of the area enclosed in the ellipse on the left hand side of the figure. The number of crossings of the elliptic curve with f changed, with β. For higher values of β there were three crossings, however for lower values there was a single one. The value of β is a materials property, which is constant for a given material, and can only be varied by changes in imperfection concentrations.

To determine the influence of the imperfection term on the potential energy surface, the total potential as a function of η was determined using equation (5.1). Figure 4.5 shows diagrams at various values of β, at a constant value of f. This figure illustrates the case of high concentrations of imperfections, and relatively large Ginzburg/random-field contributions, which results in a single solution that is a global minimizer. However, this potential well is non-parabolic, unlike the solution for the high-temperature paraelectric state. It has a flat potential minimum, where the polarization is non-zero. The value of η was changed significantly with finite values of β and f. Accordingly, the symmetry of this phase cannot truly be cubic, but rather should be considered as pseudo-cubic.

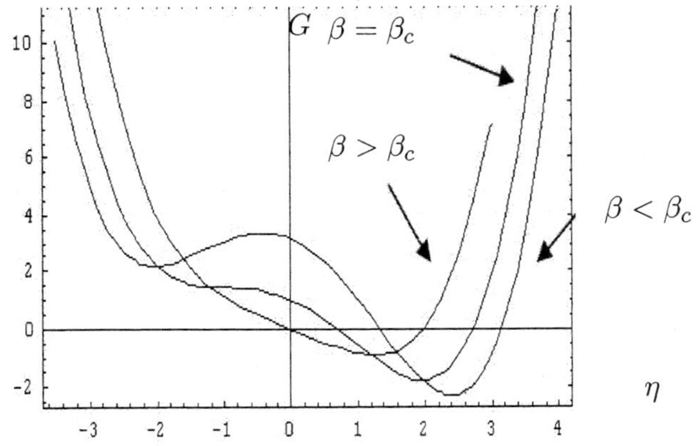

Figure 4.5. Effect of Ginzburg contribution on potential diagrams at constant random-field contribution for various values of β.

The combination of the results demonstrates that the domain of solution and the ordered state of continuous symmetry are unstable against finite imperfection, Ginzburg and random-field contributions to the total energy. All of these non-idealities place restrictions upon the phase space accessible to the solution. Clarification theorems 3 and 4 predicted that finite contributions from these terms result in domain reduction ($\Omega > \Omega_m$), restricting the phase space to subdomains solutions. This can be seen from the simple consideration that if $\beta > \beta_c$ there is only one potential, whereas if $\beta < \beta_c$ there are two. Accordingly, there must be a change in the symmetry, where the subdomains have a subgroup symmetry of that of the full domain solution. This occurs due to the breaking of continuous symmetry.

5.3 Common Sequence of Domain Transitions in Imperfect Ferroelectrics

With increasing Ginzburg/random-field and imperfection contributions to the energy density, the above results predicted a sequence of changes from a domain of solutions with a specified point group, to a restricted subdomain of solutions, and finally to a single well potential with a non-zero η which accordingly has pseudo-cubic symmetry.

We now present experimental evidence that such a sequence and changes in continuous symmetry of the ordered state exists. Transmission electron microscopy (TEM) investigations of mixed-site cation ferroelectrics have demonstrated a sequence of ferroelectric domain (i.e., spatial region having the same polarization solution) transitions with increasing Ginzburg/random-field and/or imperfection contributions [13-18]. Similar trends have been observed in modified

$$(1-x)Pb(Mg_{1/3}Nb_{2/3})O_3 - (x)PbTiO_3$$

or $PMN-PT$ [13,14], $(Pb_{1-3x/2}La_x)(Zr_{(1-x)}Ti_{(x)})O_3$ or $PLZT$ [15,16], $(Sr_{1-x}Ba_x)Nb_2O_6$ or SBN [17,18], and $(Ba_{1-3x/2}La_x)Ti_{(x)}O_3$ [19].

This sequence is illustrated for PMN-PT in Figure 6, which shows bright-field TEM micrographs of ferroelectric domains for various PT contents and aliovalent substituent concentrations. For PMN-PT 60/40, which is slightly above the morphtropic phase boundary (MPB) between ferroelectric rhombohedral (FEr) and tetragonal (FEt) phases, micron-sized domains were found (Figure 6a). Within each domain, the polarization can be considered as homogeneous. Multiple domains exist on translation across the image, representing regions with different local (global) minimizers. Ginzburg contributions occur at the boundaries between the regions, where there is a significant gradient of η. The domains for this composition are large, thus the Ginzburg contributions in the stationary case are relatively small. The symmetry is either 4mm or 3m, depending upon slight variations of composition and/or temperature.

Near the MPB (PMN-PT 65/35), the domain size was dramatically reduced as can be seen in Figure 5b. The domains for this composition were $\sim 1000\ \overset{o}{A}$ in length and $\sim 100\ \overset{o}{A}$ wide [13,14]. The small size of these domains clearly demonstrates the importance of the Ginzburg/random-field terms on the stability and potential diagrams. Upon aliovalent substitution y (at.%) in PMN-PT 65/35, the regularity in morphology and orientation of the small domains was gradually broken down for $0 < y < 0.01$ (compare Figures 6b and 6c). For $y = 0.05$, only polar clusters were found that were $\sim 100\ \overset{o}{A}$ in size (Figure 6d), and had little crystallographic relationships between each other. This condition is the so-called relaxor ferroelectric state, which is characterized by pseudo-cubic symmetry [20], however it has a non-zero η.

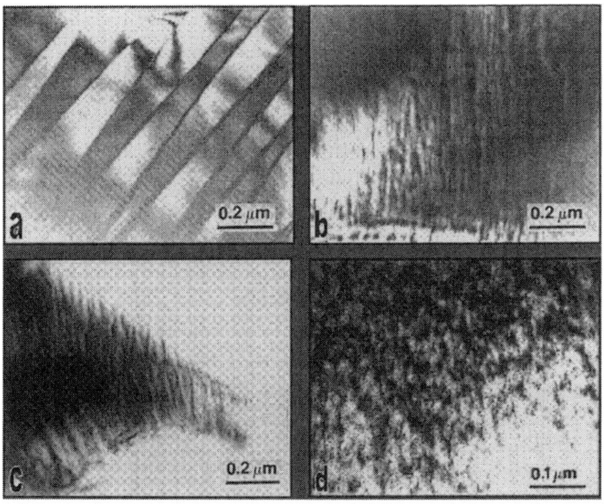

Figure 4.6. Bright field images of domain states in PMN-PT. (a) PMN-PT 60/40 on the FEt side of the MPB, (b) PMN-PT 65/35, (c) PMN-PT 65/35 modified with 1 at.% La on the Pb-site, and (d) PMN-PT 65/35 modified with 5 at.% La.

Theory predicts and experiment confirms a sequence of changes with increasing Ginzburg/random-field and imperfection contributions. This sequence is:

(i) In the presence of minimal Ginzburg/random-field and imperfection effects, large domains of a symmetry which specifies the low temperature transformed state;

(ii) to smaller sub-domains (experimentally, ~100 $\overset{o}{A}$ in width and ~1000 $\overset{o}{A}$ in length), due to an instability of η to finite Ginzburg/random-field and imperfection contributions;

(iii) to small regions which can not be classified as domains (experimentally, isotropic polar regions with an average radius ~100 $\overset{o}{A}$) that have a non-zero η and that have pseudo-cubic symmetry.

6. Summary

In this investigation, a general solution to the Landau-Ginzburg equation in the presence of finite Ginzburg, random-field, and imperfection contributions to the total energy has been achieved for the 1rst time. The solution was arrived at by use of a triality theorem, via a canonical dual transformation. The combination of the results demonstrates that domains and the ordered state of continuous symmetry are unstable against finite imperfection and Ginzburg/random-field contributions to the total energy. All of these non-idealities place restrictions upon the phase space accessible to the solution. In the case of imperfect ferroelectrics, theory predicts and experiment confirms a sequence of domain and limiting symmetry changes with increasing Ginzburg/random-field and imperfection contributions. Although the approach was only applied to ferroelectrics, it has relevance to any material having an order parameter. The approach has been applied only to the stationary solution. However, the statically admissible canonical electric field couples into equation (19) via $(\mathrm{div}\boldsymbol{\sigma} - Q_r)^2$, thus an applied field will act to enhance the Ginzburg/random-field contribution. Accordingly, non-stationary solutions will have field-induced domain and symmetry changes, rather than conventional domain boundary movement. Our theorems and confirming experimental results have significant implications to the field of real (i.e., imperfect, or inhomogeneous) materials with domains and order parameters.

Acknowledgements

This research was supported by the Office of Naval Research under grants N000140210340, N000140210126, and MURI N000140110761.

References

[1] A.P. Levanyuk and A.S. Sigov, (1988). *Defects and Structural Phase Transitions*, Gordon and Breach Science Publishers, New York, New York. (and reference therein)

[2] R. Blinc and A. P. Levanyuk Ed. (1986). *Incommensurate Phases in Dielectrics*, Elsevier Science Publisher B.V., (and reference therein)

[3] E.K.H. Salje (1993). *Phase Transitions in Ferroelastic and Coelastic Crystals* Cambridge University Press, Cambridge, United Kingdom. (and reference therein)

[4] J. Krumhansl, J. Sethna, and L. Wickham, Phys. Rev. B 52, 803 (1995).

[5] Mielke, Alexander (2002). The Ginzburg-Landau equation in its role as a modulation equation. *Handbook of dynamical systems*, Vol. 2, 759–834, North-Holland, Amsterdam.

[6] Y. Imry and S. Ma, Phys. Rev. Lett. 35 1399 (1975).

[7] Ekeland, I and Temam, R (1976), *Convex Analysis and Variational Problems,* North-Holland.

[8] Gao, DY (1998), Duality, triality and complementary extremum principles in nonconvex parametric variational problems with applications, *IMA J. of Applied Math.*, **61**, 199-235.

[9] Gao, D.Y. (1999) General analytic solutions and complementary variational principles for large deformation nonsmooth mechanics. *Meccanica*, **34**, 169-198.

[10] Gao, DY (2000a), *Duality Principles in Nonconvex Systems: Theory, Methods and Applications*, Kluwer Academic Publishers, Dordrecht /Boston / London, xvii+454pp.

[11] Gao, D.Y. (2000b). Analytic solutions and triality theory for nonconvex and nonsmooth variational problems with applications, *Nonlinear Analysis*, **42**, 1161-1193.

[12] Gao, D.Y. (2000c). Canonical dual transformation method and generalized triality theory in nonsmooth global optimization, *J. Global Optimization*, **17**, 97-126

[13] D. Viehland, Myung-Chul Kim, Z. Xu, and Jie-Fang Li, Appl. Phys. Let. 67, 2471-2473 (1995).

[14] Z. Xu, M.C. Kim, Jie-Fang Li, and Dwight Viehland, Phil. Mag. A 74, 395-406 (1996).

[15] X.H. Dai, Z. Xu and D. Viehland, Phil Mag. B 70, 33-48 (1994).

[16] X.H. Dai, PhD Dissertation, University of Illinois, Urbana, IL (1995).

[17] D. Viehland, Z. Xu, and W. Huang, Phil. Mag. A 71, 205-217 (1995).

[18] W. Huang, PhD Dissertation, University of Illinois, Urbana, IL (1996).

[19] R. Zhang, PhD Dissertation, University of Illinois, Urbana, IL (1998).

[20] D. Viehland, J. Appl. Phys. 88, 4794-4806 (2000).

Chapter 5

MATHEMATICAL MODELING OF THREE-DIMENSIONAL DELAMINATION PROCESSES OF LAMINATED COMPOSITES

Thomas C. Gasser and Gerhard A. Holzapfel
Institute for Structural Analysis, Computational Biomechanics
Graz University of Technology Schiesstattgasse 14/B, A-8010 Graz, Austria
gh@biomech.tu-graz.ac.at

Abstract The mathematical modeling of 3D delamination failure of laminated composites is discussed. Strong discontinuities are considered in the kinematically framework, which provides the basis for the embedded representation of discontinuities in finite elements. Suitable expressions for a transversely isotropic traction law in form of a displacement-energy function are derived to describe the constitutive response of the interface of laminated composites. Softening phenomena of interfaces are modeled by an isotropic damage law, while the continuous bulk material is modeled as an elastic fiber-reinforced composite. The variational formulation is based on a three-field Hu-Washizu functional which is accompanied with the enhanced assumed strain method. Three different finite element formulations are delineated. A biomechanical example investigates the dissection of the middle layer of a healthy artery, and compares the numerical results of the different finite element formulations obtained from regular and distorted meshes. Soft tissue dissection occurs, for example, during balloon angioplasty, which is a mechanical procedure frequently performed to reduce the severity of atherosclerotic stenoses. Physical and numerical analyses of delamination processes are of pressing scientific and clinical need.

Keywords: Delamination, Dissection, Failure, Discontinuous Finite Element, Cohesive Finite Element, Laminate, Composite, Biomechanics, Artery

1. Introduction

In recent years there has been a significant interest in the discontinuous modeling of strain localization phenomena. To design meaningful constitutive models for capturing localization phenomena and to implement these models in an efficient, fast and numerically accurate way is a major task in science, industrial engineering practice and in the field of biomechanics.

Traditionally, cracks have been represented in discrete or smeared techniques, for an overview see the recent articles Jirásek, 2000 and de Borst, 2001. The discrete approach is more complicated for implementations in a finite element program, and it enforces the direction of crack propagation to be along the element boundaries. Furthermore, problems arise during bandwidth minimization and mesh generation (Lotfi and Shing, 1995) . Alternative approaches, in particular for delamination analysis of composites, is the application of (cohesive) interface models, see, for example, Alfano and Crisfield, 2001. The smeared approach treats the cracked material as an equivalent continuum. A standard kinematical representation of highly localized strains, leads, however, to various problems (see Jirásek, 2000) such as overestimation of the stiffness and the strength of structures in shear dominated problems among others (see, for example, Rots, 1988).

In this communication we focus on the mathematical modeling of three-dimensional delamination of composites, in particular, we study the dissection (failure) of an arterial layer, which may be caused through non-physiological loading during, for example, the mechanical procedure of balloon angioplasty. We employ the cohesive crack concept and derive finite elements with embedded strong (displacement) discontinuities which are based on the class of mixed **E**nhanced **A**ssumed **S**train (EAS) method proposed in Simo and Rifai, 1990. The cohesive crack concept combines the strong features of the two traditional approaches (discrete and smeared) suitable to capture cracking.

In Section 2 we introduce kinematical quantities, and characterize strong discontinuities in terms of a jump of the displacement field. We continue with Section 3 by providing an anisotropic model suitable to characterize the mechanical response of fiber-reinforced composites. In addition, we propose a cohesive material model capturing transversely isotropic responses, and assume that all inelastic deformation takes place at the displacement discontinuity. The traction vector is given in terms of the gap displacement, the spatial normal vector and a scalar damage variable – associated spatial tangent moduli are provided. The variational formulation, introduced in Section 4, is based on minimizing a three-field Hu-Washizu functional, where the compatible displacement, the enhanced strain and the first Piola-Kirchhoff stress tensor are the independent fields. Additionally, we delineate three types of finite element models with embedded discontinuities, the so-called **S**tatically **O**ptimal **S**ymmetric (SOS) formulation, the **K**inematically **O**ptimal **S**ymmetric (KOS) formulation, and the **S**tatically and **K**inematically **O**ptimal **N**on-symmetric (SKON) formulations (for an overview see Jirásek, 2000). In Section 5 one representative numerical example compares results obtained from 3D finite element analyses under quasi-static conditions, which are achieved with the three (SOS, KOS, SKON) formulations, and the use of regular and distorted meshes generated with constant-strain tetrahedral elements.

2. Kinematics

In this section we provide the fundamental properties for the (finite) motion and deformation of a continuum including strong discontinuities. We consider a spatially non-continuous motion $\boldsymbol{\chi}(\mathbf{X},t)$ in the n-dimensional *Euclidean space* $\mathbb{R}^n$. As usual, $\boldsymbol{\chi}(\mathbf{X},t)$ maps a body $\mathcal{B}$, occupied by the reference configuration $\Omega_0 \in \mathbb{R}^n$, into its spatial configuration $\Omega \in \mathbb{R}^n$, where these configurations are bounded by the surfaces $\partial\Omega_0 \in \mathbb{R}^{n-1}$ and $\partial\Omega \in \mathbb{R}^{n-1}$, respectively (see, for example, Holzapfel, 2000).

A strong discontinuity separates the body $\mathcal{B}$ into the sub-bodies $\mathcal{B}^i$, $i = 1,2$, occupying the sub-regions Ω^i, $i = 1,2$, so that $\Omega = \Omega^1 \cup \Omega^2$, see Fig. 5.1 (the associated referential sub-regions are Ω_0^i). Subsequently, we denote the surfaces occupied by the discontinuity in Ω as $\partial\Omega_{\mathrm{d}}^i \in \mathbb{R}^{n-1}$, $i = 1,2$, with the associated referential surfaces $\partial\Omega_{0\mathrm{d}}^i \in \mathbb{R}^{n-1}$. The remaining boundary surfaces of Ω^i are denoted by $\partial\Omega^i \in \mathbb{R}^{n-1}$ so that $\partial\Omega = \partial\Omega^1 \cup \partial\Omega^2$, see Fig. 5.1 (the boundary surfaces of Ω_0^i read $\partial\Omega_0^i \in \mathbb{R}^{n-1}$).

As can be seen from Fig. 5.1, motion $\boldsymbol{\chi}$ maps a point $\mathbf{X}_{\mathrm{d}}$ located at the discontinuity in the reference configuration into two associated points $\mathbf{x}_{\mathrm{d}}^1$ and $\mathbf{x}_{\mathrm{d}}^2$. For characterizing the discontinuity we introduce the outward normal vectors $\mathbf{N}^i$ and $\mathbf{n}^i$, $i = 1,2$, to the discontinuities associated with the reference and spatial configurations, respectively. Since a distinction between $\mathbf{N}^1$ and $\mathbf{N}^2$ is not necessary, in the following we denote the normal vector acting at $\mathbf{X}_{\mathrm{d}}$ just as $\mathbf{N}$. In addition, we just use $\partial\Omega_{0\mathrm{d}}$ for characterizing the referential surface of the discontinuity. Note that for laminated composites normal vector fields within the whole body are defined *a priori*.

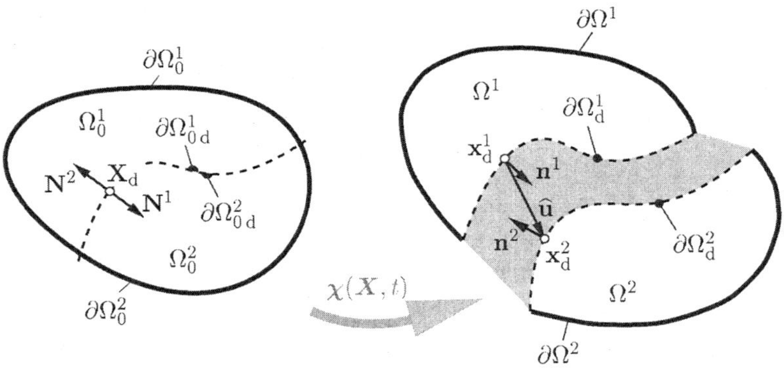

Figure 5.1. Kinematics of a body separated by a displacement discontinuity.

Following Oliver, 1996a, we represent the displacement field $\mathbf{u}(\mathbf{X},t)$ by an additive decomposition into continuous $\mathbf{u}_\mathrm{c}$ and discontinuous $\mathcal{H}_\mathrm{d}\mathbf{u}_\mathrm{e}$ parts. Thus,

$$\mathbf{u}(\mathbf{X},t) = \mathbf{u}_\mathrm{c}(\mathbf{X},t) + \mathcal{H}_\mathrm{d}(\mathbf{X})\mathbf{u}_\mathrm{e}(\mathbf{X},t) \tag{2.1}$$

holds, where

$$\mathcal{H}_\mathrm{d}(\mathbf{X}) = \begin{cases} 0, & \mathbf{X} \in \Omega_0^1, \\ 1, & \mathbf{X} \in \Omega_0^2 \end{cases} \tag{2.2}$$

denotes the *Heaviside* function (centered at the discontinuity), which is associated with the reference configuration, and which is independent of time t. Based on definition (2.1) the displacement field $\mathbf{u}_\mathrm{e}(\mathbf{X},t)$ is continuous in $\mathbf{X}$ and t, and coincides for all points $\mathbf{X}_\mathrm{d} \in \partial\Omega_{0\,\mathrm{d}}^1$ (equivalent to $\mathbf{X}_\mathrm{d} \in \partial\Omega_{0\,\mathrm{d}}^2$) on the discontinuity with the displacement jump (or gap displacement), denoted by $\widehat{\mathbf{u}}(\mathbf{X}_\mathrm{d},t) = \mathbf{x}_\mathrm{d}^2(\mathbf{X}_\mathrm{d},t) - \mathbf{x}_\mathrm{d}^1(\mathbf{X}_\mathrm{d},t)$.

Note that we have limited our considerations to two sub-bodies. The kinematical concept can, however, be easily extended to a finite number of sub-bodies, with non-intersecting discontinuities.

2.1 Discontinuous deformation gradient

Here we derive the explicit expression of the discontinuous deformation gradient $\mathbf{F}$, which is a primary measure of deformation.

We start with the derivation of the material gradient $\operatorname{Grad}\mathbf{u}(\mathbf{X},t)$ of the displacement field and use the following (decomposed) representation

$$\operatorname{Grad}\mathbf{u}(\mathbf{X},t) = \operatorname{Grad}\mathbf{u}_\mathrm{c}(\mathbf{X},t) + \overline{\operatorname{Grad}}\,\mathbf{u}_\mathrm{e}(\mathbf{X},t) + \widehat{\operatorname{Grad}}\,\mathbf{u}_\mathrm{e}(\mathbf{X},t), \tag{2.3}$$

where $\mathrm{Grad}\,\mathbf{u}_\mathrm{c}(\mathbf{X},t)$ is the 'standard' material gradient of the continuous displacement field $\mathbf{u}_\mathrm{c}$, while $\overline{\mathrm{Grad}\,\mathbf{u}_\mathrm{e}}(\mathbf{X},t)$ and $\widehat{\mathrm{Grad}\,\mathbf{u}_\mathrm{e}}$ denote the *bounded* and *unbounded* parts due to the displacement discontinuity $\mathcal{H}_\mathrm{d}\mathbf{u}_\mathrm{e}$. Using eq. (2.1) and the relation $\mathrm{Grad}\,\mathcal{H}_\mathrm{d}(\mathbf{X}) = \delta_\mathrm{d}(\mathbf{X})\mathbf{N}(\mathbf{X},t)$ of the *Heaviside* function, the unbounded contribution reads

$$\left.\begin{aligned} \overline{\mathrm{Grad}\,\mathbf{u}_\mathrm{e}}(\mathbf{X},t) &= \mathcal{H}_\mathrm{d}(\mathbf{X})\mathrm{Grad}\,\mathbf{u}_\mathrm{e}(\mathbf{X},t), \\ \widehat{\mathrm{Grad}\,\mathbf{u}_\mathrm{e}}(\mathbf{X},t) &= \delta_\mathrm{d}(\mathbf{X})[\mathbf{u}_\mathrm{e}(\mathbf{X},t)\otimes\mathbf{N}(\mathbf{X},t)], \end{aligned}\right\} \tag{2.4}$$

where

$$\delta_\mathrm{d}(\mathbf{X}) = \begin{cases} 0, & \mathbf{X} \neq \mathbf{X}_\mathrm{d}, \\ \infty, & \mathbf{X} = \mathbf{X}_\mathrm{d} \end{cases} \tag{2.5}$$

denotes the *Dirac-delta* functional centered at the discontinuity. Based on eqs. (2.3) and (2.4) the discontinuous deformation gradient is defined to be $\mathbf{F}(\mathbf{X},t) = \mathbf{I} + \mathrm{Grad}\,\mathbf{u}(\mathbf{X},t)$, where $\mathbf{I}$ denotes the unit tensor. The deformation gradient constitutes the basis for the computation of the right and left Cauchy-Green tensors, i.e. $\mathbf{C} = \mathbf{F}^\mathrm{T}\mathbf{F}$ and $\mathbf{b} = \mathbf{F}\mathbf{F}^\mathrm{T}$, as used throughout the paper.

2.2 Definition of a fictitious discontinuity

As can be seen from Fig 5.1, the opened crack is determined by two surfaces ($\partial\Omega_\mathrm{d}^1$ and $\partial\Omega_\mathrm{d}^2$) with different outward normal vectors $\mathbf{n}^1$ and $\mathbf{n}^2$ in the current configuration. In order to provide a description of the (cohesive) discontinuous constitutive model for nonlinear deformation behavior of materials, by following Wells, 2001, we define a *fictitious* discontinuity located in the current configuration. In particular, we introduce the motion $\overline{\chi}(\mathbf{X}_\mathrm{d},t)$ through the deformation gradient $\overline{\mathbf{F}}(\mathbf{X}_\mathrm{d},t) = \mathbf{I} + \mathrm{Grad}\,\mathbf{u}_\mathrm{c} + \overline{\mathrm{Grad}\,\mathbf{u}_\mathrm{e}} + (\widehat{\mathbf{u}}\otimes\mathbf{N})/2$ which maps a point $\mathbf{X}_\mathrm{d} \in \partial\Omega_{0\,\mathrm{d}}$ into a place $\overline{\mathbf{x}}_\mathrm{d}$ located on the fictitious discontinuity. The gap displacement at $\mathbf{X}_\mathrm{d}$ is described by $\widehat{\mathbf{u}}$. Hence, the normal vector $\overline{\mathbf{n}}$ at $\overline{\mathbf{x}}_\mathrm{d}$ of the fictitious discontinuity is defined via a weighted push-forward operation of the covariant vector $\mathbf{N}$ according to

$$\overline{\mathbf{n}} = \frac{\mathbf{N}\overline{\mathbf{F}}^{-1}}{|\mathbf{N}\overline{\mathbf{F}}^{-1}|}. \tag{2.6}$$

Based on definition (2.6) arbitrary cohesive constitutive relations for transversely isotropic materials can be formulated.

3. Material models

In this section we derive and particularize constitutive equations. Since we consider strong (displacement) discontinuities we distinguish between two types of material models, i.e. continuous and cohesive (discontinuous) material models.

For the continuous constitutive response of the laminated composite we follow Holzapfel et al., 2000a, and use the additive representation of the strain energy

$$\Psi(\mathbf{C}, \mathbf{A}_1, \mathbf{A}_2) = L(J) + \overline{\Psi}(\overline{\mathbf{C}}, \mathbf{A}_1, \mathbf{A}_2), \tag{3.1}$$

where $J = \det\mathbf{F} > 0$ and $\overline{\mathbf{C}} = J^{-2/3}\mathbf{F}^{\mathrm{T}}\mathbf{F}$ denotes the Jacobian determinant and the modified right Cauchy-Green tensor, respectively. The first part $L(J)$ characterizes a mathematically motivated penalty function (with $L(1) = 0$) enforcing the incompressibility constraint. The second part is the energy stored in the material when it is subjected to isochoric elastic deformations. The anisotropic structure of Ψ is described in terms of the (structural) tensors $\mathbf{A}_1 = \mathbf{a}_{0\,1} \otimes \mathbf{a}_{0\,1}$ and $\mathbf{A}_2 = \mathbf{a}_{0\,2} \otimes \mathbf{a}_{0\,2}$, where the unit vector fields $\mathbf{a}_{0\,i}(\mathbf{X})$, $|\mathbf{a}_{0\,i}| = 1$, $i = 1, 2$, define the preferential fiber directions at point $\mathbf{X} \in \Omega_0$ (for further details see Holzapfel et al., 2000a).

The cohesive (discontinues) constitutive model is based on the definition of a displacement-energy function (cohesive potential) with respect to the previous introduced fictitious discontinuity located at the current configuration. We restrict our work to transversely isotropy and introduce

$$\psi(d, \widehat{\mathbf{u}}, \overline{\mathbf{n}}) = (1 - d)\widetilde{\psi}(\widehat{\mathbf{u}}, \overline{\mathbf{n}}), \tag{3.2}$$

where $\widetilde{\psi}$ denotes an effective cohesive potential with respect to the current area $\overline{s}$ associated with the fictitious discontinuity in the current configuration. In eq. (3.2) $d \in [0, 1]$ denotes a damage variable describing the state of damage of the interface zone in an isotropic fashion. The standard Coleman-Noll procedure leads to a (constitutive) relation between the (Cauchy) traction vector $\mathbf{t}$, acting at the discontinuity, and the gap displacement $\widehat{\mathbf{u}}$. Consequently, the associated tangent moduli $\mathbf{c}_{\widehat{\mathbf{u}}}$ and $\mathbf{c}_{\overline{\mathbf{n}}}$ in spatial descriptions follow from $\mathbf{t}$ by consistent linearization. Thus,

$$\mathbf{t} = (1 - d)\widetilde{\mathbf{t}}, \qquad \mathbf{c}_{\widehat{\mathbf{u}}} = (1 - d)\widetilde{\mathbf{c}}_{\widehat{\mathbf{u}}}, \qquad \mathbf{c}_{\overline{\mathbf{n}}} = (1 - d)\widetilde{\mathbf{c}}_{\overline{\mathbf{n}}}, \tag{3.3}$$

$$\text{with} \qquad \widetilde{\mathbf{t}} = \frac{\partial \widetilde{\psi}}{\partial \widehat{\mathbf{u}}}, \qquad \widetilde{\mathbf{c}}_{\widehat{\mathbf{u}}} = \frac{\partial^2 \widetilde{\psi}}{\partial \widehat{\mathbf{u}} \partial \widehat{\mathbf{u}}}, \qquad \widetilde{\mathbf{c}}_{\overline{\mathbf{n}}} = \frac{\partial^2 \widetilde{\psi}}{\partial \widehat{\mathbf{u}} \partial \overline{\mathbf{n}}}, \tag{3.4}$$

where $\widetilde{\mathbf{t}}$ defines the effective Cauchy traction vector, and $\widetilde{\mathbf{c}}_{\widehat{\mathbf{u}}}$ and $\widetilde{\mathbf{c}}_{\overline{\mathbf{n}}}$ the effective tangent moduli. Note that the tangent modulus $\mathbf{c}_{\widehat{\mathbf{u}}}$ is symmetric, while $\mathbf{c}_{\overline{\mathbf{n}}}$ is non-symmetric, in general.

The introduced effective cohesive potential $\widetilde{\psi}(\widehat{\mathbf{u}}, \overline{\mathbf{n}})$ has to obey the standard objectivity requirement, i.e.

$$\widetilde{\psi}(\widehat{\mathbf{u}}, \overline{\mathbf{n}}) = \widetilde{\psi}(\mathbf{Q}\widehat{\mathbf{u}}, \mathbf{Q}^{-\mathrm{T}}\overline{\mathbf{n}}), \tag{3.5}$$

where $\mathbf{Q}$ is a proper orthogonal tensor, with $\det\mathbf{Q} = +1$. It describes an arbitrary rigid-body rotation superimposed onto the current configuration. Hence, we assume the effective cohesive potential of the form $\widetilde{\psi}(\widehat{\mathbf{u}} \otimes \widehat{\mathbf{u}}, \overline{\mathbf{n}} \otimes \overline{\mathbf{n}})$, and introduce the invariant-based representation

$$\widetilde{\psi}(\widehat{\mathbf{u}} \otimes \widehat{\mathbf{u}}, \overline{\mathbf{n}} \otimes \overline{\mathbf{n}}) = \widetilde{\psi}(\widehat{I}_1, \widehat{I}_2, \widehat{I}_3, \widehat{I}_4, \widehat{I}_5), \tag{3.6}$$

where the scalar invariants of the set $\{\widehat{\mathbf{u}} \otimes \widehat{\mathbf{u}}, \overline{\mathbf{n}} \otimes \overline{\mathbf{n}}\}$ of symmetric second-order tensors are defined via (see, for example, Spencer, 1984)

$$\widehat{I}_1 = \mathrm{tr}(\widehat{\mathbf{u}} \otimes \widehat{\mathbf{u}}), \qquad \widehat{I}_2 = \mathrm{tr}(\widehat{\mathbf{u}} \otimes \widehat{\mathbf{u}})^2, \qquad \widehat{I}_3 = \mathrm{tr}(\widehat{\mathbf{u}} \otimes \widehat{\mathbf{u}})^3, \tag{3.7}$$

$$\widehat{I}_4 = (\widehat{\mathbf{u}} \otimes \widehat{\mathbf{u}}) : (\overline{\mathbf{n}} \otimes \overline{\mathbf{n}}), \qquad \widehat{I}_5 = (\widehat{\mathbf{u}} \otimes \widehat{\mathbf{u}})^2 : (\overline{\mathbf{n}} \otimes \overline{\mathbf{n}}). \tag{3.8}$$

Note, that the introduced structural tensor $\overline{\mathbf{n}} \otimes \overline{\mathbf{n}}$ in eq. (3.6) is invariant with respect to the definition of the unit vector $\overline{\mathbf{n}} = -\overline{\mathbf{n}}$, which is an essential objectivity requirement.

4. Variational formulation

The variational formulation is based on a three-field Hu-Washizu functional accompanied with the EAS-method, see Simo and Rifai, 1990. According to the EAS-method we assume that

$$\mathbf{F} = \mathbf{I} + \mathrm{Grad}\,\mathbf{u}_{\mathrm{c}} + \mathbf{H}_{\mathrm{e}}, \tag{4.1}$$

where $\mathbf{u}_{\mathrm{c}}$ denotes the compatible displacement field, as introduced in eq. (2.1). The enhanced assumed strain field, denoted by $\mathbf{H}_{\mathrm{e}}$, is a two-point tensor not subject to any interelement continuity requirements.

We consider the three-field Hu-Washizu functional to be of the form $\Pi(\mathbf{u}_{\mathrm{c}}, \mathbf{H}_{\mathrm{e}}, \mathbf{P})$, where $\mathbf{u}_{\mathrm{c}}, \mathbf{H}_{\mathrm{e}}$ and $\mathbf{P}$ denote the three independent fields, the compatible displacement, the enhanced strain and the first Piola-Kirchhoff stress tensor, respectively. Note that, by satisfying angular momentum in an arbitrary point $\mathbf{X}$, i.e. $\mathbf{P}\mathbf{F}^{\mathrm{T}}(\mathbf{u}_{\mathrm{c}}, \mathbf{H}_{\mathrm{e}}) = \mathbf{F}(\mathbf{u}_{\mathrm{c}}, \mathbf{H}_{\mathrm{e}})\mathbf{P}^{\mathrm{T}}$, a constraint is enforced onto the (otherwise independent) three fields. Consequently, three stationarity conditions (variational statements) may be derived according to (Simo and Armero, 1992)

$$\left.\begin{aligned} \int_{\Omega_0} \operatorname{Grad} \delta \mathbf{u}_c : \left(2\mathbf{F} \frac{\partial \Psi(\mathbf{C})}{\partial \mathbf{C}} \right) \mathrm{d}V - \delta \Pi_{\mathrm{ext}}(\delta \mathbf{u}) &= 0, \\ \int_{\Omega_0} \delta \mathbf{H}_\mathrm{e} : \left[-\mathbf{P} + \left(2\mathbf{F} \frac{\partial \Psi(\mathbf{C})}{\partial \mathbf{C}} \right) \right] \mathrm{d}V &= 0, \\ \int_{\Omega_0} \delta \mathbf{P} : \mathbf{H}_\mathrm{e} \mathrm{d}V &= 0, \end{aligned} \right\} \tag{4.2}$$

where $\delta \mathbf{u}_c$, $\delta \mathbf{P}$ and $\delta \mathbf{H}_\mathrm{e}$ are admissible variations of $\mathbf{u}_c$, $\mathbf{P}$ and $\mathbf{H}_\mathrm{e}$, respectively. Note that in eq. (4.2) appears no variation of the (total) gradient $\mathbf{H}$ rather then the enhanced part $\mathbf{H}_\mathrm{e}$. In addition, $\delta \Pi_{\mathrm{ext}}(\delta \mathbf{u})$ and $\mathrm{d}V$ denote the variation of the potential energy of the external loading and the volume element in the reference configuration, respectively.

Similar to the introduced kinematics we assume that the enhanced strain field $\mathbf{H}_\mathrm{e}$ (and the associated first variation $\delta \mathbf{H}_\mathrm{e}$) can be represented by the additive decomposition $\mathbf{H}_\mathrm{e} = \mathbf{H} + \widehat{\mathbf{H}}$ into a bounded contribution $\mathbf{H}$ and an unbounded contribution $\widehat{\mathbf{H}}$. In order to satisfy the orthogonality condition $(4.2)_3$ (in a weak sense) we employ the Galerkin method, which leads to a two-field problem, where the independent stress field $\mathbf{P}$ disappears from the variational equation $(4.2)_2$, see Simo and Rifai, 1990. This approach leads to the SOS formulation, which is closely related to smeared approaches, which, therefore, show the typical stress-locking phenomena (for an overview see Jirásek, 2000).

In order to overcome the locking phenomena, Dvorkin et al., 1990 (with generalizations introduced by Simo and Oliver, 1994, Oliver, 1996a and Oliver, 1996b) proposed a non-symmetric formulation called the SKON formulation. In this approach the kinematics of strong discontinuities (as introduced in the Section 2) is fulfilled properly, however this approach uses the Petrov-Galerkin method in oder to satisfy eq. $(4.2)_3$. Furthermore, the stress field $\mathbf{P}$ in eq. (4.2) does not disappear, although the SKON formulation deals with a two-field approach. Hence, this approach is not consistent with the variational formulation since it is assumed that $\mathbf{H}_\mathrm{e}$ and $\delta \mathbf{H}_\mathrm{e}$ are elements of the same space, which is not the case for the Petrov-Galerkin method.

Finally the KOS formulation (see Lotfi and Shing, 1995), which is based on the Galerkin method, describes the kinematics satisfactorily but it does not satisfy the orthogonality condition $(4.2)_3$.

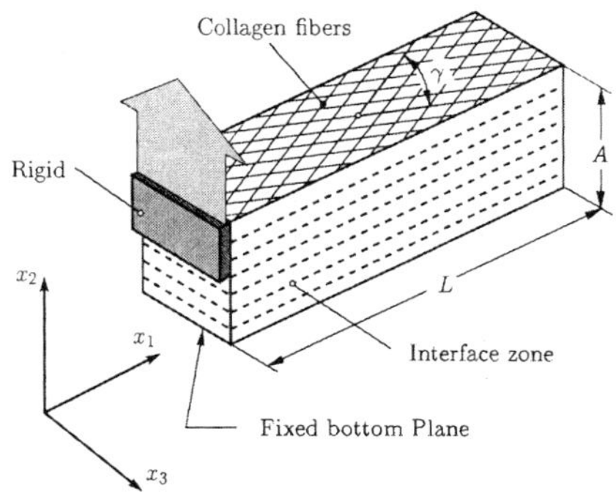

Figure 5.2. Reference geometry, boundary conditions and loading for the dissection analysis of the middle layer of an artery. The collagen fibers and the interface zone are schematically indicated. In order to initialize the crack in the middle of the specimen a rigid component transmits the load into the strip.

5. Representative numerical example

This section has the goal to investigate the efficiency of the proposed framework by comparing the numerical results obtained from the three classes of models, i.e. SOS, KOS and SKON. The introduced cohesive finite element formulation has been implemented into the multi-purpose finite element analysis program FEAP (Taylor, 2000), and constant-strain tetrahedral elements are used.

5.1 Dissection of the middle layer of an artery

In the example we perform a finite element analysis of the middle layer (the media) of an artery considered to be delaminated (dissected). The media is an important load-carrying layer in a healthy artery, and is roughly composed by a varying number of well-defined concentrically fiber-reinforced layers (lamellae). For a more detailed exposition see Holzapfel et al., 2000a, Section 2, and Holzapfel, 2002 with more references therein. Balloon dilations of arteries, frequently performed to reduce the severity of atherosclerotic stenoses, often leads to tissue dissections, which to analyze computationally is of pressing scientific and clinical need. The media is modeled as a 3D problem, where regular and distorted meshes are used for reason of comparison.

The reference geometry of the media, considered as a fiber-reinforced strip, is shown in Fig. 5.2, where $L = 10.0\,(\mathrm{mm})$ and $A = 1.0\,(\mathrm{mm})$ are the dimensions. The bottom face of the strip is fixed. The collagen fibers and the interface zones are schematically shown in Fig. 5.2. We consider a displacement-driven problem, whereby the displacement of the rigid component, as indicated in Fig. 5.2, is prescribed along the x_2-

direction. The rigid body transmits the load into the strip, which leads to a desired crack initialization in the middle of the specimen. During additional loading the stress state determines the crack propagation. The calculation is performed under 3D conditions, where symmetry is assumed with respect to the plane $x_3 = 0.0$.

The continuous model is given by the (strictly convex) penalty function $L(J) = \kappa(J-1)^2/2$, where the penalty parameter κ is chosen to be

$$100.0\,(\mathrm{mN/mm^2}) \equiv 100.0\,(\mathrm{kPa}).$$

The material parameters for the (isotropic) matrix material and the fibers are adopted from Holzapfel et al., 2000b. The angle between the two fiber directions in the reference configuration is given by $\gamma = 14.0°$.

In view of missing experimental data for the interface of the lamellar units we use a linear elastic particularization of eq. (3.2), and propose

$$\psi(d, \widehat{I}_1, \widehat{I}_4) = (1-d)\left[\frac{c_{\overline{\mathbf{n}}}}{2}\widehat{I}_4 + \frac{c_{\mathbf{t}}}{2}(\widehat{I}_1 - \widehat{I}_4)\right], \tag{5.1}$$

$$\text{with} \quad d = 1 - \exp\left(-a\,\widehat{u}_{\mathrm{H}}{}^{b}\right), \tag{5.2}$$

where $c_{\overline{\mathbf{n}}}$ and $c_{\mathbf{t}}$ are material parameters characterizing the normal and tangential responses, respectively. The evolution of the damage at the interface is described by the dimensionless material parameters a and b, while $\widehat{u}_{\mathrm{H}}(\mathbf{X}_{\mathrm{d}}, t) = \max[|\widehat{\mathbf{u}}(\mathbf{X}_{\mathrm{d}}, s)|]_{s=0}^{t}$ is the magnitude of the maximum gap displacement of the whole deformation path of a particular point $\mathbf{X}_{\mathrm{d}}$ at the interface at time t. Hence, using the particularization (5.1) and eqs. (3.3) and (3.4) it is straightforward to derive the Cauchy traction vector $\mathbf{t}$ and the associated tangent moduli $\mathbf{c}_{\widehat{\mathbf{u}}}$ and $\mathbf{c}_{\overline{\mathbf{n}}}$.

For the numerical example we used the material parameters

$$c_{\overline{\mathbf{n}}} = 1000.0\,(\mathrm{mN/mm^3}) \quad \text{and} \quad c_{\mathbf{t}} = c_{\overline{\mathbf{n}}}/2$$

for the normal and tangential responses, respectively. In addition, isotropic damage is specified by the parameters $a = 8.0\,(\mathrm{mm}^{-1})$ and $b = 0.5$. This set of parameters is associated with about $8.5\,(\mathrm{mN/mm^2}) \equiv 8.5\,(\mathrm{kPa})$ cohesive strength for mode I and about 4.25 (kPa) for mode failure II and III. In order to avoid penetration of the cohesive zone a penalty constraint is introduced, where the penalty (normal) stiffness of $1000(\mathrm{N/mm^3})$ was taken into account. The fully 3D problem is computed with regular and distorted meshes of 20580, 7500 and 1620 elements.

In Fig. 5.3 the load-displacement responses for the 3D computations are shown. On the regular meshes all three finite element formulations lead to the same result and on the distorted meshes the SOS formulation leads to the typically stress locking phenomena, see Fig. 5.3(a). The KOS formulation leads even on the distorted mesh to satisfying results, see Fig. 5.3(b). With the SKON formulation and for distorted meshes we were not able to compute the whole dissection process with the fixed load step of 0.1 (mm), see Fig. 5.3(c).

In Fig. 5.4 the maximum principal Cauchy stresses, in (mN/mm^3), are plotted onto the deformed configurations during the dissection process. Distorted meshes of 1620 and 7500 tetrahedral elements are employed and the KOS formulation is used. Note the stress concentration around the delamination front, and the good qualitatively agreement for both calculations. A second region of stress concentration can be observed around the location where the load is transmitted into the strip. More detailed investigations have shown that the maximum principal stresses in this region are aligned with the interface such that no further delamination took place.

6. Conclusion

Mathematical modeling of three-dimensional delamination processes of laminated composites was presented. In particular, the delamination (dissection) of the middle layer of a healthy artery — regarded as a highly deformable composite structure — was studied in more detail. Different EAS-based finite element formulations with embedded strong discontinuities are compared and discussed.

Dependent on the interpolation of the enhanced assumed strain field $\mathbf{H}_\mathrm{e}$ (and its associated admissible variation) one can set up three classes of finite element models, known as statically optimal symmetric (SOS), kinematically optimal non-symmetric (SKON), and kinematically optimal symmetric (KOS). While the SOS formulation satisfies a three-field Hu-Washizu variational principle, it shows pronounced stress-locking phenomena on distorted meshes. Both the SKON and KOS formulations overcome stress-locking, however, they are not consistent with the variational principle. On the basis of the delamination (dissection) of an arterial layer, modeled as a 3D problem with constant-strain tetrahedral elements, regular and distorted meshes are used to compare the associated load-displacement responses. As expected, stress locking phenomena were seen for the SOS formulation. Only the KOS formulation has led to meaningful results on distorted meshes. It was not possible to achieve similar results on distorted meshes with the SKON formulation. A comparative study of the maximum principal Cauchy stresses for distorted meshes of 1620 and 7500 tetrahedral elements have given qualitative good agreement.

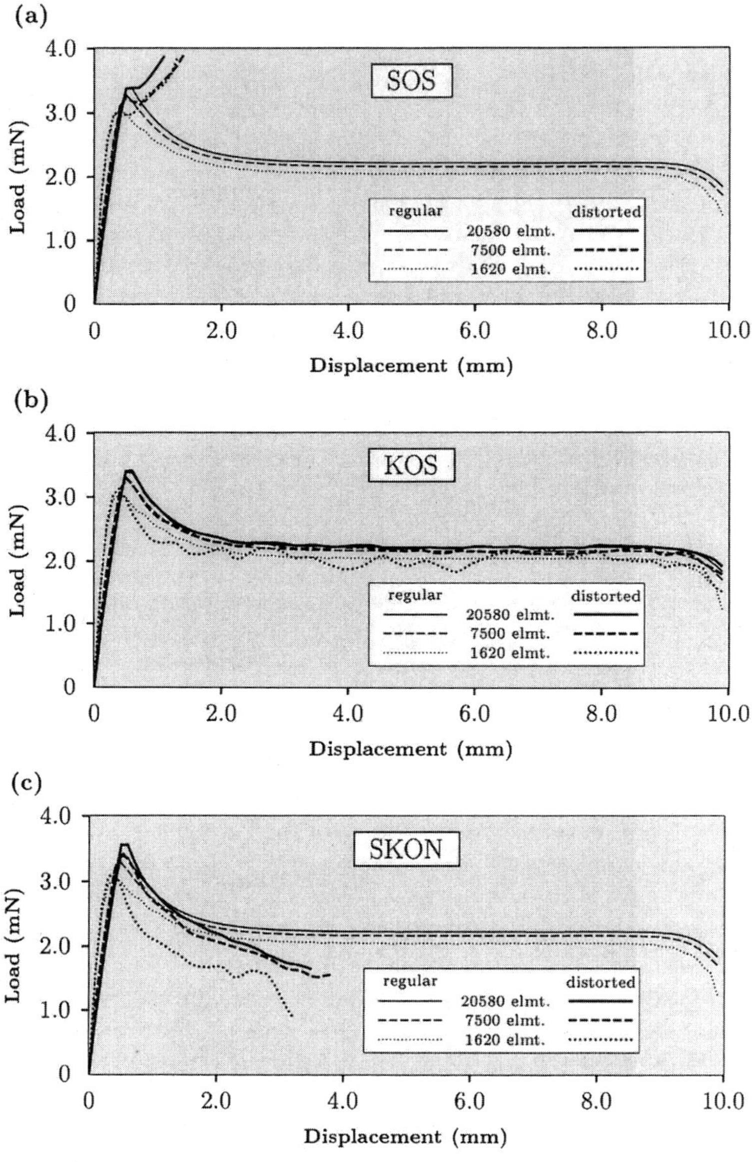

Figure 5.3. 3D dissection analysis of the middle layer of an artery. Load-displacement response for the (a) SOS, (b) KOS and (c) SKON formulations. Regular and distorted meshes using 20580, 7500 and 1620 elements are used. Note the stress locking effects accompanied with the SOS formulation for distorted meshes. With the SKON formulation and for distorted meshes no meaningful 3D results could be achieved with the fixed load step of 0.1 (mm).

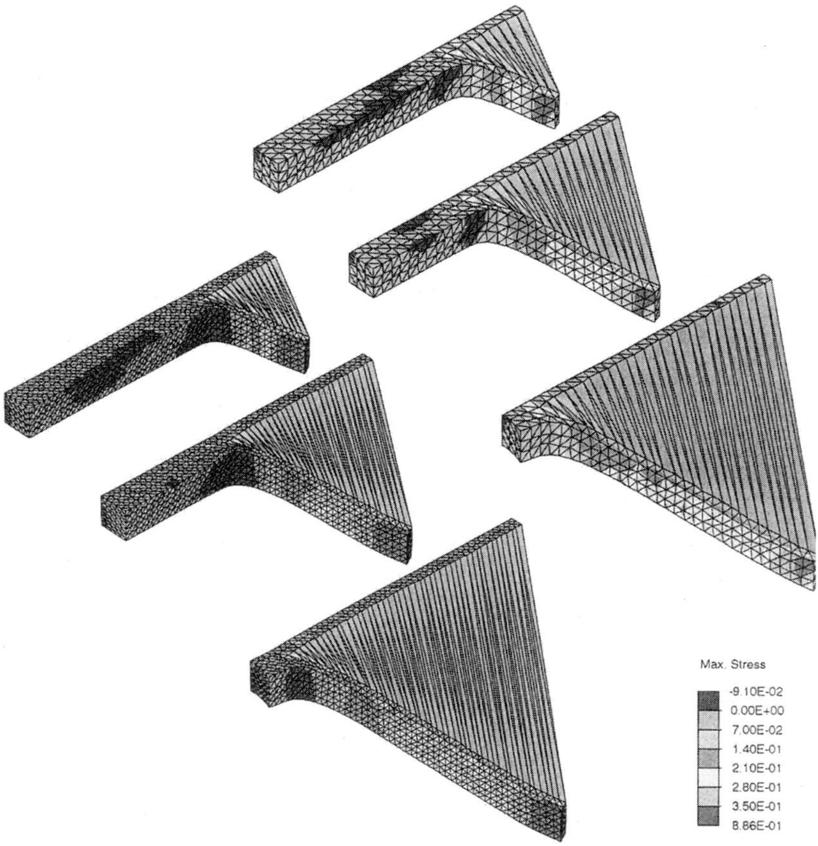

Figure 5.4. Maximum principal stresses, in (mN/mm^2), are plotted onto the deformed configurations during the dissection process. The 3D computation is based on the KOS formulation. Slightly distorted meshes with 1620 and 7500 tetrahedral elements were used.

Acknowledgments

Financial support for this research was provided by the *Austrian Science Foundation* under *START-Award Y74-TEC*.

References

Alfano, G. and Crisfield, M. A. (2001). Finite element interface models for the delamination analysis of laminated composites: Mechanical and computational issues. *Int. J. Numer. Meth. Engr.*, 50:1701–1736.

de Borst, R. (2001). Some recent issues in computational failure mechanics. *Int. J. Numer. Meth. Engr.*, 52:63–95.

Dvorkin, E. N., Cuitiño, A. M., and Gioia, G. (1990). Finite elements with displacement interpolated embedded localization lines insensitive to mesh size and distortions. *Int. J. Numer. Meth. Engr.*, 30:541–564.

Holzapfel, G. A. (2000). *Nonlinear Solid Mechanics. A Continuum Approach for Engineering.* John Wiley & Sons, Chichester.

Holzapfel, G. A. (2002). Structural and numerical models for the (visco) elastic response of arterial walls with residual stresses. In Holzapfel, G. A. and Ogden, R. W., editors, *Biomechanics of Soft Tissue in Cardiovascular Systems.* Springer-Verlag, Wien. CISM Courses and Lectures, International Centre for Mechanical Sciences.

Holzapfel, G. A., Gasser, T. C., and Ogden, R. W. (2000a). A new constitutive framework for arterial wall mechanics and a comparative study of material models. *J. Elasticity*, 61:1–48.

Holzapfel, G. A., Schulze-Bauer, C. A. J., and Stadler, M. (2000b). Mechanics of angioplasty: Wall, balloon and stent. In Casey, J. and Bao, G., editors, *Mechanics in Biology*, New York. The American Society of Mechanical Engineers (ASME). AMD-Vol. 242/BED-Vol. 46, pp. 141-156.

Jirásek, M. (2000). Comparative study on finite elements with embedded discontinuities. *Comput. Meth. Appl. Mech. Engr.*, 188:307–330.

Lotfi, H. R. and Shing, P. B. (1995). Embedded representation of fracture in concrete with mixed finite elements. *Int. J. Numer. Meth. Engr.*, 38:1307–1325.

Oliver, J. (1996a). Modelling strong discontinuities in solid mechanics via strain softening constitutive equations. Part 1: Fundamentals. *Int. J. Numer. Meth. Engr.*, 39:3575–3600.

Oliver, J. (1996b). Modelling strong discontinuities in solid mechanics via strain softening constitutive equations. Part 2: Numerical simulation. *Int. J. Numer. Meth. Engr.*, 39:3601–3624.

Rots, J. G. (1988). *Computational modeling of concrete fracture.* PhD thesis, Delft University of Technology, Netherlands.

Simo, J. C. and Armero, F. (1992). Geometrically non–linear enhanced strain mixed methods and the method of incompatible modes. *Int. J. Numer. Meth. Engr.*, 33:1413–1449.

Simo, J. C. and Oliver, J. (1994). A new approach to the analysis and simulation of strain softening in solids. In Z. P. Bažant et al., editors, *Fracture and Damage in Quasibrittle Structures*, pages 25–39. F. N. Spon, London.

Simo, J. C. and Rifai, M. S. (1990). A class of mixed assumed strain methods and the method of incompatible modes. *Int. J. Numer. Meth. Engr.*, 29:1595–1638.

Spencer, A. J. M. (1984). Constitutive theory for strongly anisotropic solids. In Spencer, A. J. M., editor, *Continuum Theory of the Mechanics of Fibre-Reinforced Composites*, pages 1–32. Springer-Verlag, Wien. CISM Courses and Lectures No. 282, International Centre for Mechanical Sciences.

Taylor, R. L. (2000). *FEAP - A Finite Element Analysis Program - Version 7.3.* University of California at Berkeley.

Wells, G. N. (2001). *Discontinuous modelling of strain localization and failure.* PhD thesis, Delft University of Technology, Netherlands.

Chapter 6

NEWTON'S AND POISSON'S IMPACT LAW FOR THE NON-CONVEX CASE OF RE-ENTRANT CORNERS

Christoph Glocker
IMES – Center of Mechanics, ETH Zentrum
CH-8092 Zürich, Switzerland
glocker@imes.mavt.ethz.ch

Abstract The paper reviews the frictionless collision problem in rigid body dynamics. Newton's and Poisson's impact laws are stated in inequality form for one collision point and extended by superposition to multi-contact configurations. One special case within this framework are impacts with global dissipation index, for which it is shown that Newton's impact law reduces to Moreau's impact rule and that both of them coincide with Poisson's law when a certain kinematic compatibility condition is met. A geometrical interpretation of this impact law is given for a tangentially regular boundary and then extended to re-entrant corners.

Keywords: rigid body, impact, Newton, Poisson, projection, reflection.

1. Introduction

There are at least two historic approaches to impacts: Newton's kinematic impact law which reverses the sign of the relative velocity at the impact and takes into account dissipation by a coefficient $0 \leq \varepsilon \leq 1$,

$$\varepsilon\,(v - V) = (C - c), \tag{1.1}$$

and Poisson's impact law which requires a decomposition of the impact process into a compression phase $(-)$ and a decompression phase $(+)$ in order to define the restitution coefficient ε by the ratio of the corresponding impulsive forces,

$$\Lambda^{(+)} = \varepsilon\,\Lambda^{(-)}. \tag{1.2}$$

Both impact laws in their original versions were certainly not formulated with the intention to be used for a multibody system as a general rule on how to treat velocity jumps: Newton performed experiments with balls on a string in order to verify primarily his lex tertia [11]. Poisson investigated already the impact between two rigid bodies by using Newton's second law and Euler's axiom, with (1.2) as an additional relation to determine the "normal force" at the impact. This is already close to the today's methods of treating impacts, where the impact law is regarded as an independent equation, which can generally not be derived from any mechanical principle when dealing with *rigid* bodies.

The goal of this paper is to adapt the historic impact laws (1.1), (1.2) to the multi-impact problem in rigid body dynamics, and to study their geometric meaning within the framework of analytical mechanics. One of the main impacts for treating such problems in finite-degree-of-freedom dynamics was given by Moreau's non-smooth dynamical equations. In our setting they read for the frictionless case

$$\begin{aligned} \mathcal{C} &:= \{\, \mathbf{q} \,|\, g_i(\mathbf{q}) \geq 0, \quad i = 1, \ldots, m \,\} \\ \mathcal{H} &:= \{\, i \,|\, g_i(\mathbf{q}_0) = 0 \,\} \\ \mathcal{T}_\mathcal{C}(\mathbf{q}_0) &= \{\, \mathbf{y} \,|\, -\mathbf{w}_i^T \mathbf{y} \leq 0,\ \mathbf{w}_i^T := \nabla g_i(\mathbf{q}_0),\ \forall i \in \mathcal{H} \,\} \\ \mathbf{M}(\mathbf{q}_0)\,(\mathbf{u}^+ - \mathbf{u}^-) &= \mathbf{R} \\ -\mathbf{R} &\in \mathcal{N}_{\mathcal{T}_\mathcal{C}(\mathbf{q}_0)}(\mathbf{v}) \\ \mathbf{v} &= \tfrac{1}{1+\varepsilon}\,\mathbf{u}^+ + \tfrac{\varepsilon}{1+\varepsilon}\,\mathbf{u}^-, \end{aligned} \tag{1.3}$$

see [9] for the precise interrelation to Moreau's original formulation in [8]. Only the impact itself is treated in this paper, and nothing will be said about adjoint periods of impact-free motion.

The first three equations in (1.3) concern the kinematics of the simple unilaterally constrained system and define the feasible region $\mathcal{C}$ of the system, the active set $\mathcal{H}$ formed by the indices of the unilateral constraints $g_i \geq 0$ which hold as an equality at the point of evaluation $\mathbf{q}_0 \in \mathcal{C}$, and the convex tangent cone $\mathcal{T}_\mathcal{C}(\mathbf{q}_0)$ which approximates $\mathcal{C}$ in the neighborhood of $\mathbf{q}_0$. The remaining three equations concern non-smooth dynamics at the impact time t_0. The impact equations in the fourth line of (1.3) represent the Newton-Euler equations in integrated form where $\mathbf{M}(\mathbf{q}_0)$ denotes the mass matrix, $\mathbf{u}^-$ and $\mathbf{u}^+$ the pre- and post-impact velocity, and $\mathbf{R}$ the generalized impulsive force. The last two lines state an impact law which makes use of a global restitution coefficient ε. This law will be discussed in full detail in section 5 to enlighten its geometric meaning for the impact and its correlation to the classical approaches (1.1) and (1.2).

The paper is organized as follows: In section 2 we state the impact equations of a mechanical system within the framework of perfect collisions. This set of equations consists of the Newton-Euler equations which are integrated over a singleton in time, the definition of the perfect unilateral constraints, and the relative velocities which are instantaneously changed by the impact. To get a complete description of the dynamics at the impact, one has to introduce additional impact laws which is done by two alternative methods via inequalities in section 3 for Newtonian, and in section 4 for Poisson impacts. For a further analysis we restrict ourselves to a special subclass of impacts. We call them impacts with global dissipation index, occurring in multi-contact configurations for which all coefficients of restitution are equal to each other. For such systems it is shown in section 3 that Newton's impact law for simple unilaterally constrained systems agrees with Moreau's non-smooth dynamical equations. We also prove equivalence with Poisson's law in section 5 when a certain kinematic compatibility condition is met, and we present a geometric interpretation of this class of impacts. In section 6, this geometric approach is extended to the non-convex case of re-entrant corners in the feasible domain of displacements, reflecting, for example, the corner-corner impact of two colliding rectangular blocks. The methods used in this paper are based on the geometry of cones and go far beyond simple unilateral constraints.

2. Impact equations

Throughout the paper we treat a finite-dimensional holonomic rheonomic mechanical system. We denote by t time and by $\mathbf{q}(t) \in \mathbb{R}^f$ the Lagrangian coordinates of the system, i.e. a set of local coordinates of the breathing f-dimensional configuration manifold. The velocities of the system are denoted by $\mathbf{u}(t)$ and are assumed to be functions of bounded variations with $\dot{\mathbf{q}}(t) = \mathbf{u}(t)$ almost everywhere. A general framework for non-smooth dynamics is provided by the Newton-Euler equations stated as an equality of measures [8], from which we only investigate the atomic part in a form obtained by integration over a singleton t_0,

$$\mathbf{M}(\mathbf{q}_0, t_0)\,(\mathbf{u}^+(t_0) - \mathbf{u}^-(t_0)) = \mathbf{R}(t_0), \tag{2.1}$$

where $\mathbf{q}_0 = \mathbf{q}(t_0)$. These are the *impact equations* of the system. $\mathbf{M}(\mathbf{q}_0, t_0)$ denotes the $f \times f$ symmetric and positive definite inertia operator which will later serve as the kinetic metric when the geometry of impacts is discussed, $\mathbf{u}^-(t_0)$ and $\mathbf{u}^+(t_0)$ are the left and right limits of the velocities at t_0, i.e. the pre- and post-impact velocities that are different in the case of an impact, and $\mathbf{R}(t_0)$ any generalized impulsive

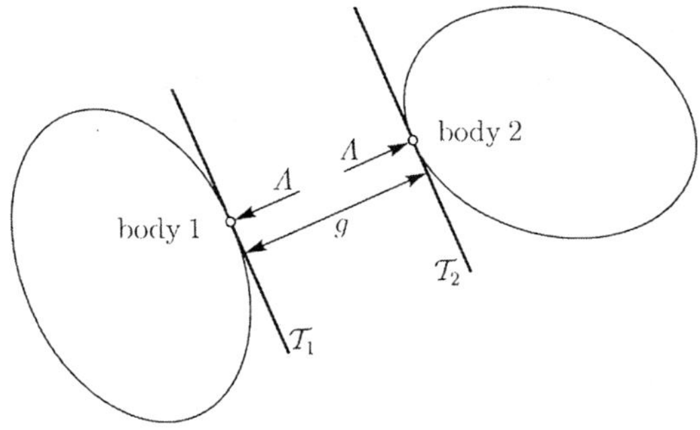

Figure 6.1. Gap function g and impulsive impact forces Λ.

force acting on the system. Any non-atomic kinetic entity like finite forces or gyroscopical accelerations does not contribute to (2.1).

We further assume m simple unilateral constraints $g_i(\mathbf{q}, t) \geq 0$ which define, for any fixed time t, the set of admissible displacements

$$\mathcal{C}(t) := \{\, \mathbf{q} \,|\, g_i(\mathbf{q}, t) \geq 0, \quad i = 1, \ldots, m\}, \tag{2.2}$$

and hence a non-smooth moving boundary in the configuration manifold. In collision problems, the unilateral constraints are set up by the gaps between the individual bodies as depicted in figure 6.1. A vanishing gap $g_i = 0$ at the time t_0 thus indicates that the associated bodies are in contact, which may be expressed by the active set of inequality constraints

$$\mathcal{H}(t_0) := \{\, i \,|\, g_i(\mathbf{q}_0, t_0) = 0\,\}. \tag{2.3}$$

In order to obtain the relative velocity γ_i between two contact partners, we differentiate the gap function $g_i(\mathbf{q}, t)$ with respect to time. This yields, evaluated at time t_0,

$$\gamma_i^{\pm}(t_0) = \mathbf{w}_i^T(\mathbf{q}_0, t_0)\, \mathbf{u}^{\pm}(t_0) + \hat{w}_i(\mathbf{q}_0, t_0) \tag{2.4}$$

with $\mathbf{w}_i^T = \frac{\partial g_i}{\partial \mathbf{q}}$ the spatial and $\hat{w}_i = \frac{\partial g_i}{\partial t}$ the temporal coordinates of the differential of g_i. Finally, let us specify the type of impulsive forces $\mathbf{R}$ and thus the type of impacts that we study in this paper. We restrict ourselves to pure collision problems. Impulsive forces applied on the bodies from outside, or impulsive forces as reactions due to a non-smooth configuration manifold as treated in [8] are excluded. We assume frictionless collisions to stay within the framework of perfect unilateral constraints [3], which causes $\mathbf{R}$ to be of the form

$$\mathbf{R}(t_0) = \sum_{i \in \mathcal{H}} \mathbf{w}_i(\mathbf{q}_0, t_0)\, \Lambda_i(t_0) \tag{2.5}$$

with Lagrangian impact multipliers Λ_i which are the impulsive forces at the collision points. These multipliers are later used together with the pre- and post-impact relative velocities $\gamma_i^{\pm}$ from (2.4) to set up the impact laws (1.1) and (1.2) in inequality form. For a detailed discussion of the physical meaning and the mathematical formulation of all the terms in (2.1)–(2.5) we refer especially to [9], where rigid body collision systems are treated. Note also that we entered by the summation in (2.5) already the multi-contact collision problem. Multi-contact configurations are, indeed, one of the standard problems in the dynamics of machines.

Equations (2.1)–(2.5) incorporate already one part of the impact law, i.e. the trivial case of an open contact at which no impulsive force can occur. This is taken into account in (2.5) by summing up only the impulsive forces of the closed contacts $i \in \mathcal{H}$, a concept that is later required when the geometrical meaning of the impact laws is investigated. Note that $\mathbf{u}^+ = \mathbf{u}^-$ for $\mathcal{H} = \emptyset$, i.e. no velocity jump. The impact laws for the closed contacts $i \in \mathcal{H}$ are still missing. In sections 3 and 4 we discuss two different approaches obtained from Newton's impact law (1.1) and Poisson's impact law (1.2) by a proper re-formulation as inequalities.

3. Newton's impact law

In this section we complete the impact equations (2.1)–(2.5) by Newton's impact law in inequality form for the closed contact $g_i(\mathbf{q}_0, t_0) = 0$ and prove equivalence with Moreau's equations (1.3) for scleronomic systems.

The unilateral version of *Newton's impact law* (1.1) in terms of the pre- and post-impact normal relative velocity $\gamma_i^{\pm}(t_0) = \mathbf{w}_i^T(\mathbf{q}_0, t_0)\,\mathbf{u}^{\pm}(t_0) + \hat{w}_i(\mathbf{q}_0, t_0)$ and associated impulsive force $\Lambda_i(t_0)$ is

$$\gamma_i^+ + \varepsilon_i\gamma_i^- \geq 0, \quad \Lambda_i \geq 0, \quad (\gamma_i^+ + \varepsilon_i\gamma_i^-)\,\Lambda_i = 0, \tag{3.1}$$

with the following physical behavior: The impulsive force, if there is any, should act as a compressive magnitude, $\Lambda_i \geq 0$. In the case of a non-vanishing impulse ($\Lambda_i > 0$) we apply Newton's impact law as usual, i.e. $\gamma_i^+ = -\varepsilon_i\gamma_i^-$, which is expressed by the third condition in the second line in (3.1). The magnitude ε_i denotes Newton's coefficient of restitution with the usual values $0 \leq \varepsilon_i \leq 1$. The case $\varepsilon_i = 0$ corresponds to a completely inelastic impact with vanishing post-impact relative velocity ($\gamma_i^+ = 0$) and may later be recognized to be a projection, whereas $\varepsilon_i = 1$ represents a completely elastic impact at which the relative velocity γ_i is inverted ($\gamma_i^+ = -\gamma_i^-$) to be later interpreted as a reflection. Suppose now that, for any reason, the contact does *not* participate in the impact, i.e. that the value of the impulsive force is zero, although the contact is

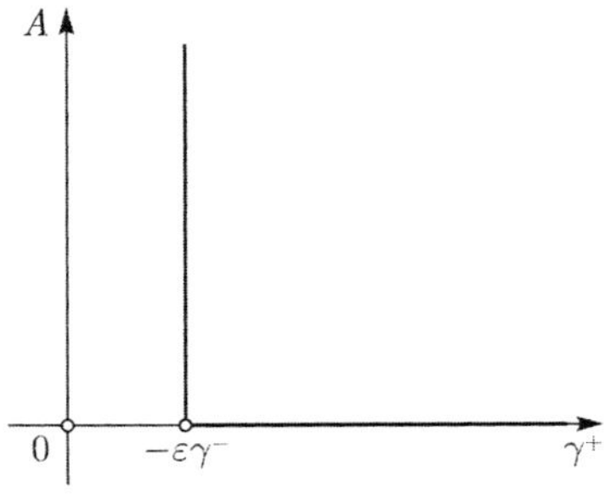

Figure 6.2. Newton's impact law.

closed. This happens normally for multi-contact situations as described in [9]. For this case we allow post-impact relative velocities *higher* than what is prescribed by Newton's impact law in the case of a non-vanishing impulse, $\gamma_i^+ \geq -\varepsilon_i \gamma_i^-$, in order to express that the contact is superfluous and could be removed without changing the contact-impact process. The graph of Newton's impact law in inequality form is depicted in figure 6.2 and is the horizontal translate of a unilateral primitive expressed by the complementarity conditions $\Lambda_i \geq 0$, $\gamma_i^+ \geq 0$, $\Lambda_i \gamma_i^+ = 0$.

In the following we rewrite the impact equations (2.1)–(2.5) together with Newton's impact law (3.1) to obtain finally a relation to Moreau's rule (1.3). Within these calculations, all coefficients of restitution are assumed to be equal and denoted by ε with possible values $0 \leq \varepsilon \leq 1$. This ε serves as the *global dissipation index* of the system at the impact. We first set

$$\begin{aligned} \xi_i &:= \tfrac{1}{1+\varepsilon}\left(\gamma_i^+ + \varepsilon \gamma_i^-\right) \\ &= \mathbf{w}_i^T\left(\tfrac{1}{1+\varepsilon}\,\mathbf{u}^+ + \tfrac{\varepsilon}{1+\varepsilon}\,\mathbf{u}^-\right) + \hat{w}_i \end{aligned} \tag{3.2}$$

with $\gamma_i^\pm$ according to (2.4). We abbreviate the term in brackets by $\mathbf{v}$ and rewrite (2.1), (2.5) and (3.1) as

$$\begin{gathered} \mathbf{M}(\mathbf{u}^+ - \mathbf{u}^-) = \mathbf{R}, \quad \mathbf{v} = \tfrac{1}{1+\varepsilon}\,\mathbf{u}^+ + \tfrac{\varepsilon}{1+\varepsilon}\,\mathbf{u}^- \\ \xi_i = \mathbf{w}_i^T \mathbf{v} + \hat{w}_i, \quad \mathbf{R} = \sum_{i \in \mathcal{H}} \mathbf{w}_i \Lambda_i, \\ \xi_i \geq 0, \quad \Lambda_i \geq 0, \quad \xi_i \Lambda_i = 0 \end{gathered} \tag{3.3}$$

with the unilateral constraints g_i and the index set $\mathcal{H}$ as defined in (2.2) and (2.3).

We are now going to reformulate the complementarity conditions of the third line of (3.3). We start by writing them as a variational inequality,

$$-\sum_{i\in\mathcal{H}} \Lambda_i(\xi_i^\star - \xi_i) \le 0, \quad \xi_i \ge 0, \quad \forall \xi_i^\star \ge 0 \qquad (i \in \mathcal{H}), \tag{3.4}$$

and continue by expressing the two inequalities $\xi_i \ge 0$ and $\forall \xi_i^\star \ge 0$ again in variational form,

$$\begin{aligned} \xi_i \ge 0, \quad i \in \mathcal{H} \quad &\Leftrightarrow \quad \xi_i \text{ such that } -\sum_{j\in\mathcal{H}} \alpha_j \xi_j \le 0 \quad \forall \alpha_j \ge 0, \\ \forall \xi_i^\star \ge 0, \quad i \in \mathcal{H} \quad &\Leftrightarrow \quad \forall \xi_i^\star \text{ such that } -\sum_{j\in\mathcal{H}} \alpha_j \xi_j^\star \le 0 \quad \forall \alpha_j \ge 0. \end{aligned} \tag{3.5}$$

Next, regard the variations $\xi_i^\star$ as affine functions of some velocity variations $\mathbf{v}^\star \in \mathbb{R}^f$ and make use of the kinematic equation in (3.3)

$$\xi_i^\star := \mathbf{w}_i^T \mathbf{v}^\star + \hat{w}_i, \quad \xi_i = \mathbf{w}_i^T \mathbf{v} + \hat{w}_i \tag{3.6}$$

to eliminate ξ_i and $\xi_i^\star$ from (3.4) and (3.5). The first inequality in (3.4) then becomes

$$-\sum_{i\in\mathcal{H}} \Lambda_i \, \mathbf{w}_i^T (\mathbf{v}^\star - \mathbf{v}) \le 0, \tag{3.7}$$

where $(\mathbf{v}^\star - \mathbf{v})$ constitutes the *virtual velocities* at the impact. The conditions (3.5) on ξ_i and $\xi_i^\star$ are now replaced by corresponding conditions on $\mathbf{v}$ and $\mathbf{v}^\star$,

$$\begin{aligned} \mathbf{v} \text{ such that } \quad &-\sum_{j\in\mathcal{H}} \alpha_j(\mathbf{w}_j^T \mathbf{v} + \hat{w}_j) \le 0 \quad \forall \alpha_j \ge 0, \\ \forall \mathbf{v}^\star \text{ such that } \quad &-\sum_{j\in\mathcal{H}} \alpha_j(\mathbf{w}_j^T \mathbf{v}^\star + \hat{w}_j) \le 0 \quad \forall \alpha_j \ge 0, \end{aligned} \tag{3.8}$$

which is equivalent to (3.5) when $\{\mathbf{w}_i \,|\, i \in \mathcal{H}\}$ is a linearly independent set.

Denote further by $\mathbf{b}$ a solution of the linear system of equations $-\mathbf{w}_j^T \mathbf{b} = \hat{w}_j$ $(j \in \mathcal{H})$ which exists for linearly independent $\{\mathbf{w}_i\}$, but also for linearly dependent $\{\mathbf{w}_i\}$ when the kinematics of the system has been properly formulated. With such a $\mathbf{b}$ and (3.7), (3.8) we can now state Newton's impact problem (3.3) as

$$\begin{aligned} &\mathbf{M}(\mathbf{u}^+ - \mathbf{u}^-) = \mathbf{R}, \quad \mathbf{v} = \tfrac{1}{1+\varepsilon}\,\mathbf{u}^+ + \tfrac{\varepsilon}{1+\varepsilon}\,\mathbf{u}^-, \\ &\mathbf{R} = \sum_{i\in\mathcal{H}} \mathbf{w}_i \Lambda_i, \quad -\sum_{i\in\mathcal{H}} \Lambda_i \, \mathbf{w}_i^T (\mathbf{v}^\star - \mathbf{v}) \le 0, \\ &\mathbf{v} \text{ such that } \quad -\sum_{j\in\mathcal{H}} \alpha_j \mathbf{w}_j^T (\mathbf{v} - \mathbf{b}) \le 0 \quad \forall \alpha_j \ge 0, \\ &\forall \mathbf{v}^\star \text{ such that } \quad -\sum_{j\in\mathcal{H}} \alpha_j \mathbf{w}_j^T (\mathbf{v}^\star - \mathbf{b}) \le 0 \quad \forall \alpha_j \ge 0. \end{aligned} \tag{3.9}$$

Finally, we take into account that the elements of the tangent cone of a dynamical system with simple unilateral constraints are characterized according to (0.15). This changes (3.9) to

$$\begin{gathered}\mathbf{M}(\mathbf{u}^+ - \mathbf{u}^-) = \mathbf{R}, \quad \mathbf{v} = \tfrac{1}{1+\varepsilon}\,\mathbf{u}^+ + \tfrac{\varepsilon}{1+\varepsilon}\,\mathbf{u}^-, \\ -\mathbf{R}^T(\mathbf{v}^\star - \mathbf{v}) \leq 0, \\ \mathbf{v} \in \mathcal{T}_C(\mathbf{q}_0) + \mathbf{b}, \quad \forall \mathbf{v}^\star \in \mathcal{T}_C(\mathbf{q}_0) + \mathbf{b}.\end{gathered} \tag{3.10}$$

Note that both, the velocity $\mathbf{v}$ and their variation $\mathbf{v}^\star$ are taken from a *translate* of the tangent cone, because we are dealing with *moving* sets ($g_i(\mathbf{q}, t) \leq 0$) caused by external kinematic excitation. In contrast, for the *virtual velocities* $\delta\mathbf{v} := \mathbf{v}^\star - \mathbf{v}$ it holds that $\delta\mathbf{v} \in \mathcal{T}_{\mathcal{T}_C(\mathbf{q}_0)}(\mathbf{v} - \mathbf{b}) \supset \mathcal{T}_C(\mathbf{q}_0)$. Newton's impact problem (3.10) may thus be stated in the following variational form: For

$$\mathbf{v} = \tfrac{1}{1+\varepsilon}\,\mathbf{u}^+ + \tfrac{\varepsilon}{1+\varepsilon}\,\mathbf{u}^- \tag{3.11}$$

and given $\mathbf{u}^-$ find $\mathbf{u}^+$ such that

$$-\delta\mathbf{v}^T\mathbf{M}(\mathbf{u}^+ - \mathbf{u}^-) \leq 0, \quad \mathbf{v} \in \mathcal{T}_C(\mathbf{q}_0) + \mathbf{b}, \quad \forall \delta\mathbf{v} \in \mathcal{T}_{\mathcal{T}_C(\mathbf{q}_0)}(\mathbf{v} - \mathbf{b}). \tag{3.12}$$

These are by (0.7) exactly the conditions for an element $-\mathbf{M}(\mathbf{u}^+ - \mathbf{u}^-)$ to belong to the polar of $\mathcal{T}_{\mathcal{T}_C(\mathbf{q}_0)}(\mathbf{v} - \mathbf{b})$,

$$-\mathbf{M}(\mathbf{u}^+ - \mathbf{u}^-) \in \mathcal{N}_{\mathcal{T}_C(\mathbf{q}_0)}(\mathbf{v} - \mathbf{b}), \quad (\mathbf{v} - \mathbf{b}) \in \mathcal{T}_C(\mathbf{q}_0), \tag{3.13}$$

which expresses Newton's impact problem as an inclusion. We have thus proven the following:

Proposition 6.1 *The impact problem* (3.11), (3.12) *or* (3.11), (3.13) *is equivalent to the formulation* (2.1)–(3.1) *of Newton's multi-contact impact law for simple unilaterally constrained systems with global dissipation index in the sense that for any* $\mathbf{u}^+$ *satisfying* (3.11), (3.12) *or* (3.11), (3.13) *values of* Λ_i *and* γ_i *can be found such that the local impact laws* (3.1) *hold.*

Note, however, that (3.11)–(3.13) can be regarded as a *generalization* of (2.1)–(3.1) because these formulations hold beyond simple unilateral constraints. They are still valid for (tangentially regular) sets C which can not be expressed by the intersection of a finite number of inequalities $g_i(\mathbf{q}, t) \geq 0$ with non-vanishing gradients $\mathbf{M}^{-1}\mathbf{w}_i$ at their boundaries. An example in $\mathbb{R}^3$ of such a set is already the circular cone $\{\lambda(\mathbf{e} + \mathcal{D}) \mid \lambda \geq 0\}$ with $0 \neq \mathbf{e} \perp \mathcal{D}$ and $\mathcal{D}$ the closed unit disc in $\mathbb{R}^2$ with respect to the 2-norm.

By proposition 6.1 we have also proven that Moreau's impact problem stated in the form (1.3) agrees with Newtonian impacts for scleronomic systems ($\hat{w}_i = 0$, hence $\mathbf{b} = 0$ as a possible choice): The first two equations in (1.3) are met again in (2.2) and (2.3), the tangent cone from the third line in (1.3) is just an alternative representation of (0.15), and the last three lines in (1.3) are simply (3.11) and (3.13).

Along with the impact equations, the impact law $-\mathbf{R} \in \mathcal{N}_{\mathcal{T}_C(\mathbf{q}_0)}(\mathbf{v} - \mathbf{b})$ for systems with global dissipation index singles out one particular element $-\mathbf{R}$ of $\mathcal{N}_C(\mathbf{q}_0)$. This element is unique as we will see in section 5 and is always contained in $\mathcal{N}_C(\mathbf{q}_0)$, because $\mathcal{N}_{\mathcal{T}_C(\mathbf{q}_0)}(\mathbf{v}-\mathbf{b}) \subset \mathcal{N}_C(\mathbf{q}_0)$ for $(\mathbf{v}-\mathbf{b}) \in \mathcal{T}_C(\mathbf{q}_0)$ by (*iii*) in the proof of proposition A.5 and (0.8) which connects $\mathcal{T}^{\perp}_{(\cdot)}$ and $\mathcal{N}_{(\cdot)}$. There is nothing axiomatic behind this impact rule, and one can not expect this law to hold for any impact event in a physical system. For a detailed discussion of this topic see [9]. The only prediction which may be taken for granted is that $-\mathbf{R}$ lies in the normal cone $\mathcal{N}_C(\mathbf{q}_0)$, when the displacements $\mathbf{q}_0$ and the normal cone $\mathcal{N}_C(\mathbf{q}_0)$ are constant at the impact. This is the classical assumption for impact events in dynamics, see e.g. [9] for additional comments.

We will finally state equation (3.13) in the tangent space at point $\mathbf{q}_0$. With $\mathcal{T}^{\perp}_{\mathcal{T}_C(\mathbf{q}_0)}(\mathbf{v} - \mathbf{b}) = \mathbf{M}^{-1}\mathcal{N}_{\mathcal{T}_C(\mathbf{q}_0)}(\mathbf{v} - \mathbf{b})$ the cone orthogonal to $\mathcal{T}_{\mathcal{T}_C(\mathbf{q}_0)}(\mathbf{v} - \mathbf{b})$ as a subset of the tangent space we obtain

$$-(\mathbf{u}^+ - \mathbf{u}^-) \in \mathcal{T}^{\perp}_{\mathcal{T}_C(\mathbf{q}_0)}(\mathbf{v} - \mathbf{b}), \quad (\mathbf{v} - \mathbf{b}) \in \mathcal{T}_C(\mathbf{q}_0), \tag{3.14}$$

which can be rewritten according to proposition A.5, (0.18), (0.19) as

$$\mathcal{T}_C^{\perp}(\mathbf{q}_0) \ni -(\mathbf{u}^+ - \mathbf{u}^-) \perp (\mathbf{v} - \mathbf{b}) \in \mathcal{T}_C(\mathbf{q}_0), \tag{3.15}$$

where $\perp$ means orthogonal with respect to the kinetic metric. This formulation of Newton's impact law will further be used to gain a geometric interpretation of the impact and to show equivalence with Poisson's law.

4. Poisson's impact law

As an alternative to (3.1) we may use a unilateral formulation of *Poisson's impact law* (1.2) to complete the impact equations (2.1)–(2.5). This requires a decomposition of the impact process (2.1), (2.5) into two succeeding phases, the *phase of compression*

$$\mathbf{M}(\mathbf{u}^\circ - \mathbf{u}^-) = \mathbf{R}^{(-)}, \quad \mathbf{R}^{(-)} = \sum_{i \in \mathcal{H}} \mathbf{w}_i \Lambda_i^{(-)} \tag{4.1}$$

and the *phase of decompression*

$$\mathbf{M}(\mathbf{u}^+ - \mathbf{u}^\circ) = \mathbf{R}^{(+)}, \quad \mathbf{R}^{(+)} = \sum_{i \in \mathcal{H}} \mathbf{w}_i \Lambda_i^{(+)} \tag{4.2}$$

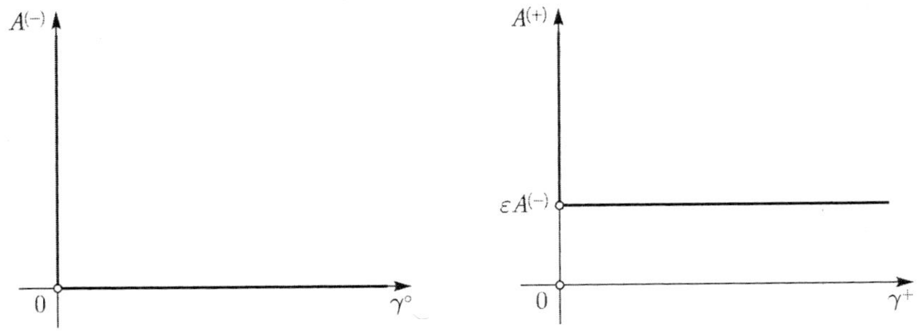

Figure 6.3. Poisson's impact law: Compression and decompression.

and was first formulated as an inequality law in [6], see also [7] and [9]. These two phases describing the impact process work as follows:

Compression starts with the pre-impact velocities which are denoted by upper indices − as before, and terminates with a velocity state indicated by upper indices ∘. The impulsive force acting at compression is denoted by $\Lambda_i^{(-)}$. The end of compression is defined by the demand that the approaching process ($\gamma_i < 0$) of the bodies has to be finished, thus $\gamma_i^\circ \geq 0$. We take, as before, $\Lambda_i^{(-)} \geq 0$ to ensure the compressive character of the impulsive force, and propose complementary behavior of γ_i° and $\Lambda_i^{(-)}$. This yields the impact law for the compression phase

$$\gamma_i^\circ \geq 0, \quad \Lambda_i^{(-)} \geq 0, \quad \gamma_i^\circ \Lambda_i^{(-)} = 0 \tag{4.3}$$

where we recognize that compression corresponds to a Newtonian inelastic impact, i.e. (3.1) applies with $\varepsilon_i = 0$. The graph of the impact law for compression is depicted in the right diagram of figure 6.3 and constitutes a unilateral primitive expressed by the complementarities in (4.3).

Decompression starts with the velocity state at the end of compression (upper index ∘) and terminates with the post-impact velocities (upper index +). The impulse at decompression is denoted by $\Lambda_i^{(+)}$ and is determined by a modified Poisson's hypothesis which originally is (1.2), $\Lambda_i^{(+)} = \varepsilon_i \Lambda_i^{(-)}$. In order to get a complete description of the impact process, even this relation has to be written in a unilateral manner, in particular with respect to multi-contact configurations. After the impact phase has terminated the relative velocity γ_i must show non-negative values. We therefore demand $\gamma_i^+ \geq 0$. If the original Poisson impulse $\Lambda_i^{(+)}$ from (1.2) is too small to ensure this, it must be enlarged enough such that at least $\gamma_i^+ = 0$ can be satisfied. We thus allow $\Lambda_i^{(+)} \geq \varepsilon_i \Lambda_i^{(-)}$ and propose to set $\gamma_i^+ = 0$ whenever $\Lambda_i^{(+)} > \varepsilon_i \Lambda_i^{(-)}$. On the other hand, if $\Lambda_i^{(+)}$ from (1.2) is high enough to admit separation we allow any value $\gamma_i^+ \geq 0$. This behavior can be expressed by the impact law

$$\gamma_i^+ \geq 0, \quad \Lambda_i^{(+)} - \varepsilon_i \Lambda_i^{(-)} \geq 0, \quad \gamma_i^+ (\Lambda_i^{(+)} - \varepsilon_i \Lambda_i^{(-)}) = 0 \tag{4.4}$$

which corresponds to a shifting of the compression impact characteristic as depicted in the right diagram of figure 6.3. Here, ε_i denotes the Poisson's coefficient of restitution. The impact terminates already after compression when $\varepsilon_i = 0 \; \forall i \in \mathcal{H}$ in the sense that decompression does not alter the post-compression velocity $\mathbf{u}^\circ$. For the case $\varepsilon_i = 1$ (and one single contact, of course), an impulse of the same magnitude as at compression is applied, leading finally again to an inversion of the relative velocity. For $\varepsilon_i \geq 0$, the overall value of Poisson's impulsive force $\Lambda_i^{(-)} + \Lambda_i^{(+)}$ satisfies by (4.3) and (4.4) the inequality $\Lambda_i \geq 0$ as for the Newtonian impact (3.1) and thus ensures, again, its compressive character.

The goal is now to state Poisson's impact process in a form similar to Newton's problem as done in the previous section. As before, we assume that all coefficients of restitution are equal and denote them by ε, to enter again the framework of impacts with global dissipation index. Since Poisson's phase of compression coincides with a completely inelastic Newtonian impact ($\varepsilon = 0$), the four representations corresponding to (3.12)–(3.15) are immediately available,

$$\begin{gathered}
-\delta \mathbf{u}^{\circ T} \mathbf{M}(\mathbf{u}^\circ - \mathbf{u}^-) \leq 0, \quad \mathbf{u}^\circ \in \mathcal{T}_\mathcal{C}(\mathbf{q}_0) + \mathbf{b}, \quad \forall \delta \mathbf{u}^\circ \in \mathcal{T}_{\mathcal{T}_\mathcal{C}(\mathbf{q}_0)}(\mathbf{u}^\circ - \mathbf{b}), \\
-\mathbf{M}(\mathbf{u}^\circ - \mathbf{u}^-) \in \mathcal{N}_{\mathcal{T}_\mathcal{C}(\mathbf{q}_0)}(\mathbf{u}^\circ - \mathbf{b}), \quad (\mathbf{u}^\circ - \mathbf{b}) \in \mathcal{T}_\mathcal{C}(\mathbf{q}_0), \\
-(\mathbf{u}^\circ - \mathbf{u}^-) \in \mathcal{T}^\perp_{\mathcal{T}_\mathcal{C}(\mathbf{q}_0)}(\mathbf{u}^\circ - \mathbf{b}), \quad (\mathbf{u}^\circ - \mathbf{b}) \in \mathcal{T}_\mathcal{C}(\mathbf{q}_0), \\
\mathcal{T}_\mathcal{C}^\perp(\mathbf{q}_0) \ni -(\mathbf{u}^\circ - \mathbf{u}^-) \perp (\mathbf{u}^\circ - \mathbf{b}) \in \mathcal{T}_\mathcal{C}(\mathbf{q}_0)
\end{gathered} \tag{4.5}$$

with $\mathbf{v} \equiv \mathbf{u}^\circ$ from (3.11) and $\delta \mathbf{u}^\circ = \mathbf{u}^{\circ\star} - \mathbf{u}^\circ$.

For decompression we set

$$\Delta_i := \Lambda_i^{(+)} - \varepsilon \Lambda_i^{(-)} \tag{4.6}$$

and denote by $\mathbf{G} := \sum_{i \in \mathcal{H}} \mathbf{w}_i \Delta_i$ the corresponding impulsive force in $\mathbb{R}^f$. We then obtain the representation

$$\begin{gathered}
\mathbf{M}(\mathbf{u}^+ - \mathbf{u}^\circ) = \mathbf{R}^{(+)}, \quad \mathbf{R}^{(+)} = \mathbf{G} + \varepsilon \mathbf{R}^{(-)} \\
\gamma_i^+ = \mathbf{w}_i^T \mathbf{u}^+ + \hat{w}_i, \quad \mathbf{G} = \sum_{i \in \mathcal{H}} \mathbf{w}_i \Delta_i, \\
\gamma_i^+ \geq 0, \quad \Delta_i \geq 0, \quad \gamma_i^+ \Delta_i = 0,
\end{gathered} \tag{4.7}$$

where $\mathbf{R}^{(-)}$ is the compression impulse $\sum_{i\in\mathcal{H}} \mathbf{w}_i \Lambda_i^{(-)}$. Equations (4.2) and (4.4) may be immediately recognized from (4.7) when $\mathbf{R}^{(-)}$ and $\mathbf{G}$ are eliminated and (4.6) is re-substituted.

By performing the same steps as in (3.4)–(3.9) for Newtonian impacts, the complementarities from the last line in (4.7) are now rewritten as

$$\begin{aligned} & -\sum_{i\in\mathcal{H}} \Delta_i \mathbf{w}_i^T(\mathbf{u}^{+\star} - \mathbf{u}^+) \le 0, \\ \mathbf{u}^+ \text{ such that } & -\sum_{j\in\mathcal{H}} \alpha_j \mathbf{w}_j^T(\mathbf{u}^+ - \mathbf{b}) \le 0 \quad \forall \alpha_j \ge 0, \\ \forall \mathbf{u}^{+\star} \text{ such that } & -\sum_{j\in\mathcal{H}} \alpha_j \mathbf{w}_j^T(\mathbf{u}^{+\star} - \mathbf{b}) \le 0 \quad \forall \alpha_j \ge 0, \end{aligned} \tag{4.8}$$

where $\gamma_i^+ = \mathbf{w}_i^T \mathbf{u}^+ + \hat{w}_i$ and $\gamma_i^{+\star} = \mathbf{w}_i^T \mathbf{u}^{+\star} + \hat{w}_i$ have been used. This yields together with the first two lines in (4.7) Poisson's impact problem in the form

$$\begin{gathered} \mathbf{M}(\mathbf{u}^+ - \mathbf{u}^\circ) = \mathbf{R}^{(+)}, \quad \mathbf{R}^{(+)} = \mathbf{G} + \varepsilon \mathbf{R}^{(-)} \\ -\mathbf{G}^T(\mathbf{u}^{+\star} - \mathbf{u}^+) \le 0, \\ \mathbf{u}^+ \in \mathcal{T}_\mathcal{C}(\mathbf{q}_0) + \mathbf{b}, \quad \forall \mathbf{u}^{+\star} \in \mathcal{T}_\mathcal{C}(\mathbf{q}_0) + \mathbf{b}. \end{gathered} \tag{4.9}$$

With $\delta\mathbf{u}^+ := \mathbf{u}^{+\star} - \mathbf{u}^+$ and the compression impulse $\mathbf{R}^{(-)} = \mathbf{M}(\mathbf{u}^\circ - \mathbf{u}^-)$ we obtain from (4.9) the four representations corresponding to (4.5) for decompression:

$$\begin{gathered} -\delta\mathbf{u}^{+T}\mathbf{M}\mathbf{u}_\Delta \le 0, \quad \mathbf{u}^+ \in \mathcal{T}_\mathcal{C}(\mathbf{q}_0) + \mathbf{b}, \quad \forall \delta\mathbf{u}^+ \in \mathcal{T}_{\mathcal{T}_\mathcal{C}(\mathbf{q}_0)}(\mathbf{u}^+ - \mathbf{b}), \\ -\mathbf{M}\mathbf{u}_\Delta \in \mathcal{N}_{\mathcal{T}_\mathcal{C}(\mathbf{q}_0)}(\mathbf{u}^+ - \mathbf{b}), \quad (\mathbf{u}^+ - \mathbf{b}) \in \mathcal{T}_\mathcal{C}(\mathbf{q}_0), \\ -\mathbf{u}_\Delta \in \mathcal{T}^\perp_{\mathcal{T}_\mathcal{C}(\mathbf{q}_0)}(\mathbf{u}^+ - \mathbf{b}), \quad (\mathbf{u}^+ - \mathbf{b}) \in \mathcal{T}_\mathcal{C}(\mathbf{q}_0), \\ \mathcal{T}_\mathcal{C}^\perp(\mathbf{q}_0) \ni -\mathbf{u}_\Delta \perp (\mathbf{u}^+ - \mathbf{b}) \in \mathcal{T}_\mathcal{C}(\mathbf{q}_0) \\ \text{and } \mathbf{u}_\Delta = (\mathbf{u}^+ - \mathbf{u}^\circ) - \varepsilon(\mathbf{u}^\circ - \mathbf{u}^-). \end{gathered} \tag{4.10}$$

In particular, the fourth formulation in (4.10) will be used in the next section to show equivalence with Newton's problem.

5. A geometric interpretation of impacts

In this section we give a geometric interpretation of Newton's impact problem for tangentially regular sets and show equivalence with Poisson's problem under a certain kinematic condition. The same nomenclature as in the previous sections is used. In particular, $\mathcal{T}_\mathcal{C}(\mathbf{q}_0) + \mathbf{b}$ denotes the translate of the tangent cone at $\mathbf{q}_0$ in the moving set $\mathcal{C}$, and $\mathcal{T}_\mathcal{C}^\perp(\mathbf{q}_0)$ the cone orthogonal to $\mathcal{T}_\mathcal{C}(\mathbf{q}_0)$ with respect to the inner product on the

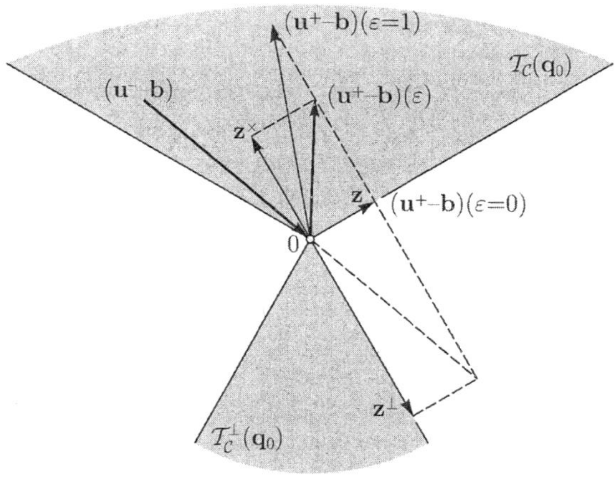

Figure 6.4. The geometry of impacts with global dissipation index.

tangent space as before. With $(\mathbf{u}^- - \mathbf{b})$ being the pre-impact velocity of the system *relative* to the tangent cone $\mathcal{T}_C(\mathbf{q}_0)$, we propose the following geometric construction of the impact process:

According to theorem 6, perform an orthogonal decomposition of $(\mathbf{u}^- - \mathbf{b})$ into

$$(\mathbf{u}^- - \mathbf{b}) = \mathbf{z} + \mathbf{z}^\perp \tag{5.1}$$

such that the vectors $\mathbf{z}$ and $\mathbf{z}^\perp$ satisfy

$$\mathcal{T}_C^\perp(\mathbf{q}_0) \ni \mathbf{z}^\perp \perp \mathbf{z} \in \mathcal{T}_C(\mathbf{q}_0). \tag{5.2}$$

This decomposition is unique. The term $\mathbf{z}$ plays the role of the *tangential component* of $(\mathbf{u}^- - \mathbf{b})$ which remains unchanged by the impact, whereas the *normal component* $\mathbf{z}^\perp$ is "inverted" by the impact rule

$$\mathbf{z}^\times := -\varepsilon \mathbf{z}^\perp \quad (0 \le \varepsilon \le 1), \tag{5.3}$$

where ε is the coefficient of restitution as before. The post-impact velocity relative to $\mathcal{T}_C(\mathbf{q}_0)$ is then set to be

$$(\mathbf{u}^+ - \mathbf{b}) := \mathbf{z} + \mathbf{z}^\times \tag{5.4}$$

according to the construction shown in figure 6.4.

Proposition 6.2 *For* $(\mathbf{u}^- - \mathbf{b}) \in -\mathcal{T}_C(\mathbf{q}_0)$ *the impact law* (5.1)–(5.4) *yields physically admissible post-impact velocities, i.e.* $(\mathbf{u}^+ - \mathbf{b}) \in \mathcal{T}_C(\mathbf{q}_0)$.

Proof. Denote by $\mathcal{L}(\mathbf{q}_0)$ any arbitrary two-dimensional subspace containing the triple $\{(\mathbf{u}^- - \mathbf{b}), \mathbf{z}, \mathbf{z}^\perp\}$. The subspace $\mathcal{L}(\mathbf{q}_0)$ is uniquely

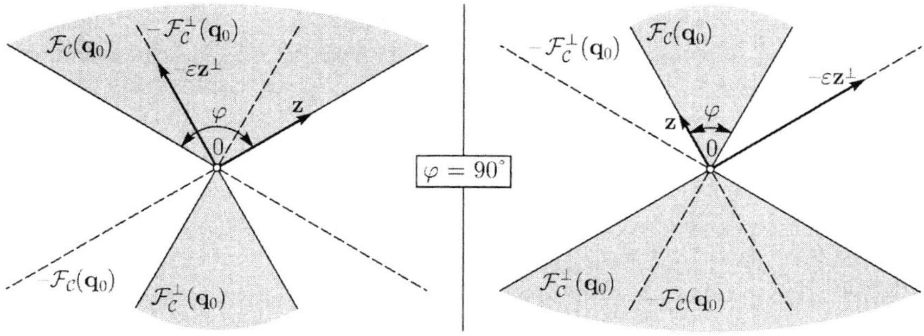

Figure 6.5. On the kinematic compatibility of Newton's impact law.

generated by this set, as soon as two out of the three members are linearly independent. Let further be $\mathcal{F}_\mathcal{C}(\mathbf{q}_0) := \mathcal{L}(\mathbf{q}_0) \cap \mathcal{T}_\mathcal{C}(\mathbf{q}_0)$ and $\mathcal{F}_\mathcal{C}^\perp(\mathbf{q}_0) := \mathcal{L}(\mathbf{q}_0) \cap \mathcal{T}_\mathcal{C}^\perp(\mathbf{q}_0)$ which are two closed convex cones orthogonal to each other such that one of the two configurations in figure 6.5 is met. Consider first the case $\varphi \geq 90°$ for which $-\mathcal{F}_\mathcal{C}^\perp(\mathbf{q}_0) \subset \mathcal{F}_\mathcal{C}(\mathbf{q}_0)$. In this case it holds for any $\varepsilon \geq 0$ and any $\mathbf{z} \in \mathcal{F}_\mathcal{C}(\mathbf{q}_0)$, $\mathbf{z}^\perp \in \mathcal{F}_\mathcal{C}^\perp(\mathbf{q}_0)$ that $(\mathbf{u}^+ - \mathbf{b}) = \mathbf{z} - \varepsilon\mathbf{z}^\perp \in \mathcal{F}_\mathcal{C}(\mathbf{q}_0)$, i.e. admissible post-impact velocities are assured independent of the value of $(\mathbf{u}^- - \mathbf{b})$. Consider now $\varphi < 90°$. We have $-\mathcal{F}_\mathcal{C}(\mathbf{q}_0) \subset \mathcal{F}_\mathcal{C}^\perp(\mathbf{q}_0)$. For $(\mathbf{u}^- - \mathbf{b}) \in -\mathcal{F}_\mathcal{C}(\mathbf{q}_0)$ by assumption and a proper choice of $\mathcal{L}(\mathbf{q}_0)$ if any, we have thus $(\mathbf{u}^- - \mathbf{b}) \in \mathcal{F}_\mathcal{C}^\perp(\mathbf{q}_0)$, and the orthogonal decomposition yields $\mathbf{z} = 0$, $\mathbf{z}^\perp = (\mathbf{u}^- - \mathbf{b})$ with $\mathbf{z}^\perp \in -\mathcal{F}_\mathcal{C}(\mathbf{q}_0)$. This means by (5.3), (5.4) that $(\mathbf{u}^+ - \mathbf{b}) = -\varepsilon\mathbf{z}^\perp \in \mathcal{F}_\mathcal{C}(\mathbf{q}_0) \subset \mathcal{T}_\mathcal{C}(\mathbf{q}_0)$ and completes the proof, because this or the other situation applies for any possible $\mathcal{L}(\mathbf{q}_0)$. ■

Note that the impact law (5.1)–(5.4) *may* produce non-compatible post-impact velocities $\mathbf{u}^+$ when the pre-impact velocities $\mathbf{u}^-$ are not from $\mathcal{T}_\mathcal{C}(\mathbf{q}_0) + \mathbf{b}$, for instance when initial conditions are arbitrarily chosen or as a result of numerical round-offs. For $\varphi < 90°$ as in the right diagram of figure 6.5 one has in general $(\mathbf{u}^+ - \mathbf{b}) \in -\mathcal{F}_\mathcal{C}^\perp(\mathbf{q}_0)$ which might, of course, violate $(\mathbf{u}^+ - \mathbf{b}) \in \mathcal{F}_\mathcal{C}(\mathbf{q}_0)$. In contrast, Poisson's impact law (4.5), (4.10) provides *always* compatible post-impact velocities and so Newton's law for *completely inelastic impacts* ($\varepsilon = 0$).

From figure 6.4 one recognizes immediately a lot of properties of the impact law: It is always energetically consistent, because for systems without external excitation ($\mathbf{b} = 0$) the kinetic energy T satisfies $2\,T^+ = \|\mathbf{u}^+\|^2 \leq \|\mathbf{u}^-\|^2 = 2\,T^-$, where equality holds for $\varepsilon = 1$ and maximal dissipation is achieved for $\varepsilon = 0$. For systems *with* external excitation energetical consistency has to be understood as to satisfy $\|\mathbf{u}^+ - \mathbf{b}\|^2 \leq \|\mathbf{u}^- - \mathbf{b}\|^2$. A completely elastic impact ($\varepsilon = 1$) can be interpreted as a *reflection* on a hyperplane $\mathcal{H}$ with normal $\mathbf{z}^\perp$, whereas a completely inelastic impact ($\varepsilon = 0$) corresponds to an *orthogonal projection* of $(\mathbf{u}^- - \mathbf{b})$ on $\mathcal{H}$ to give $\mathbf{z}$. In terms of a minimization problem are $\mathbf{z}$ and $\mathbf{z}^\perp$ the nearest points to $(\mathbf{u}^- - \mathbf{b})$ in the sets $\mathcal{T}_C(\mathbf{q}_0)$ and $\mathcal{T}_C^\perp(\mathbf{q}_0)$, respectively. The corresponding maps are called *proximations* and denoted by

$$\mathbf{z} = \mathrm{prox}_{\mathcal{T}_C(\mathbf{q}_0)}(\mathbf{u}^- - \mathbf{b}), \quad \mathbf{z}^\perp = \mathrm{prox}_{\mathcal{T}_C^\perp(\mathbf{q}_0)}(\mathbf{u}^- - \mathbf{b}). \tag{5.5}$$

For example, the impact law (5.1)–(5.4) might equivalently be stated in terms of proximations as

$$(\mathbf{u}^+ - \mathbf{b}) = (1+\varepsilon)\,\mathrm{prox}_{\mathcal{T}_C(\mathbf{q}_0)}(\mathbf{u}^- - \mathbf{b}) - \varepsilon(\mathbf{u}^- - \mathbf{b}) \tag{5.6}$$

when the first equation in (5.5) is used. Further, we recognize that the proximation in (5.6) becomes the identity whenever $(\mathbf{u}^- - \mathbf{b}) \in \mathcal{T}_C(\mathbf{q}_0)$. In this case $(\mathbf{u}^+ - \mathbf{b}) \equiv (\mathbf{u}^- - \mathbf{b})$, thus no impact occurs.

Proposition 6.3 *The construction given by equations* (5.1)–(5.4) *and in figure* 6.4 *is the geometric version of Newton's impact law* (3.15), (3.11) *for systems with global dissipation index.*

Proof. We take (5.1), (5.3), (5.4) and solve them for $\mathbf{z}$ and $\mathbf{z}^\perp$,

$$\begin{aligned}(\mathbf{u}^- - \mathbf{b}) &= \mathbf{z} + \mathbf{z}^\perp,\\ (\mathbf{u}^+ - \mathbf{b}) &= \mathbf{z} - \varepsilon\mathbf{z}^\perp,\end{aligned} \quad\Leftrightarrow\quad \begin{aligned}\mathbf{z} &= \tfrac{1}{1+\varepsilon}(\mathbf{u}^+ + \varepsilon\mathbf{u}^-) - \mathbf{b},\\ \mathbf{z}^\perp &= \tfrac{-1}{1+\varepsilon}(\mathbf{u}^+ - \mathbf{u}^-),\end{aligned} \tag{5.7}$$

in order to express the orthogonal decomposition (5.2) as

$$\mathcal{T}_C^\perp(\mathbf{q}_0) \ni \left(\tfrac{-1}{1+\varepsilon}(\mathbf{u}^+ - \mathbf{u}^-)\right) \perp \left(\tfrac{1}{1+\varepsilon}(\mathbf{u}^+ + \varepsilon\mathbf{u}^-) - \mathbf{b}\right) \in \mathcal{T}_C(\mathbf{q}_0). \tag{5.8}$$

By substituting $\mathbf{v}$ from (3.11) we obtain

$$\mathcal{T}_C^\perp(\mathbf{q}_0) \ni -(\mathbf{u}^+ - \mathbf{u}^-)\tfrac{1}{1+\varepsilon} \perp (\mathbf{v} - \mathbf{b}) \in \mathcal{T}_C(\mathbf{q}_0). \tag{5.9}$$

This is exactly (3.15), because we may cancel $\frac{1}{1+\varepsilon} > 0$ from (5.9) since $\mathcal{T}_C^\perp(\mathbf{q}_0)$ is a cone. ■

We may further recognize a relation expressing that, indeed, some relative velocity has been "inverted" by the impact law. From the upper right equation in (5.7) and $\mathbf{v}$ in (3.11) we have

$$\mathbf{z} = \mathbf{v} - \mathbf{b} \tag{5.10}$$

which we use to eliminate $\mathbf{z}$ from the left two equations in (5.7). This yields

$$(\mathbf{u}^+ - \mathbf{v}) = -\varepsilon\,(\mathbf{u}^- - \mathbf{v}), \tag{5.11}$$

even for systems with kinematic excitation ($\mathbf{b} \neq 0$).

Proposition 6.4 *For a common coefficient of restitution ε and a common pre-impact velocity $(\mathbf{u}^- - \mathbf{b}) \in -\mathcal{T}_C(\mathbf{q}_0)$, the post-impact velocities $\mathbf{u}_N^+$ obtained from Newton's law* (3.15), (3.11) *agree with $\mathbf{u}_P^+$ obtained from Poisson's law* (4.5), (4.10).

Proof. For the representation of Poisson's phase of decompression we take the last equation in (4.10), i.e.

$$\begin{aligned} \mathcal{T}_C^{\perp}(\mathbf{q}_0) \ni -\mathbf{u}_\Delta \ \perp \ (\mathbf{u}_P^+ - \mathbf{b}) \in \mathcal{T}_C(\mathbf{q}_0) \\ \text{and } \mathbf{u}_\Delta = (\mathbf{u}_P^+ - \mathbf{u}^\circ) - \varepsilon(\mathbf{u}^\circ - \mathbf{u}^-). \end{aligned} \tag{5.12}$$

In a first step we express $\mathbf{u}^\circ$ in terms of the Newtonian post-impact velocity $\mathbf{u}_N^+$ which can be done as follows: Suppose that all terms in (5.1)–(5.4) of the Newtonian impact have already been determined. We know that Poisson's phase of compression coincides with a completely inelastic Newtonian impact ($\mathbf{z}^\times = 0$). Thus, the Poisson velocity $(\mathbf{u}^\circ - \mathbf{b})$ after compression coincides with $\mathbf{z}$ by (5.4) or figure 6.4, and $\mathbf{u}^\circ \equiv \mathbf{v}$ due to (5.10). With this result, (5.11) becomes

$$(\mathbf{u}_N^+ - \mathbf{u}^\circ) = -\varepsilon\,(\mathbf{u}^- - \mathbf{u}^\circ), \tag{5.13}$$

and $\mathbf{u}_\Delta$ in (5.12) reduces to

$$\mathbf{u}_\Delta = \mathbf{u}_P^+ - \mathbf{u}_N^+. \tag{5.14}$$

Poisson's law (5.12) may thus be written as

$$\mathcal{T}_C^{\perp}(\mathbf{q}_0) \ni -(\mathbf{u}_P^+ - \mathbf{u}_N^+) \ \perp \ (\mathbf{u}_P^+ - \mathbf{b}) \in \mathcal{T}_C(\mathbf{q}_0), \tag{5.15}$$

which are the necessary and sufficient conditions of the orthogonal decomposition problem (0.17): Decompose $(\mathbf{u}_N^+ - \mathbf{b})$ uniquely into

$$(\mathbf{u}_N^+ - \mathbf{b}) = (\mathbf{u}_N^+ - \mathbf{u}_P^+) + (\mathbf{u}_P^+ - \mathbf{b}) \tag{5.16}$$

such that (5.15) holds. Since we know by proposition 6.2 that $(\mathbf{u}_N^+ - \mathbf{b}) \in \mathcal{T}_C(\mathbf{q}_0)$, the (unique) solution of (5.15), (5.16) is $(\mathbf{u}_N^+ - \mathbf{b}) = (\mathbf{u}_P^+ - \mathbf{b})$ and $(\mathbf{u}_N^+ - \mathbf{u}_P^+) = 0$, hence $\mathbf{u}_N^+ = \mathbf{u}_P^+$. ∎

Equations (5.15), (5.16) express nothing else than a (second) orthogonal projection of the Newtonian post-impact velocity $(\mathbf{u}_N^+ - \mathbf{b})$ on the cone $\mathcal{T}_C(\mathbf{q}_0)$ to obtain Poisson's velocity $(\mathbf{u}_P^+ - \mathbf{b})$, see figure 6.6. Thus, if $(\mathbf{u}_N^+ - \mathbf{b})$ is already contained in $\mathcal{T}_C(\mathbf{q}_0)$, then this second projection does not alter anything, and one obtains $\mathbf{u}_P^+ \equiv \mathbf{u}_N^+$ as, for example, in the situation described in proposition 6.4. Otherwise, for incompatible

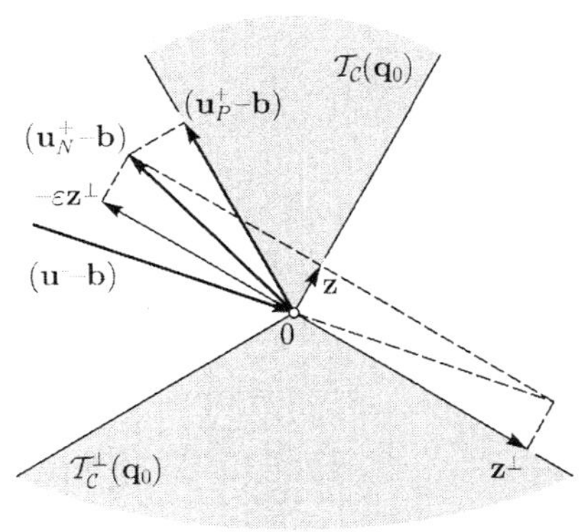

Figure 6.6. On the difference of Newton's and Poisson's impact law.

Newtonian post-impact velocities $(\mathbf{u}_N^+ - \mathbf{b}) \notin \mathcal{T}_C(\mathbf{q}_0)$, this second projection changes $\mathbf{u}_N^+$ to $\mathbf{u}_P^+$ and ensures that $(\mathbf{u}_P^+ - \mathbf{b}) \in \mathcal{T}_C(\mathbf{q}_0)$. As a simple consequence on the orthogonal projection properties one also has the inequality $\|\mathbf{u}_P^+ - \mathbf{b}\|^2 \leq \|\mathbf{u}_N^+ - \mathbf{b}\|^2$ which relates the energetical properties of both impact laws. Note also that the concept of applying a second orthogonal projection may easily be extended to configurations with re-entrant corners as treated in the next section for Newtonian impacts.

6. On impacts at re-entrant corners

We finally consider the case that the boundary of the admissible displacements is not tangentially regular at the point of impact $\mathbf{q}_0$, but can be approximated well by the (non-convex) contingent cone $\mathcal{K}_C(\mathbf{q}_0)$. Note that such a configuration can *not* be generated by the intersection of smooth simple unilateral constraints, but occurs in practice when, for example, two rectangular blocks hit each other at their corners. We try to find a reasonable extension of Moreau's impact law which we do by just replacing the tangent cone with the contingent cone in the appropriate equations. The set of admissible virtual velocities is now defined by the contingent cone, whereas the impulsive forces at the impact are still taken from the normal cone with the corresponding set $\mathcal{T}_C^\perp(\mathbf{q}_0)$ in the tangent space. We thus rewrite Newton's impact law (3.14) now as

$$-(\mathbf{u}^+ - \mathbf{u}^-) \in \mathcal{T}^\perp_{\mathcal{K}_C(\mathbf{q}_0)}(\mathbf{v} - \mathbf{b}), \quad (\mathbf{v} - \mathbf{b}) \in \mathcal{K}_C(\mathbf{q}_0) \tag{6.1}$$

with $\mathbf{v}$ as defined in (3.11) and note that the inclusion $\mathcal{T}^\perp_{\mathcal{K}_C(\mathbf{q}_0)}(\mathbf{v} - \mathbf{b}) \subset \mathcal{T}_C^\perp(\mathbf{q}_0)$ still holds. Of course, the orthogonal decomposition performed

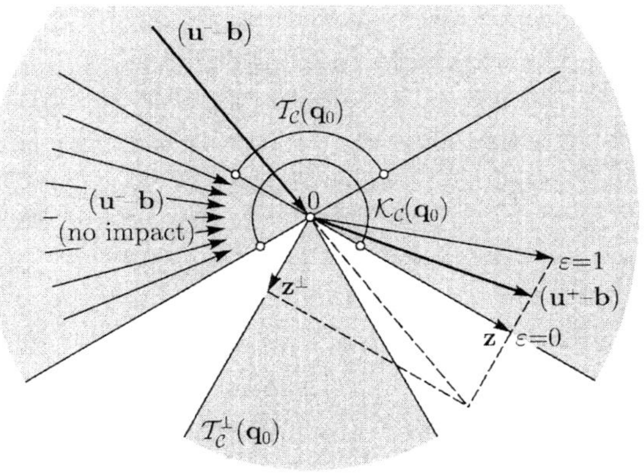

Figure 6.7. Impact at a re-entrant corner.

in (3.15) does not longer apply, because convexity of the two participating cones is required in theorem 6.

By using the same notation for the decomposition of $\mathbf{u}^-$ and $\mathbf{u}^+$ as in (5.7),

$$(\mathbf{u}^- - \mathbf{b}) =: \mathbf{z} + \mathbf{z}^\perp, \qquad (\mathbf{u}^+ - \mathbf{b}) =: \mathbf{z} - \varepsilon \mathbf{z}^\perp, \tag{6.2}$$

the impact law (6.1) becomes

$$\mathbf{z}^\perp \in \mathcal{T}^\perp_{\mathcal{K}_\mathcal{C}(\mathbf{q}_0)}(\mathbf{z}), \quad \mathbf{z} \in \mathcal{K}_\mathcal{C}(\mathbf{q}_0) \tag{6.3}$$

and is the version of (5.2) extended to re-entrant corners with the geometrical meaning as sketched in figure 6.7. A representation in terms of proximal points also fails, because due to the lack of convexity of $\mathcal{K}_\mathcal{C}(\mathbf{q}_0)$ one is no longer able to express (6.3) as a *minimization* problem. The *extended* statement of (5.5) is: Find the critical points of the function

$$\Phi(\mathbf{z}) = \tfrac{1}{2}\|\mathbf{z} - (\mathbf{u}^- - \mathbf{b})\|^2 + I_{\mathcal{K}_\mathcal{C}(\mathbf{q}_0)}(\mathbf{z}), \tag{6.4}$$

where $I_\mathcal{N}(\mathbf{z})$ denotes the value of the indicator function of $\mathcal{N}$ at $\mathbf{z}$, i.e. $I_\mathcal{N}(\mathbf{z}) = 0$ for $\mathbf{z} \in \mathcal{N}$ and $I_\mathcal{N}(\mathbf{z}) = +\infty$ for $\mathbf{z} \notin \mathcal{N}$. The solution set of this problem contains, among others, the proximal points to $(\mathbf{u}^- - \mathbf{b})$ in $\mathcal{K}_\mathcal{C}(\mathbf{q}_0)$ and is set up by the zeros of the generalized gradient of (6.4). These are the values of $\mathbf{z}$ which satisfy

$$\begin{aligned} 0 \in \bar{\partial}\Phi(\mathbf{z}) &= \mathbf{M}(\mathbf{z} - (\mathbf{u}^- - \mathbf{b})) + \bar{\partial} I_{\mathcal{K}_\mathcal{C}(\mathbf{q}_0)}(\mathbf{z}) \\ &= \mathbf{M}(\mathbf{z} - (\mathbf{u}^- - \mathbf{b})) + \mathcal{N}_{\mathcal{K}_\mathcal{C}(\mathbf{q}_0)}(\mathbf{z}). \end{aligned} \tag{6.5}$$

Or, with $\mathbf{z}^\perp = (\mathbf{u}^- - \mathbf{b}) - \mathbf{z}$ and by taking the terms in (6.5) from the cotangent space to the tangent space, we obtain

$$\mathbf{z}^\perp \in \mathcal{T}^\perp_{\mathcal{K}_\mathcal{C}(\mathbf{q}_0)}(\mathbf{z}), \tag{6.6}$$

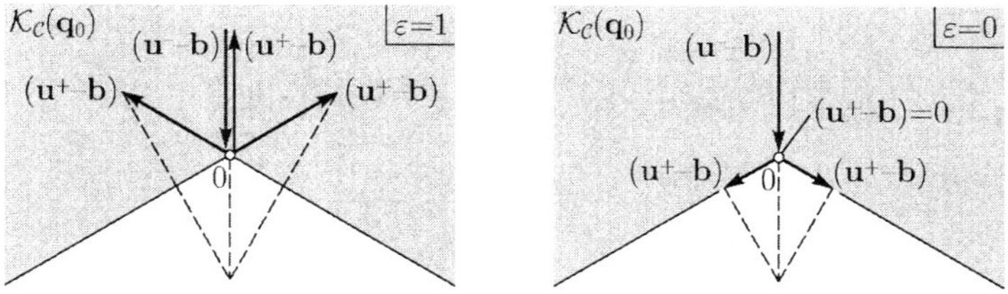

Figure 6.8. Non-uniqueness of the impact law at re-entrant corners.

which is together with the restriction $\mathbf{z} \in \mathcal{K}_C(\mathbf{q}_0)$ again the impact law (6.3).

Due to the non-convexity of $\mathcal{K}_C(\mathbf{q}_0)$ one can *not* expect the solution set of (6.3) to consist of one element only. Non-unique post-impact velocities *must* be accepted to occur for certain configurations, such as for the symmetric case depicted in figure 6.8 for $\varepsilon = 1$ and $\varepsilon = 0$, where three different solutions are met. An impact law for re-entrant corners was also introduced in [7] in the sense that (6.4) is taken as a *minimization* problem, leading to a restricted set of solutions.

7. Conclusion

In this paper we have presented the today's approach to frictionless collisions in rigid body systems. In principle there are two different methods: The first is based on a kinematic assumption and called the Newtonian impact law, the second is based on an impulse ratio and usually understood as Poisson's law. Both approaches coincide under certain conditions and were analyzed in a geometrical setting with the help of variational calculus. Although clearly originating in the collision problem of rigid bodies under simple unilateral constraints, the presented methods must be looked upon as a theoretical framework useful for many impact problems in general finite-degree-of-freedom dynamics, with even a clear geometric meaning relating to projections and reflections. Most of the ideas carried out and collected together in this paper, however, have their roots in the article [8] written by Moreau.

Appendix

The concept of linearization plays a crucial role in analysis. However linearization fails, if the investigated objects become non-smooth. The next step is to look for convexification. Instead of working with linear spaces one has to deal with cones which are convex at best. The cones introduced in the following characterize sets in the neighborhood of a chosen point and indicate whether convexification is possible or not. As a standard tool for inequality systems, these cones have already been used in sections 3–6; the related representations, variational formulations and decomposition rules as far as needed are presented in this section.

Cones and convexification

A subset $\mathcal{R}$ of $\mathbb{R}^f$ is called a *cone* if $\lambda \mathbf{v} \in \mathcal{R}$ for $\mathbf{v} \in \mathcal{R}$ and $\lambda > 0$. Cones are therefore collections of rays emanating from the origin. Let $\mathcal{C} \subset \mathbb{R}^f$ be a given set. The *contingent cone* to $\mathcal{C}$ at a point $\mathbf{q}$ is defined in [2] to be

$$\mathcal{K}_{\mathcal{C}}(\mathbf{q}) = \bigcap_{\delta>0} \bigcap_{\alpha>0} \bigcup_{0<\mu\leq\alpha} \left(\tfrac{1}{\mu}\left(\mathcal{C}-\mathbf{q}\right)+\delta\mathcal{B}\right) = \limsup_{\mu\downarrow 0} \tfrac{1}{\mu}\left(\mathcal{C}-\mathbf{q}\right), \tag{0.1}$$

where $\delta\mathcal{B}$ denotes the closed ball at center zero with radius δ. Equivalently,

$$\begin{aligned} \mathbf{v} \in \mathcal{K}_{\mathcal{C}}(\mathbf{q}) \iff & \forall\delta>0 \quad \forall\alpha>0 \quad \exists\mu\in(0,\alpha] \quad \exists\tilde{\mathbf{v}}\in\mathbf{v}+\delta\mathcal{B} \\ & \text{satisfying } \mathbf{q}+\mu\tilde{\mathbf{v}}\in\mathcal{C}, \end{aligned} \tag{0.2}$$

or stated in terms of sequences

$$\mathbf{v} \in \mathcal{K}_{\mathcal{C}}(\mathbf{q}) \iff \exists\mu_n\downarrow 0 \quad \exists\mathbf{v}_n\to\mathbf{v} \quad \text{satisfying } \mathbf{q}+\mu_n\mathbf{v}_n\in\mathcal{C} \quad \forall n. \tag{0.3}$$

There is a second cone, called the *tangent cone*, which is always convex in contrast to $\mathcal{K}_{\mathcal{C}}$. It is defined by

$$\mathcal{T}_{\mathcal{C}}(\mathbf{q}) = \bigcap_{\delta>0} \bigcup_{\substack{\alpha>0\\ \beta>0}} \bigcap_{\substack{0<\mu\leq\alpha\\ \tilde{\mathbf{q}}\in\mathcal{B}_{\mathcal{C}}(\mathbf{q},\beta)}} \left(\tfrac{1}{\mu}\left(\mathcal{C}-\tilde{\mathbf{q}}\right)+\delta\mathcal{B}\right) = \liminf_{\substack{\mu\downarrow 0\\ \tilde{\mathbf{q}}\,\vec{_{\mathcal{C}}}\,\mathbf{q}}} \tfrac{1}{\mu}\left(\mathcal{C}-\tilde{\mathbf{q}}\right), \tag{0.4}$$

where $\mathcal{B}_{\mathcal{C}}(\mathbf{q},\beta) := \mathcal{C}\cap(\mathbf{q}+\beta\mathcal{B})$ and $\tilde{\mathbf{q}}\,\vec{_{\mathcal{C}}}\,\mathbf{q}$ denotes the convergence of $\tilde{\mathbf{q}}$ to $\mathbf{q}$ in $\mathcal{C}$. As above, equivalent definitions are

$$\begin{aligned} \mathbf{v} \in \mathcal{T}_{\mathcal{C}}(\mathbf{q}) \iff & \forall\delta>0 \quad \exists\alpha>0 \quad \exists\beta>0 \quad \text{such that} \\ & \forall\tilde{\mathbf{q}}\in\mathcal{B}_{\mathcal{C}}(\mathbf{q},\beta) \quad \forall\mu\in(0,\alpha] \quad \exists\tilde{\mathbf{v}}\in\mathbf{v}+\delta\mathcal{B} \\ & \text{satisfying } \tilde{\mathbf{q}}+\mu\tilde{\mathbf{v}}\in\mathcal{C} \end{aligned} \tag{0.5}$$

or, in terms of sequences

$$\begin{aligned} \mathbf{v} \in \mathcal{T}_{\mathcal{C}}(\mathbf{q}) \iff & \forall\mathbf{q}_n\,\vec{_{\mathcal{C}}}\,\mathbf{q} \quad \forall\mu_n\downarrow 0 \quad \exists\mathbf{v}_n\to\mathbf{v} \\ & \text{satisfying } \mathbf{q}_n+\mu_n\mathbf{v}_n\in\mathcal{C} \quad \forall n. \end{aligned} \tag{0.6}$$

Note that $\mathcal{T}_{\mathcal{C}}(\mathbf{q}) \subset \mathcal{K}_{\mathcal{C}}(\mathbf{q})$ which is obvious when setting $\mathbf{q}_1 = \mathbf{q}_2 = \ldots = \mathbf{q}$ in (0.6) and comparing it with (0.3). We set $\mathcal{K}_\emptyset(\mathbf{q}) = \emptyset$, $\mathcal{T}_\emptyset(\mathbf{q}) = \emptyset$ and we observe that both cones are closed subsets of $\mathbb{R}^f$. For $\mathcal{C}$ being convex we have $\mathcal{K}_{\mathcal{C}}(\mathbf{q}) = \mathcal{T}_{\mathcal{C}}(\mathbf{q})$. A set is called to be *tangentially regular* at a point $\mathbf{q}$ if $\mathcal{K}_{\mathcal{C}}(\mathbf{q}) = \mathcal{T}_{\mathcal{C}}(\mathbf{q})$. Tangential regularity holds especially for convex sets but not generally. For nonconvex sets we may have $\mathcal{K}_{\mathcal{C}}(\mathbf{q}) = \mathcal{T}_{\mathcal{C}}(\mathbf{q})$ but also cases where this equality does not apply, the latter being called a *re-entrant corner point.*

We may further introduce an object playing the role of normal vectors to $\mathcal{C}$ at $\mathbf{q}$. This object is called the *normal cone* to $\mathcal{C}$ at $\mathbf{q}$ and is defined via the variational inequality

$$\mathcal{N}_{\mathcal{C}}(\mathbf{q}) = \{\, \mathbf{r} \mid \mathbf{r}^T\mathbf{v} \leq 0 \quad \forall\mathbf{v}\in\mathcal{T}_{\mathcal{C}}(\mathbf{q}) \,\}, \tag{0.7}$$

providing the set of all linear maps acting on $\mathbf{v}$ with values in $\mathbb{R}_0^-$. We set $\mathcal{N}_\emptyset(\mathbf{q}) = \emptyset$ and note that $\mathcal{N}_{\mathcal{C}}(\mathbf{q})$ is a closed convex cone. One observes that the cones $\mathcal{T}_{\mathcal{C}}(\mathbf{q})$ and

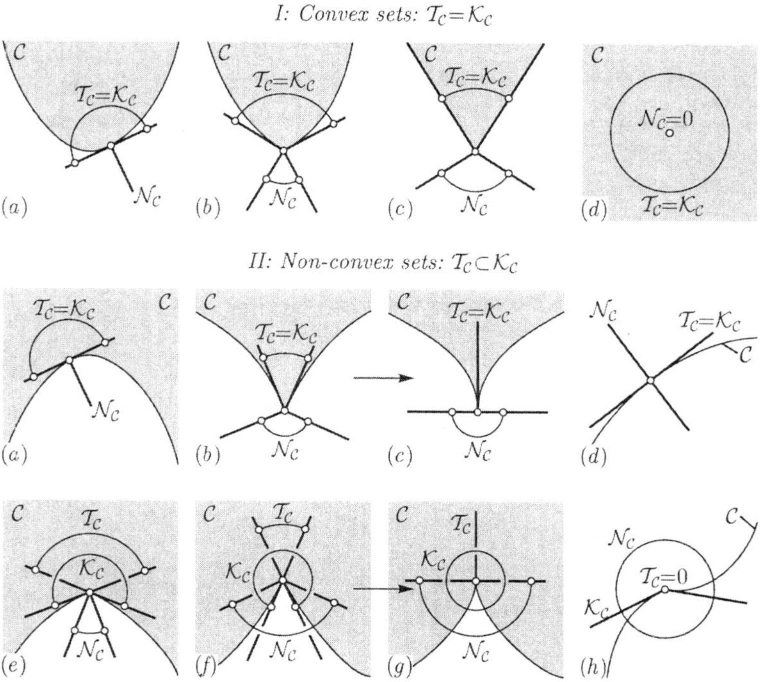

Figure 6.A.1. The cones $\mathcal{K}_C$, $\mathcal{T}_C$ and $\mathcal{N}_C$ at different points of $\mathcal{C}$.

$\mathcal{N}_C(\mathbf{q})$ are mutually polar which allows to express the orthogonality of spaces in terms of polarity of cones.

For $\mathcal{C}$ a subset of a differentiable manifold $\mathcal{M}$ the cones $\mathcal{K}_C(\mathbf{q})$ and $\mathcal{T}_C(\mathbf{q})$ at some point $\mathbf{q} \in \mathcal{C}$ are contained in the tangent space of $\mathcal{M}$ at $\mathbf{q}$, whereas $\mathcal{N}_C(\mathbf{q})$ is a subset of the cotangent space. We denote by $\mathbf{M}$ the natural isomorphism induced by the metric on $\mathcal{M}$ in use which takes elements from the tangent to the cotangent space and introduce, in addition, the image and pre-image of the cones $\mathcal{T}_C(\mathbf{q})$ and $\mathcal{N}_C(\mathbf{q})$, respectively,

$$\mathcal{N}_C^{\perp}(\mathbf{q}) := \mathbf{M}(\mathbf{q})\, \mathcal{T}_C(\mathbf{q}), \quad \mathcal{T}_C^{\perp}(\mathbf{q}) := \mathbf{M}^{-1}(\mathbf{q})\, \mathcal{N}_C(\mathbf{q}). \tag{0.8}$$

By equations (0.4), (0.7), (0.8) we have thus obtained two pairs of orthogonal cones $(\mathcal{T}_C(\mathbf{q}), \mathcal{T}_C^{\perp}(\mathbf{q}))$ and $(\mathcal{N}_C(\mathbf{q}), \mathcal{N}_C^{\perp}(\mathbf{q}))$ characterized by

$$\begin{aligned} \mathbf{v} \cdot \mathbf{v}^{\perp} \leq 0 \quad \forall \mathbf{v} \in \mathcal{T}_C(\mathbf{q}),\ \forall \mathbf{v}^{\perp} \in \mathcal{T}_C^{\perp}(\mathbf{q}), \\ \mathbf{r} \cdot \mathbf{r}^{\perp} \leq 0 \quad \forall \mathbf{r} \in \mathcal{N}_C(\mathbf{q}),\ \forall \mathbf{r}^{\perp} \in \mathcal{N}_C^{\perp}(\mathbf{q}), \end{aligned} \tag{0.9}$$

where the dots denote the associated inner products on the tangent and the cotangent space, respectively.

In order to get a better sense on how the three cones $\mathcal{K}_C(\mathbf{q})$, $\mathcal{T}_C(\mathbf{q})$ and $\mathcal{N}_C(\mathbf{q})$ are arranged, we refer to figure 6.A.1 where different situations for convex and nonconvex sets are shown, also including submanifolds as a special case. For convex sets $\mathcal{C}$

tangential regularity holds at every point $\mathbf{q} \in \mathcal{C}$, see configurations $I(a)$–$I(d)$. In particular, $\mathcal{T}_\mathcal{C} = \mathcal{K}_\mathcal{C}$ is a half-space and $\mathcal{N}_\mathcal{C}$ degenerates to a ray when the boundary is smooth ($I(a)$), and $\mathcal{K}_\mathcal{C} = \mathcal{T}_\mathcal{C} = \mathbb{R}^f$ and $\mathcal{N}_\mathcal{C} = \{0\}$ for $\mathbf{q} \in \text{int}\,\mathcal{C}$, see $I(d)$. Non-convex sets (II) may be subdivided up into two classes depending on the shape of the boundary. Configurations for which the boundary is tangentially regular at some point $\mathbf{q}$ are depicted in $II(a)$–$II(d)$. For these cases we have $\mathcal{T}_\mathcal{C} = \mathcal{K}_\mathcal{C}$, including, for example, smooth submanifolds ($II(d)$) for which $\mathcal{T}_\mathcal{C} = \mathcal{K}_\mathcal{C}$ is recognized to be the classical tangent space to $\mathcal{C}$ at $\mathbf{q}$ and $\mathcal{N}_\mathcal{C}$ its orthogonal complement. Re-entrant corners ($\mathcal{T}_\mathcal{C} \subset \mathcal{K}_\mathcal{C}$) at $\mathbf{q}$ are depicted in $II(e)$–$II(h)$. Note in particular the case $II(h)$ of a non-smooth submanifold, at which $\mathcal{T}_\mathcal{C}$ may even reduce to 0 and $\mathcal{N}_\mathcal{C}$ may become the whole $\mathbb{R}^f$.

Representations of the normal cone

The cones introduced in the preceding section are, in general, quite hard to get. We will discuss only two cases which are of relevance for the impact theory as far as it has been developed in sections 3–6: Firstly, if the set $\mathcal{C}$ is convex and secondly, if a manifold is restricted by simple unilateral constraints. For convex sets $\mathcal{C}$ the normal cone to $\mathcal{C}$ at $\mathbf{q}$ becomes

$$\mathcal{N}_\mathcal{C}(\mathbf{q}) = \{\, \mathbf{r} \,|\, \mathbf{r}^T(\mathbf{q}^\star - \mathbf{q}) \le 0, \quad \mathbf{q} \in \mathcal{C}, \quad \forall \mathbf{q}^\star \in \mathcal{C} \,\}, \tag{0.10}$$

see e.g. [10], and consists of all vectors $\mathbf{r}$ which do not make an acute "angle" with any line segment in $\mathcal{C}$ with $\mathbf{q}$ as endpoint. The other cones are then easily derivable through (0.8) and the "inverse" of (0.7),

$$\mathcal{T}_\mathcal{C}(\mathbf{q}) = \{\, \mathbf{v} \,|\, \mathbf{r}^T \mathbf{v} \le 0 \quad \forall \mathbf{r} \in \mathcal{N}_\mathcal{C}(\mathbf{q}) \,\}, \tag{0.11}$$

which hold generally and do not require convexity of $\mathcal{C}$.

For the second case suppose that the (non-convex) set of feasible displacements $\mathcal{C}$ is represented by m simple unilateral smooth constraints $g_i(\mathbf{q}) \ge 0$ as depicted in figure 6.A.2,

$$\mathcal{C} := \{\, \mathbf{q} \,|\, g_i(\mathbf{q}) \ge 0, \quad i = 1, \ldots, m \,\}. \tag{0.12}$$

Denote by $\mathbf{w}_i$ the differential of g_i and by $\mathcal{H}$ the indices of the active set of constraints,

$$\mathbf{w}_i^T(\mathbf{q}) := \tfrac{\partial g_i}{\partial \mathbf{q}}, \qquad \mathcal{H} := \{\, i \,|\, g_i(\mathbf{q}) = 0 \,\}, \tag{0.13}$$

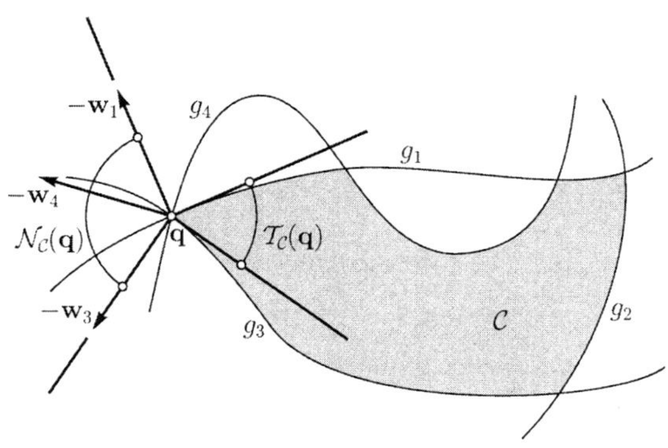

Figure 6.A.2. Normal and tangent cone for simple unilateral constraints.

and assume that $\mathbf{w}_i \neq 0$ when $i \in \mathcal{H}$. The normal cone $\mathcal{N}_\mathcal{C}(\mathbf{q})$ to $\mathcal{C}$ at $\mathbf{q}$ is then generated by the vectors $\mathbf{w}_i$ of the active set,

$$\mathcal{N}_\mathcal{C}(\mathbf{q}) = \{\, \mathbf{r} \,|\, \mathbf{r} = -\sum_{i\in\mathcal{H}} \alpha_i\, \mathbf{w}_i, \quad \alpha_i \geq 0 \,\}, \tag{0.14}$$

and the elements of the tangent cone are according to (0.11) characterized by

$$\mathbf{v} \in \mathcal{T}_\mathcal{C}(\mathbf{q}) \quad \Leftrightarrow \quad -\sum_{i\in\mathcal{H}} \alpha_i\, \mathbf{w}_i^T \mathbf{v} \leq 0, \quad \forall \alpha_i \geq 0. \tag{0.15}$$

It should be noted that the normal and the tangent cone associated with simple unilateral constraints are already polyhedral; more general situations occur if the set $\mathcal{C}$ of admissible displacements may not be obtained by the intersection of simple smooth inequality constraints, such as the circular cone $\{\, \lambda(\mathbf{e}_3 + \mathcal{D}_{12}),\ \lambda \geq 0 \,\}$ in $\mathbb{R}^3$, where $\mathbf{e}_3$ denotes the unit vector in the 3-direction and $\mathcal{D}_{12}$ the closed unit disc centered at 0 in the 1-2-plane.

Orthogonal decompositions

In this section we review briefly the decomposition of a vector with respect to a pair of orthogonal convex cones and discuss some other related formulations. The following statement is contained as a special case in Moreau's theorem, see e.g. [10] for the full version and the proof, and generalizes the classical vector decomposition with respect to orthogonal linear spaces.

Theorem 6 *Let $\mathcal{U}$ be a finite-dimensional real inner product space, $\mathcal{R}$ a closed convex cone in $\mathcal{U}$ and $\mathcal{R}^\perp$ the closed convex cone in $\mathcal{U}$ orthogonal to $\mathcal{R}$ defined by*

$$\mathcal{R}^\perp = \{\, \mathbf{v}^\perp \,|\, \mathbf{v}^\perp \cdot \mathbf{v} \leq 0 \quad \forall \mathbf{v} \in \mathcal{R} \,\}, \tag{0.16}$$

where $\cdot$ denotes the inner product. Any $\mathbf{u} \in \mathcal{U}$ can then be expressed uniquely as a sum $\mathbf{u} = \mathbf{v} + \mathbf{v}^\perp$ such that

$$\mathbf{v} \in \mathcal{R}, \quad \mathbf{v}^\perp \in \mathcal{R}^\perp, \quad \mathbf{v} \cdot \mathbf{v}^\perp = 0. \tag{0.17}$$

This decomposition is depicted in the left part of figure 6.A.3. For $\mathbf{u} \notin \mathcal{R} \cup \mathcal{R}^\perp$ one obtains for the decomposition $\mathbf{v} \neq 0$, $\mathbf{v}^\perp \neq 0$, whereas $\mathbf{v}^\perp = 0$ or $\mathbf{v} = 0$ as soon as

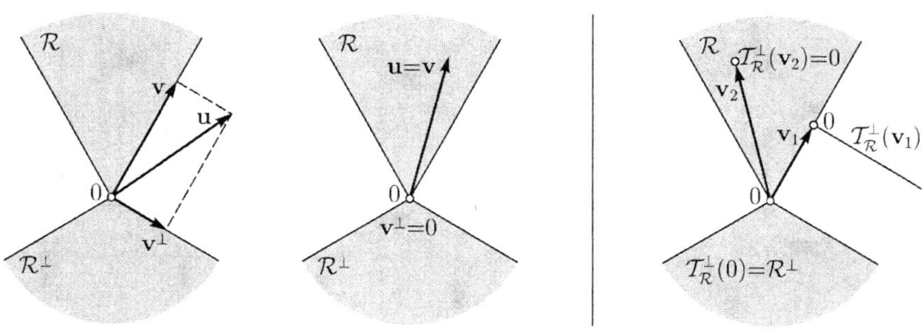

Figure 6.A.3. Orthogonal pair of cones $(\mathcal{R}, \mathcal{R}^\perp)$, and the cone $\mathcal{T}_\mathcal{R}^\perp(\cdot)$ orthogonal to the tangent cone of $\mathcal{R}$ at $(\cdot)$.

$\mathbf{u} \in \mathcal{R}$ or $\mathbf{u} \in \mathcal{R}^\perp$, respectively. Apparently is $\mathbf{v}$ the nearest point to $\mathbf{u}$ in the set $\mathcal{R}$, which is usually denoted by $\mathbf{v} = \text{prox}_\mathcal{R}(\mathbf{u})$. The corresponding map $\mathbf{u} \to \text{prox}_\mathcal{R}(\mathbf{u})$ is called a proximation which is, in fact, a projection because $\text{prox}^2_\mathcal{R}(\mathbf{u}) = \text{prox}_\mathcal{R}(\mathbf{u})$. In addition, this projection is orthogonal since $\text{prox}_\mathcal{R}(\mathbf{u}) \cdot (\mathbf{u} - \text{prox}_\mathcal{R}(\mathbf{u})) = 0$. Of course, the same properties apply for the second term $\mathbf{v}^\perp$ in the decomposition due to symmetry, i.e. $\mathbf{v}^\perp = \text{prox}_{\mathcal{R}^\perp}(\mathbf{u})$.

Proposition A.5 *Let $\mathcal{U}$, $\mathcal{R}$ and $\mathcal{R}^\perp$ be defined as in theorem* 6. *Then the following conditions are equivalent to each other:*

$$\mathbf{v} \in \mathcal{R}, \quad \mathbf{v}^\perp \in \mathcal{R}^\perp, \quad \mathbf{v} \cdot \mathbf{v}^\perp = 0 \tag{0.18}$$

$$\mathbf{v} \in \mathcal{R}, \quad \mathbf{v}^\perp \in \mathcal{T}^\perp_\mathcal{R}(\mathbf{v}) \tag{0.19}$$

$$\mathbf{v}^\perp \in \mathcal{R}^\perp, \quad \mathbf{v} \in \mathcal{T}^\perp_{\mathcal{R}^\perp}(\mathbf{v}^\perp) \tag{0.20}$$

Here, $\mathcal{T}^\perp_\mathcal{R}(\mathbf{v})$ and $\mathcal{T}^\perp_{\mathcal{R}^\perp}(\mathbf{v}^\perp)$ denote the cones orthogonal to the tangent cones to $\mathcal{R}$ at $\mathbf{v}$ and $\mathcal{R}^\perp$ at $\mathbf{v}^\perp$, respectively.

The cone $\mathcal{T}^\perp_\mathcal{R}(\mathbf{v})$ is depicted in the right part of figure 6.A.3 for some elements $\mathbf{v} \in \mathcal{R}$.

Proof. In order to show the equivalence of (0.18) and (0.19) we first recall that the cone $\mathcal{T}^\perp_\mathcal{R}(\mathbf{v})$ can be expressed via the second equation in (0.8) and (0.10) as

$$\mathcal{T}^\perp_\mathcal{R}(\mathbf{v}) = \{\,\mathbf{u}\,|\,\mathbf{u} \cdot (\mathbf{v}^\star - \mathbf{v}) \leq 0, \quad \mathbf{v} \in \mathcal{R}, \quad \forall \mathbf{v}^\star \in \mathcal{R}\}. \tag{0.21}$$

Further, we note that

(*i*) $\mathcal{R}^\perp = \{\,\mathbf{u}\,|\,\mathbf{u} \cdot \mathbf{v}^\star \leq 0, \quad \forall \mathbf{v}^\star \in \mathcal{R}\}$

(*ii*) $\mathcal{T}^\perp_\mathcal{R}(\mathbf{v}) = \{\,\mathbf{u}\,|\,\mathbf{u} \cdot \mathbf{z}^\star \leq 0, \quad \mathbf{v} \in \mathcal{R}, \quad \forall \mathbf{z}^\star \in \mathcal{R} - \mathbf{v}\}$

(*iii*) $\mathcal{T}^\perp_\mathcal{R}(\mathbf{v}) \subset \mathcal{R}^\perp$ with $\mathcal{T}^\perp_\mathcal{R}(0) = \mathcal{R}^\perp$

where (*i*) is the definition of $\mathcal{R}^\perp$ from (0.16), (*ii*) is obtained from (0.21) when setting $\mathbf{z}^\star := \mathbf{v}^\star - \mathbf{v}$, and (*iii*) holds because (*ii*) is more restrictive on $\mathbf{u}$ than (*i*): The set $\mathcal{R} - \mathbf{v}$ is the translate of the closed convex cone $\mathcal{R}$ in the direction $-\mathbf{v}$ with $\mathbf{v} \in \mathcal{R}$, hence $\mathcal{R} - \mathbf{v} \supset \mathcal{R}$.

Suppose now that (0.18) holds. The second condition, $\mathbf{v}^\perp \in \mathcal{R}^\perp$, means by (*i*) that $\mathbf{v}^\perp \cdot \mathbf{v}^\star \leq 0\ \forall \mathbf{v}^\star \in \mathcal{R}$. Subtracting from this expression the third condition $\mathbf{v}^\perp \cdot \mathbf{v} = 0$ in (0.18) gives $\mathbf{v}^\perp \cdot (\mathbf{v}^\star - \mathbf{v}) \leq 0\ \forall \mathbf{v}^\star \in \mathcal{R}$. This is by (0.21) together with $\mathbf{v} \in \mathcal{R}$ already $\mathbf{v}^\perp \in \mathcal{T}^\perp_\mathcal{R}(\mathbf{v})$ in (0.19).

To show the converse we rewrite $\mathbf{v}^\perp \in \mathcal{T}^\perp_\mathcal{R}(\mathbf{v})$ in (0.19) with the help of (0.21) as a variational inequality, $\mathbf{v}^\perp \cdot (\mathbf{v}^\star - \mathbf{v}) \leq 0$, which has to hold for any $\mathbf{v}^\star \in \mathcal{R}$ when $\mathbf{v} \in \mathcal{R}$. Choose now $\mathbf{v}^\star = 2\mathbf{v} \in \mathcal{R}$ and $\mathbf{v}^\star = 0 \in \mathcal{R}$ which is possible because $\mathcal{R}$ is a closed cone and $\mathbf{v} \in \mathcal{R}$, and evaluate the variational expression for both cases. This gives $\mathbf{v}^\perp \cdot \mathbf{v} \leq 0$ on the one hand and $\mathbf{v}^\perp \cdot \mathbf{v} \geq 0$ on the other hand, hence $\mathbf{v}^\perp \cdot \mathbf{v} = 0$ which is the third condition in (0.18). It remains to show that $\mathbf{v}^\perp \in \mathcal{R}^\perp$ which has already been done in (*iii*).

To prove the equivalence of (0.18) and (0.20) one proceeds exactly in the same manner. ■

References

[1] Aubin, J., and Ekeland, I. (1984). *Applied nonlinear analysis*, New York: John Wiley & Sons.

[2] Ballard, P. (2000). The dynamics of discrete mechanical systems with perfect unilateral constraints. *Arch. Rational Mech. Anal.* **154** 3, 199–274.

[3] Glocker, Ch. (1995). Multiple impacts with friction in rigid multibody systems. *Nonlinear Dynamics* **7**, 471–497.

[4] Glocker, Ch. (1995). Dynamik von Starrkörpersystemen mit Reibung und Stößen. VDI-Fortschrittberichte Mechanik/Bruchmechanik, Reihe 18, Nr. 182, VDI-Verlag, Düsseldorf.

[5] Glocker, Ch. (1999). Velocity jumps induced by C^0-constraints. In *Proc. of the ASME 1999 Design Engineering Technical Conferences, Las Vegas, Nevada, 12–15 September 1999*, pp. 1–8, CD-ROM.

[6] Glocker, Ch. (2001). On Frictionless Impact Models in Rigid-Body Systems. *Phil. Trans. Royal. Soc. Lond. A* **359**, 2385–2404.

[7] Kane, C., Repetto, E.A., Ortiz, M., and Marsden, J.E. (1999). Finite element analysis of nonsmooth contact. *Comp. Methods Appl. Mech. Eng.* **180**, 1–26.

[8] Moreau, J. J. (1988). Unilateral contact and dry friction in finite freedom dynamics. In *Non-Smooth mechanics and applications, CISM Courses and Lectures Vol. 302* (eds J. J. Moreau & P. D. Panagiotopoulos), pp. 1–82. Wien: Springer.

[9] Pfeiffer, F., and Glocker, Ch. (1996). *Multibody dynamics with unilateral contacts*, New York: John Wiley & Sons.

[10] Rockafellar, R.T. (1972). *Convex analysis*, Princeton, New Jersey: Princeton Univ. Press.

[11] Szabó, I. (1979). *Geschichte der mechanischen Prinzipien*, 2nd edn. Basel: Birkhäuser.

Chapter 7

DUALITY IN KINEMATIC APPROACHES OF LIMIT AND SHAKEDOWN ANALYSIS OF STRUCTURES

Nguyen-Dang Hung, Aimin Yan and Vu Duc Khoi

LTAS-Fracture Mechanics, University of Liege

1 Chemin des Chevreuils, Bat.52

4000 Liege, Belgium

h.nguyendang@ulg.ac.be, am.yan@ulg.ac.be, *vdkhoi@hotmail.com*

Abstract: The paper describes the duality in limit and shakedown analysis of structures. Starting from Koiter's kinematic theorem, the primal-dual optimization condition is established by adopting static terms as complementary variables. A highly effective algorithm is developed with the use of displacement-based finite elements. Newton-like iterations are performed to give both upper and lower bounds of the load factor at every iteration, which converge rapidly to the accurate solution of shakedown limit.

Keywords: Duality, shakedown analysis, limit analysis, nonlinear optimization

1. Introduction

Among the methods of plasticity employed in structural design, limit analysis and shakedown analysis are two useful tools because they provide realistic assessment of safety margin with respect to plastic failure. Limit analysis is used when loads are applied in a monotonic and proportional way, while shakedown analysis should be performed when loads vary arbitrarily or cyclically. It is clear that the former may be included in the latter as a special case. Traditionally, the shakedown load calculation was formulated using either static or kinematic theorems [7]. Based on Melan's static theorem one looks for an optimal residual stress field to find a maximum loading factor, while by Koiter's kinematic theorem one minimizes the loading factors by optimizing the kinematical mechanisms [6]. The two inadaptation load limits (the incremental plasticity or ratchetting limit and the alternating plasticity or low-cycle fatigue limit) may be calculated separately. The calculation of the former is most important in practice and it may be transformed into an equivalent limit analysis [19-20]. Alternatively, a unique shakedown solution may be determined as the minimum of the two inadaptation limits [2, 11, 21].

The fundamental theorems of limit and shakedown analysis were established long time ago. Some practical calculating methods were developed, see for example [2, 7, 12, 16-21]. However in most existing algorithms, the duality between the static lower bound and the kinematic upper bound of the load factor was not effectively used in numerical implementation. The optimization variables are either purely static or purely kinematic. This impeded further improvement of the calculating efficiency. From both physical and mathematical point of views, the duality is a very attractive property of shakedown theory connecting the upper and lower bound theorems. It has been showed in Maier [8] that by using mathematical programming method the precise duality of the static and kinematic approaches holds. In [4-5], a convex-analysis approach is used to establish a dual kinematic safety factor based on Melan's condition. The practical application of the duality was explored in limit analysis by Zouain *et al.* [22, 10]. Specially, an efficient primal-dual interior-point algorithm for minimizing a sum of Euclidean norms has been developed by Andersen *et al.* [1]. They showed that the application of the duality indeed leads to very accurate results in limit analysis with high efficiency.

The present work studies the duality in shakedown analysis of structures in discretized kinematic formulation. On this basis, an efficient algorithm of shakedown limits is proposed and implemented with displacement-based finite elements [14-15].

2. Shakedown analysis as a non-linear programming

2.1 Kinematic formulation

Let consider a structure made of elastic-perfectly plastic material and subjected to n time-dependent loads $\overline{P}_k^0(t)$ which vary independently within a given range:

$$\overline{P}_k^0(t) \in I_k^0 = \left[\overline{P}_k^-, \overline{P}_k^+\right] = \left[\mu_k^-, \mu_k^+\right] P_k^0 \quad k = \overline{1,n} \tag{2.1}$$

These loads form a convex polyhedral domain D of n dimensions with 2^n vertices in load space. The load domain can be represented in the following linear form [7, 19]:

$$P(t) = \sum_{k=1}^{n} \mu_k(t) P_k^0 \tag{2.2}$$

$$\mu_k^- \le \mu_k(t) \le \mu_k^+ \qquad k = \overline{1,n} \tag{2.3}$$

where P_k^0 is k-th nominal load and μ_k^-, μ_k^+ are its varying bounds. In many cases, it is useful to describe the load domain in stress space. To this end, we use here the notion of a fictitious elastic response of the structure under the same loading. The fictitious elastic stress $\sigma_{ij}^E(\mathbf{x}, t)$ is written in a form similar to (2.2):

$$\sigma_{ij}^E(\mathbf{x}, t) = \sum_{k=1}^{n} \mu_k(t) \sigma_{ij}^{Ek}\left(\mathbf{x}, P_k^0\right) \tag{2.4}$$

where $\sigma_{ij}^{Ek}\left(\mathbf{x}, P_k^0\right)$ denotes the stress field of the structure when subjected to k-th nominal load P_k^0.

Based on the kinematic theorem of Koiter [6], the shakedown load multiplier may be formulated in the following form:

$$\alpha^+ = \min \sum_{k \in I_D} \int_V D^p(\dot{\varepsilon}_{ij}^k) dV \tag{a}$$

$$s.t.\begin{cases}\Delta\varepsilon_{ij}=\sum_{k\in I_D}\dot{\varepsilon}_{ij}^{k} & & (b)\\ \Delta\varepsilon_{ij}=\frac{1}{2}\left(\frac{\partial\Delta u_i}{\partial x_j}+\frac{\partial\Delta u_j}{\partial x_i}\right) & \text{in } V & (c)\\ \Delta u_i=0 & \text{on } A_u & (d)\\ \sum_{k\in I_D}\int_V\sigma_{ij}^{E}\left(\mathbf{x},P_k^0\right)\dot{\varepsilon}_{ij}^{k}dV=1 & & (e)\end{cases} \tag{2.5}$$

where α^+ denotes the shakedown load multiplier; u is the displacement at point $\mathbf{x}$ of volume V; $\dot{\varepsilon}_{ij}^{k}$ is the corresponding strain rate at load vertex k; $\Delta\varepsilon_{ij}$ is the total plastic strain increment after a loading cycle; I_D is the set of all load vertices and $D^p\left(\dot{\varepsilon}_{ij}^{k}\right)$ denotes the plastic dissipation rate. By using von Mises yield criterion we have:

$$D^p=\sigma_p\left(\frac{3}{2}\dot{e}_{ij}^{p}\dot{e}_{ij}^{p}\right)^{1/2} \tag{2.6}$$

where $\dot{e}_{ij}^{p}$ denotes the deviatoric part of $\dot{\varepsilon}_{ij}^{p}$, and σ_p is the yield stress of material.

The discretized form of (2.5) by means of the finite element method can be expressed as follows:

$$\alpha^+=\min\sum_{k=1}^{m}\sum_{i=1}^{NG}\sqrt{2}w_ik_v\sqrt{\dot{\boldsymbol{\varepsilon}}_{ik}^{T}\mathbf{D}\dot{\boldsymbol{\varepsilon}}_{ik}+\varepsilon_0^2}$$

$$s.t:\begin{cases}\sum_{k=1}^{m}\dot{\boldsymbol{\varepsilon}}_{ik}=\mathbf{B}_i\mathbf{q} & \forall i=\overline{1,NG}\\ \mathbf{D}_v\dot{\boldsymbol{\varepsilon}}_{ik}=\mathbf{0} & \forall i=\overline{1,NG}\\ \sum_{k=1}^{m}\sum_{i=1}^{NG}w_i\dot{\boldsymbol{\varepsilon}}_{ik}^{T}\boldsymbol{\sigma}_{ik}^{E}=1\end{cases} \tag{2.7}$$

where $\dot{\boldsymbol{\varepsilon}}_{ik}$ and $\boldsymbol{\sigma}_{ik}^{E}$ denote the vector of deformation rate and the vector of the fictitious elastic stress at Gauss point i and load vertex k; $\mathbf{q}$ is the nodal displacement vector; $\mathbf{B}_i$ is the strain matrix; $m=2^n$; NG denotes the total number of Gauss points of the whole structure with integration weight w_i at Gauss point i; ε_0 is a small value for avoiding the singularity [1].

2.2 Static formulation

Based on the static theorem of Melan, the shakedown load multiplier can be written in a lower bound formation:

$$\alpha^{-} = \max \alpha \qquad (a)$$

$$\text{s.t.} \begin{cases} \partial_j \bar{\rho}_{ij} = 0 & \text{in } V \qquad (b) \\ \bar{n}_j \bar{\rho}_{ij} = 0 & \text{on } A_\sigma \qquad (c) \\ f\left(\alpha \sigma_{ij}^{E}\left(\mathbf{x}, \hat{P}_k\right) + \bar{\rho}_{ij}\right) \leq 0 & \forall k = \overline{1, m} \qquad (d) \end{cases}$$

(2.8)

To have a strict lower bound, one should use statically admissible (SA) elements to discretize (2.8). However, as we know, the SA elements are not widely available. By using kinematically admissible elements instead, the shakedown load multiplier may be discretized as follows:

$$\alpha^{-} = \max \alpha \qquad (a)$$

$$s.t: \begin{cases} \sum_{i=1}^{NG} w_i \mathbf{B}_i^T \bar{\boldsymbol{\rho}}_i = \mathbf{B}^T \bar{\boldsymbol{\rho}} = \mathbf{0} & (b) \\ f\left(\alpha \boldsymbol{\sigma}_{ik}^{E} + \bar{\boldsymbol{\rho}}_i\right) \leq 0 \qquad \forall k = 1 \ldots m \quad \forall i = 1 \ldots NG & (c) \end{cases} \qquad (2.9)$$

The fictitious elastic stress $\boldsymbol{\sigma}_{ik}^{E}$ in this formulation is a discretized field with its values computed at Gauss points like the residual stress field $\bar{\boldsymbol{\rho}}_i$. Note that the non-linear mathematical optimization problem (2.9) has $(Nsc \times NG + 1)$ variables: the global residual stress $\bar{\boldsymbol{\rho}}$ and load multiplier α. Here Nsc is the number of stress components.

The formulation (2.9) is an approximation because we can only have a quasi-equilibrium state as well as yield criterion satisfied at discrete Gauss points of the structure. In strict meaning, the load multiplier obtained by (2.9) loses its bounding characteristic: with a coarse FM meshing and finite iterations, it is no more a true lower bound of the shakedown limit load multiplier.

If the von Mises yield criterion is used, the requirement (2.9c) represents a system of NG nonlinear inequality constraints. These nonlinear constraints are the major obstacle in implementing the von Mises yield criterion in shakedown analysis. In order to overcome this difficulty, the Tresca or linearized von Mises yield criterion may be used. Trying to avoid the use of linearization, Stein *et al.* [12] proposed a sequential quadratic procedure with a base-reduction technique. This technique permits us to consider only some small number of residual stress vectors as active variables at a time step. Based on this set of active variables, the global residual stress vector is improved, a new set of active variables is sought for and the procedure continues until a solution is found.

Experiences showed that the difficulties resulting from the non-linear constraints of (2.9) may also be avoided if we do not directly maximize the load multiplier but consider it as a complementary problem of the kinematic formulation (2.7). In order to do so, in the following section the duality between formulations (2.7) and (2.9) will be discussed in details.

3. The duality between kinematic and static formulations

By restricting ourselves to a polyhedral form of load domain, we show in this section that the static lower bound based on Melan theorem (2.9) may be viewed as the dual form of the kinematic upper bound (2.7). For the sake of simplicity, let us rewrite the upper bound limit (2.7) in a simpler form by setting:

- The new strain rate vector $\mathbf{e}_{ik}$ (the dot mark denoting time derivative is omitted for simplicity):

$$\mathbf{e}_{ik} = w_i \mathbf{D}^{1/2} \dot{\boldsymbol{\varepsilon}}_{ik} \tag{3.1}$$

- The new fictitious elastic stress field $\mathbf{t}_{ik}$:

$$\mathbf{t}_{ik} = \mathbf{D}^{-1/2} \boldsymbol{\sigma}_{ik}^{E} \tag{3.2}$$

- The new deformation matrix $\hat{\mathbf{B}}_i$:

$$\hat{\mathbf{B}}_i = w_i \mathbf{D}^{1/2} \mathbf{B}_i \tag{3.3}$$

In the above definition $\mathbf{D}^{1/2}$ and $\mathbf{D}^{-1/2}$ are a symmetric matrices such that:

$$\begin{aligned} \mathbf{D}^{-1/2} &= \left(\mathbf{D}^{1/2}\right)^{-1} \\ \mathbf{D} &= \mathbf{D}^{1/2}\mathbf{D}^{1/2} \end{aligned} \tag{3.4}$$

With these definitions, the objective function of (2.7) becomes:

$$\sum_{k=1}^{m} \sum_{i=1}^{NG} \sqrt{2} w_i k_v \sqrt{\dot{\boldsymbol{\varepsilon}}_{ik}^T \mathbf{D} \dot{\boldsymbol{\varepsilon}}_{ik} + \varepsilon_0^2} = \sqrt{2} k_v \sum_{k=1}^{m} \sum_{i=1}^{NG} \sqrt{\mathbf{e}_{ik}^T \mathbf{e}_{ik} + \varepsilon^2} \tag{3.5}$$

In the formulation (3.5), ε^2 is a small positive number suitably chosen to avoid the singularity of the objective function. By substituting (3.1-3.5) into (2.7) one obtains a simplified version for upper bound of shakedown limit load multiplier:

$$\alpha^+ = \min \sum_{i=1}^{NG}\sum_{k=1}^{m} \sqrt{2}k_v \sqrt{\mathbf{e}_{ik}^T\mathbf{e}_{ik} + \varepsilon^2} \qquad (a)$$

$$s.t: \begin{cases} \sum_{k=1}^{m} \mathbf{e}_{ik} - \hat{\mathbf{B}}_i\mathbf{q} = \mathbf{0} & \forall i = \overline{1, NG} \qquad (b) \\ \frac{1}{3}\mathbf{D}_v\mathbf{e}_{ik} = \mathbf{0} & \forall i = \overline{1, NG},\ \forall k = \overline{1,\ m} \qquad (c) \\ \sum_{i=1}^{NG}\sum_{k=1}^{m} \mathbf{e}_{ik}^T\mathbf{t}_{ik} - 1 = \mathbf{0} & \qquad (d) \end{cases} \qquad (3.6)$$

where factor (1/3) is added in (3.6c) for a technical reason. As it is well-known, limit analysis may be considered as a special case of shakedown analysis. Andersen *et al.* [1], while considering a problem of minimizing a sum of Euclidean norms, found that in case of limit analysis there exists a dual form for (3.6). A generalization of this dual form is proposed hereafter through the following propositions [14-15]:

Proposition 1:

If there exists a finite solution α^+ for the kinematic shakedown load multiplier (3.6) with $\varepsilon^2 = 0$ then α^+ has its dual form as:

$$\alpha^- = \max_{\boldsymbol{\gamma}_{ik}, \boldsymbol{\beta}_i, \alpha} \alpha \qquad (a)$$

$$s.t: \begin{cases} \left\| \boldsymbol{\gamma}_{ik} + \boldsymbol{\beta}_i + \mathbf{t}_{ik}\alpha \right\| \le \sqrt{2}k_v & (b) \\ \sum_{i=1}^{NG} \hat{\mathbf{B}}_i^T \boldsymbol{\beta}_{ik} = 0 & (c) \end{cases} \qquad (3.7)$$

where: $\|.\|$ *denotes Euclidean vector norm.*

Proof:

By setting $\varepsilon^2 = 0$, let us write the Lagrange dual function of (3.6) as:

$$F_L = \sum_{i=1}^{NG}\left\{\sum_{k=1}^{m}\sqrt{2}k_v\left(\sqrt{\mathbf{e}_{ik}^T\mathbf{e}_{ik}}\right) - \sum_{k=1}^{m}\frac{1}{3}\left(\boldsymbol{\gamma}_{ik}^T\mathbf{D}_v\mathbf{e}_{ik}\right) - \boldsymbol{\beta}_i^T\left(\sum_{k=1}^{m}\mathbf{e}_{ik} - \hat{\mathbf{B}}_i\mathbf{q}\right)\right\}$$
$$-\alpha\left(\sum_{i=1}^{NG}\sum_{k=1}^{m}\mathbf{e}_{ik}^T\mathbf{t}_{ik} - 1\right) \quad (3.8)$$

where $\boldsymbol{\gamma}_{ik}, \boldsymbol{\beta}_i, \alpha$ are Lagrange multipliers. Note that $\boldsymbol{\gamma}_{ik}, \boldsymbol{\beta}_i$ are vectors at Gauss point i for each load vertex k while α is merely a scalar. The dual problem of (3.6) is now:

$$\max_{\gamma_{ik}, \boldsymbol{\beta}_i, \alpha}\left(\min_{\mathbf{e}_{ik}, \mathbf{q}} F_L\right) \quad (3.9)$$

Because a finite solution of (3.6) exists, the constraint system (3.6b-d) is affine and the objective function is convex, then the duality theorem states that there exists no dual gap between the primal and dual solutions:

$$\min_{\mathbf{h}(\mathbf{e}_{ik}, \mathbf{q})=\mathbf{0}} \sum_{i=1}^{NG}\sum_{k=1}^{m}\sqrt{2}k_v\sqrt{\mathbf{e}_{ik}^T\mathbf{e}_{ik}} = \max_{\gamma_{ik}, \boldsymbol{\beta}_i, \alpha}\left(\min_{\mathbf{e}_{ik}, \mathbf{q}} F_L\right) \quad (3.10)$$

where $\mathbf{h}(\mathbf{e}_{ik}, \mathbf{q}) = \mathbf{0}$ stands for linear constraint system (3.6b-d).

The Lagrange dual function (3.8) may be written in another form:

$$F_L = \sum_{i=1}^{NG}\sum_{k=1}^{m}\left(\frac{\sqrt{2}k_v\mathbf{e}_{ik}}{\sqrt{\mathbf{e}_{ik}^T\mathbf{e}_{ik}}} - \boldsymbol{\gamma}_{ik} - \boldsymbol{\beta}_i - \mathbf{t}_{ik}\alpha\right)^T\mathbf{e}_{ik} + \sum_{i=1}^{NG}\boldsymbol{\beta}_i^T\hat{\mathbf{B}}_i\mathbf{q} + \alpha \quad (3.11)$$

By writing (3.8) in the form (3.11) we adopt here the convention that if the vector norm of strain rate $\|\mathbf{e}_{ik}\|$ is equal to zero then: $\dfrac{\sqrt{2}k_v\mathbf{e}_{ik}^T}{\sqrt{\mathbf{e}_{ik}^T\mathbf{e}_{ik}}}\mathbf{e}_{ik} = 0$.

Due to the existence of a dual solution α^- having no gap to the primal α^+, it is required that for any solution set of Lagrange multipliers $(\gamma_{ik}, \boldsymbol{\beta}_i, \alpha)$ the function $\left(\min_{\mathbf{e}_{ik}, \mathbf{q}} F_L\right)$ must have a finite value. To this end, the following system must be satisfied:

$$\begin{cases} \left(\dfrac{\sqrt{2}k_v\mathbf{e}_{ik}}{\sqrt{\mathbf{e}_{ik}^T\mathbf{e}_{ik}}} - \boldsymbol{\gamma}_{ik} - \boldsymbol{\beta}_i - \mathbf{t}_{ik}\alpha\right)^T\mathbf{e}_{ik} \geq 0 \quad \forall\mathbf{e}_{ik} & (a) \\ \displaystyle\sum_{i=1}^{NG}\boldsymbol{\beta}_i^T\hat{\mathbf{B}}_i\mathbf{q} = 0 \quad \forall\mathbf{q} & (b) \end{cases} \quad (3.12)$$

otherwise we may show that:

$$\min_{\mathbf{e}_{ik},\mathbf{q}} F_L \to -\infty \tag{3.13}$$

According to (3.13), the function of $\left(\min_{\mathbf{e}_{ik},\mathbf{q}} F_L\right)$ is bounded from below:

$$\min_{\mathbf{e}_{ik},\mathbf{q}} F_L \geq \alpha \tag{3.14}$$

It reaches α when, for example, all strain rates and displacement are equal to zero. This fact leads to the conclusion:

$$\min_{\mathbf{e}_{ik},\mathbf{q}} F_L = \alpha \tag{3.15}$$

The condition (3.12b) is equivalent to:

$$\sum_{i=1}^{NG} \mathbf{B}_i^T \boldsymbol{\beta}_i = \mathbf{0} \tag{3.16}$$

Further more, we may point out that the condition (3.12a) is equivalent to restriction on only multipliers $\boldsymbol{\gamma}_{ik}, \boldsymbol{\beta}_i, \alpha$:

$$\left\| \boldsymbol{\gamma}_{ik} + \boldsymbol{\beta}_i + \mathbf{t}_{ik}\alpha \right\| \leq \sqrt{2}k_v \qquad \forall i,k \tag{3.17}$$

Equalities (3.15), (3.16) and inequality (3.17) conclude our proof.

Despite the fact that the dual form of (3.6) exists, the formulation (3.7) does not appear as the one resulting from discretization of Melan's condition. In order to have von Mises condition in the dual form, the Lagrange multiplier $\boldsymbol{\gamma}_{ik}$ in the above system should be eliminated and we have:

$$\begin{aligned} &\alpha^- = \max_{\boldsymbol{\beta}_i,\alpha} \alpha && (a) \\ &s.t: \begin{cases} f(\mathbf{D}^{1/2}\boldsymbol{\beta}_i + \alpha\boldsymbol{\sigma}_{ij}^E) \leq 0 & (b) \\ \sum_{i=1}^{NG} w_i \mathbf{B}_i^T \mathbf{D}^{1/2} \boldsymbol{\beta}_i = \mathbf{0} & (c) \end{cases} \end{aligned} \tag{3.18}$$

In this case, (3.18) is a discretized form of the shakedown load multiplier resulting from Melan's theorem. In the formulation (3.18) the vector $\mathbf{D}^{1/2}\boldsymbol{\beta}_i$ can be interpreted as the stress vector of a time-independent residual stress field: the value of this vector is calculated at each Gauss point, independently of load vertices or, in other word, independently of time. Of course Lagrange multiplier α represents the shakedown load factor. To arrive above duality of (3.18), we propose the following theorem:

Proposition 2:

If there exists a finite solution α^+ *for the kinematic shakedown load multiplier (3.6) with* $\varepsilon^2 = 0$*, and if the incompressibility is automatically satisfied, then the kinematic formulation has its dual form as the static one resulted from Melan's theorem:*

$$\min_{\mathbf{h}(\mathbf{e}_{ik},\mathbf{q})=\mathbf{0}} \sum_{i=1}^{NG}\sum_{k=1}^{m} \sqrt{2}k_v \sqrt{\mathbf{e}_{ik}^T \mathbf{e}_{ik}} = \max_{\begin{cases}\mathbf{B}^T\overline{\boldsymbol{\rho}}=\mathbf{0}\\ f\left(\alpha\overline{\boldsymbol{\sigma}}_{ik}^E+\overline{\boldsymbol{\rho}}_{ik}\right)\le 0\end{cases}} \alpha \tag{3.19}$$

where f is the von Mises yield function.

The two propositions show implicitly that that if kinematically admissible elements and von Mises yield criterion are used, the limit load multipliers formulated by static and kinematic theorems are actually the approximation of the true upper bound.

4. Primal-dual algorithm for shakedown analysis

Using the Lagrange multipliers described in §3, the stationary conditions of (3.6) can be written as follows:

$$\begin{aligned}
&\frac{\sqrt{2}k_v \mathbf{e}_{ik}}{\sqrt{\mathbf{e}_{ik}^T \mathbf{e}_{ik}}} - \left(\boldsymbol{\gamma}_{ik} + \boldsymbol{\beta}_i + \alpha \mathbf{t}_{ik}\right) &= \mathbf{0} \quad &(a)\\
&\mathbf{D}_v \mathbf{e}_{ik} &= \mathbf{0} \quad &(b)\\
&\sum_{k=1}^{m} \mathbf{e}_{ik} - \hat{\mathbf{B}}_i \mathbf{q} &= \mathbf{0} \quad &(c)\\
&\sum_{i=1}^{NG} \left(\hat{\mathbf{B}}_i^T \boldsymbol{\beta}_i\right) &= \mathbf{0} \quad &(d)\\
&\sum_{i=1}^{NG}\sum_{k=1}^{m} \mathbf{e}_{ik}^T \mathbf{t}_{ik} - 1 &= \mathbf{0} \quad &(e)
\end{aligned} \tag{4.1}$$

Theoretically the shakedown load factor may be found by solving the system (4.1). Unfortunately, solving directly this system is not a good idea because it leads to a system of equations much bigger than that in the case of purely elastic computation. The resulted system thus requires large amount of computer memory as well as computational effort to solve. Trying to keep our problem size as small as possible (as an elastic analysis), we use here the penalty method to handle the equality conditions (4.1b-c) with Lagrange multipliers playing intermediate roles. A similar technique, which showed great efficiency in large-scale problems, was successfully applied in limit analysis by Andersen *et al.* [1].

Numerically speaking, the stationary conditions (4.1) are very difficult to satisfy due to their singular property. Specially we lack an appropriate criterion to stop optimization process or to assess the exactness of the solution. However, if a strictly lower bound is found at the same time with a strictly upper bound, we possess a very useful tool to control the obtained results. Although those strict bounds are hard to find, in the following proposed algorithm we will try to build some of their approximations while employing Newton method and making use of (3.7) to solve (4.1).

Primal-dual algorithm **[15]**

1. Initialize displacement and strain rate vectors $\mathbf{q}^0$ and $\mathbf{e}^0$ such that the normalized condition is satisfied:

$$\sum_{i=1}^{NG}\sum_{k=1}^{m}\mathbf{t}_{ik}^T\mathbf{e}_{ik}^0 = 1 \tag{4.2}$$

Set all stress vectors to nulls:

$$\boldsymbol{\gamma}_{ik}^0 = \mathbf{0} \quad \boldsymbol{\beta}_i^0 = \mathbf{0} \quad \forall i = \overline{1, NG}, \quad k = \overline{1, m} \tag{4.3}$$

Set up values for penalty parameter c and for ε. Set up convergence criteria.

2. Calculate incremental vectors $d\mathbf{q}$, $d\mathbf{e}_{ik}$ of displacement, deformations and $(d\alpha + \alpha)$ at the current values of $\mathbf{q},\mathbf{e}$ by solving the following system:

$$\begin{cases} d\mathbf{e}_{ik} = \sqrt{\mathbf{e}_{ik}^T\mathbf{e}_{ik} + \varepsilon^2}\,\hat{\mathbf{M}}_{ik}^{-1}\left(-\mathbf{g}_{ik} + d\boldsymbol{\beta}_i + \mathbf{t}_{ik}d\alpha\right) - \hat{\mathbf{M}}_{ik}^{-1}\mathbf{f}_{ik} \\ \sum_{i=1}^{NG}\sum_{k=1}^{m}\mathbf{t}_{ik}^T\left(\mathbf{e}_{ik} + d\mathbf{e}_{ik}\right) = 1 \\ d\mathbf{q} = -\mathbf{q} + \hat{\mathbf{S}}^{-1}\hat{\mathbf{f}}_1 + (\alpha + d\alpha)\hat{\mathbf{S}}^{-1}\hat{\mathbf{f}}_2 \end{cases} \tag{4.4}$$

where:

$$\hat{\mathbf{S}} = \sum_{i=1}^{NG} \hat{\mathbf{B}}_i^T \hat{\mathbf{K}}_i^{-1} \hat{\mathbf{B}}_i$$

$$\hat{\mathbf{f}}_1 = \sum_{i=1}^{NG} \hat{\mathbf{B}}_i^T \hat{\mathbf{K}}_i^{-1} \left(\sum_{k=1}^{m} \mathbf{e}_{ik} - \sum_{k=1}^{m} \hat{\mathbf{M}}_{ik}^{-1} \mathbf{e}_{ik} - c \sum_{k=1}^{m} \sqrt{\mathbf{e}_{ik}^T \mathbf{e}_{ik} + \varepsilon^2}\, \hat{\mathbf{M}}_{ik}^{-1} \mathbf{D}_v \mathbf{e}_{ik} \right)$$

$$\hat{\mathbf{f}}_2 = \sum_{i=1}^{NG} \hat{\mathbf{B}}_i^T \hat{\mathbf{K}}_i^{-1} \sum_{k=1}^{m} \sqrt{\mathbf{e}_{ik}^T \mathbf{e}_{ik} + \varepsilon^2}\, \hat{\mathbf{M}}_{ik}^{-1} \mathbf{t}_{ik}$$

$$\mathbf{f}_{ik} = \mathbf{e}_{ik} - \sqrt{\mathbf{e}_{ik}^T \mathbf{e}_{ik} + \varepsilon^2} \left(\boldsymbol{\gamma}_{ik} + \boldsymbol{\beta}_i + \alpha \mathbf{t}_{ik} \right)$$

$$\mathbf{g}_{ik} = \boldsymbol{\gamma}_{ik} + c \mathbf{D}_v \mathbf{e}_{ik}$$

$$\mathbf{h}_i = \boldsymbol{\beta}_i + c \left(\sum_{k=1}^{m} \mathbf{e}_{ik} - \hat{\mathbf{B}}_i \mathbf{q} \right)$$

$$\hat{\mathbf{K}}_i = \left[\mathbf{I} + c \sum_{k=1}^{m} \sqrt{\mathbf{e}_{ik}^T \mathbf{e}_{ik} + \varepsilon^2}\, \hat{\mathbf{M}}_{ik}^{-1} \right]$$

$$\hat{\mathbf{M}}_{ik} = \frac{1}{2} \left(\mathbf{M}_{ik} + \mathbf{M}_{ik}^T \right) + \sqrt{\mathbf{e}_{ik}^T \mathbf{e}_{ik} + \varepsilon^2}\, c \mathbf{D}_v$$

$$\mathbf{M}_{ik} = \left[\mathbf{I} - \frac{\left(\boldsymbol{\gamma}_{ik} + \boldsymbol{\beta}_i + \alpha \mathbf{t}_{ik} \right) \mathbf{e}_{ik}^T}{\sqrt{\mathbf{e}_{ik}^T \mathbf{e}_{ik} + \varepsilon^2}} \right] \tag{4.5}$$

3. Perform a line-search to find $\hat{\lambda}_q$ such that:

$$\hat{\lambda}_q = \min F_P \left(\mathbf{q} + \lambda d\mathbf{q}, \mathbf{e} + \lambda d\mathbf{e} \right) \tag{4.6}$$

where F_P is the penalty function:

$$F_P = \sum_{i=1}^{NG} \left\{ \sqrt{2} k_v \sum_{k=1}^{m} \sqrt{\mathbf{e}_{ik}^T \mathbf{e}_{ik} + \varepsilon^2} + \frac{c}{2} \sum_{k=1}^{m} \mathbf{e}_{ik}^T \mathbf{D}_v \mathbf{e}_{ik} \right.$$
$$\left. + \frac{c}{2} \left(\sum_{k=1}^{m} \mathbf{e}_{ik} - \mathbf{B}_i \mathbf{q} \right)^T \left(\sum_{k=1}^{m} \mathbf{e}_{ik} - \mathbf{B}_i \mathbf{q} \right) \right\} \tag{4.7}$$

4. Update the displacement and strain rate vectors as:

$$\mathbf{q} = \mathbf{q} + \hat{\lambda}_q d\mathbf{q}, \quad \mathbf{e}_{ik} = \mathbf{e}_{ik} + \hat{\lambda}_q d\mathbf{e}_{ik} \tag{4.8}$$

5. Calculate the incremental vectors of stress vectors $\boldsymbol{\gamma}_{ik}, \boldsymbol{\beta}_i$:

$$d\boldsymbol{\gamma}_{ik} = -c\mathbf{D}_v d\mathbf{e}_{ik} - \mathbf{g}_{ik}, \quad d\boldsymbol{\beta}_i = -c\left(\sum_{k=1}^{m} d\mathbf{e}_{ik} - \hat{\mathbf{B}}_i d\mathbf{q}\right) - \mathbf{h}_i \quad (4.9)$$

Perform a line-search to find $\hat{\lambda}_s$ such that:

$$\begin{aligned} &\hat{\lambda}_s = \max \lambda \\ s.t: \quad &\left\| \left(\boldsymbol{\gamma}_{ik} + \boldsymbol{\beta}_i + \mathbf{t}_{ik}\alpha\right) + \lambda\left(d\boldsymbol{\gamma}_{ik} + d\boldsymbol{\beta}_i + \mathbf{t}_{ik} d\alpha\right) \right\| \leq 1 \end{aligned} \quad (4.10)$$

Update stress vectors $\boldsymbol{\gamma}_{ik}, \boldsymbol{\beta}_i$ and shakedown limit α (with a chosen parameter τ: $0 < \tau \leq 1$):

$$\boldsymbol{\gamma}_{ik} = \boldsymbol{\gamma}_{ik} + \tau\hat{\lambda}_s d\boldsymbol{\gamma}_{ik}, \; \boldsymbol{\beta}_i = \boldsymbol{\beta}_i + \tau\hat{\lambda}_s d\boldsymbol{\beta}_i, \; \alpha = \alpha + \tau\hat{\lambda}_s d\alpha \quad (4.11)$$

6. Check the convergence criteria: if they are all satisfied then stop, otherwise repeat steps 2-5.

Regardless of some possible numerical obstacles, we can show that the algorithm converges to a solution set $(\overline{\mathbf{q}}, \overline{\mathbf{e}}, \overline{\alpha})$. The convergence property of this primal-dual algorithm is proved by the following theorem (the proof of this theorem can be found [14]):

Theorem 1: *If a feasible set* $(\mathbf{q}^0, \mathbf{e}^0)$ *can be found such that:* $\sum_{i=1}^{NG}\sum_{k=1}^{m} \mathbf{e}_{ik}^{0T}\mathbf{t}_{ik} \neq 0$, *the primal-dual algorithm leads to solution of (2.2).*

Because limit analysis may be considered as a special case of shake down when the structure is loaded with only one monotonic load, i.e. $[\mu_0, \mu_0]P_0$, the above algorithm is expected to give accurate solutions in limit analysis. Numerical examples hereafter shows that such requirement is fairly satisfied.

5. Numerical examples

5.1 Square plate with a central hole

Let consider first a simple application where we can compare numerical results with the existing analytic ones to verify the efficiency of the developed algorithm. A square plate with a central hole is subjected to two loads p_1 and p_2 varying independently ($p_1 \in [0,\ p_0^1]$, $p_2 \in [0,\ p_0^2]$), see Fig. 1.

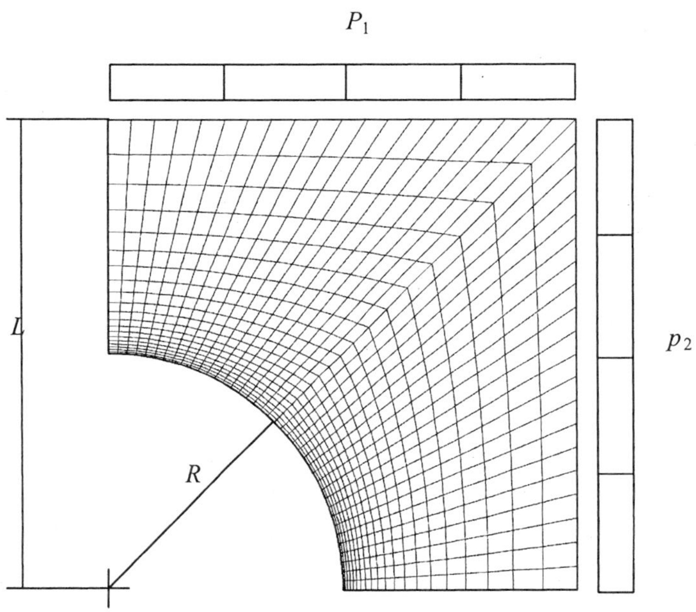

Fig. 1 Square plate with a central hole (FE mesh)

The analytical lower and upper bounds of limit load (using Von Mises yield criterion) can be found in the work of Gaydon and McCrum [3]. Numerical investigations are carried to estimate the plastic collapse limit and the shakedown limit for different ratios D/L. Due to the symmetry, one fourth of the plate is modelled by 800 plane quadrilateral 8-node elements. As an example, we present in Fig 2 the convergence of solutions in the case of $R/L = 0.2$ and $p_1 \neq 0$, $p_2 = 0$. In this special geometry, the exact collapse limit load is $p_{\lim} = 0.8\sigma_o$ (σ_o is the yield stress). Our corresponding numerical values obtained are $0.79924\,\sigma_o$ as lower bound and $0.80038\,\sigma_o$ as upper bound after 10 iterations. The shakedown load (alternating plasticity limit) can be estimated as $0.59947\,\sigma_o$ by an elastic analysis. The present dual algorithm gives $0.59947\sigma_o$ as lower bound and $0.59949\sigma_o$ as upper bound (Fig. 2). In Fig. 3, we present a comparison of limit analysis in the case of two proportional loads with different geometry ration of *R/L*. We note the solutions of Gayon and McCrum are exact in the range 0.483 <*R/L*≤ 1 where analytical lower bound coincides with upper bound. Indeed, the numerical comparisons with this example show excellent precision and efficiency of the developed algorithm.

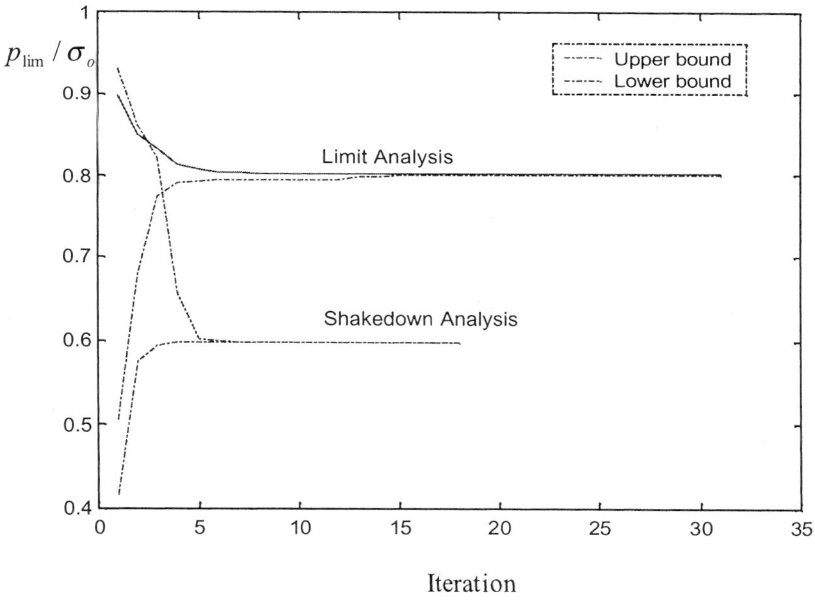

Fig. 2 Limit and shakedown bounds converge rapidly to the exact solution ($R/L = 0.2, p_0^1 \neq 0, p_0^2 = 0$)

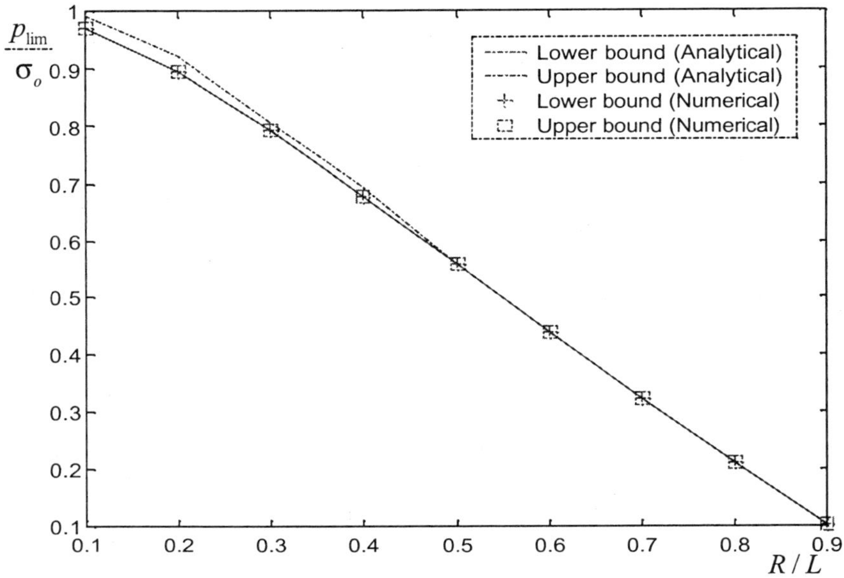

Fig. 3 Limit analysis ($p_1 = p_2$)

5.2 Pipe-junction subjected to internal pressure

The problem was examined by Staat and Heitzer [11]. In their studies, 125 solid 27-node hexahedron elements were used to model one fourth of the structure because of its symmetric property. In our analysis, the FE mesh is presented in Fig 4: it contains 720 solid 20-node hexahedron elements with 12479 DOF. The dimensions of the pipe-junction are chosen as described in [11]. The structure is subjected to internal pressure p varying within $[0, p_0]$, the ends of the junction are closed (so the axial tension exists). Numerical integration is realized with 2×2×2 Gauss points.

Numerical analysis presented in Fig 5 and Table 1 leads to the plastic collapse pressure of $0.14434\sigma_0$ (upper bound) and $0.14424\sigma_0$ (lower bound) compared with lower bound of $0.134\sigma_0$ obtained by Staat and Heitzer. Shakedown analysis gives $0.11044\sigma_0$ as upper bound and $0.10983\sigma_0$ as lower bound. The alternating limit calculated by using elastic analysis is $0.10983\sigma_0$. Staat and Heitzer [11] obtained a lower bound of $0.0952\sigma_0$.

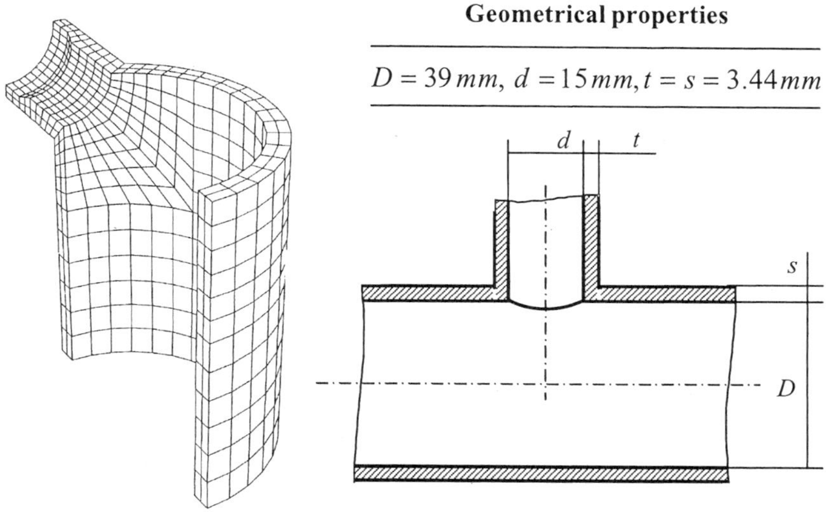

Fig. 4 FE mesh and geometrical properties

Table 1. Limit analysis of pipe-junction

Iteration	Upper bound	Lower bound
1	0.18317	0.89147
2	0.15565	0.11970
3	0.14715	0.13265
4	0.14524	0.13693
5	0.14469	0.13994
6	0.14449	0.14136
7	0.14440	0.14265
8	0.14436	0.14348
9	0.14434	0.14362
10	0.14434	0.14381

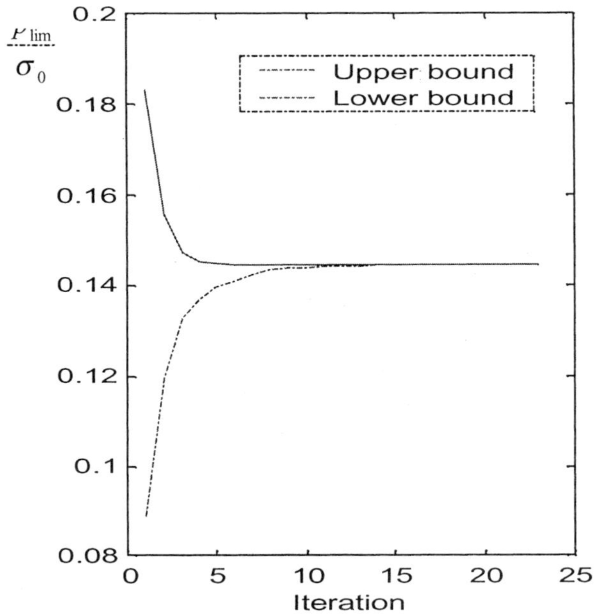

Fig. 5 Convergence of limit bounds of pipe-junction

Table 2. Shakedown analysis of pipe-junction

Iteration	Upper bound	Lower bound
1	0.51474	0.06288
2	0.24996	0.09484
3	0.17092	0.09817
4	0.11751	0.09916
5	0.10391	0.09937
6	0.10116	0.09939
7	0.10046	0.09939
8	0.10033	0.09939
9	0.10032	0.09939
10	0.10031	0.09939

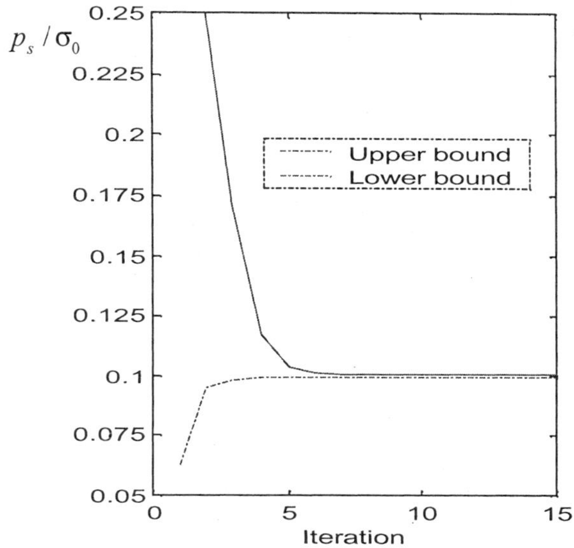

Fig. 6 Convergence of shakedown bounds of pipe-junction

Note that in this example the structure fails due to the alternating plasticity, thus the shakedown load factor could be evaluated more precisely if we use a set of 3×3×3 Gauss points in numerical integration. The shakedown limit load factor in this case is $0.10031\,\sigma_0$ (upper bound) and $0.09939\sigma_0$ (lower bound). The alternating limit calculated by using elastic stresses is $0.09939\,\sigma_0$. The result depicted in Fig 6 is obtained with 3×3×3 Gauss points. It is important to note that whatever the number of Gauss points, the two bounds converge rapidly to the alternating limit. Finally we note that an iteration in the present calculation is about equivalent to 1.1~1.2 elastic calculation. The limit and shakedown analysis by the presented method is much more efficient than an elastic-plastic finite element analysis of structures.

6. Conclusions

Duality between kinematic and static formulation of shakedown analysis is demonstrated in this paper. Starting from the kinematic theorem of Koiter, a new primal-dual algorithm for shakedown analysis of structures is developed. The penalty and Lagrange methods are used to handle the incompressibility, compatibility and normalization conditions.

Consequently, the method may be implemented with any displacement-based finite elements. The residual stresses and hydrostatic pressure are introduced as complementary variables conjugated respectively with the compatibility and incompressibility conditions. Newton iterative method is used to solve the resulted system of equations. The problem size is successfully reduced to the size of linear elastic analysis, thus there exists no size limit in practical application as in non-linear optimization. Line search is performed to improve the current kinematic solution, the plastic admissible condition is used to regulate the static variables. At every iteration, the upper bound is given by the obtained kinematic field and the lower bound is given by the obtained static field. The numerical examples show high calculating efficiency: both upper and lower bounds converge to the accurate solution after few iterations. Although no proof is found to guarantee that the lower bound increases after every iteration, it is an interesting feature observed in all of our examinations.

Acknowledgements

The work is supported by the European Community under the Brite-EuRam project LISA BE97-4547, Contract n° BRPR-CT97-0595

References

[1] Andersen K.D., Christiansen E., Conn A.R., Overton M.L. (2000): An efficient primal-dual interior-point method for minimizing a sum of Euclidean norms. *SIAM J. Sci. Comput.* **22**. 243-262.

[2] Carvelli V., Cen Z.Z., Liu Y., Maier G. (1999): Shakedown analysis of defective pressure vessels by a kinematic approach *Archive of Applied Mechanics* **69**. 751-764.

[3] Gaydon F.A., MacCrum A.W. (1954): A theoretical investigation of the yield point loading of a square plate with a circular hole. *Journal of the mechanics and physics of solids* **2**, 156-169.

[4] Kamenjarzh J., Merzljakov A. (1994): On kinematic method in shakedown theory: I. Duality of extremum problems. II. Modified kinematic method *Int. J. Plasticity* **10**. 363-380 and 381-392

[5] Kamenjarzh J., Weichert D. (1992): On kinematic upper bounds for the safety factor in shakedown theory. *Int. J. Plasticity* **8**. 827-837.

[6] Koiter W.T.(1960): General theorems for elastic plastic solids. In *Progress in Solid Mechanics,* ed. By Sneddon I.N.& Hill R. Nord-Holland, Amsterdam, 165-221.

[7] Konig J.A. (1987): *Shakedown of Elastic-plastic Structure.* PWN – Polish Scientific Publishers, Elsevier.

[8] Maier G. (1973): A shakedown matrix theory allowing for work hardening and second-order geometric effects. In: *Foundations of plasticity* , edited by Sawczuk A., North-Holland, Amsterdam. 417-433.

[9] Polizzotto C. (1993): On the condition to prevent plastic shakedown of structures: Part II – The plastic shakedown limit load. *J. Appl. Mech. ASME* **60.** 20-25.

[10] Silveira J.L., Zouain N. (1997). *On extremum principles and algorithms for shakedown analysis. Eur. J. Mech., A/Solids* **16** (N° 5). 757-778.

[11] Staat M., Heitzer M. (1997): Limit and shakedown analysis for plastic safety of complex structures. *Transactions of the 14th International Conference on Structural Mechanics in Reactor Technology (SMiRT 14),* Lon, France, August 17-22.

[12] Stein E., Zhang G. (1993): Theoretical and numerical shakedown analysis for kinematic hardening materials. In *Progress in computational analysis of inelastic structures.* Springer Verlage.

[13] Stein E., Zhang G., Huang Y. (1993): Modelling and computation of shakedown problems for nonlinear hardening materials. *Computer Methods in Applied Mechanics and Engineering* **103**. 247-272.

[14] Vu D.K., Yan A.M., Nguyen-Dang H. (2001). A new dual shakedown analysis of structures. *Proceedings of the ECCM-2001 European Conference on Computational Mechanics*, Cracow, Poland, June 26-29.

[15] Vu D.K., Yan A.M., Nguyen-Dang H. (2003): A dual form for discretized kinematic formulation in shakedown analysis. In print in *International journal of Solids and Structures.*

[16] Yan A.M., Nguyen-Dang H. (1999): Limit analysis of cracked structures by mathematical programming and finite element technique. *Computational Mechanics* **24** (N°5). 319-333.

[17] Yan A.M., Jospin R.J., Nguyen-Dang H(1999): An enhanced elbow element – application in limit state analysis. *Int. Journal of Num. Meth. Engng.* **46**. 409-431.

[18] Yan A.M., Nguyen-Dang H. (2000). Direct finite element kinematical approaches in limit and shakedown analysis of shells and elbows. *Inelastic Analysis of Structure under Variable Loads. Theory and Engineering Applications.* D. Weichert and G. Maier (eds.). Kluwer Academic Publishers. 233-254.

[19] Yan A.M., Nguyen-Dang H. (2001). Non-linear upper bound shakedown analysis with temperature dependent yield limit. *Int. Journal of Num. Meth. Engng,* **50**. 1145-1168

[20] Yan A.M., Vu D.K., Nguyen-Dang H. (2001): Incremental plasticity limit of structures by finite elements analysis. *ERMESC'III, July 25-28* Shanghai.

[21] Zhang Y.G. (1995): An iteration algorithm for kinematic shakedown analysis. *Computer Methods Appl. Mech. Engrg.* **126**. 333-341.

[22] Zouain N., Herskovits J., Borges L.A., Feijoo R.A. (1993) An iterative algorithm for limit analysis with non-linear yield function. *Int. J. Solids Structures,* **39**. 1397-1417.

Chapter 8

BIFURCATION ANALYSIS OF SHALLOW SPHERICAL SHELLS WITH MERIDIONALLY NONUNIFORM LOADING

Charles G. Lange* & Frederic Y.M. Wan

Department of Mathematics

University of California, Irvine, CA92697-3875, U. S. A.

fwan@math.uci.edu

Abstract Numerical solutions of the relevant boundary value problem suggest that the first bifurcation from the basic solution for a spherical cap under a class of meridionally nonuniform loading is to a dimple state. Delicate asymptotic analysis of the linearized buckling problem confirm this observation.

1. Introduction

Under favorable conditions, dome-shaped, thin elastic shell of revolution is known to undergo a predominantly inextensional bending type deformation in the form of a finite axisymmetric dimple centered at the apex of the shells. The possibility of polar dimpling was suggested in [1] and investigated in [2-9] and elsewhere. Consider for example a very thin shallow clamped spherical cap subject to a quadratically varying pressure loading $p_n = p_o(1 - c^2x^2)$, where $x = r/r_o$ is the radial distance from the axis of revolution of the cap's middle surface normalized by its value at the outer edge. It was shown in [9] that polar dimpling is a possible mode of deformation if $c > 1$ and $p_o \geq p_{cr}$ for some critical load magnitude p_{cr}. Moreover, for very thin shells, we see from the results of [9] that p_{cr} is much smaller than the classical buckling load for a complete spherical shell p_s with the same radius of curvature, i.e., $p_{cr} \ll p_s$. This gives rise to the possibility of axisymmetric buckling at

*Deceased

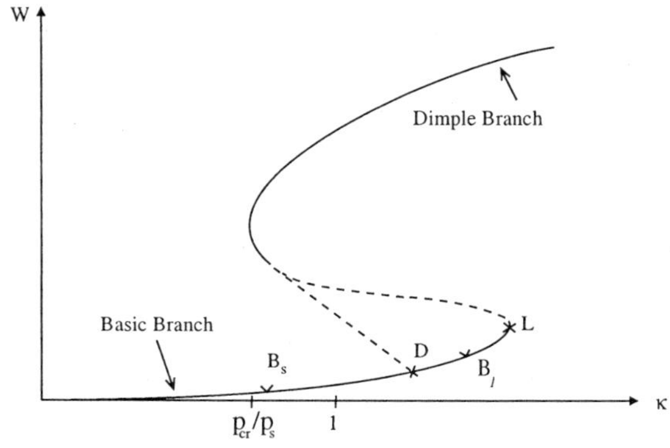

Figure 8.1. LOAD DEFORMATION CURVE

load level comparable to, or well below the critical load for unsymmetrical buckling. An investigation of axisymmetric buckling for meridionally nonuniform pressure loading is therefore of interest.

With $p_o = p_{cr}$, the size of the resulting dimple was found in [9] to be given by the dimensionless dimple (base) radius $x_{cr} = \sqrt{2/3}c^{-1}$. For larger values of p_o, two dimples of different sizes are possible, one with dimensionless dimple radius $x_G > x_{cr}$ and the other with dimensionless dimple radius $x_L < x_{cr}$. As p_o increases, x_G increases toward unity or $x_t = \sqrt{2}/c$ whichever is smaller, while x_L decreases to zero. Whether the dimple mode is possible or not, a non-dimple mode of deformation (called the basic branch hereafter) is also possible up to a certain threshold load magnitude. (A more precise description will be given later.) These results may be summarized by an overall load-deformation plot such as the one shown in Figure 8.1 giving the L_2 norm, W, of the axial displacement of the shell's midsurface as a function of a dimensionless load magnitude parameter $\kappa = p_o/p_s$. Accurate numerical solutions of the relevant nonlinear boundary value problem governing the elastostatics of the spherical cap give the solid curves in Figure 8.1 as (a qualitative description of) the overall load-deformation relation corresponding to a typical load-geometric configuration. The qualitative features of the shell behavior described above for a spherical cap subject to a particular pressure distribution are typical for many dome-shaped thin elastic shells of revolution subject to more general axisymmetric loading. They give rise to several questions which are important in design studies: Is the *dimple branch* of the load-deformation relation in Figure (1) the result of a bifurcation from the basic branch at some point D, or the consequence of the presence of a limit point L along an S shape load-deformation curve as indicated by the dotted portion connecting the top of the basic branch to the dimple branch? Does bifurcation (buckling) occur elsewhere along the basic branch, e.g., at some B_ℓ and B_s? Is the buckling load above (as in the case of B_ℓ) or below (as in the case of B_s) p_s? The answers to these important questions for the specific problem of a spherical cap as well as similar questions for other dome-shaped thin shells of revolution with more general axisymmetric loading evidently cannot be obtained from numerical solutions of the relevant boundary value problems, however accurate these solutions may be. In this paper, we show that it is possible to gain some insight to the actual shell behavior by obtaining asymptotic solutions of the relevant boundary value problems and eigenvalue problems. For simplicity, we consider here only the spherical cap problem and obtain an asymptotic solution for the basic branch behavior and for the *axisymmetric* buckling load of the "linearized" buckling theory. The basic mathematical problem in our axisymmetric buckling analysis is an eigenvalue problem which is nonlinear in the eigenvalue parameter. It is of some mathematical interest that a matched asymptotic solution of this eigenvalue problem requires an intermediate layer solution to connect the inner solution near the apex of the shell to a WKBJ type solution away from the apex. To the authors, the presence of such an intermediate layer solution is novel and rather unexpected.

2. Formulation

Consider an elastic shell of revolution which has undergone infinitesimal strain, and axisymmetric finite deformations. The elastostatics of such a shell may be formulated as a boundary value problem for two simultaneous second order nonlinear ordinary differential equations for a stress function and a meridional slope variable [10,11]. For a shallow spherical cap with an undeformed meridional slope $\xi_o x$, subject to a normal[1] pressure distribution $p_n = p_o p(x)$ where p_o is chosen so that $p(x) \leq 1$, we may take these equations in the following dimensionless form [11]:

$$\varepsilon^2 x[\Psi'' + \frac{1}{x}\Psi' - \frac{1}{x^2}\Psi] + \frac{1}{2}(\Phi^2 - x^2) = 0 \qquad (2.1)$$

$$(0 < x < 1)$$

$$\varepsilon^2 x[\Phi'' + \frac{1}{x}\Phi' - \frac{1}{x^2}\Phi] - \Phi\Psi = 4\kappa x P(x) \qquad (2.2)$$

with

$$P(x) = \frac{1}{x} \int_o^x p(t)\, t dt. \tag{2.3}$$

In (2.1)-(2.3), a prime indicates differentiation with respect to the dimensionless radial coordinate x (as defined in section (1)) and $\xi_o \Phi(x)$ denotes the meridional slope of the deformed middle surface of the shell. In terms of radial coordinate r_o of the shell edge, the shell thickness h, Young's modulus E and Poisson's ratio ν, we also

$$\varepsilon^2 = \frac{ha}{2r_o^2\sqrt{3(1-\nu^2)}}, \qquad \kappa = \frac{p_o}{p_s}, \qquad p_s = \frac{2Eh^2}{a^2\sqrt{3(1-\nu^2)}} \tag{2.4}$$

where $a = r_o/\xi_o$ is the radius of the spherical middle surface and p_s is the classical buckling load of a complete spherical shell under uniform pressure. The dimensionless stress function Ψ is the conventional stress function normalized by $p_s a r_o/4$ with $p_s a\Psi/4x$ and $p_s a\Psi'/4$ being the radial and hoop stress resultant of the shell, respectively.

The fourth order system of the two ODE, (2.1)-(2.2), is supplemented by four boundary conditions. Two are regularity conditions at the apex of the shell:

$$x = 0: \qquad \Phi = 0, \qquad \Psi = 0, \tag{2.5}$$

(with the second being a consequence of the symmetry condition of no radial displacement at the apex). The other two come from the edge constraints at $x = 1$. For a clamped edge, we have

$$x = 1: \qquad \Phi = 1, \qquad \Psi' - \nu\Psi = 0. \tag{2.6}$$

They correspond to the condition of no change in meridional slope and no radial displacement at the edge.

With $H = r_o^2/2a$ being the rise of the shell edge above the base plane tangent to the apex, the parameter $\varepsilon^2 = O(h/H)$ is a measure of the thickness-to-rise ratio. In this paper, we consider only the range $\varepsilon^2 \ll 1$ so that the spherical cap behaves as a shell (rather than a flat plate). We will also focus our discussion on the case of a quadratically varying normal load with

$$p(x) = 1 - c^2x^2, \qquad P(x) = \frac{1}{2}x(1 - \frac{1}{2}c^2x^2). \tag{2.7}$$

It should be clear from the subsequent development that a similar analysis may be carried out for other loadings and edge supports.

3. The Basic Branch Solution

It is not difficult to see that the system (2.1) and (2.2) with $P(x)$ given by (2.7) has an exact solution

$$\Phi_p = Ax, \qquad \Psi_p = \frac{\kappa}{A}x(c^2x^2 - 2) \tag{3.1}$$

where A is a root of

$$A^3 - A + 16\varepsilon^2 c^2 \kappa = 0 \tag{3.2}$$

The cubic equation has only one negative root if $16\varepsilon^2c^2\kappa > 2/(3\sqrt{3})$ or $\kappa > \kappa_T \equiv [24\sqrt{3}c^2\varepsilon^2]^{-1}$, one negative and one poistive (double) root if $\kappa = \kappa_T$, and one negative and two poistive roots if $0 < \kappa < \kappa_T$. For $0 < 16\varepsilon^2c^2\kappa \ll 1$, the three roots may be obtained in the form of parametric series in powers of ε^2:

$$A = \begin{cases} 1 - 8\kappa c^2\varepsilon^2 + \ldots\ldots\ldots\ldots\ldots\ldots\ldots\ldots \equiv A_b \\ -1 - 8\kappa c^2\varepsilon^2 + \ldots\ldots\ldots\ldots\ldots\ldots\ldots \equiv A_d \\ 16\kappa c^2\varepsilon^2[1 + (16\kappa c^2\varepsilon^2)^2 + \ldots\ldots] \equiv A_f \end{cases} \tag{3.3}$$

Note that the larger positive root is slightly less than unity, the smaller positive root is slightly larger than zero and the negative root is slightly less than -1. For fixed values of c and ε, both A_b and A_d decrease while A_f increases with $\kappa(\geq 0)$.

The exact solution (3.1) of the ODE also satisfies the regularity conditions (2.5) at $x = 0$, but does not satisfy the clamped edge conditions (2.6) at $x = 1$. With $\varepsilon^2 \ll 1$, this exact solution of (2.1), (2.2) and (2.5) may be modified by a boundary layer correction so that the modified solution also satisfies the clamped edge conditions (2.6) asymptotically and therefore is an asymptotic solution of the boundary value problem for Φ and Ψ. We omit the routine calculations for the method of matched asymptotic expansions and merely give here the final composite basic branch solution, denoted by Φ_b and Ψ_b, corresponding to the root A_b of the cubic equation (3.2):

$$\Phi_b = A_b x + \frac{\kappa\varepsilon}{\lambda_p - \lambda_m}[(3-\nu)c^2 - 2(1-\nu)][e^{\lambda_p\bar{x}} - e^{\lambda_m\bar{x}}] + O(\varepsilon^2)$$

$$\begin{aligned} \Psi_b \ = \ & \frac{\kappa}{A_b}x(c^2x^2 - 2) + \frac{\kappa\varepsilon}{\lambda_p - \lambda_m}[(3-\nu)c^2 - 2(1-\nu)][(\lambda_p^2 - \bar{\kappa})e^{\lambda_p\bar{x}} \\ & -(\lambda_m^2 - \bar{\kappa})e^{\lambda_m\bar{x}}] + O(\varepsilon^2) \end{aligned} \tag{3.4}$$

where

$$\bar{x} = \frac{1-x}{\varepsilon}, \qquad \bar{\kappa} = \frac{1}{2}\kappa(c^2-2)$$

$$\lambda_p = -\sqrt{\bar{\kappa} + \sqrt{\bar{\kappa}^2 - 1}}, \qquad \lambda_p^2 = \bar{\kappa} + \sqrt{\bar{\kappa}^2-1} \tag{3.5}$$

$$\lambda_m = -\sqrt{\bar{\kappa} - \sqrt{\bar{\kappa}^2 - 1}}, \qquad \lambda_m^2 = \bar{\kappa} - \sqrt{\bar{\kappa}^2-1}$$

The matched asymptotic expansion solution (3.4) was obtained under the assumption $8\kappa c^2\varepsilon^2 \ll 1$ and is therefore valid only for sufficiently small values of ε. To the same order of approximation, we may set $A_b = 1$ in (3.4).

The asymptotic solution (3.4) is also inappropriate when $\bar{\kappa}$ is nearly unity, say $\bar{\kappa} = 1 + O(\varepsilon)$, for we have $\lambda_p - \lambda_m = O(\varepsilon)$ at most in that case. It is not difficult to verify that an appropriate basic solution for $\bar{\kappa} = 1 + O(\varepsilon)$ is given by

$$\Phi_B \sim A_b x - \varepsilon\left[2\bar{x}\left(\nu - \frac{3c^2-2}{c^2-2}\right)e^{-\bar{x}}\right] \tag{3.6}$$

$$\Psi_B \sim \frac{\kappa x}{A_b}(c^2x^2-2) + \varepsilon\left[(4+2\bar{x})\left(\nu - \frac{3c^2-2}{c^2-2}\right)e^{-\bar{x}}\right]$$

where A_b may be approximated by unity and

$$\kappa = \frac{2}{c^2-2}[1+0(\varepsilon)]. \tag{3.7}$$

Both the basic branch solution (3.5) for $\bar{\kappa}$ away from unity and the basic branch solution (3.6) for $\bar{\kappa} = 1 + 0(\varepsilon)$ will be needed in the buckling analysis in a later section.

Numerical solutions for different combinations of c, ε and κ accurate to at least five significant figures have been generated by the general boundary value problem solver COLSYS (or COLNEW) [12, 13] to confirm the adequacy of the asymptotic solution (3.4) for $8\kappa c^2\varepsilon^2 \ll 1$. The difference between the numerical and asymptotic solution is at most $O(16\kappa c^2\varepsilon^2)$ in all cases. Beyond the range of parameter values allowed by the asymptotic solution (3.4), COLSYS has also been used to generate the basic branch solution as well as the corresponding portion of the overall load-deformation curve in Figure (1) for different combinations of $c(>1)$ and $\varepsilon(\ll 1)$ and for κ beyond the range $8\kappa c^2\varepsilon^2 \ll 1$. We could not obtain a basic branch solution for $\kappa \geq \bar{\kappa}_T$ for some upper bound $\bar{\kappa}_T$ which depends on the value of c and ε. For $c = 2$, we found that COLSYS failed to converge for sufficiently large κ for $\varepsilon = 0.01, 0.04$ and

0.1 with the upper bound $\bar{\kappa}_T$ for each ε given in Table (1). The corresponding value of $\kappa_T \equiv [24\sqrt{3}c^2\varepsilon^2]^{-1}$ beyond which there cannot be a positive root for (3.2) (and hence not a basic branch solution) are also listed there for comparison.

Table (1) The Values of κ_T and $\bar{\kappa}_T$ for $c = 2$

ε	0.01	0.04	0.1
$\bar{\kappa}_T$	60	3.1	0.48
κ_T	$60.14\cdots$	$3.759\cdots$	$0.6014\cdots$

Also, COLSYS encountered convergence difficulty for several κ values well below the bound $\bar{\kappa}_T$ ($< \kappa_T$ in all cases). It raises the possibility of multiple equilibrium states associated with a bifurcation phenomenon at those κ values. As we shall see in the next section, bifurcation into another axisymmetric equilibrium state does occur at these κ values.

4. Bifurcation from the Basic Branch Solution

Bifurcation from the basic branch solution occurs when there are more than one solution of the boundary value problem (2.1), (2.2), (2.5) and (2.6) in the neighborhood of the bifurcation point. Let ϕ and ψ be the deviation of a second solution

$$\Phi = \Phi_b + \phi, \qquad \Psi = \Psi_b + \psi \tag{4.1}$$

from the basic branch solution Φ_b and Ψ_b, respectively. The boundary value problem for Φ and Ψ can then expressed in terms of ϕ and ψ as

$$\varepsilon^2[\phi'' + \frac{1}{x}\phi' - \frac{1}{x^2}\phi] - \Phi_b\psi - \Psi_b\phi = \phi\psi$$

$$\varepsilon^2 x[\psi'' + \frac{1}{x}\psi' - \frac{1}{x^2}\psi] + \Phi_b\phi = -\frac{1}{2}\phi^2 \tag{4.2}$$

with

$$x = 0: \qquad \phi = 0, \qquad \psi = 0 \tag{4.3}$$

$$x = 1: \qquad \phi = 0, \qquad \psi' - \nu\psi = 0 \tag{4.4}$$

The problem (4.2)-(4.4) is a nonlinear eigenvalue problem with κ as the eigenvalue parameter. Near the point of bifurication, we expect $|\phi|$ and $|\psi|$ generally to be relatively small of higher order compared to the basic branch solution and the ODE (4.2) can be linearized. Upon omitting the nonlinear terms on the right hand side of (4.2), we are left with the linearized eigenvalue problem

$$\varepsilon^2 x[u'' + \frac{1}{x}u' - \frac{1}{x^2}u] - \Phi_b v - \Psi_b u = 0 \tag{4.2'}$$

$$\varepsilon^2 x[v'' + \frac{1}{x}v' - \frac{1}{x^2}v] + \Phi_b u = 0$$

with

$$x = 0: \qquad u = v = 0 \tag{4.3'}$$

$$x = 1: \qquad u = v' - \nu v = 0 \tag{4.4'}$$

Note that the eigenvalue parameter κ appears nonlinearly in the basic branch solution Φ_b and Ψ_b. For $8\kappa\varepsilon^2 c^2 \ll 1$, the rather complex dependence of Φ_b and Ψ_b on κ can be seen from the asymptotic solution (3.4) and (3.5).

With Φ_b and Ψ_b already determined from section (3), the solution of the nonlinear eigenvalue problem (4.2)-(4.4) or of the linearized eigenvalue problem (4.2′)-(4.4′) may be obtained by various approximate methods. We confine ourselves to obtaining an approximate solution of the linearized problem in this article by three different methods. The first method is to discretize the continuous eigenvalue problem by approximating it by its finite difference analogue. To leading order in ε, Φ_b and Ψ_b are linear in κ, the difference analogue is, to the same order in ε, a matrix eigenvalue problem. The eigenvalues for the (NxN) matrix problem provide an approximate solution for the first N eigenvalues of the continuous problem. A second method of solution (which does not require a leading term approximation for Φ_b and Ψ_b) consists of solving the boundary value problem (4.2′)-(4.4′) for a fixed κ with the boundary condition $\phi(1)$=0 replaced by $\phi'(0)$=1. Starting with $0 < \kappa \ll 1$, repeat the solution process for an increasing sequence of κ values and, for each κ, check for a change in the sign of $\phi(1)$ from the solution for the previous value of κ. For every sign change, there is an eigenvalue between two successive values of κ.

With either method, the approximate eigenvalues found may be made more accurate by finding a COLSYS solution of the inhomogeneous linearized boundary value problem defined by (4.2′)-(4.4′) and

$$\kappa' = 0, \qquad \phi'(0) = 1 \tag{4.5}$$

with the additional boundary condition serving only to normalize the eigenfunctions. The approximate eigenvalues and eigenfunctions found by either of the above methods provide the initial guesses for the iteration scheme of COLSYS. In this way, two eigenvalues are found near unity (to five significant figure accuracy) for $c = 2$, $\nu = 0.3$ and for two different values of ε; they are recorded in Table (2).

Table (2) Eigenvalues for Linearized Buckling Theory for $c = 2$ & $\nu = 0.3$ by COLSYS

ε	κ_1	κ_2
0.04	$0.96465\cdots$	$0.98533\cdots$
0.01	$0.993806\cdots$	1.0138

Accurate numerical solutions were also obtained by COLSYS for the dimple branch solutions [9]. It is significant that, for the same two values of ε, the smaller of the two dimples shrinks as κ increases from κ_{cr} and eventually disappear around $\kappa \simeq 1$. This suggests that the dimple branch solution actually bifurcates from the basic branch solution at the buckling load. Unfortunately, the eigenvalues obtained by COLSYS in Table (2) may or may not be the lowest eigenvalues of the linearized eigenvalue problem as the iterative solution scheme in COLSYS may have converged to these larger eigenvalues. To find the lowest branch point, we consider in the next section an asymptotic solution of the linearized eigenvalue problem. Away from the apex of the spherical cap, an adequate approximate solution of (4.2′)-(4.4′) for $\varepsilon \ll 1$ may be obtained by the WKBJ method. Such an asymptotic solution is not uniformly valid for $0 \leq x \leq 1$, and the regularity conditions at $x = 0$ cannot be applied to the WKBJ solution. It is necessary to obtain also an asymptotic solution valid in the neightborhood of $x = 0$. Such a solution would be matched to the WKBJ solution in an intermediate region and then the clamped edge conditions would be used to determine the eigenvalues and therefore the bifurcation points along the overall load-deformation curve of the basic branch solution [14].

For $\bar{\kappa}$ away from unity and $c = 2$, (so that $\bar{\kappa} = \kappa$) the above method gives the following condition for the admissable values of κ for a non-trivial solution for ϕ and ψ:

$$\left[\frac{A^2 - \kappa}{2A}\right]^{\frac{1}{2}} sinh\left[\frac{1}{\varepsilon}(\frac{\kappa}{A})^{\frac{1}{2}}\{\sqrt{1+\gamma} + \gamma \ell n(\frac{1+\sqrt{1+\gamma}}{\sqrt{\gamma}})\}\right] \tag{4.6}$$

$$= \left[\frac{A^2 + \kappa}{2A}\right]^{\frac{1}{2}} sin\left[\frac{1}{\varepsilon}(\frac{\kappa}{A})^{\frac{1}{2}}\{\sqrt{\tau - 1} + \frac{1}{2}\tau sin^{-1}(\frac{1}{\sqrt{\tau}}) - \frac{3}{2}\pi\}\right]$$

where

$$\gamma = \frac{A^2 - \kappa}{2A\kappa}, \qquad \tau = \frac{A^2 + \kappa}{2A\kappa} \tag{4.7}$$

and A is a function of κ (and vice versa) by the cubic equation (3.2). For $0 < \varepsilon \ll 1$, the roots of (4.6) must be such that we have $A^2 \approx \kappa$ ($=\bar{\kappa}$ as we have taken $c = 2$ in (4.6) and (4.7)); otherwise the exponential growth of hyperbolic sine term on the left cannot be matched by the (at most) algebraic growth on the right. We therefore conclude that, for $0 < \varepsilon \ll 1$, the linearized eigenvalue problem (4.2′)-(4.4′) for the determination of bifurcation from the basic branch admits only the trivial solution unless $\kappa \approx A^2$. From (3.2), we have for $\kappa = A^2$ (and $c = 2$),

$$A_b = \sqrt{1 + (32\varepsilon^2)^2} - 32\varepsilon^2 \cong 1 - 32\varepsilon^2 + O(\varepsilon^4)$$

$$\kappa = A^2 = 1 + 2(32\varepsilon^2)^2 - 64\varepsilon^2\sqrt{1 + (32\varepsilon^2)^2} \simeq 1 - 64\varepsilon^2 + O(\varepsilon^4)$$

It follows that, for $0 < \varepsilon \ll 1$ and $c = 2$, branching from the basic solution does not occur if the pressure magnitude p_0 is <u>not</u> nearly the classical buckling pressure p_s of the complete spherical shell, say $\kappa = 1 + O(\varepsilon)$.

We note that no conclusion on bifurcation in the range $\kappa = 1+O(\varepsilon) \approx A^2$ can be drawn from (4.6)-(4.7) as they were obtained with Φ_B and Ψ_B taken from (3.4). This asymptotic solution for the basic branch is not valid for $\kappa = 1 + O(\varepsilon)$; we should use the expressions given by (3.6) instead. When we applied the same solution process with Φ_B and Ψ_B taken from (3.6), we found that it is not possible to match the WKBJ solution (valid away from $x = 0$) with the Bessel function solution (valid near the apex of the shell) for the linearized eigenvalue problem (4.2′)-(4.4′). As we shall see in the next section, an intermediate layer solution is required to bridge the two solutions.

5. Asymptotic Analysis of the Linearized Eigenvalue Problem

5.1 Bessel Function Solution Near the Apex

For $\varepsilon \ll 1$, an asymptotic solution of the linearized eigenvalue problem (4.2′)-(4.4′) is complicated by the singularity of the differential equations at $x = 0$. In the neighborhood of the apex of the cap, we set $y = x/\varepsilon$ and write the ODE(4.2′) as

$$L_y[u] - [A_b + O(e^{-1/\varepsilon})]v + \frac{2\kappa}{A_b}[(1 - \frac{1}{2}C^2\varepsilon^2y^2) + O(e^{-1/\varepsilon})]u = 0 \quad (5.1)$$

$$L_y[v] + [A_b + O(e^{-1/\varepsilon})]u = 0$$

with

$$L_y[f] \equiv \frac{d^2f}{dy^2} + \frac{1}{y}\frac{df}{dy} - \frac{1}{y^2}f \quad (5.2)$$

For the basic branch solution, we have $A_b = 1+O(\kappa c^2\varepsilon^2) = 1+O(\varepsilon^2)$ as we consider in this section only $\kappa \approx A^2$. We also take $c = 2$ to simplify the discussion. In that case we have $\kappa = \bar{\kappa} = [1 + \alpha_1\varepsilon + \alpha_2\varepsilon^2 + \cdots]$ and omit the exponentially small terms to simplify (5.1) to

$$L_y[u] - [1+O(\varepsilon^2)]v + \frac{2[1+\alpha_1\varepsilon + O(\varepsilon^2)][1-2\varepsilon^2y^2]}{[1+O(\varepsilon^2)]}u = O(e^{-1/\varepsilon}) \quad (5.1')$$

$$L_y[v] + [1+O(\varepsilon^2)]u = O(e^{-1/\varepsilon})$$

We may now (tentatively) consider a regular perturbation solution for u and v in powers of ε. The leading term u_0 and v_0 of this perturbation solution is of course given by

$$L_y[u_0] - v_0 + 2u_0 = 0 \quad (5.3)$$

$$L_y[v_0] + u_0 = 0$$

with the regularity conditions (see(4.3$'$))

$$x = 0: \qquad u_0(0) = v_0(0) = 0. \quad (5.4)$$

It is not difficult to verify that the exact solution of (5.3) and (5.4) is

$$v_0 = a_0 J_1(y) + b_0 y J_1^\bullet(y), \quad (5.5)$$

$$u_0 = (a_0 + 2b_0)J_1(y) + b_0 y J_1^\bullet(y)$$

where $(\ \ )^\bullet \equiv d(\ \ )/dy$ and a_0 and b_0 are arbitrary real constants of integration to be determined by matching to an asymptotic solution for $x \gg \varepsilon$. For matching, we need the following asymptotic behavior of $J_1(y)$ and $yJ_1^\bullet(y)$ for $y \gg 1$:

$$J_1(y) \sim \left(\frac{2}{\pi y}\right)^{\frac{1}{2}} cos\left(y - \frac{3\pi}{4}\right)\{1 + O(y^{-3/2})\} \quad (5.6)$$

$$yJ_1^\bullet(y) \sim -\left(\frac{2y}{\pi}\right)^{\frac{1}{2}} sin\left(y - \frac{3\pi}{4}\right)\{1 + O(y^{-1/2})\}$$

5.2 WKBJ Solution Away from the Apex

Away from $x = 0$, an asymptotic solution of the linealized ODE (4.2$'$) may be obtained by the standard WKBJ method. The solution process is complicated by the fact that the boundary layer terms in the basic branch solution Φ_b and Ψ_b (which appear as coefficients of the ODE for u and v) are no longer exponentially small near $x = 1$, and therefore cannot be neglected in the solution process. Fortunately, the contribution of these boundary layer terms to the final WKBJ solution is nevertheless negligibly small (so that they can in fact be deleted from the ODE before the WKBJ calculations). We omit the details and simply write down the leading term WKBJ solution:

$$\begin{aligned} v &\sim c_0 e^{i[xs+sin^{-1}x]/2\varepsilon} e^{\frac{\alpha_1}{4}ixs} / \sqrt{x(1-\tfrac{\alpha_1}{2})s(1-s^4)} \\ &+ d_0 e^{-i[xs+sin^{-1}x]/2\varepsilon} e^{-\frac{\alpha_1}{4}ixs} / \sqrt{x(1+\tfrac{\alpha_1}{2})s(1-s^4)} \\ &+ \text{(complex conjugate of the two preceding terms)} \\ &\equiv v_0 + v_0^* \end{aligned}$$

$$u \sim s^2 v_0 + (s^2 v_0)^*$$

with

$$s = \sqrt{1-x^2} + ix, \qquad 1 - s^4 = 8x^2(1-x^2) + i4x(2x^2-1)\sqrt{1-x^2} \tag{5.7}$$

where c_0 and d_0 are arbitrary complex constants. For simplicity, we have given v and u in (5.7) only for the case $c = 2$ in the pressure distribution (2.7) for which extensive numerical results have been obtained and recorded in this article and elsewhere. We will henceforth focus our discussion on this case though a similar analysis applies to the general case.

Two of the four real constants in (5.7) may be determined by the two boundary conditions at $x = 1$. The remaining two constants are to be determined by matching the WKBJ solution with the Bessel function solution of section (5.1). However, for $x = z\delta_1(\varepsilon)$ where $\delta_1(\varepsilon) \to 0$ but $\delta_1(\varepsilon)/\varepsilon \to \infty$ as $\varepsilon \to 0$, we have

$$v \sim \frac{c_0 i^{-3/2} e^{iz\delta_1/\varepsilon - (\delta_1 z)^2/2\varepsilon}}{2[\delta_1 z]^{1-\alpha_1/4}} [1 + O(z\delta_1)] \tag{5.7'}$$

$$+ \frac{d_0 i^{-3/2} e^{-iz\delta_1/\varepsilon + (\delta_1 z)^2/2\varepsilon}}{2[\delta_1 z]^{1+\alpha_1/4}} [1 + O(z\delta_1)]$$

$$+ \text{ complex conjugate of the preceding terms}$$

It is simply not possible to match v as given by (5.7′) and v_0 as given by (5.5) with $y = z/\delta_0(\varepsilon)$ where $\delta_0(\varepsilon) \to 0$ as $\varepsilon \to 0$, keeping in mind the asymptotic behavior of J_1 and $J_1^\bullet$ given by (5.6). This suggests an intermediate layer solution exists to bridge the WKBJ solution and the Bessel function solution.

5.3 An Intermediate Layer Solution

The expression (5.7′) for v must remain bounded for $x = O(\delta_1(\varepsilon))$. This suggests a stretched variable $z = x/\sqrt{\varepsilon}$ for the intermediate layer region. In terms of z, the ODE (4.2′) may be written as

$$\varepsilon L_z[u] - [1 + O(\varepsilon^2)]v + \frac{2[1 + \alpha_1\varepsilon + O(\varepsilon^2)][1 - \frac{1}{2}c^2\varepsilon z^2]}{[1 + O(\varepsilon^2)]}u = O(e^{-1/\varepsilon})$$

$$\varepsilon L_z[v] + [1 + O(\varepsilon^2)]u = O(e^{-1/\varepsilon})$$

Up to terms of order ε^2 or smaller, we have $(u, v) \sim (\tilde{u}, \tilde{v})$ with

$$\varepsilon L_z[\tilde{u}] - \tilde{v} + 2(1 + \alpha_1\varepsilon)(1 - \frac{1}{2}\varepsilon c^2 z^2)\tilde{u} = 0$$

$$\varepsilon L_z[\tilde{v}] - \tilde{u} = 0 \tag{5.8}$$

We seek an asymptotic solution of (5.9) in the form

$$\begin{pmatrix} \tilde{u} \\ \tilde{v} \end{pmatrix} \sim e^{iz/\sqrt{\varepsilon}} \sum_{n=0}^{\infty} \begin{pmatrix} U_n \\ V_n \end{pmatrix} \varepsilon^{n/2} \tag{5.9}$$

It is not difficult to verify that the ODE for $\tilde{u}$ and $\tilde{v}$ are satisfied identically in ε to tems of order ε if

$$U_0 = V_0, \qquad U_1 = V_1 - 2iV_0{}' - \frac{i}{z}V_0,$$

$$U_2 = V_2 - 2iV_1{}' - \frac{i}{z}V_1 + (\frac{3}{4z^2} + \frac{\alpha_1}{2} - z^2)V_0 \tag{5.10}$$

and

$$V_0{}'' + \frac{1}{z}V_0{}' - (\frac{1}{4z^2} - \frac{\alpha_1}{2} + z^2)V_0 = 0 \tag{5.11}$$

with $(\quad)' = d(\quad)/dz$. The ODE (5.12) for V_0 may be transformed into

$$w'' + (\frac{\alpha_1}{2} - z^2)w = 0 \tag{5.12}$$

where $w = \sqrt{z}V_0$. The solutions for (5.13) are parabolic cylinder functions $D_\nu(t)$ and $D_{-\nu-1}$ (it) where $\nu = (\alpha_1 - 2)/4$ and $t = \sqrt{2}z$ [13]. Altogether, we have the following asymptotic solution for $\tilde{u}$ and $\tilde{v}$:

$$\tilde{u} \sim \tilde{v} \sim \frac{e^{iz/\sqrt{\varepsilon}}}{\sqrt{z}}[c_1 D_{-\frac{1}{2}+\frac{\alpha_1}{4}}(\sqrt{2}z) + d_1 D_{-\frac{1}{2}-\frac{\alpha_1}{4}}(i\sqrt{2}z)][1+O(\sqrt{\varepsilon})] \tag{5.13}$$

+complex conjugate of the preceding expression.

For large values of the argument, we have the following asymptotic behavior for$D_\nu(t)$ and $D_{-\nu-1}$(it):

$$\begin{aligned} D_\nu(t) &\sim t^\nu e^{-t^2/4} \textstyle\sum_{n=0}^{\infty}(-1)^n \gamma_n t^{-2n}, \\ D_{-\nu-1}(it) &\sim (it)^{-\nu-1} e^{t^2/4} \textstyle\sum_{n=0}^{\infty} \gamma_n t^{-2n} \end{aligned} \tag{5.14}$$

for $t \to \infty, |arg(t)| < 3\pi/4$ with $\gamma_0 = 1$, $\gamma_1 = \nu(\nu-1)/2$, etc. They show that matching of $(\tilde{u}, \tilde{v})$ and (u, v) as given by (5.7) may be accomplished by taking

$$c_1 = \frac{2^{-(\alpha_1+6)/8}}{i^{3/2}\varepsilon^{-(\alpha_1-4)/8}} c_0, \qquad d_1 = \frac{2^{(\alpha_1-6)/8} i^{\alpha_1/4}}{\varepsilon^{(\alpha_1+4)/8}} d_0^* \tag{5.15}$$

For small values of the argument, we have from the Taylor series representation of the parabolic cylinder functions

$$D_\nu(t) = \frac{\sqrt{2^\nu \pi}}{\Gamma(\frac{1}{2}-\frac{1}{2}\nu)}[1 - \frac{\Gamma(\frac{1}{2}-\frac{1}{2}\nu)}{\Gamma(\frac{-1}{2}\nu)}\sqrt{2}t + O(t^2)] \tag{5.16}$$

which allows to match the parabolic cylinder function solution with the Bessel function solution in an intermediate (overlapping) region of validity. The matching is accomplished to order ε with

$$\begin{aligned} c_{1r} + \frac{\Gamma(\frac{3}{4}-\frac{\alpha_1}{8})}{2^{\alpha_1/4}\Gamma(\frac{3}{4}+\frac{\alpha_1}{8})} d_{1r} = c_{1i} + \frac{\Gamma(\frac{3}{4}-\frac{\alpha_1}{8})}{2^{\alpha_1/4}\Gamma(\frac{3}{4}+\frac{\alpha_1}{8})} d_{1i} = -\frac{\varepsilon^{1/4}\Gamma(\frac{3}{4}-\frac{\alpha_1}{8})}{\pi 2^{(\alpha_1+6)/8}} a_0 \\ -c_{1r} + \frac{2^{\alpha_1/4}\Gamma(\frac{1}{4}-\frac{\alpha_1}{8})}{\Gamma(\frac{1}{4}+\frac{\alpha_1}{8})} d_{li} = c_{1i} + \frac{2^{\alpha_1/4}\Gamma(\frac{1}{4}-\frac{\alpha_1}{8})}{\Gamma(\frac{1}{4}+\frac{\alpha_1}{8})} d_{lr} = \frac{\varepsilon^{-1/4}\Gamma(\frac{1}{4}-\frac{\alpha_1}{8})}{\pi 2^{(\alpha+14)/8}} \bar{b}_0 \end{aligned} \tag{5.17}$$

where (c_{1r}, d_{1r}) and (c_{1i}, d_{1i}) are the real and imaginary parts of (c_1, d_1) respectively, and $b_0 = \bar{b}_0\sqrt{\varepsilon}$.

5.4 Matching and Solutions of Eigenvalue Problem

The four conditions (5.17) determine the two complex constants c_1 and d_1 in terms of the two real constants a_0 and $\bar{b}_0$. It follows from (5.15) that the two complex constants c_0 and d_0, which appear in the WKBJ solution, are also determined in terms of the two real constants a_0 and $\bar{b}_0$. Finally, these two unknown real constants must be chosen so that clamped edge conditions (4.4′) are satisfied by the WKBJ solution (5.7). The two clamped edge conditions at $x = 1$ are two homogeneous linear equations for a_0 and $\bar{b}_0$. For a nontrivial solution, the determinant of the 2x2 coefficient matrix must vanish. We omit the lengthy expression for the determinant obtained from this tedious but straightforward solution procedure, and merely report that, except for exponentially small terms, the determinant vanishes only if $\alpha_1 = 0$. Therefore, the eigenvalues of the linearized problem (4.2′)-(4.4′) is given by $\bar{\kappa} = \kappa = 1 + 0(\varepsilon^2)$ (for $c = 2$). The eigenfunctions are given

- by (5.7) with $\alpha_1 = 0$ away from the apex,
- by (5.5) in the neighborhood of the apex, and
- by (5.14) with $\alpha_1 = 0$ for an intermediate region.

The complex constants c_0, d_0, c_1 and d_1 are determined in terms of the two real constants a_0 and $b_0 \equiv \bar{b}_0\sqrt{\varepsilon}$ by (5.16) and (5.18). The constant a_0 is determined in terms of $\bar{b}_0$ by either one of the clamped edge conditions. The result $\alpha_1 = 0$ is consistent with the COLSYS solutions given in Table (2). To determine the $O(\varepsilon^2)$ correction for the eigenvalues, we will need higher order terms in the various asymptotic expansions including the basic branch solution.

6. Concluding Remarks

By the asymptotic analysis of sections (3)-(5), we have established that, for $c = 2$ and $\varepsilon \ll 1$, branching from the basic solution (of section (3) or (4)) is not possible for p_o not nearly p_s. Branching when $(\bar{\kappa} =)\kappa = [1 + 0(\varepsilon^2)]$ remains a possibility but requires a separate analysis. With the smaller size dimple shrinking to zero as κ tends to 1-, it would seem that the lowest mode axisymmetric bifurcation is to the dimple branch. In any event, axisymmetric bifurcation first occurs at a rather high pressure magnitude, namely $p_0 \cong p_s$ for $c = 2$ and $\varepsilon \ll 1$, compared to the critical magnitude p_{cr} for which the appearance of a dimple is possible; the linear buckling load is therefore not an important design consideration (unless polar dimpling is the preferred mode) compared to asymmetric bifurcation, say. For this reason, we did not attempt to describe in more details the eigenvalues and eigenfunctions than we did in sections (4) and (5) or the branching analysis for $\kappa = O(1)$. On the other hand, the analysis of post-buckling behavior and imperfection sensitivity should be of interest.

Notes

1. Within the framework of a shallow shell theory, there is no distinction between normal and axial loads.

References

[1] K.O. Friedrich, On the minimum buckling load for spherical shells , *Th. von Karman Anniversary Volume*, Caltech (Pasadena), 1941, 258-272.

[2] D.G. Ashwell, On the large deflexion of a spherical shell with an inward point load , *Proc. IUTAM Symp. on Thin Shells (Delft),* Ed. W.T. Koiter, North-Holand, Amsterdam, 1960, 43-63.

[3] C.V. Ranjan and C.R. Steele, Large deflection of deep spherical shells under concentrated load , *Proc. AIAA/ASME 18th Structures, Struc. Dynamics and Materials Conf.* (San Diego), 1977, 269-278.

[4] F.Y.M. Wan, The dimpling of spherical caps , *Mechanics Today* **5** (The Eric Reissner Anniversary Volume), 1980, 495-508.

[5] F.Y.M. Wan, Polar dimpling of complete spherical shells , *Proc. (3rd) IUTAM Sym. on Thin Shells (Tibilisi): Theory of Shells,* Ed. W.T. Koiter & G.K. Mikhailov, North-Holland, 1980, 589-605.

[6] F.Y.M. Wan & U. Ascher, Horizontal and flatpoints in shallow shell dimpling , *Proc. BAIL I Conf.* Ed. J. Miller, Boole Press, Dublin, 1980, 659-679.

[7] L.A. Taber, Large deflection of a fluid-filled spherical shell under a point load , *J. Appl. Mech.,* **49**, 1982, 121-128.

[8] F.Y.M. Wan, Shallow caps with a localized pressure distribution centered at the apex , *Proc. EUROMECH Colloq. (No. 165) on Flexible Shells (Munich)*, Ed. E. Axelrad & F. Emmerling, Springer-Verlag, Berlin-New York-Heidelberg, 1984, 124-145.

[9] D.F. Parker & F.Y.M. Wan, Finite polar dimpling of shallow caps under sub-buckling pressure loading , *SIAM J. of Appl. Math.*, **44** 1984, 301-326.

[10] E. Reissner, On axisymmetric deformations of thin shells of revolution , *Proc. Symp. in Appl. Math.,* Vol.III, Amer. Math. Soc., Providence, 1950, 27-52.

[11] E. Ressner, Symmetric bending of shallow shells of revolution , *J. Math Mech.,* **1**, 1958, 121-140.

[12] U. Ascher, J. Christiansen and R.D. Russell, A collocation solver for mixed order systems of boundary value problems , *Math. Comp.*, **33**, 1979, 659-679.

[13] U. Ascher, R.M.M. Mattheij & R.D. Russell, *Numerical Solutions of Boundary Value Problems for Ordinary Differential Equations*, (Classics in Applied Mathematics, vol. 13) SIAM Publications, Philadelphia, 1995.

[14] M. Abramowitz & I.A. Stegun, *Handbook of Mathematical Functions*, AMS No. 55 of National Bureau of Standards, 1964, U.S. Government Printing Office, Washington D.C.

[15] C.G. Lange & G.A. Kriegsmann, The axisymmetric branching behavior of complete spherical shells , *Quart. Appl. Math.*, **39**, 1981, 145-178.

Chapter 9

DUALITY FOR ENTROPY OPTIMIZATION AND ITS APPLICATIONS

Xingsi Li and Shaohua Pan
Department of Engineering Mechanics and Department of Applied Mathematics, Dahlia University of Technology, Dahlia, China, 116024.
Lixs@dlut.edu.cn pshcoral12@sina.com

Abstract This paper describes dual formulations of two entropy optimization principles, Jaynes' maximum entropy and Kullback-Leibler's minimum cross-entropy principles. Particular emphases are given to their applications in various optimization problems such as minimax, complementarity and nonlinear programming problems.

Keywords: Entropy optimization, Duality, Min-Max, Smoothing technique, Lagrangian perturbation.

1. Introduction

The concept of entropy was originated from thermodynamics and later extended to statistical mechanics and information theory. A significant movement for promoting the entropy to wide applications is the link of it with probability. Both of Jaynes' maximum entropy [1] and Kullback-Leibler's minimum cross-entropy [2] principles establish certain criteria to determine the probabilities based on "partial" (incomplete) information. The maximum entropy principle claims: "in making inference on the basis of partial information, we must use that probability distribution which has maximum entropy subject to whatever is known. This is the only unbiased assignment we can make; to use any other would amount to arbitrary assumption of information which by hypothesis we do not have." Mathematically, it is stated as an optimization problem

(E1):

$$
\begin{aligned}
\max\ S(\mathbf{p}) &:= -\sum_{i=1}^{n} p_i \ln p_i \\
s.t.\ \sum_{i=1}^{n} p_i f_{ji}(x) &= \mathrm{E}[f_j],\ j = 1,2,\ldots,m \\
\sum_{i=1}^{n} p_i &= 1,\ p_i \geq 0,\ i = 1,2,\ldots,n
\end{aligned}
\tag{1.1}
$$

where $\mathbf{p} = (p_1, p_2, \ldots, p_n)$ stands for the probability vector to be determined, $\mathrm{E}[f_j]$ denotes the jth moment obtained from some probabilistic experiments and $S(\mathbf{p})$ represents the Shannon entropy measure. It can be easily verified that this principle will choose a uniform distribution in case of no information (without any moment constraint). As a matter of fact, it always chooses a probability distribution as close as possible to a uniform distribution. If a prior probability $\mathbf{q} = (q_1, q_2, \ldots, q_n)$ is known, in addition to the moment constraints in (E1), one should determine the probability $\mathbf{p}$ based on the minimum cross-entropy principle that chooses $\mathbf{p}$ as close as possible to $\mathbf{q}$. Mathematically, it leads to the following entropy optimization problem (E2):

$$
\begin{aligned}
\min\ D(\mathbf{p}, \mathbf{q}) &:= \sum_{i=1}^{n} p_i \ln \frac{p_i}{q_i} \\
s.t.\ \sum_{i=1}^{n} p_i f_{ji}(x) &= \mathrm{E}[f_j],\ j = 1,2,\ldots,m \\
\sum_{i=1}^{n} p_i &= 1,\ p_i \geq 0,\ i = 1,2,\ldots,n
\end{aligned}
\tag{1.2}
$$

where $D(\mathbf{p}, \mathbf{q})$ stands for Kullback-Leibler's cross-entropy or relative entropy. At present, these two entropy optimization principles have found their applications in many disciplines throughout science and technology, which are directly or indirectly involved with randomness and uncertainty.

In this paper, we present their dual formulations. Fortunately, it turns out that these dual problems are simpler unconstrained optimization problems. It is this feature to motivate our study of the applications in establishing efficient algorithms for solving various optimization problems, including minimax, complementarity and nonlinear programming problems.

2. Duality for Entropy Optimization Problems

For the problem (E1), if define Lagrange multipliers ω for the normality constraint $\sum_{i=1}^{n} p_i = 1$ and u_j, $j = 1, 2, \ldots, m$ for the m expectation constraints, the Lagrange function $L(\mathbf{p}, \mathbf{u}, \omega)$ has the following form:

$$L(\mathbf{p}, \mathbf{u}, \omega) := -\sum_{i=1}^{n} p_i \ln p_i + \sum_{j=1}^{m} u_j \Big(\sum_{i=1}^{n} p_i f_{ji} - \mathrm{E}[f_j] \Big) + \omega \Big(\sum_{i=1}^{n} p_i - 1 \Big) \tag{2.1}$$

Accordingly, the dual problem (DE1) of (E1) is defined as

$$\min_{\mathbf{u}, \omega} \max_{\mathbf{p}} L(\mathbf{p}, \mathbf{u}, \omega) \tag{2.2}$$

Since $L(\mathbf{p}, \mathbf{u}, \omega)$ is a convex function of variable $\mathbf{p}$, the sufficient and necessary conditions for $\mathbf{p}^*$ being an optimal solution to the maximum problem in (2.2) are:

$$\frac{\partial L(\mathbf{p}, \mathbf{u}, \omega)}{\partial p_i} = -1 - \ln p_i + \sum_{j=1}^{m} u_j f_{ji} + w = 0, \quad i = 1, 2, \ldots, n$$

i.e. ,

$$p_i = \exp \Big(\sum_{j=1}^{m} u_j f_{ji} + w - 1 \Big) = 0, \quad i = 1, 2, \ldots, n \tag{2.3}$$

Consequently, substituting it into the objective function $L(\mathbf{p}, \mathbf{u}, \omega)$ of (2.2) to eliminate $\mathbf{p}$, we get the dual problem (DE1)

$$\min_{\mathbf{u}, \omega} D(\mathbf{u}, \omega) := \exp(\omega - 1) \Big[\sum_{i=1}^{n} \exp \Big(\sum_{j=1}^{m} u_j f_{ji} \Big) \Big] - w - \sum_{j=1}^{m} u_j \mathrm{E}[f_j] \tag{2.4}$$

Considering the convexity of $D(\mathbf{u}, \omega)$ in variable ω again, the optimal condition for ω^* being a solution of this dual problem

$$\frac{\partial D(\mathbf{u}, \omega)}{\partial \omega} = \exp(\omega - 1) \Big[\sum_{i=1}^{n} \exp \Big(\sum_{j=1}^{m} u_j f_{ji} \Big) \Big] - 1 = 0 \tag{2.5}$$

that is,

$$\omega = -\ln \Big\{ \sum_{i=1}^{n} \exp \Big(\sum_{j=1}^{m} u_j f_{ji} \Big) \Big\} + 1 \tag{2.6}$$

By this relationship, the dual problem (DE1) can be simplified as

$$\min D(\mathbf{u}) := \ln \Big\{ \sum_{i=1}^{n} \exp \Big(\sum_{j=1}^{m} u_j f_{ji} \Big) \Big\} - \sum_{j=1}^{m} u_j \mathrm{E}[f_j] \tag{2.7}$$

Moreover, combining (2.3) and (2.6) yields

$$p_i = \frac{\exp(\sum_{j=1}^m u_j f_{ji})}{\sum_{i=1}^n \exp(\sum_{j=1}^m u_j f_{ji})}, \quad i = 1, 2, \ldots, n \tag{2.8}$$

Similarly, we can derive the dual problem (DE2) of (E2)

$$\max D(\mathbf{u}, \mathbf{q}) := -\ln\Big\{\sum_{i=1}^n q_i \exp\Big(\sum_{j=1}^m u_j f_{ji}\Big)\Big\} + \sum_{j=1}^m u_j \mathrm{E}[f_j] \tag{2.9}$$

Note that, in this case,

$$p_i = \frac{q_i \exp(\sum_{j=1}^m u_j f_{ji})}{\sum_{i=1}^n q_i \exp(\sum_{j=1}^m u_j f_{ji})}, \quad i = 1, 2, \ldots, n \tag{2.10}$$

and the priori probability $\mathbf{q}$ is regarded as a parameter vector only.

From the dual theorem of convex program, the above discussions demonstrate that the primal entropy optimization problems could be solved by simply unconstrained dual programs, that is, substituting the solution $\mathbf{u}^*$ of (DE1) and (DE2) into the formula (2.8) and (2.10) respectively generates an optimal solution $\mathbf{p}^*$ to (E1) and (E2). In addition, it is very interesting to note that the object function of dual problem (2.7) has a formal implication in mechanics, in which the first term corresponds to the "strain energy", while the second term to "potential energy" of external forces.

3. The Applications of Duality for Entropy Optimization

From last section, we see, the solution of entropy optimization problems (E1) and (E2) could be obtained by solving their dual programs. The feature mainly stems from the ability for the primal counterparts being explicitly solved, and therefore lends themselves to various applications. In this section, we present several applications in establishing efficient algorithms for solving some deterministic optimization problems. These algorithms are developed by introducing the entropy function as a regularization technique and constructing certain artificial entropy optimization problems. The presentation is divided into two parts: one is based on some smoothing approximations to max function and the other comes from Lagrangian perturbations.

3.1 Smoothing technique for the finite min-max problem

Now, we derive a smooth approximation to the objective function of the finite min-max problem (MMP):

$$\min_{x} \varphi(x) := \max_{1\leq i\leq m} \{g_1(x), g_2(x), \ldots g_m(x)\} \tag{3.1}$$

where all component functions $g_i : \mathbb{R}^n \to \mathbb{R}$, $i = 1, 2, \ldots, m$ are assumed to be continuously differentiable. Nevertheless, this is a typical non-smooth optimization problem due to the non-differentiability of the max function $\varphi(x)$.

Since the min-max problem (MMP) frequently arises in numerical analysis and engineering design [3, 4, 5], and especially the max-type functions are often involved in many formulations of various optimization problems, a large number of algorithms have been devised to solve this problem, of which smooth approximations are worth of special attention. Observing the Lagrange function of (MMP)

$$L(x, \lambda) := \sum_{i=1}^{m} \lambda_i g_i(x), \quad \forall x \in \mathbb{R}^n, \ \lambda \in \Delta$$

where $\Delta := \left\{\lambda \in \mathbb{R}^m | \sum_{i=1}^{m} \lambda_i = 1, \lambda_i \geq 0, i = 1, 2, \ldots, m\right\}$, we find that the Lagrange multipliers have natural implications of probability and the max-valued function $\varphi(x)$ could be denoted as

$$\varphi(x) = \max_{\lambda\in\Delta} L(x, \lambda)$$

So, we introduce the Shannon's entropy and Kullback-Leibler's cross-entropy into the Lagrangian function $L(x, \lambda)$, respectively, and construct the following entropy optimization problems (PE1) and (PE2):

$$\max_{\lambda\in\Delta} L_p(x, \lambda) := \sum_{i=1}^{m} \lambda_i g_i(x) - p^{-1} \sum_{i=1}^{m} \lambda_i \ln \lambda_i \tag{3.2}$$

and

$$\max_{\lambda\in\Delta} L_p(x, \lambda, \mu) := \sum_{i=1}^{m} \lambda_i g_i(x) - p^{-1} \sum_{i=1}^{m} \lambda_i \ln \frac{\lambda_i}{\mu_i} \tag{3.3}$$

where $p > 0$ is a control parameter and $\mu \in int(\Delta)$ stands for the Lagrange multiplier vector obtained in last iteration. In virtue of the derivation in Section 2, it is easily shown that their dual problems transform the original minimax problem (MMP) into smooth unconstrained optimization problems:

$$\min_x \varphi_p(x) := p^{-1} \ln \Big\{ \sum_{i=1}^{m} \exp[pg_i(x)] \Big\} \tag{3.4}$$

$$\min_x \varphi_p(x, \mu) := p^{-1} \ln \Big\{ \sum_{i=1}^{m} \mu_i \exp[pg_i(x)] \Big\} \tag{3.5}$$

Furthermore, the following properties established in [6]

1. **error bound:** $\varphi(x) \leq \varphi_p(x) \leq \varphi(x) + \ln(m)/p$

$$\varphi(x) + p^{-1} \ln \textstyle\sum_{i \in I(x)} \mu_i \leq \varphi_p(x, \mu) \leq \varphi(x)$$

2. **monotonicity:** $\varphi_p(x) \geq \varphi_q(x), \quad p \leq q$

$$\varphi_p(x, \mu) \leq \varphi_q(x, \mu), \quad p \leq q$$

3. **approximation:** $\lim_{p \to +\infty} \varphi_p(x) = \varphi(x)$

$$\lim_{p \to +\infty} \varphi_p(x, \mu) = \varphi(x)$$

4. **convexity:** $\varphi_p(x)$ **and** $\varphi_p(x, \mu)$ **are convex if all** g_i **are convex**

where $I(x) := \{i \mid g_i(x) = \varphi(x),\ i = 1, 2, \ldots, m\}$, illustrate that smooth functions $\varphi(x)$ and $\varphi_p(x)$ uniformly approximate to the non-differentiable max function $\varphi(x)$ from above and below, respectively, that is,

$$\varphi_p(x, \mu) \leq \varphi(x) \leq \varphi_p(x) \quad \forall x \in \mathbb{R}^n, \ \mu \in \Delta$$

As a result, to solve the smooth unconstrained optimization problem (3.4) or (3.5) yields an approximate solution of the original non-smooth problem (MMP).

As the max function $\varphi(x)$ frequently appears in various formulations or reformulations of numerical analysis and optimization, there have left wide space for applications of the smooth functions $\varphi_p(x)$ and $\varphi_p(x, \mu)$. Compared with the former, the latter has an advantage of avoiding the ill-conditioning, with the cost of extra multiplier updates. To our experiences, however, there exists a "zero" pitfall of using the smooth function $\varphi_p(x, \mu)$, because updating of Lagrange multipliers is based upon

$$\hat{\mu}_i = \frac{\mu_i \exp[pg_i(x)]}{\sum_{i=1}^{m} \mu_i \exp[pg_i(x)]}, \quad i = 1, 2, \ldots, m$$

which may give rise to the instability even premature termination of algorithms at a non-optimal solution. To overcome this drawback needs very sophisticated care for algorithmic implementation. In the following, we restrict ourselves to the applications of $\varphi_p(x)$ partially due to these arguments. The appealing properties of this function have made it indispensable for certain smoothing methods. Now, we consider its applications in the following aspects:

3-1-1 Nonlinear programming problem

Consider the following nonlinear programming problem (NLP):

$$\begin{aligned} &\min_x \ f(x) \\ &s.t. \ \ g_i(x) \leq 0, \ i = 1, 2, \ldots, m \end{aligned} \tag{3.6}$$

where $f(x)$ and $g_i(x)$, $i = 1, 2, \ldots, m$ are continuously differentiable. The presence of inequality constraints brings the main difficulty to its solution.

A direct way is to exploit penalty functions to convert it into an unconstrained one. Here, we focus on two exact penalty functions, i.e., L_1 and L_∞ exact penalty functions. For the L_1 exact penalty method, (NLP) is transformed into the following non-smooth unconstrained optimization problem:

$$\min_x \ f(x) + \alpha \sum_{i=1}^{m} \max\{0, g_i(x)\}$$

whose solution becomes exactly optimal for the original problem as long as the penalty parameter $\alpha > \max_{1 \leq i \leq m} \lambda_i^*$, where λ_i^*, $i = 1, 2, \ldots, m$ denote the optimal Lagrange multipliers of original problem (3.6). For the L_∞ exact penalty method, (NLP) is converted to

$$\min_x \ f(x) + \alpha \max\left\{0, g_1(x), \cdots, g_m(x)\right\}$$

and the penalty factor $\alpha > \sum_{i=1}^m \lambda_i^*$ is required in this case. Replacing the above max-valued functions with their smooth functions through $\varphi_p(x)$, we can get an approximate solution of (NLP) by solving the smooth unconstrained optimization problem

$$\min_x \ f(x) + (\alpha/p) \sum_{i=1}^{m} \ln\left[1 + \exp(pg_i(x))\right] \tag{3.7}$$

or

$$\min_x \ f(x) + (\alpha/p) \ln\left\{1 + \sum_{i=1}^{m} \exp[pg_i(x)]\right\} \tag{3.8}$$

Thus, the smooth function $\varphi_p(x)$ makes it possible to solve the nonlinear programming problem with these non-differentiable exact penalty functions in practice. Following this line, we have proposed a quasi-exact penalty function method for solving (NLP) in [7].

Another way to resolve the difficulty of inequality constraints is that all constraints of the original problem are equivalently replaced, through the maximum operator, by singly constrained one

$$\begin{aligned} &\min_x \ f(x) \\ &s.t. \ \ \varphi(x) := \max_{1\leq i\leq m} \{g_i(x)\} \leq 0 \end{aligned} \tag{3.9}$$

and then the non-smooth $\varphi(x)$ is replaced by its uniform approximation $\varphi_p(x)$. As a consequence, the problem (NLP) with multiple constraints can be approximately solved by solving the following smooth and singly-constrained one:

$$\begin{aligned} &\min_x \ f(x) \\ &s.t. \ \ \varphi_p(x) \leq 0 \end{aligned} \tag{3.10}$$

In [8], such an approach is termed as the aggregate function method.

3-1-2 Complementarity problem

Consider the vertical nonlinear complementarity problem (VNCP):

$$x \geq 0, \ F_1(x) \geq 0, \cdots, F_m(x) \geq 0, \ x_j \prod_{i=1}^{m} F_i^j(x) = 0$$

where $F_i(x), \ i = 1, 2, \ldots, m$ are functions from $\mathbb{R}^n$ into itself and $F_i^j(x)$ denotes the jth component of $F_i(x)$. Obviously, in the special case of $m = 1$, the problem (VNCP) reduces to a general nonlinear complementarity problem (NCP):

$$x \geq 0, \ F_1(x) \geq 0, \ x^{\mathsf{T}} F_1(x) = 0$$

It is easy to verify that the problem (VNCP) is equivalent to the following non-smooth equations:

$$\begin{aligned} &\min\{x_j, F_1^j(x), \ldots, F_m^j(x)\} = \\ &-\max\{-x_j, -F_1^j(x), \ldots, -F_m^j(x)\} = 0 \end{aligned}$$

Once again, one can replace the maximum operator here by smoothing approximation with $\varphi_p(x)$ and convert it into a smooth system of equations. A non-interior continuation method for generalized linear complementarity problems is developed in [9].

3-1-3 Box constrained variational inequality problem (BVIP)

This problem is to find an $x \in [l, u]$ such that

$$(y - x)^{\mathsf{T}} F(x) \geq 0, \quad \forall y \in [l, u]$$

where $[l, u]$ is a box constraint in $\mathbb{R}^n$ with $l \leq u$. It is easy to show that the problem (BVIP) is equivalent to the system of equations:

$$x - mid\{l, u, x - F(x)\} = mid\{x - l, x - u, F(x)\} = 0$$

where the mid operator $mid\{a, b, c\}$ can be represented by

$$mid\{a, b, c\} = a + b + c - \min\{a, b, c\} - \max\{a, b, c\}$$

Here the max and min operators could be replaced by smoothing approximations with $\varphi_p(x)$ in proper forms.

3-1-4 Global optimization

The smooth approximations $\varphi_p(x)$ can be generalized for infinite case, i.e.,

$$\sup_{x \in X} f(x) \approx p^{-1} \ln \int_{x \in X} \exp[pf(x)] dx \tag{3.11}$$

which provides a framework for devising global optimization algorithms. This result probably dates back to Laplace, and which nowadays is explained and used in the context of "large deviation" in Probability and Statistics. In fact, it can be derived with continuous entropy function by a similar procedure to discrete case.

Of course, the applications of function $\varphi_p(x)$ are not limited as above. For instance, it has been extensively used in multi-objective optimization, bi-level programming, random programming, etc.

3.2 Lagrangian perturbation

The Lagrange function has played important role both in theoretical and algorithmic developments of optimization. This section employs Lagrangian perturbed by entropy function to derive the familiar exponential penalty functions, whereby revealing the dually equivalent relationship between them.

Consider the NLP problem (3.6), with resort to the Lagrangian

$$L(x, \lambda) := f(x) + \sum_{i=1}^{m} \lambda_i g_i(x), \quad \forall x \in \mathbb{R}^n, \ \lambda \in \mathbb{R}_+^m$$

it can be rewritten as

$$\min_x \sup_{\lambda \geq 0} L(x, \lambda) \tag{3.12}$$

because of the following fact

$$\sup_{\lambda \geq 0} L(x, \lambda) = \begin{cases} f(x) & x \in X \\ +\infty & x \notin X \end{cases}$$

where $X := \{x \in \mathbb{R}^n | \ g_i(x) \leq 0, \ i = 1, 2, \ldots, m\}$, or simplified as

$$\min_x \ \Phi(x)$$

by defining

$$\Phi(x) := \sup_{\lambda \geq 0} L(x, \lambda) \tag{3.13}$$

If the maximization contained in last equation could be analytically carried out in multiplier space, it would be suffice to perform an unconstrained minimization of $\Phi(x)$ in primal space only. Unfortunately, the linear form of $L(x, \lambda)$ in dual variables $\lambda_i, \ i = 1, 2, \ldots, m$ prevents from a continuous and analytical solution $\lambda(x)$.

To overcome this drawback of $L(x, \lambda)$, we employ entropy function $\xi(\lambda) := \sum_{i=1}^{m} \lambda_i \ln \lambda_i$ as a perturbing term to construct the following perturbed maximization problem

$$\sup_{\lambda \geq 0} L_p(x, \lambda) := f(x) + \sum_{i=1}^{m} \lambda_i g_i(x) - p^{-1} \xi(\lambda) \tag{3.14}$$

where $p > 0$ is a controlling parameter. As entropy function $\xi(\lambda)$ is strict convex in the closed convex set $\mathbb{R}_+^m$ and coercive, i.e., for every sequence $\{\lambda_k\}$ such that $\|\lambda_k\| \to +\infty$, it always holds

$$\lim_{k \to \infty} \xi(\lambda_k) = \infty,$$

the maximization problem (3.14) has a unique analytic solution $\lambda(x, p)$ for any $x \in \mathbb{R}^n$ and the objective function $L_p(x, \lambda)$ attains a finite value at this point. In view of (3.13), this maximization actually defines a perturbed function of $\Phi(x)$

$$\Phi_p(x) := L(x, \lambda(x, p)) = \sup_{\lambda \geq 0} L_p(x, \lambda) \tag{3.15}$$

Moreover, it can be proved that $\Phi_p(x)$ approximate to $\Phi(x)$ for any fixed $x \in \mathbb{R}^n$. By a simple computation, we obtain

$$\lambda_i(x,p) = \exp[pf_i(x) - 1], \; i = 1, 2, \ldots, m \tag{3.16}$$

On substituting it into the objective function $L_p(x, \lambda)$ to eliminate λ, we recover the familiar exponential penalty function

$$\Phi_p(x) = f(x) + p^{-1} \sum_{i=1}^{m} \exp[pg_i(x) - 1] \tag{3.17}$$

If the relative entropy function $\xi(\lambda, \mu) := \sum_{i=1}^{m} \lambda_i \ln \frac{\lambda_i}{\mu_i}$ is adopted as the perturbing term, the exponential multiplier penalty function

$$\Phi_p(x, \mu) = f(x) + p^{-1} \sum_{i=1}^{m} \mu_i \exp[pg_i(x) - 1] \tag{3.18}$$

can be derived in a very similar manner. As is well known, this corresponds to the multiplier methods of $\Phi_p(x)$ and the vector μ (≥ 0) of Lagrange multipliers has to be updated through iterations.

The above derivation demonstrates that the traditional exponential penalty functions have incorporated estimations of Lagrange multipliers and they are conjugate to some specific perturbed Lagrangian functions in nature. As a matter of fact, we can replace the entropy functions with general convex function $\xi(\lambda)$ or $\xi(\lambda, \mu)$ as perturbing functions to derive other penalty functions.

4. Concluding and Remarks

This paper derives the dual formulations of entropy optimization problems and presents their several applications in developing efficient algorithms, which reflect the important role of duality for entropy optimization in the field of mathematical programming. Of course, since entropy optimization problem itself originates from many different fields, the potential of the duality should not be limited in this presentation. Particularly, as entropy optimization problem is always interrelated with randomness and uncertainty, how to make use of it to devise algorithms for deterministic optimization problems awaits for further research.

References

[1] E. T. Jaynes. Where do we stand on maximum entropy. In R. D. Levine and M. Tribus, Editor, *The Maximum Entropy Formalism*, pages 15-118. MIT Press, Cambridge MA, 1978.

[2] S. Kullback. *Information theory and statistics.* John Wiley, New York, 1959.

[3] E. Polak. On the mathematical foundations of nondifferentiable optimization. *SIAM Review*, 29: 21-89, 1987.

[4] R. T. Rockafellar. Computational schemes for large-scale problems in extended linear-quadratic programming. *Mathematical Programming*, 48: 447-474, 1990.

[5] E. Polak, J. E. Higgins, and D. Q. Maynes. A barrier function method for minimax problems. *Mathematical Programming*, 54: 155-176,1992.

[6] X. S. Li. On the Entropic regularization method for solving minmax problem with applications. *Mathematical Methods of Operations Research*, 46: 119-130, 1997.

[7] X. S. Li. A smooth quasi-exact penalty function for nonlinear programming. *Chinese Science Bulletin*, 37: 806-609, 1991.

[8] X. S. Li. An aggregate function method for nonlinear programming. *Science China Series A*, 34: 1467-1473, 1991.

[9] J. M. Peng and Z. H. Lin. A non-interior-point continuation method for generalized linear complementarity problems. *Mathematical Programming*, 86: 533-563, 1999.

Chapter 10

DUAL VARIATIONAL PRINCIPLES FOR THE FREE-BOUNDARY PROBLEM OF CAVITATED BEARING LUBRICATION

Gao-Lian Liu

Shanghai Institute of Appl. Maths. & Mechanics, Shanghai University, Shanghai 200072, China

E-mail: *liu_gaolian@hotmail.com*

Abstract: The free-boundary problem in cavitated bearing lubrication is handled in a systematic way via the powerful mathematical tool -- the functional variation with variable domain, resulting in a pair of dual variational principle families for it. In this way, a new rigorous and sound theoretical foundation for the finite element analysis of the cavitated bearing lubrication problem is founded.

Keywords: free-boundary value problem, dual variational principle, lubrication

1. Introduction

Since the early 1900s, tremendous progress has been made in fluid-film lubrication [1,2]. This activity in hydrodynamic lubrication theory stems from the fact that every machine with moving parts can not successfully operate under unsuccessful performance of each bearing. The rapid developments in computers and computational fluid dynamics since 1960 have made possible the prediction of bearing performance via solving the relevant Reynolds equation by numerical methods such as finite difference method, finite element method etc. [2,4]. However, this prediction can be done successfully only for cases with full oil films of positive pressure and does not hold true when film rupture occurs due to cavitation. The Reynolds equation is of elliptic type and its unique solution requires boundary conditions of Dirichlet or Neumann type for the pressure. Sometimes, this mathematically properly-posed boundary-value problem may predict negative pressure in some parts of the oil film. This is physically impossible and we observe the formation of cavitation bubbles filled with oil vapor. This phenomenon is called partial lubrication and the main problem is that the position and shape of the cavities are unknown a priori (free boundary). The existence of the free boundary makes the problem nonlinear and very difficult to solve even numerically. On the other hand, we know [5-8] that from the viewpoint of calculus of variations every unknown (free) boundary/interface can be regarded as a variable-domain boundary and hence can be treated conveniently via the powerful tool: variation with variable domain. Such an approach is just employed in the present paper. It should be noted that in the last two decades the free boundary problem is handled mostly by means of variational inequalities [3,4]. It is anticipated that the present approach based on variable-domain variational principles joined with the deforming finite element [5] will be a powerful competitor of it.

2. BASIC EQUATIONS FOR HYDRODYNAMIC LUBRICATION

The governing equation is the Reynolds equation in the following dimensionless form (Fig. 1):

$$\frac{\partial}{\partial x}\left(h^3 \frac{\partial p}{\partial x}\right)+\frac{\partial}{\partial z}\left(h^3 \frac{\partial p}{\partial z}\right)=\frac{dh}{dx} \tag{1}$$

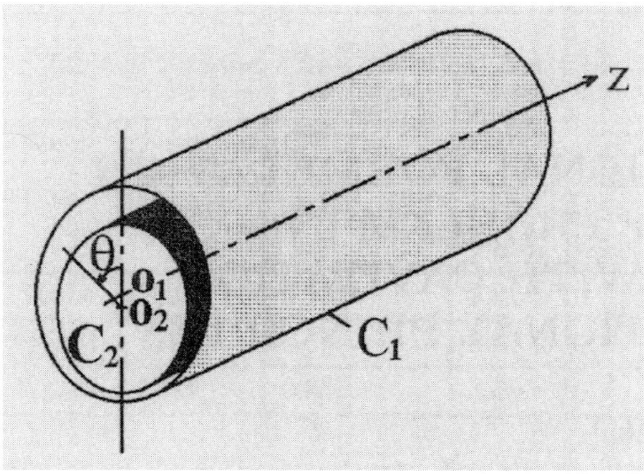

Fig. 1 Journal bearing

where

$x=\theta r_2 / D$, $z=z'/D$, $l=o_1 o_2$, $D=2\pi r_2$, $U=\omega r_2$, $\varepsilon=l/(r_1-r_2)$,
$h=h'/(r_1-r_2)=1+\varepsilon\cos 2\pi x$, $p=(p'-p'_c)(r_1-r_2)^2/(6\mu UD)$,
and p'_c is pressure in cavitated region, h' is oil film thickness, the superscript (') denotes the dimensional quantities.

Introducing a vector $\vec{q}$ defined by

$$\vec{q}=h^3\nabla p \tag{2}$$

Eq. (1) can be rewritten as

$$\nabla\cdot\vec{q}=dh/dx$$

or

$$\frac{\partial(q_1-h)}{\partial x}+\frac{\partial q_2}{\partial z}=0 \tag{3}$$

Boundary conditions (B.C.)

Assuming that the boundary $\partial\Omega$ of the solution domain Ω ($\Omega = \Omega_0 \cup \Omega_c$) is composed of 2 segments: $\partial\Omega = \partial\Omega_0 \cup \partial\Omega_c$ where $\partial\Omega_c$ is the boundary of the cavitated subdomain Ω_c with the pressure $p_c = 0$, the boundary conditions can be formulated as follows [3,4]

$$\text{on } \partial\Omega_0\text{:} \quad p = 0 \tag{4}$$

$$\text{on } \partial\Omega_c\text{:} \quad p = 0 \tag{5a}$$

$$\text{and } \partial p / \partial n = 0 \tag{5b}$$

where $\partial\Omega_c$ is a free boundary. Thus, the present lubrication problem is fully formulated by the Eqs. (1), (4) and (5a, b) or by Eqs. (2), (3), (4) and (5a, b).

3. PRIMARY VARIATIONAL PRINCIPLE (VP) FAMILY

3.1 VP-I

Starting from Eqs. (1), (4) and (5a, b) the following variable-domain VP for the entitled problem can be established by means of the inverse deduction method suggested in Ref. [8]: $\delta J_I = 0$ and

$$J_I(p, \partial\Omega_c) = \iint_{(\Omega_0)} \left\{ \frac{h^3}{2} (\nabla p)^2 + p \frac{dh}{dx} \right\} dxdz \tag{6}$$

To show this, it is necessary to employ the following formula for the functional variation with variable domain [5,6]

$$\delta J = \iint_{(\Omega_0)} \left(\frac{\partial F}{\partial p} - \nabla \cdot \vec{G} \right) \tilde{\delta} p dxdz + \oint_{\partial\Omega} G_n \delta p ds + \int_{\partial\Omega_c} \left\{ \left(F - G_n \frac{\partial p}{\partial n} \right) \delta e_n - G_n \frac{\partial p}{\partial s} \delta e_s \right\}$$

(7)

where F is the function under integration in Eq. (6), $\vec{e}$ is the position vector of the boundary $\partial\Omega_c$, $\tilde{\delta}$ is the fixed-domain variation,

$$\vec{G} = \frac{\partial F}{\partial p_x}\vec{i} + \frac{\partial F}{\partial p_z}\vec{j} = h^3 \nabla p, \quad \frac{\partial F}{\partial p} = \frac{dh}{dx}, \quad p_x = \frac{\partial p}{\partial x}.$$

Upon substituting in Eq. (7) we obtain

$$\delta J_I = \iint_{(\Omega_0)} \left[\frac{dh}{dx} - \frac{\partial}{\partial x}\left(h^3 \frac{\partial p}{\partial x} \right) - \frac{\partial}{\partial z}\left(h^3 \frac{\partial p}{\partial z} \right) \right] \tilde{\delta} p dxdz + \oint_{\partial\Omega} h^3 \frac{\partial p}{\partial n} \delta p ds$$

$$+ \int_{\partial\Omega_c} \left\{ \left(\frac{1}{2} h^3 \left[\left(\frac{\partial p}{\partial s} \right)^2 - \left(\frac{\partial p}{\partial n} \right)^2 \right] + \frac{dh}{dx} p \right) \delta e_n - h^3 \frac{\partial p}{\partial n} \frac{\partial p}{\partial s} \delta e_s \right\} ds \qquad (8)$$

where the subscripts s, n denote the tangential and normal components to the boundary respectively.

Putting $\delta J_I = 0$, we get the following set of stationary conditions of J_I :

Euler equation: Eq. (1)

Natural BC:

On $\partial\Omega_0$: Essential BC: $p = 0$

On $\partial\Omega_c$: There may be two alternative approaches:

(i) Using $p = 0$ as essential BC, we obtain the natural BC: $\partial p / \partial n = 0$

(ii) From the integral term along $\partial\Omega$ in Eqs. (8) we obtain the natural BC: $\partial p / \partial s = 0$, then the integral term along $\partial\Omega_c$ yields another natural BC:

$$\frac{1}{2} h^3 \left(\frac{\partial p}{\partial s} \right)^2 + \frac{dh}{dx} p = 0 \qquad (9)$$

Noting that at the two endpoints of $\partial\Omega_c$ the pressure is zero, so the solution of Eq. (9) must be $p = 0$. Thus, we have a natural BC: $p = 0$.

Obviously, both 2 approaches give the required BC Eqs. (5A, B), so either of them can be employed in practice. It seems simpler to use the second approach, but in the finite element solution the natural BC $p = 0$ is satisfied with less accuracy than the first approach.

This VP-I was also obtained in Ref. [10] but in an alternative way.

3.2 GENERALIZED VP-II

Let us apply the constraint-removing transformation [8,9] to VP-I. The following augmented functional is constructed by adjoining Eq. (2) to J_I in Eq. (6) by Lagrange multipliers λ_1 and λ_2:

$$J'_I(p,\vec{q},\lambda_1,\lambda_2,\partial\Omega_c)= \iint_{(\Omega_0)}\left\{\frac{1}{2h^3}\left(q_1^2+q_2^2\right)+\lambda_1\left(h^3\frac{\partial p}{\partial x}-q_1\right)+\lambda_2\left(h^3\frac{\partial p}{\partial z}-q_2\right)+\frac{dh}{dx}p\right\}dxdz \tag{10}$$

From $\delta J'_I=0$, the multipliers can be identified:

$$\lambda_1=\frac{1}{h^3}q_1,\qquad \lambda_2=\frac{1}{h^3}q_2 \tag{11}$$

Inserting Eq. (11) in Eq. (10) we obtain:

Generalized VP-II: $\delta J_{II}=0$ with independent variation of p, $\vec{q}$ and free boundary $\partial\Omega_c$, and

$$J_{II}(p,\vec{q},\partial\Omega_c)=\iint_{(\Omega_0)}\left\{-\frac{1}{2h^3}\left(q_1^2+q_2^2\right)+\vec{q}\cdot\nabla p+\frac{dh}{dx}p\right\}dxdz \tag{12}$$

Applying the general formula for variable-domain variation Eq. (7) to Eq. (12), it is easy to show that the stationary conditions for J_{II} are:

Euler equations: Eqs. (2) and (3).

Natural BC: the same as in VP-I.

3.3 SUB-GENERALIZED VP-III

Applying the constraint-recovering transformation [8] to the generalized VP-II, we obtain a series of sub-generalized VP, of which only one is given here:

$\delta J_{III}=0$, where

$$J_{III}(p,q,\partial\Omega_c)=\iint_{(\Omega_0)}\left\{-\frac{1}{2h^3}q_1^2+q_1\frac{\partial p}{\partial x}+\frac{h^3}{2}\left(\frac{\partial p}{\partial z}\right)^2+\frac{dh}{dx}p\right\}dxdz \tag{13}$$

It is easy to show that applying Eq. (7) and taking the independent variations of p, q_1 and the free boundary $\partial\Omega_c$, we arrive at the following stationary conditions for J_{III}:

Euler equations:

$$q_1 = h^3 \frac{\partial p}{\partial x}$$

and
$$\frac{\partial (q_1 - h)}{\partial x} + \frac{\partial}{\partial z}\left(h^3 \frac{\partial p}{\partial z} \right) = 0,$$

which are equivalent to Eq. (1).

Natural BC: the same as in VP-I.

4. DUAL VARIATIONAL PRINCIPLE FAMILY

Another family of variational principles which is dual with respect to the primary one obtained above can be derived via a systematic approach suggested in Ref. [8] as follows.

4.1 Dual VP-IV

To derive the dual VP of the VP-I, first of all, we integrate Eq. (12) by parts and then apply the involutory transformation to it, disregarding the boundary integral term, we obtain

$$J'_{II}\left(p, \vec{q}, \partial\Omega_c\right) = -\iint_{(\Omega_0)} \left\{ \frac{1}{2h^3}\left(q_1^2 + q_2^2\right) + p\left(\frac{\partial q_1}{\partial x} + \frac{\partial q_2}{\partial z} - \frac{dh}{dx} \right) \right\} dxdz \tag{14}$$

If Eq. (3) is used as a variational constraint and let $q^2 = q_1^2 + q_2^2$, the Eq. (14) can be reduced to

$$J'_{III}\left(\vec{q}, \partial\Omega_c\right) = -\iint_{(\Omega_0)} \frac{1}{2h^3} q^2 \cdot dxdz \tag{15}$$

It should be noted that in this equation the pressure has disappeared. If a new unknown function $\Pi(x, y)$ called the dual variable to p is introduced by

$$\frac{\partial \Pi}{\partial x} = q_2, \qquad \frac{\partial \Pi}{\partial z} = h - q_1 \tag{16}$$

and the vorticity of Eq. (2) is taken

$$\frac{\partial}{\partial x}\left(\frac{q_2}{h^3} \right) - \frac{\partial}{\partial z}\left(\frac{q_1}{h^3} \right) = 0 \tag{17a}$$

The following partial differential equation for the dual variable Π can be obtained

$$\frac{\partial}{\partial x}\left(\frac{1}{h^3}\frac{\partial \Pi}{\partial x}\right)+\frac{\partial}{\partial z}\left(\frac{1}{h^3}\frac{\partial \Pi}{\partial z}\right)=0 \tag{17b}$$

Employing Eq. (16) as the constraints of Eq. (15) in place of Eq. (3) and inserting in Eq. (15) to eliminating q, the following VP-IV which is dual to VP-I is obtained: $\delta J_{IV}=0$

$$J_{IV}(\Pi,\partial\Omega_c)=-\iint_{(\Omega_0)}\frac{1}{2h^3}\left\{\left(h-\frac{\partial \Pi}{\partial z}\right)^2+\left(\frac{\partial \Pi}{\partial x}\right)^2\right\}dxdz \tag{18}$$

Upon applying the variable-domain variation formula Eq. (7) to J_{IV} and taking the variable-domain variation of Eq. (18), we get

$$\begin{aligned}\delta J_{IV} = &\iint_{(\Omega_0)}\left\{\frac{\partial}{\partial x}\left(\frac{1}{h^3}\frac{\partial \Pi}{\partial x}\right)+\frac{\partial}{\partial z}\left(\frac{1}{h^3}\frac{\partial \Pi}{\partial z}\right)\right\}\widetilde{\delta\Pi}dxdz\\ &+\int_{(\partial\Omega_0)}\frac{1}{h^3}\left(hn_z-\frac{\partial \Pi}{\partial n}\right)\delta\Pi ds\\ &+\int_{\partial\Omega_c}\left\{\begin{aligned}&\frac{1}{h^3}\left(hn_z-\frac{\partial \Pi}{\partial n}\right)\delta\Pi+\left[\frac{-1}{2h^3}q^2+\frac{1}{h^3}\left(\frac{\partial \Pi}{\partial n}-n_z h\right)\frac{\partial \Pi}{\partial n}\right]\delta e_n\\ &+\frac{1}{h^3}\left(\frac{\partial \Pi}{\partial n}-n_z h\right)\frac{\partial \Pi}{\partial s}\delta e_s\end{aligned}\right\}ds\end{aligned} \tag{19}$$

and from $\delta J_{IV}=0$ the following stationary conditions for J_{IV} result :

Euler equation: Eq. (17b)

Natural BC:

on $\partial\Omega_0$: $\frac{\partial \Pi}{\partial n}=hn_z$, leading to $\frac{\partial p}{\partial s}=0$ (20a)

on $\partial\Omega_c$:

1) $\delta\Pi$: $\frac{\partial \Pi}{\partial n}=hn_z$, leading to $\frac{\partial p}{\partial s}=0$

2) δe_n: $q = 0$, i.e. $|\nabla p| = 0$, leading to Eq. (5B). (20b)

3) δe_s: $\dfrac{\partial p}{\partial s} = 0$

It should be pointed out at this point that:

(1) Owing to the fact that the dual variable Π appears in all equations (16)-(18) exclusively in form of partial derivatives, so, to ensure the uniqueness of the solution, it is necessary and sufficient to assign a value of Π at some selected point of the flow field.

(2) The natural BC Eqs. (20a, b) show only that the pressure is constant along the boundaries, so it seems that they might not be identical with the original BC Eqs. (4) & (5a). However, this result is correct and reasonable, because this VP-IV includes the dual variable Π (but not the primary variable p). After $\Pi(x, y)$ is computed, the q-field can be obtained from Eq. (16). Then the p-field should be solved from Eq. (2) by using the BC Eqs. (4) & (5a).

Thus, we can see that this dual VP-IV is really capable of giving the solution to the present lubrication problem.

4.2 Dual Generalized VP-V

Applying the constraint-removing transformation to Eq. (15) by employing two Lagrange multipliers λ_3 and λ_4 to join the constraint Eq. (16), we obtain the following augmented functional J'_{IV}

$$J'_{IV}\left(\Pi, \vec{q}, \lambda_3, \lambda_4, \partial\Omega_c\right) = \iint_{(\Omega_0)} \left\{ \frac{-1}{2h^3} q^2 + \lambda_3 \left(\frac{\partial \Pi}{\partial x} - q_2 \right) + \lambda_4 \left(\frac{\partial \Pi}{\partial z} + q_1 - h \right) \right\} dxdz \tag{21}$$

Taking the first variation of J'_{IV} and from $\delta J'_{IV} = 0$ the multipliers λ_3 and λ_4 can be identified as follows:

$$\delta q_{1,2}:\ \lambda_3 = \frac{-q_2}{h^3}, \qquad \lambda_4 = \frac{q_1}{h^3} \tag{22}$$

$\delta\Pi$: $\dfrac{\partial \lambda_3}{\partial x} + \dfrac{\partial \lambda_4}{\partial z} = 0$, leading to Eq. (17A).

Upon substituting Eq. (22) in Eq. (21) the following dual generalized VP results:

$$J_V\left(\Pi, \vec{q}, \partial\Omega_c\right) = \iint_{(\Omega_0)} \left\{ \frac{1}{2h^3} \left[q^2 - 2q_2 \frac{\partial \Pi}{\partial x} + 2q_1 \left(\frac{\partial \Pi}{\partial z} - h \right) \right] \right\} dxdz \tag{23}$$

Using the formula for the variable-domain variation Eq. (7) to handle the free boundary $\partial\Omega_c$, it is easy to show that the stationary conditions of J_V have the form:

Euler equations: Eqs. (17A) and (16).

Natural BC: the same as in VP-IV

5. CONCLUDING REMARKS

The free-boundary problem in cavitated lubrication has been successfully handled via the powerful mathematical tool – the functional variation with variable domain, leading to a pair of dual VP and the related families of sub-generalized and generalized VP. Thus, a rigorous and sound theoretical basis for the finite element solution of the lubrication problem is founded. As for the numerical realization and implementation of the variable-domain variation at the free boundary it is necessary to employ the special self-deforming finite element models and the interested reader is advised to refer to the Refs. [5,7] for details.

Acknowledgement: The support of the National Natural Science Foundation of China is gratefully acknowledged.

REFERENCES

[1] Cameron, A., Principle of Lubrication. Longmans, London, 1966

[2] Dowson, D. & Taylor, C.M., Cavitation in Bearings. Annual Rev. Fluid Mechanics, **11** (1979): 35-60

[3] Capriz, G. & Cimatti, G., Free boundary problems in the theory of hydrodynamic lubrication: A survey. Free boundary problems: Theory

and applications, Vol. II, Fasano, A. & Primicesio, M. (eds), Pitman, Boston, 1983 (Research Notes in Maths. 79) 613-635

[4] Cuvelier, C., A free boundary problem in hydrodynamic lubrication including surface tension. Proc. 6th Intl. Conf. Num. Methods in Fluid Dynamics, Cabannes, H. et al. (eds.), Springer, Berlin, 1979 (Lecture Notes in Physics 90) 143-148

[5] Liu, Gao-Lian. Variable-domain variational finite element method: A general approach to free/moving boundary problems in heat and fluid flow, Nonlinear Analysis, **30** (1997) 5229-5239

[6] Liu, Gao-Lian, Variational principles (VP) and generalized VP for fully 3-D transonic flow with shocks in a turbo-rotor (I), Acta Mechanica, **95** (1992): 117-130

[7] Liu, Gao-Lian & Yan, Shan, A unified variable-domain variational approach to hybrid problems of compressible blade-to-blade flow. Intl. J. Turbo & Jet Engines, **10** (1993): 273-284

[8] Liu, Gao-Lian, Derivation and transformation of VP with emphasis on inverse and hybrid problems in fluid dynamics: A systematic approach, Acta Mechanica, **140** (2000): 73-89 [or: Chinese J. Engrg. Thermophysics, **11** (1990) 136-142]

[9] Chien, Wei-Zang, Generalized VP, Intelligence Press, Shanghai, 1985

[10] Rohde, S.M. & McAllister, G.T., A variational formulation for a class of free boundary problems arising in hydrodynamic lubrication. Int. J. Engrg. Sci., **13** (1975): 841-850

[11] Courant R. & Hilbert D., Methods of Mathematical Physics, Vol. I, Interscience, 1953

[12] Liu, Gao-Lian, Variable-domain variational formulation of free-boundary problem in bearing lubrication with cavitation, Pt. I, Computational Fluid Dynamics Journal, **10** (2002) 539-541

Chapter 11

FINITE DIMENSIONAL FRICTIONAL CONTACT QUASI–STATIC *RATE* AND *EVOLUTION* PROBLEMS REVISITED

J.A.C. Martins
Instituto Superior Técnico, Dep. Eng. Civil e Arquitectura and I.C.I.S.T., Av. Rovisco Pais, 1049–001 Lisboa, Portugal
jmartins@civil.ist.utl.pt

A. Pinto da Costa
Instituto Superior Técnico, Dep. Eng. Civil e Arquitectura and I.C.I.S.T., Av. Rovisco Pais, 1049–001 Lisboa, Portugal
apcosta@civil.ist.utl.pt

Abstract In this paper we revisit the quasi-static *rate* and the quasi-static *evolution* problems for finite dimensional nonlinear elastic systems, in frictional contact with curved obstacles. Our objectives are: (i) to present complementarity formulations for the quasi-static rate problem; (ii) to discuss conditions for existence and uniqueness of solution for that problem; (iii) to present a differential inclusion formulation and related mathematical results for the quasi–static evolution problem; (iv) to present numerical results for a specific finite element example of (non–) uniqueness of solution for the quasi–static rate problem.

Keywords: friction, bifurcation, quasi–static trajectory, rate problem, complementarity, finite elements

Introduction

The problems addressed in this paper are the quasi-static *rate* and the quasi–static *evolution* problems for finite dimensional nonlinear elastic systems, in frictional contact with curved obstacles. The quasi–static evolution problem consists of finding the time evolution of the system configuration and contact reactions, for given variation in time of the applied forces. The quasi-static rate problem consists of finding, for the equilibrium configuration that exists at some time and for a given external loading rate at that time, the first order right rate of change of configuration and reactions. Inertia effects are not taken into account in both problems.

Our purposes in this paper are the following:

(*i*) To present complementarity formulations for the quasi–static (first order) rate problem at equilibrium configurations of a finite dimensional frictional contact system. The unilateral frictional contact conditions are characterized by imposing that the static and the kinematic variables (or some transformed quantities of them) belong to some admissible mutually dual cones, and satisfy appropriate complementarity (orthogonality) conditions. This kind of formulation has much in common with the one adopted earlier in (De Saxce and Feng, 1991) for (non-associative) friction and elastic–plastic problems. Similarly to (Vola et al., 1999), the present formulation of the rate problem takes into account not only the particles in contact with impending slip but also those in grazing contact. Moreover, the complementarity form adopted for the unilateral frictional contact conditions naturally leads to NCP (Nonlinear Complementarity Problem) formulations (mixed *explicit* complementarity problems, (Hyers et al., 1997)) for the first order rate problem. Other related, but *different*, formulations can be found in (Martins et al., 2002). Early complementarity formulations for the same problem can be found in (Klarbring, 1999).

(*ii*) To discuss conditions for existence and uniqueness of solution for the above problem. Whenever possible, and similarly to (Martins et al., 2002), these conditions are presented in terms of properties of some *effective* stiffness matrices that incorporate the elastic and the geometrically nonlinear load effects in usual tangent stiffness matrices as well as obstacle curvature and friction effects. Similarly to what happens in the 2 degree–of–freedom example of Klarbring (Klarbring, 1990), and with some similarity also to the "instantaneous" dynamic problem studied in (Trinkle et al., 1997), we emphasize here the different role that the particles in grazing contact and in impending sliding have on (non–) existence and (non–) uniqueness of solution (Martins et al., 2002).

(*iii*) To present a differential inclusion formulation for the quasi-static evolution problem, which generalizes, to the case of arbitrary generalized coordinates and curved obstacles, the one recently proposed in (Andersson et al., 1999). The present formulation allows a discussion of the mathematical issue of existence of solution that is shorter than the one in (Andersson et al., 1999); here a straightforward application of general mathematical results in e.g. (Deimling, 1992) becomes possible.

(*iv*) To present numerical results for a specific finite element example of (non–) uniqueness of solution for the quasi–static rate problem. These numerical studies are performed with the algorithm of de Moor (De Moor et al., 1992), which is capable of computing all the solutions of a generalized linear complementarity problem.

1. Kinematics and quasi–static evolutions

We consider a finite dimensional nonlinear elastic system with plane motion. At each time its configuration is described by the vector of generalized coordinates $\mathbf{X} \in \mathbb{R}^N$. A finite number of particles of the system may have unilateral contact with smooth curved fixed obstacles. The set $\mathcal{P}_C \subset \mathbb{N}$ groups the labels (p) of the Contact candidate particles ($\#\mathcal{P}_C = n_C$). For each contact candidate particle p, the corresponding obstacle is identified by the set of vectors $\mathbf{X} \in \mathbb{R}^N$ such that $\Phi^p(\mathbf{X}) = 0$. We assume that in the absence of unilateral constraints the components X_i of $\mathbf{X}$ are independent. The components of the velocity and of the reaction vectors of all contact candidate particles in local orthonormal bases $(\mathbf{n}^p, \mathbf{t}^p)$ are grouped in the vectors $\mathbf{v}$ and $\mathbf{r}$, which belong to $\mathbb{R}^{2n_C}$; the vectors $\mathbf{n}^p$ point towards the interior of the corresponding obstacle. The relations between the vectors $\mathbf{v}$ and $\mathbf{r}$ and the generalized velocity and reaction vectors $\mathbf{V}$ and $\mathbf{R}$ are expressed by

$$\mathbf{v} = \mathbf{G}\mathbf{V}, \qquad \mathbf{R} = \mathbf{G}^T\mathbf{r}. \tag{1.1}$$

It is assumed that

$$\textit{the } 2n_C \times N \textit{ matrix } \mathbf{G} \textit{ has maximal rank.} \tag{1.2}$$

In the finite element case the matrix $\mathbf{G}$ has the form $\mathbf{G} = [\mathbf{0} \quad \mathbf{Q}]$, where $\mathbf{0}$ is the $2n_C \times (N - 2n_C)$ null matrix and $\mathbf{Q}$ is a $2n_C \times 2n_C$ block diagonal matrix whose 2×2 diagonal blocks are orthogonal matrices that relate the normal and tangential velocity and reaction components of the contact candidate nodes with their components along global cartesian axes.

For any contact candidate particle $p \in \mathcal{P}_C$, the classical *unilateral contact conditions* and the *friction law of Coulomb* at some time $t \geq 0$

are

$$\Phi^p(\mathbf{X}(t)) \leq 0, \qquad r_n^p(t) \leq 0, \qquad \Phi^p(\mathbf{X}(t))\, r_n^p(t) = 0 \tag{1.3}$$

and

$$r_t^p(t) \in \mu r_n^p(t)\, \sigma[v_t^p(t)], \tag{1.4}$$

respectively; $\mu \geq 0$ is the coefficient of friction, $\sigma[.]$ denotes the multi–valued application such that, for each $x \in \mathbb{R}$,

$$\sigma[x] = \begin{cases} x/|x|, & \text{if } x \neq 0, \\ [-1, +1], & \text{if } x = 0. \end{cases} \tag{1.5}$$

The friction law can be alternatively written in the equivalent form of an inequality plus an equality:

$$|r_t^p(t)| + \mu r_n^p(t) \leq 0, \qquad r_t^p(t) v_t^p(t) - \mu r_n^p(t) |v_t^p(t)| = 0. \tag{1.6}$$

For each $\mathbf{X} \in \mathbb{R}^N$ the decomposition of the set $\mathcal{P}_C$ of contact candidate particles is introduced

$$\mathcal{P}_C = \mathcal{P}_f(\mathbf{X}) \cup \mathcal{P}_c(\mathbf{X}),$$

$$\mathcal{P}_f(\mathbf{X}) \doteq \{p \in \mathcal{P}_C : \Phi^p(\mathbf{X}) < 0\}$$
$$\text{[particles currently not in contact } (\textit{free})],$$

$$\mathcal{P}_c(\mathbf{X}) \doteq \{p \in \mathcal{P}_C : \Phi^p(\mathbf{X}) \geq 0\}$$
$$\text{[particles currently in } \textit{contact}].$$

The *set of admissible configurations* is defined by

$$\mathcal{K}_{\mathbf{X}} \doteq \{\mathbf{X} \in \mathbb{R}^N : \Phi^p(\mathbf{X}) \leq 0, \text{ for all } p \in \mathcal{P}_C\}. \tag{1.7}$$

The *(configuration dependent) set of admissible right velocities* of the contact candidate particles is

$$\mathcal{K}_{\mathbf{v}}(\mathbf{X}) = \left\{\mathbf{v} \in \mathbb{R}^{2n_C} : v_n \leq 0, \text{ in } \mathcal{P}_c(\mathbf{X})\right\}. \tag{1.8}$$

In what concerns the reaction forces, we define, for each $\mathbf{X} \in \mathbb{R}^N$, the *(configuration dependent) set of admissible reaction forces*:

$$\begin{aligned} \mathcal{K}_{\mathbf{r}}(\mathbf{X}) \doteq \{\mathbf{r} \in \mathbb{R}^{2n_C} : r_n = r_t = 0, \text{ in } \mathcal{P}_f(\mathbf{X}); \\ r_n \leq 0 \text{ and } |r_t| + \mu r_n \leq 0, \text{ in } \mathcal{P}_c(\mathbf{X})\}. \end{aligned} \tag{1.9}$$

As done in (1.8) and (1.9), the dependency of some (local) contact related quantities on the contact particle p may be omitted in some circumstances, i.e. v_n^p, v_t^p, r_n^p and r_t^p are sometimes abbreviated to v_n, v_t, r_n and r_t, respectively.

The quasi–static evolution problem consists of finding the time evolution of the system configuration and contact reactions, for given variation in time of the applied forces. $\mathbf{F}^U(\mathbf{X})$ is the vector of the generalized elastic forces, $\mathbf{F}^\Omega(\mathbf{X},\Lambda)$ is the vector of the generalized external forces, Λ is a loading parameter, and the sum of $\mathbf{F}^U$ with $\mathbf{F}^\Omega$ is denoted by $\mathbf{F}$: $\mathbf{F}(\mathbf{X},\Lambda) = \mathbf{F}^U(\mathbf{X}) + \mathbf{F}^\Omega(\mathbf{X},\Lambda)$. For initial configuration $(\mathbf{X}_0)$ and reaction $(\mathbf{r}_0)$ satisfying (see (1.7) and (1.9))

$$\mathbf{X}_0 \in \mathcal{K}_\mathbf{X}, \qquad \mathbf{r}_0 \in \mathcal{K}_\mathbf{r}(\mathbf{X}_0) \tag{1.10}$$

$$\mathbf{F}(\mathbf{X}_0, \Lambda(0)) + \mathbf{G}^T(\mathbf{X}_0)\mathbf{r}_0 = \mathbf{0}, \tag{1.11}$$

a quasi–static evolution $(\mathbf{X}(t), \mathbf{r}(t))$ is characterized by absolutely continuous functions $\mathbf{X}(t)$ and $\mathbf{r}(t)$ such that the equilibrium equations

$$\mathbf{F}(\mathbf{X}(t), \Lambda(t)) + \mathbf{G}^T(\mathbf{X}(t))\mathbf{r}(t) = \mathbf{0}, \tag{1.12}$$

and the unilateral contact conditions (1.3) are satisfied for all $t \geq 0$, the Coulomb friction law (1.4) is verified for almost all $t \geq 0$ and the initial conditions

$$\mathbf{X}(0) = \mathbf{X}_0, \qquad \mathbf{r}(0) = \mathbf{r}_0 \tag{1.13}$$

are satisfied at $t = 0$.

2. The rate problem and the tangent stiffness matrix

For given equilibrium state and load rate at some time, the quasi–static rate problem consists of determining the right rate of change of state at that time due to that rate of change of load.

A decomposition of the set of particles currently in contact that is useful in this context is the following. For each $\mathbf{X} \in \mathbb{R}^N$ and each $\mathbf{r} \in \mathcal{K}_\mathbf{r}(\mathbf{X})$, we let

$$\mathcal{P}_c(\mathbf{X}) = \mathcal{P}_z(\mathbf{X},\mathbf{r}) \cup \mathcal{P}_d(\mathbf{X},\mathbf{r}) \cup \mathcal{P}_s(\mathbf{X},\mathbf{r}),$$

where the above disjoint subsets are

$$\mathcal{P}_z(\mathbf{X},\mathbf{r}) \doteq \{p \in \mathcal{P}_c : r_n^p = r_t^p = 0\}$$

[particles in contact with *zero* reaction, grazing contact] ,

$$\mathcal{P}_d(\mathbf{X},\mathbf{r}) \doteq \{p \in \mathcal{P}_c : r_n^p < 0 \text{ and } |r_t^p| < -\mu r_n^p\}$$

[particles in contact with reaction strictly inside the friction cone and consequent vanishing (right) *displacement* rate],

$$\mathcal{P}_s(\mathbf{X},\mathbf{r}) \doteq \{p \in \mathcal{P}_c : r_n^p < 0 \text{ and } |r_t^p| = -\mu r_n^p\}$$

[particles in contact with non–vanishing reaction on the boundary of the friction cone and consequent possible *slip* in the near future].

For $\mathbf{X} \in \mathbb{R}^N$ and $\mathbf{r} \in \mathcal{K}_\mathbf{r}(\mathbf{X})$ we define the *(configuration and reaction dependent) set of admissible right velocities,*

$$\begin{aligned}\mathcal{K}_\mathbf{v}(\mathbf{X},\mathbf{r}) \doteq \{\mathbf{v} \in \mathbb{R}^{2n_C} : \; & v_n \le 0, \text{ in } \mathcal{P}_z; \\ & v_n = 0 \text{ and } \sigma[r_t]v_t \le 0, \text{ in } \mathcal{P}_s; \\ & v_n = v_t = 0, \text{ in } \mathcal{P}_d\} \subset \mathcal{K}_\mathbf{v}(\mathbf{X}).\end{aligned} \tag{2.1}$$

Having in mind (1.9), the *(configuration and reaction dependent) set of admissible right reaction rates* is defined by

$$\begin{aligned}\mathcal{K}_\mathbf{w}(\mathbf{X},\mathbf{r}) \doteq \{\mathbf{w} \in \mathbb{R}^{2n_C} : \; & w_n = w_t = 0, \text{ in } \mathcal{P}_f; \\ & w_n \le 0, |w_t| + \mu w_n \le 0, \text{ in } \mathcal{P}_z; \\ & \sigma[r_t]w_t + \mu w_n \le 0, \text{ in } \mathcal{P}_s\}.\end{aligned} \tag{2.2}$$

The definitions of the sets $\mathcal{K}_\mathbf{v}(\mathbf{X},\mathbf{r})$ and $\mathcal{K}_\mathbf{w}(\mathbf{X},\mathbf{r})$ take into account the conditions that involve solely the kinematic variables $\mathbf{v}$ and solely the static variables $\mathbf{w}$. For known configurations and reactions, the right displacement and reaction rates have to satisfy other conditions than those expressed in the definitions of $\mathcal{K}_\mathbf{v}(\mathbf{X},\mathbf{r})$ and $\mathcal{K}_\mathbf{w}(\mathbf{X},\mathbf{r})$. Such conditions are complementarity conditions: $v_n w_n = 0$ and $w_t v_t - \mu w_n |v_t| = 0$ in $\mathcal{P}_z$ and $(\sigma[r_t]v_t)(\sigma[r_t]w_t + \mu w_n) = 0$ in $\mathcal{P}_s$ that involve both velocity and reaction rates. These conditions may be written in the form

$$\mathbf{w} \cdot g(\mathbf{v}) = 0 \tag{2.3}$$

where the map $g : \mathbb{R}^{2n_C} \to \mathbb{R}^{2n_C}$ results from the maps $g^p : \mathbb{R}^2 \to \mathbb{R}^2$ given, for each contact candidate particle p, by

$$\begin{aligned} g^p(\mathbf{v}^p) &= \begin{Bmatrix} v_n^p - \mu|v_t^p| \\ v_t^p \end{Bmatrix}, \text{ for } p \in \mathcal{P}_z(\mathbf{X},\mathbf{r}) \cup \mathcal{P}_s(\mathbf{X},\mathbf{r}), \\ g^p(\mathbf{v}^p) &= \mathbf{v}^p, \text{ for } p \in \mathcal{P}_f(\mathbf{X}). \end{aligned} \tag{2.4}$$

Note that $|v_t^p|$ in $\mathcal{P}_s(\mathbf{X},\mathbf{r})$ is equal to $-\sigma[r_t^p]v_t^p$. Figure 11.1(b) illustrates the duality (Hyers et al., 1997) between the restrictions $\mathcal{K}_\mathbf{w}^p(\mathbf{X},\mathbf{r})$ and $g^p(\mathcal{K}_\mathbf{v}^p(\mathbf{X},\mathbf{r}))$ of the cones $\mathcal{K}_\mathbf{w}(\mathbf{X},\mathbf{r})$ and $g(\mathcal{K}_\mathbf{v}(\mathbf{X},\mathbf{r}))$ to particles in contact with zero reaction ($p \in \mathcal{P}_z$).

The rate equilibrium equations corresponding to (1.12) are then (Martins and Pinto da Costa, 2000):

$$\dot{\mathbf{F}}^U + \dot{\mathbf{F}}^\Omega + \dot{\mathbf{R}} = \mathbf{0} \tag{2.5}$$

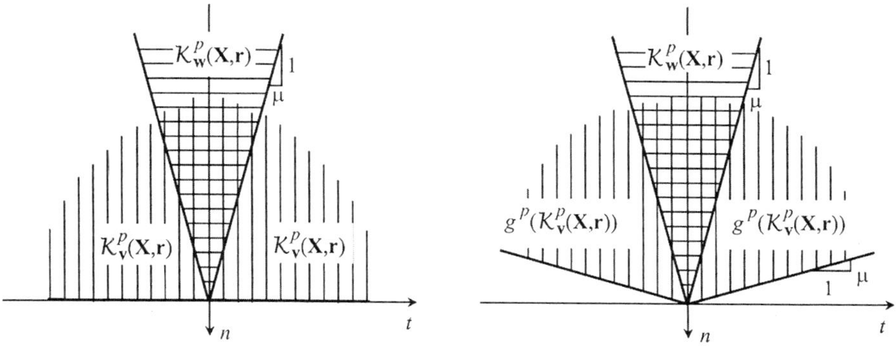

Figure 11.1. (a) The (configuration and reaction dependent) sets of admissible right velocities, $\mathcal{K}^p_{\mathbf{v}}(\mathbf{X},\mathbf{r})$, and admissible right reaction rates, $\mathcal{K}^p_{\mathbf{w}}(\mathbf{X},\mathbf{r})$, for a particle currently in contact with zero reaction, $p \in \mathcal{P}_z$. (b) The mutually dual cones $\mathcal{K}^p_{\mathbf{w}}(\mathbf{X},\mathbf{r})$ and $g^p(\mathcal{K}^p_{\mathbf{v}}(\mathbf{X},\mathbf{r}))$, for $p \in \mathcal{P}_z$.

where, as usual, $(\dot{\ })$ denotes the time derivative $\mathrm{d}()/\mathrm{d}t$,

$$\dot{\mathbf{F}}^U = -\mathbf{K}^U(\mathbf{X})\dot{\mathbf{X}}, \qquad \dot{\mathbf{F}}^\Omega = -\mathbf{K}^\Omega(\mathbf{X},\Lambda)\dot{\mathbf{X}} + \dot{\Lambda}\mathbf{P}(\mathbf{X},\Lambda), \tag{2.6}$$

$$\dot{\mathbf{R}} = \dot{\mathbf{G}}^T(\mathbf{X})\mathbf{r} + \mathbf{G}^T(\mathbf{X})\dot{\mathbf{r}} = -\mathbf{K}^C(\mathbf{X},\mathbf{r})\dot{\mathbf{X}} + \mathbf{G}^T(\mathbf{X})\dot{\mathbf{r}}, \tag{2.7}$$

$$\mathbf{K} = \mathbf{K}^U(\mathbf{X}) + \mathbf{K}^\Omega(\mathbf{X},\Lambda) + \mathbf{K}^C(\mathbf{X},\mathbf{r}). \tag{2.8}$$

$\mathbf{P} = \mathbf{P}(\mathbf{X},\Lambda)$ is the vector of the load rate direction, $\mathbf{K} = \mathbf{K}(\mathbf{X},\mathbf{r},\Lambda)$ is the *tangent stiffness matrix*, which has contributions from the elastic forces, the external forces and the contact related nonlinearities. The rate equilibrium equations are equivalently written in the form

$$\mathbf{K}\dot{\mathbf{X}} = \dot{\Lambda}\mathbf{P} + \mathbf{G}^T\dot{\mathbf{r}}, \tag{2.9}$$

and the rate problem for an equilibrium state $(\mathbf{X},\mathbf{r})$ may be stated as the following *mixed explicit complementarity problem* (Hyers et al., 1997):

Find $\mathbf{V} \in \mathbb{R}^N, \mathbf{w} \in \mathbb{R}^{2n_C}$, *such that*

$$\mathbf{K}\mathbf{V} = \dot{\Lambda}\mathbf{P} + \mathbf{G}^T\mathbf{w} \tag{2.10}$$

$$\mathcal{K}_{\mathbf{w}}(\mathbf{X},\mathbf{r}) \ni \mathbf{w} \perp g(\mathbf{G}\mathbf{V}) \in (\mathcal{K}_{\mathbf{w}}(\mathbf{X},\mathbf{r}))^* = g(\mathcal{K}_{\mathbf{v}}(\mathbf{X},\mathbf{r})) \tag{2.11}$$

3. Solvability and multiplicity in the rate problem

For simplicity of the presentation, we assume that a change of variables has been performed such that the vector $\mathbf{v}$ of normal and tangential velocities of the contact candidate particles is a sub–vector of the vector $\mathbf{V}$ of generalized velocities of the system. Both degrees–of–freedom (d) of the stick particles and the normal degrees of freedom (ns) of the particles in impending slip are then eliminated and: (i) the currently free contact candidate particles as well as all the (free) degrees–of–freedom of the system that are not subjected to any kinematic constraint will be jointly denoted by the subscript f; (ii) the subscript c will denote jointly the tangential degrees of freedom (ts) of the particles in $\mathcal{P}_s$ and the normal and tangential (z) degrees–of–freedom of the particles in $\mathcal{P}_z$. The first order rate equilibrium equations are then

$$\begin{bmatrix} \mathbf{K}_{f,f} & \mathbf{K}_{f,c} \\ \mathbf{K}_{c,f} & \mathbf{K}_{c,c} \end{bmatrix} \begin{Bmatrix} \mathbf{v}_f \\ \mathbf{v}_c \end{Bmatrix} = \begin{Bmatrix} \mathbf{0} \\ \mathbf{w}_c \end{Bmatrix} + \dot{\Lambda} \begin{Bmatrix} \mathbf{P}_f \\ \mathbf{P}_c \end{Bmatrix}. \tag{3.1}$$

Upon elimination of the free degrees of freedom, the rate equations become

$$\bar{\mathbf{K}}_{c,c}\mathbf{v}_c = \mathbf{w}_c + \dot{\Lambda}\bar{\mathbf{P}}_c \tag{3.2}$$

where $\bar{\mathbf{K}}_{c,c} = \mathbf{K}_{c,c} - \mathbf{K}_{c,f}(\mathbf{K}_{f,f})^{-1}\mathbf{K}_{f,c}$ and $\bar{\mathbf{P}}_c = \mathbf{P}_c - \mathbf{K}_{c,f}(\mathbf{K}_{f,f})^{-1}\mathbf{P}_f$, provided that the matrix $\mathbf{K}_{f,f}$ is *non–singular*. Then we perform the transformation of variables

$$\boldsymbol{\Xi}_c \doteq \begin{Bmatrix} \boldsymbol{\Xi}_{st} \\ \boldsymbol{\Xi}_{zn} \\ \boldsymbol{\Xi}_{zt} \end{Bmatrix} = \begin{Bmatrix} \mathbf{S}_s\mathbf{v}_{st} \\ -\mathbf{v}_{zn} \\ \mathbf{v}_{zt} \end{Bmatrix}, \qquad \boldsymbol{\Psi}_c \doteq \begin{Bmatrix} \boldsymbol{\Psi}_{st} \\ \boldsymbol{\Psi}_{zn} \\ \boldsymbol{\Psi}_{zt} \end{Bmatrix} = \begin{Bmatrix} \mathbf{S}_s\mathbf{w}_{st} - \mu_s\mathbf{w}_{sn} \\ -\mathbf{w}_{zn} \\ \mathbf{w}_{zt} \end{Bmatrix}. \tag{3.3}$$

With these variables the rate equations become

$$\bar{\mathbf{K}}^*_{c,c}\boldsymbol{\Xi}_c = \boldsymbol{\Psi}_c + \dot{\Lambda}\boldsymbol{\Upsilon}_c \tag{3.4}$$

where

$$\boldsymbol{\Upsilon}_c = \begin{Bmatrix} \mathbf{S}_s\bar{\mathbf{P}}_{st} - \mu_s\bar{\mathbf{P}}_{sn} \\ -\bar{\mathbf{P}}_{zn} \\ \bar{\mathbf{P}}_{zt} \end{Bmatrix} \tag{3.5}$$

and

$$\begin{aligned} \bar{\mathbf{K}}^*_{c,c} = & \begin{bmatrix} \mathbf{S}_s\bar{\mathbf{K}}_{st,st}\mathbf{S}_s & -\mathbf{S}_s\bar{\mathbf{K}}_{st,zn} & \mathbf{S}_s\bar{\mathbf{K}}_{st,zt} \\ -\bar{\mathbf{K}}_{zn,st}\mathbf{S}_s & \bar{\mathbf{K}}_{zn,zn} & -\bar{\mathbf{K}}_{zn,zt} \\ \bar{\mathbf{K}}_{zt,st}\mathbf{S}_s & -\bar{\mathbf{K}}_{zt,zn} & \bar{\mathbf{K}}_{zt,zt} \end{bmatrix} \\ & - \mu_s \begin{bmatrix} \bar{\mathbf{K}}_{sn,st}\mathbf{S}_s & -\bar{\mathbf{K}}_{sn,zn} & \bar{\mathbf{K}}_{sn,zt} \\ \mathbf{0} & \mathbf{0} & \mathbf{0} \\ \mathbf{0} & \mathbf{0} & \mathbf{0} \end{bmatrix}. \end{aligned} \tag{3.6}$$

We *assume* now that the effective stiffness matrix $\bar{\mathbf{K}}^*_{c,c}$ is non–singular. The rate problem is then written as a *mixed complementarity–inclusion* problem:

$$\begin{gathered}
\textit{Find } (\mathbf{\Psi}_c, \mathbf{\Xi}_c) \in \mathbb{R}^{n_s+2n_z} \times \mathbb{R}^{n_s+2n_z} \textit{ such that} \\
\mathbf{C}^* \mathbf{\Psi}_c = \mathbf{\Xi}_c + \dot{\Lambda} \mathbf{\Gamma}_c, \\
\mathbf{0} \leq \mathbf{\Psi}_{st} \perp \mathbf{\Xi}_{st} \geq \mathbf{0}, \\
\mathbf{0} \leq \mathbf{\Psi}_{zn} \perp \mathbf{\Xi}_{zn} \geq \mathbf{0}, \\
\psi_{zt}^p \in -\mu_z \psi_{zn}^p \sigma[\Xi_{zt}^p],\ p \in \mathcal{P}_z,
\end{gathered} \tag{3.7}$$

where $\mathbf{C}^* = (\bar{\mathbf{K}}_{c,c}^*)^{-1}$ and $\mathbf{\Gamma}_c = -\mathbf{C}^* \mathbf{\Upsilon}_c$.

Existence and uniqueness of solution to the problem (3.7) is discussed when *the matrices* $\mathbf{K}_{f,f}$ *and* $\bar{\mathbf{K}}_{c,c}^*$ (3.6) *are non–singular.* We define the following closed convex cones of admissible values of $\mathbf{\Psi}_c$ and $\mathbf{\Xi}_c$:

$$\begin{aligned}
\mathcal{K}_{\mathbf{\Psi}_c} = \{\mathbf{\Psi}_c \in \mathbb{R}^{n_s+2n_z} :\ & \psi_t \geq 0, \text{ in } \mathcal{P}_s; \\
& \psi_n \geq 0 \text{ and } -|\psi_t| + \mu_z \psi_n \geq 0, \text{ in } \mathcal{P}_z\},
\end{aligned} \tag{3.8}$$

$$\mathcal{K}_{\mathbf{\Xi}_c} = \{\mathbf{\Xi}_c \in \mathbb{R}^{n_s+2n_z} : \ \Xi_t \geq 0, \text{ in } \mathcal{P}_s;\ \Xi_n \geq 0 \text{ in } \mathcal{P}_z\}. \tag{3.9}$$

Observing that the dual cone of $\mathcal{K}_{\mathbf{\Psi}_c}$ is the set

$$\begin{aligned}
(\mathcal{K}_{\mathbf{\Psi}_c})^* = \{\mathbf{\Xi}_c \in \mathbb{R}^{n_s+2n_z} :\ & \Xi_t \geq 0, \text{ in } \mathcal{P}_s; \\
& \Xi_n - \mu_z |\Xi_t| \geq 0 \text{ in } \mathcal{P}_z\} = g(\mathcal{K}_{\mathbf{\Xi}_c})
\end{aligned} \tag{3.10}$$

where, in the spirit of (2.4), we have this time

$$g(\mathbf{\Xi}_c) = \mathbf{\Xi}_c + \begin{Bmatrix} \mathbf{0} \\ \mu_z |\mathbf{\Xi}_{zt}| \\ \mathbf{0} \end{Bmatrix} = \begin{Bmatrix} \mathbf{\Xi}_{st} \\ \mathbf{\Xi}_{zn} + \mu_z |\mathbf{\Xi}_{zt}| \\ \mathbf{\Xi}_{zt} \end{Bmatrix}, \tag{3.11}$$

the problem (3.7) is equivalent to the *explicit complementarity problem*:

$$\begin{gathered}
\textit{Find } \mathbf{\Psi}_c \in \mathcal{K}_{\mathbf{\Psi}_c} \textit{ such that} \\
f(\mathbf{\Psi}_c) = g(\mathbf{C}^* \mathbf{\Psi}_c - \dot{\Lambda} \mathbf{\Gamma}_c) \in (\mathcal{K}_{\mathbf{\Psi}_c})^* = g(\mathcal{K}_{\mathbf{\Xi}_c}) \\
\textit{and } \mathbf{\Psi}_c \cdot f(\mathbf{\Psi}_c) = 0.
\end{gathered} \tag{3.12}$$

According to Corollary 4.13 in (Hyers et al., 1997), page 64, a sufficient condition for existence of solution to this problem is that the function $f(\mathbf{\Psi}_c)$ be *completely continuous* (which holds here because it is continuous and the problem is finite dimensional) and *coercive*, in the sense that:

$$
\begin{aligned}
&\lim_{\|\boldsymbol{\Psi}_c\|\to\infty,\ \boldsymbol{\Psi}_c\in\mathcal{K}_{\boldsymbol{\Psi}_c}} \frac{\boldsymbol{\Psi}_c \cdot f(\boldsymbol{\Psi}_c)}{\|\boldsymbol{\Psi}_c\|} \\
&= \lim_{\|\boldsymbol{\Psi}_c\|\to\infty,\ \boldsymbol{\Psi}_c\in\mathcal{K}_{\boldsymbol{\Psi}_c}} \frac{1}{\|\boldsymbol{\Psi}_c\|} \left(\boldsymbol{\Psi}_c \cdot \left(\mathbf{C}^*\boldsymbol{\Psi}_c - \dot{\Lambda}\boldsymbol{\Gamma}_c\right)\right. \\
&\qquad\qquad \left. + \sum_{\mathcal{P}_z} \mu_z \psi_{zn} \left| \left(\mathbf{C}^*\boldsymbol{\Psi}_c - \dot{\Lambda}\boldsymbol{\Gamma}_c\right)_{zt} \right| \right) \\
&\qquad\qquad = +\infty
\end{aligned}
\tag{3.13}
$$

Since $\psi_{zn} \geq 0$, it follows that the above summation in $\mathcal{P}_z$ is necessarily non–negative. A sufficient condition for coercivity is then that the matrix $\mathbf{C}^*$ is strictly co–positive in the closed convex cone $\mathcal{K}_{\boldsymbol{\Psi}_c}$, i.e. that

$$\boldsymbol{\Psi}_c \cdot \mathbf{C}^*\boldsymbol{\Psi}_c > 0, \quad \forall \boldsymbol{\Psi}_c \in \mathcal{K}_{\boldsymbol{\Psi}_c}, \boldsymbol{\Psi}_c \neq \mathbf{0}.$$

It is important to observe that, in the case of a properly constrained linearly elastic system, the matrix $\mathbf{C}^*$ is *PD* (positive definite) for $\mu_s = 0$, because it is the inverse of a matrix $\bar{\mathbf{K}}^*_{c,c}$, which (for $\mu_s = 0$) is similar to the matrix $\bar{\mathbf{K}}_{c,c}$. This matrix is in turn the Schur complement of the *PD* stiffness matrix in (3.1) in its (c, c) principal sub–matrix. Then, the affine (hence, continuous) dependence of $\bar{\mathbf{K}}^*_{c,c}$ on μ_s implies that $\mathbf{C}^* = (\bar{\mathbf{K}}^*_{c,c})^{-1}$ is *PD* for sufficiently small μ_s.

To study uniqueness of solution we rewrite the conditions on the reaction rates and on the contact velocities in $\mathcal{P}_z$ in the form

$$
\begin{aligned}
\boldsymbol{\Psi}_{zt} &= \mathbf{w}_{zt} = \mu_z \mathbf{S}_z \mathbf{w}_{zn} = -\mu_z \mathbf{S}_z \boldsymbol{\Psi}_{zn} \\
\mathbf{S}_z &= diag(s^p), s^p \in [-1, +1], p \in \mathcal{P}_z \\
\Xi^p_{zt} > 0 &\Rightarrow s^p = +1; \Xi^p_{zt} < 0 \Rightarrow s^p = -1, p \in \mathcal{P}_z.
\end{aligned}
\tag{3.14}
$$

Then we assume the existence of two solutions $(\boldsymbol{\Psi}^1_c, \boldsymbol{\Xi}^1_c)$ and $(\boldsymbol{\Psi}^2_c, \boldsymbol{\Xi}^2_c)$, for the same load rate vector $\boldsymbol{\Gamma}_c$. Writing the governing equations (3.7) for each of these solutions, subtracting them, and observing that

$$\boldsymbol{\Psi}^1_{zt} - \boldsymbol{\Psi}^2_{zt} = -\mu_z \mathbf{S}^1_z \boldsymbol{\Psi}^1_{zn} + \mu_z \mathbf{S}^2_z \boldsymbol{\Psi}^2_{zn} = -\mu_z \mathbf{S}^1_z (\boldsymbol{\Psi}^1_{zn} - \boldsymbol{\Psi}^2_{zn}) - \mu_z (\mathbf{S}^1_z - \mathbf{S}^2_z) \boldsymbol{\Psi}^2_{zn}$$

we obtain

$$\left\{ \begin{array}{c} \boldsymbol{\Xi}^1_{st} - \boldsymbol{\Xi}^2_{st} \\ \boldsymbol{\Xi}^1_{zn} - \boldsymbol{\Xi}^2_{zn} \\ \boldsymbol{\Xi}^1_{zt} - \boldsymbol{\Xi}^2_{zt} \end{array} \right\} = \mathbf{C}^* \left[\begin{array}{ccc} \mathbf{I} & \mathbf{0} & \mathbf{0} \\ \mathbf{0} & \mathbf{I} & \mathbf{0} \\ \mathbf{0} & -\mu_z \mathbf{S}^1_z & \mathbf{I} \end{array} \right] \left\{ \begin{array}{c} \boldsymbol{\Psi}^1_{st} - \boldsymbol{\Psi}^2_{st} \\ \boldsymbol{\Psi}^1_{zn} - \boldsymbol{\Psi}^2_{zn} \\ -\mu_z (\mathbf{S}^1_z - \mathbf{S}^2_z) \boldsymbol{\Psi}^2_{zn} \end{array} \right\}. \tag{3.15}$$

It is possible to show that

$$\begin{aligned}(\Xi^1_{st} - \Xi^2_{st})^p(\Psi^1_{st} - \Psi^2_{st})^p \leq 0\\ (\Xi^1_{zn} - \Xi^2_{zn})^p(\Psi^1_{zn} - \Psi^2_{zn})^p \leq 0\\ (\Xi^1_{zt} - \Xi^2_{zt})^p\left(-\mu_z(\mathbf{S}^1_z - \mathbf{S}^2_z)\Psi^2_{zn}\right)^p \leq 0\end{aligned} \tag{3.16}$$

so that, if all the matrices constructed from $\mathbf{C}^*$ as indicated in the next proposition are of class P, the changes of sign in (3.16) are possible only if $(\Psi^1_c, \Xi^1_c) = (\Psi^2_c, \Xi^2_c)$.

Proposition 1 (Existence and uniqueness) *Let the matrices* $\mathbf{K}_{f,f}$ *(3.1) and* $\bar{\mathbf{K}}^*_{c,c}$ *(3.6) be non–singular. Then a sufficient condition for existence of solution is that the effective compliance matrix* $\mathbf{C}^*(\mu_s)$ *(3.7) is strictly co–positive in the closed convex cone* $\mathcal{K}_{\Psi_c}$ *(3.8). A sufficient condition of uniqueness is that, for any matrix* $\mathbf{S}_z = diag(s^p, p \in \mathcal{P}_z), s^p \in [-1, +1]$, *the matrix below is of class P:*

$$\mathbf{C}^*(\mu_s)\begin{bmatrix}\mathbf{I} & \mathbf{0} & \mathbf{0}\\ \mathbf{0} & \mathbf{I} & \mathbf{0}\\ \mathbf{0} & -\mu_z\mathbf{S}_z & \mathbf{I}\end{bmatrix}. \tag{3.17}$$

From this proposition it is clear that, similarly to the well known single particle example of Klarbring (Klarbring, 1990): – *the particles in impending slip may be responsible for non–existence of solution of the rate problem, for sufficiently large friction* μ_z; – *both the particles in impending slip and the particles in grazing contact may be responsible for multiple solutions, for sufficiently large friction,* μ_z and μ_s.

As in (Klarbring, 1999), (Klarbring, 2000) and (Martins et al., 2002), full Linear Complementarity Problems (LCP) can be obtained from (3.7) by introducing non–negative complementarity variables for the tangential direction of the particles in $\mathcal{P}_z$. Their numerical solution can then be attempted with well–known algorithms, like the Lemke method (Cottle et al., 1992). Here we study cases with multiple rate solutions or absence of rate solution for some load rates. This is done by using an algorithm due to De Moor (De Moor et al., 1992), which is capable of computing *all* the solutions of (3.7), rewritten as a Generalized Linear Complementarity Problem (GLCP) (Martins et al., 2002).

4. Existence of solution to the quasi–static evolution problem

The reaction rate – velocity form of the frictional contact conditions motivates the formulation given here for the quasi–static evolution problem. The unilateral contact and friction law conditions are satisfied provided the configuration at time 0 is admissible ($\mathbf{X}(0) \in \mathcal{K}_{\mathbf{X}}$) and the reactions and velocities satisfy:

$$\mathbf{r}(t) \in \mathcal{K}_{\mathbf{r}}(\mathbf{X}(t)) \text{ for all } t \in [0, T[, \tag{4.1}$$

$$\mathbf{G}(\mathbf{X}(t))\dot{\mathbf{X}}(t) \in \mathcal{K}_{\mathbf{v}}(\mathbf{X}(t), \mathbf{r}(t)) \text{ a.e. } t \in [0, T[. \tag{4.2}$$

In what concerns the unilateral contact law, and similarly to the Proposition 2.4 in (Moreau, 1988), we may prove the following.

Lemma 1 *Let the absolutely continuous function* $[0, T] \ni t \to \mathbf{X}(t) \in \mathbb{R}^N$ *be such that, for all* $p \in \mathcal{P}_C$*:*

$$\Phi^p(\mathbf{X}(0)) \leq 0,$$

and, for almost every $t \in [0, T]$*,*

$$v_n^p(t) = \left(\mathbf{G}(\mathbf{X}(t))\dot{\mathbf{X}}(t)\right)_n^p \leq 0, \text{ if } \Phi^p(\mathbf{X}(t)) > 0.$$

Then $\Phi^p(\mathbf{X}(t)) \leq 0$ *for all* $t \in [0, T]$ *and all* $p \in \mathcal{P}_C$*.*

Note that the slight extension of this result relative to the one in (Moreau, 1988) is that we impose the velocity constraint $v_n^p \leq 0$ only when the unilateral contact is strictly violated $\phi^p(\mathbf{X}(t)) > 0$.

The rate equilibrium equations are (2.9) where $\mathbf{K} = \mathbf{K}(\mathbf{X}, \mathbf{r}, \Lambda)$, $\mathbf{G}^T = \mathbf{G}^T(\mathbf{X})$, $\mathbf{P} = \mathbf{P}(\mathbf{X}, \Lambda)$, and $\mathbf{X}$, $\mathbf{r}$ and Λ are assumed to be absolutely continuous functions of time t. Assuming now that the matrix $\mathbf{K}$ is invertible and that the lines of the matrix $\mathbf{G}$ are linearly independent (1.2), we may rewrite (2.9) and (4.2) in the form

$$\frac{d}{dt}\begin{Bmatrix}\mathbf{X}\\ \mathbf{r}\end{Bmatrix} \in \mathbf{E}(\mathbf{X}, \mathbf{r}, \Lambda(t))\begin{Bmatrix}\dot{\Lambda}(t)\mathbf{P}(\mathbf{X}, \Lambda(t))\\ -\mathcal{K}_{\mathbf{v}}(\mathbf{X}, \mathbf{r})\end{Bmatrix}, \tag{4.3}$$

where the matrix $\mathbf{E}$ is given by

$$\mathbf{E} = \begin{bmatrix}\mathbf{K}^{-1}[\mathbf{I} - \mathbf{G}^T(\mathbf{G}\mathbf{K}^{-1}\mathbf{G}^T)^{-1}\mathbf{G}\mathbf{K}^{-1}] & -\mathbf{K}^{-1}\mathbf{G}^T(\mathbf{G}\mathbf{K}^{-1}\mathbf{G}^T)^{-1}\\ -(\mathbf{G}\mathbf{K}^{-1}\mathbf{G}^T)^{-1}\mathbf{G}\mathbf{K}^{-1} & -(\mathbf{G}\mathbf{K}^{-1}\mathbf{G}^T)^{-1}\end{bmatrix}, \tag{4.4}$$

and, in view of (4.1), the unknown pair $(\mathbf{X}, \mathbf{r})$ is assumed to belong to the set

$$\mathcal{D} = \{(\mathbf{X}, \mathbf{r}) \in \mathbb{R}^N \times \mathbb{R}^{2n_C} : \mathbf{r} \in \mathcal{K}_{\mathbf{r}}(\mathbf{X})\}. \tag{4.5}$$

In order to obtain an upper semi–continuous closed multi–valued right–hand side in the inclusion above, and since the matrix $\mathbf{K}$ is non–singular and the matrix $\mathbf{G}$ has full rank (1.2), it suffices to obtain, at each $(\mathbf{X}, \mathbf{r}) \in$

$\mathcal{D}$, the closure of the convex hull of the limit values of $\mathcal{K}_{\mathbf{v}}(\mathbf{X}', \mathbf{r}')$ as $(\mathbf{X}', \mathbf{r}') \in \mathcal{D}$ approach $(\mathbf{X}, \mathbf{r})$. Since the set $\mathcal{K}_{\mathbf{v}}(\mathbf{X}, \mathbf{r})$ is convex,

$$\overline{co}\, \mathcal{K}_{\mathbf{v}}(\mathbf{X}, \mathbf{r}) = \overline{\mathcal{K}}_{\mathbf{v}}(\mathbf{X}, \mathbf{r})$$
$$= \mathcal{K}_{\mathbf{v}}(\mathbf{X}, \mathbf{r}) \cup \left\{\mathbf{v} \in \mathbb{R}^{2n_C} : v_n^p \geq 0, \text{ for } p \in \mathcal{P}_z \text{ and } \Phi^p(\mathbf{X}) = 0\right\}, \quad (4.6)$$

and the differential inclusion (4.3) becomes

$$\frac{d}{dt}\begin{Bmatrix}\mathbf{X}\\ \mathbf{r}\end{Bmatrix} \in \mathcal{F}\left(\begin{Bmatrix}\mathbf{X}\\ \mathbf{r}\end{Bmatrix}, t\right), \quad (4.7)$$

where, for each $(\mathbf{X}, \mathbf{r}, t) \in \mathcal{D} \times [0, T[$:

$$\mathcal{F}(\mathbf{X}, \mathbf{r}, t) = \mathbf{E}(\mathbf{X}, \mathbf{r}, \Lambda(t)) \begin{Bmatrix}\dot{\Lambda}(t)\mathbf{P}(\mathbf{X}, \Lambda(t))\\ -\overline{\mathcal{K}}_{\mathbf{v}}(\mathbf{X}, \mathbf{r})\end{Bmatrix}. \quad (4.8)$$

Existence of solution for the differential inclusion (4.7) in $\mathcal{D}$ requires that $\mathcal{F}(\mathbf{X}, \mathbf{r}, t) \cap \mathcal{T}_{\mathcal{D}}(\mathbf{X}, \mathbf{r}) \neq \emptyset$ on $\mathcal{D} \times [0, T[$ (Deimling, 1992), where, for each $(\mathbf{X}, \mathbf{r}) \in \mathcal{D}$,

$$\mathcal{T}_{\mathcal{D}}(\mathbf{X}, \mathbf{r}) = \left\{(\mathbf{V}, \mathbf{w}) \in \mathbb{R}^N \times \mathbb{R}^{2n_C} : \liminf_{\lambda \to 0^+} \frac{\rho((\mathbf{X}, \mathbf{r}) + \lambda(\mathbf{V}, \mathbf{w}), \mathcal{D})}{\lambda} = 0\right\} \quad (4.9)$$

is the tangent cone to $\mathcal{D}$ at $(\mathbf{X}, \mathbf{r})$. In the previous definition, $\rho(x, \mathcal{D}) = \min_{y \in \mathcal{D}} \|y - x\|$ is the distance from x to the set $\mathcal{D}$. It is easy to compute

$$\begin{aligned}\mathcal{T}_{\mathcal{D}}(\mathbf{X}, \mathbf{r}) = \{\ & (\mathbf{V}, \mathbf{w}) \in \mathbb{R}^N \times \mathbb{R}^{2n_C} \text{ such that}\\ & (\mathbf{v}, \mathbf{w}) = (\mathbf{G}\mathbf{V}, \mathbf{w}) \in \mathbb{R}^{2n_C} \times \mathbb{R}^{2n_C} \text{ satisfy :}\\ & w_n^p = w_t^p = 0, \text{ if } p \in \mathcal{P}_f;\\ & |w_t^p| + \mu w_n^p \leq 0, \text{ if } p \in \mathcal{P}_z, \text{ with } v_n^p \geq 0 \text{ if } \Phi^p = 0\\ & \ \text{and } w_n^p < 0;\\ & \sigma[r_t^p] w_t^p + \mu w_n^p \leq 0, \text{ if } p \in \mathcal{P}_s, \text{ with } v_n^p \geq 0 \text{ if } \Phi^p = 0;\\ & v_n^p \geq 0, \text{ if } p \in \mathcal{P}_d \text{ and } \Phi^p = 0\},\end{aligned} \quad (4.10)$$

which is equivalently written as

$$\begin{aligned}\mathcal{T}_{\mathcal{D}}(\mathbf{X}, \mathbf{r}) = \{\ & (\mathbf{V}, \mathbf{w}) \in \mathbb{R}^N \times \mathbb{R}^{2n_C} \text{ such that}\\ & \mathbf{w} \in \mathcal{K}_{\mathbf{w}}(\mathbf{X}, \mathbf{r}) \text{ and } \mathbf{v} = \mathbf{G}(\mathbf{X})\mathbf{V} \text{ satisfies}\\ & v_n^p \geq 0, \text{ if } \Phi^p = 0, \text{ and } p \in \mathcal{P}_s \cup \mathcal{P}_d\\ & \text{or } (p \in \mathcal{P}_z \text{ and } w_n^p < 0)\}.\end{aligned} \quad (4.11)$$

Denoting by $\mathcal{F}_{\mathbf{V}}$ the sub–vector of $\mathcal{F}$ associated to the velocities $d\mathbf{X}/dt$, it should be clear that, by construction, $\mathbf{G}(\mathbf{X})\mathcal{F}_{\mathbf{V}}(\mathbf{X},\mathbf{r},t) = \overline{\mathcal{K}}_{\mathbf{v}}(\mathbf{X},\mathbf{r})$ for all $(\mathbf{X},\mathbf{r},t) \in \mathcal{D} \times [0,T[$. From (4.8) and (4.11) it follows then that

$$\begin{aligned}\mathcal{F}(\mathbf{X},\mathbf{r},t) \cap \mathcal{T}_{\mathcal{D}}(\mathbf{X},\mathbf{r}) = \\ = \{\, (\mathbf{V},\mathbf{w}) \in \mathbb{R}^N \times \mathbb{R}^{2n_C} \text{ such that} \\ \mathbf{K}(\mathbf{X},\mathbf{r},\mathit{\Lambda}(t))\mathbf{V} - \mathbf{G}^T(\mathbf{X})\mathbf{w} = \dot{\mathit{\Lambda}}(t)\mathbf{P}(\mathbf{X},\Lambda(t)), \\ \mathbf{w} \in \mathcal{K}_{\mathbf{w}}(\mathbf{X},\mathbf{r}), \mathbf{G}(\mathbf{X})\mathbf{V} \in \mathcal{K}_{\mathbf{v}}(\mathbf{X},\mathbf{r}) \cup \{\mathbf{v} \in \mathbb{R}^{2n_C} : \\ v_n^p \geq 0, \text{ if } \mathit{\Phi}^p = 0,\ p \in \mathcal{P}_z \text{ and } w_n^p < 0\}\}. \end{aligned} \tag{4.12}$$

The above proves the following

Proposition 2 (Existence of solution to the evolution problem) *Let initial conditions (1.10) and (1.11) be given and let $\mathit{\Lambda}(t)$ be a given absolutely continuous function in $[0,T]$ such that, for some $c \in L^1(0,T)$*

$$\begin{aligned}\|\mathcal{F}(\mathbf{X},\mathbf{r},t)\| &= \sup\{\|(\mathbf{V},\mathbf{w})\| : (\mathbf{V},\mathbf{w}) \in \mathcal{F}(\mathbf{X},\mathbf{r},t)\} \\ &\leq c(t)(1 + \|(\mathbf{X},\mathbf{r})\|) \text{ on } \mathcal{D} \times [0,T].\end{aligned} \tag{4.13}$$

Assume also that for all $(\mathbf{X},\mathbf{r},t) \in \mathcal{D} \times [0,T]$ the first order (right) rate frictional contact problem (2.10)-(2.11) is feasible, *i.e. that there exist $\mathbf{V}$ and $\mathbf{w}$ such that (2.10) holds with $\mathbf{w} \in \mathcal{K}_{\mathbf{w}}(\mathbf{X},\mathbf{r})$, and $\mathbf{G}(\mathbf{X})\mathbf{V} \in \mathcal{K}_{\mathbf{v}}(\mathbf{X},\mathbf{r})$. Then there exists a solution to the differential inclusion problem (4.7), which solves the quasi–static evolution problem (1.12,1.3,1.4,1.1*

Note that only the feasibility of the rate problem is required in Proposition 2: the complementarity conditions (2.3) of the rate problem need not to be assumed to hold. Note however that complementarity conditions involving $\mathbf{v}$ and $\mathbf{r}$ are indeed satisfied by the solution of the evolution problem: they are incorporated in the definition of $\overline{\mathcal{K}}_{\mathbf{v}}(\mathbf{X},\mathbf{r})$ in $\mathcal{F}(\mathbf{X},\mathbf{r},t)$.

5. Numerical examples

We consider a solid with the geometry indicated in Figure 11.2. The body is assumed to have linear elastic behavior and to be in a state of plane stress. The boundary segments *CD* and *EF* are fixed. The particles in the segment *AB* may have contact with an horizontal rigid flat obstacle. The friction coefficient is $\mu = 2$. The elastic properties are: modulus of elasticity $E = 5\ MPa$ and Poisson's ratio $\nu = 0.48$. The self–weight of the structure is neglected. We let all the particles in segment *AB* in grazing contact ($\mathcal{P}_z$). We use the algorithm of De Moor (De

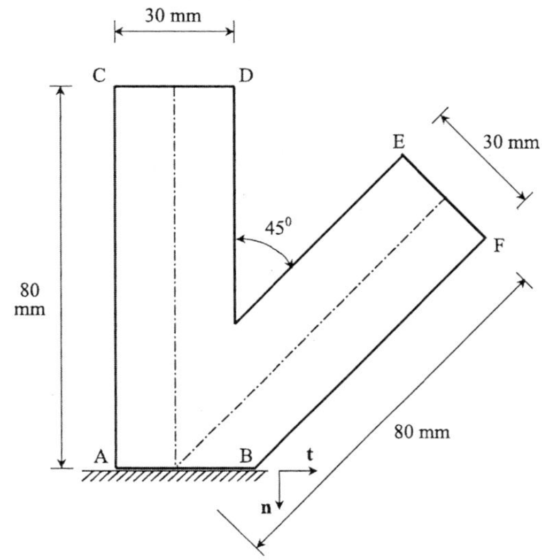

Figure 11.2. Dimensions of the finite element version of the example of Klarbring (Klarbring, 1990).

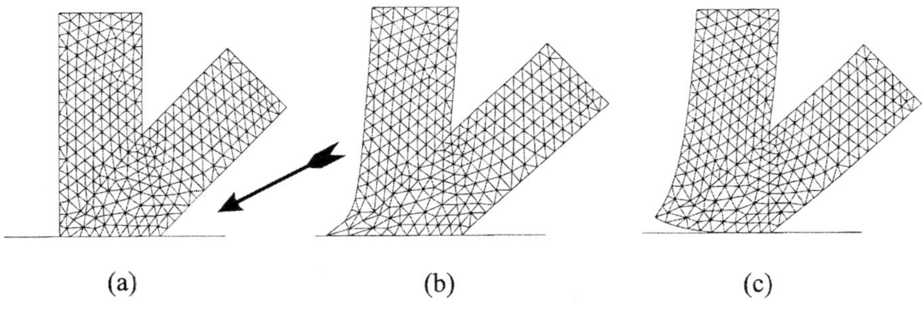

Figure 11.3. Three solutions of the rate problem, for the (undeformed) equilibrium state of the structure in contact with zero reaction at all contact nodes: (a) rate solution involving stick of all contact nodes; (b) rate solution involving slip towards the left of all contact nodes; (c) rate solution involving loss of contact and slip towards the left. Angle $\beta = 296.70^o$; coefficient of friction $\mu = 2$.

Moor et al., 1992) to solve the rate problem, for a uniformly distributed inclined load rate applied also on segment AB. The solid is discretized with 547 linear P1 finite elements, as represented in Figure 11.3(a). The

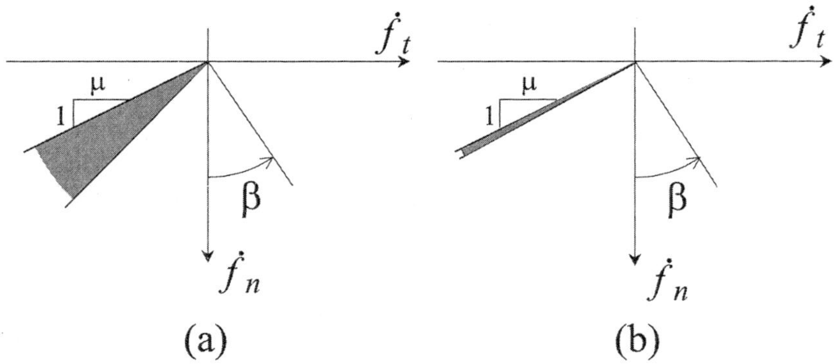

Figure 11.4. Regions in the $(\dot{f}_n, \dot{f}_t)$ plane where the first order rate problem has three solutions for an equilibrium state in grazing contact with $\mu = 2$: (a) Two degree-of-freedom example ($\beta \in]296.57^o, 315^o[$); (b) Finite element version with eleven contact nodes ($\beta \in]296.57^o, 298.09^o[$).

number of contact candidate nodes is 11. The equilibrium configuration is the undeformed configuration presented in Figure 11.3(a). For sufficiently large coefficients of friction there is a sector of load rate vectors for which there are three rate solutions. For load rate vectors outside that sector, the rate problem has one solution. Contrarily to what happens in the impending slip case treated in (Martins et al., 2002), the sector in the $(\dot{f}_n, \dot{f}_t)$ plane where there are multiple rate solutions for the finite element discretization is much smaller than the sector where there are multiple rate solutions in the single contact particle example. Figure 11.4 illustrates that fact for $\mu = 2$. The three rate solutions obtained for a load rate in the direction $cos(296.70^o)\mathbf{n}+sin(296.70^o)\mathbf{t}$ are represented in Figure 11.3: the solutions (a) and (b) involve the evolution of all the contact nodes towards stick and towards left slip, respectively, while the solution (c) involves the evolution of the six contact nodes on the left towards loss of contact and the five contact nodes on the right towards left slip.

We finally observe that one boundary line (with slope μ) of the sector of load rate directions for which non–unique rate solutions occur coincides with the limit load rate direction for sticking. Because the loading rate is uniformly distributed on the contact segment AB of the finite element model, that boundary line is the same for both the finite element model and the single particle example. It is the limit load rate direction for loss of contact, which coincides in the single particle example with the other boundary line of the non-uniqueness sector, that is not preserved when we consider the finite element model: in this case, the contact nodes *progressively* loose contact as the loading direction changes; this has no parallel in the single particle example, where that occurs at a single load rate direction; this is the probable reason for the discrepancy between the non-uniqueness region in the two models.

References

Andersson, L.–E., (1999). *Quasi–static Frictional Contact Problems with Finitely Many Degrees of Freedom.* LiTH-MAT-R-1999-22, Linköping University, Department of Mathematics.

R.W. Cottle, J.–S. Pang and R.E. Stone, (1992) *The Linear Complementarity Problem.* Academic Press, Boston.

Deimling, K. (1992). *Multivalued Differential Equations.* W. de Gruyter, Berlin–New York.

De Moor, B., Vandenberghe, L. and Vandewalle, J. (1992) The generalized linear complementarity problem and an algorithm to find all its solutions. *Mathematical Programming, Series A*, **57**(3), 415–426.

De Saxce, G. and Feng, Z.Q. (1991). New inequation and functional for contact with friction: the implicit standard material approach. *Mech. Struct. and Mach.*, **19**, 3, 301–325.

Hyers, D.H., Isac, G. and Rassias, T.M. (1997). Topics in Nonlinear Analysis and Applications, World Scientific Publishing, Singapore, New Jersey, London.

Klarbring, A. (1990) Examples of non–uniqueness and non–existence of solutions to quasistatic contact problems with friction. *Ingenieur–Archiv.*, **56**, 529–541.

A. Klarbring (1999) Contact, friction, discrete mechanical structures: analogies and dynamic problems. In Wriggers, P. and Panagiotopoulos, P., eds., *New Developments in Contact Problems*, CISM Courses and Lectures, Springer–Wien, **384**, 55–100.

A. Klarbring (2000) Contact, friction, discrete mechanical structures: analogies and dynamic problems. In Pfeiffer, F. and Glocker, C., eds., *Multibody Dynamics with Unilateral Contacts*, CISM Courses and Lectures, Springer–Wien, **421**, 147–174.

Martins, J.A.C. and Pinto da Costa, A. (2000) Stability of finite dimensional nonlinear elastic systems with unilateral contact and friction. *Int. J. Solids Structures*, **37**, 18, 2519-2564.

Martins, J.A.C., Pinto da Costa, A. and Simões, F.M.F. (2002). Some notes on friction and instabilities. *Friction and Instabilities*, CISM

Courses and Lectures, **457**, 65–136. Edited by J.A.C. Martins and M. Raous, Springer, Wien, New York.

J.J. Moreau (1988) Unilateral contact and dry friction in finite freedom dynamics. *Nonsmooth Mechanics and Applications.* CISM courses and lectures, **302**, 1–82. Edited by J.J. Moreau and P.D. Panagiotopoulos, Springer–Verlag, Wien, New York.

Trinkle, J.C., Pang, J.–S., Sudarsky, S. and Lo, G. (1997) On dynamic multi–rigid–body contact problems with Coulomb friction. *ZAMM*, **77**(4), 267–279.

Vola, D., Pinto da Costa, A., Barbarin, S., Martins, J.A.C. and Raous, M. (1999). Bifurcations and instabilities in some finite dimensional frictional contact problems. In Pfeiffer, F. and Glocker, Ch., eds., *IUTAM Symposium on Unilateral Multibody Contacts, Solids Mechanics and its Applications*, Kluwer Academic Publishers, Dordrecht, Boston, London, **72**, 179-190.

Chapter 12

MINIMAX THEORY, DUALITY AND APPLICATIONS

D. Motreanu
University of Perpignan
Department of Mathematics
66860 Perpignan, France
motreanu@univ-perp.fr

Abstract The paper presents recent minimax results in the critical point theory of functionals equal to sums of locally Lipschitz and convex, proper, lower semicontinuous terms. The applications involve semilinear elliptic boundary value problems with constraints and nonlinear discontinuities as well as nonsmooth problems in the unilateral mechanics.

Keywords: critical point, minimax value, nonsmooth analysis, nonlinear elliptic boundary value problems, unilateral mechanics

1. General Setting and Abstract Results

In this paper we are concerned with a general minimax principle for the class of nonsmooth functionals $f : X \to]-\infty, +\infty]$ on a real reflexive Banach space X satisfying the structure hypothesis

(H_f) $f = \Phi + \alpha$, *where* $\Phi : X \to \mathbb{R}$ *is locally Lipschitz continuous and* $\alpha : X \to]-\infty, +\infty]$ *is convex, proper, lower semicontinuous.*

In the sequel we use the following notation:

$$f_r = \{x \in X : f(x) \le r\}, \quad f^r = \{x \in X : r \le f(x)\} \quad (\text{for } r \in \mathbb{R}),$$

$$d(x, S) = \inf_{z \in S} \|x - z\|, \quad N_r(S) = \{z \in X : d(z, S) \le r\} \quad (\text{for } S \subset X),$$

$$B_r = \{x \in X : \|x\| \le r\} \quad (\text{for } r > 0), \quad D_{\partial\alpha} = \{x \in X : \partial\alpha(x) \neq \emptyset\},$$

where $\partial\alpha$ is the subdifferential in the sense of convex analysis of the convex, proper, lower semicontinuous function α entering hypothesis (H_f).

First we recall two basic definitions from Motreanu and Panagiotopoulos [6], Chapter 3.

Definition 1. An element $u \in X$ is called a critical point of the functional $f : X \to]-\infty, +\infty]$ satisfying assumption (H_f) if

$$\Phi^0(u; x-u) + \alpha(x) - \alpha(u) \geq 0 \quad \forall x \in X,$$

where the notation Φ^0 stands for the generalized directional derivative in the sense of Clarke [3].

We note that if Φ is continuously differentiable one obtains the definition of critical point given in Szulkin [10], while if $\alpha = 0$ one gets the definition of critical point in the sense of Chang [2].

Given a real number c, we denote

$$K_c(f) = \{u \in X : f(u) = c,\ u \text{ is a critical point of } f\}.$$

Definition 2. The function $f : X \to]-\infty, +\infty]$ satisfying (H_f) is said to verify the Palais-Smale condition around the set $S \subset X$ at the level $c \in I\!R$ if the following property holds

$(\mathrm{PS})_{f,S,c}$ *Each sequence* $\{x_n\} \subset X$ *verifying* $d(x_n, S) \to 0$, $f(x_n) \to c$,

$$\Phi^0(x_n; x - x_n) + \alpha(x) - \alpha(x_n) \geq -\varepsilon_n \|x_n - x\| \quad \forall n \in I\!N,\ x \in X,$$

where $\varepsilon_n \to 0^+$, *possesses a strongly convergent subsequence.*

The functional $f : X \to]-\infty, +\infty]$ satisfying (H_f) verifies the Palais-Smale condition $(\mathrm{PS})_{f,c}$ at the level $c \in I\!R$ if one has

$(\mathrm{PS})_{f,c}$ *Each sequence* $\{x_n\} \subset X$ *such that* $f(x_n) \to c$ *and*

$$\Phi^0(x_n; x - x_n) + \alpha(x) - \alpha(x_n) \geq -\varepsilon_n \|x_n - x\| \quad \forall n \in I\!N,\ x \in X,$$

where $\varepsilon_n \to 0^+$, *possesses a strongly convergent subsequence.*

Given a compact topological submanifold Q of X with nonempty boundary ∂Q we introduce the minimax value

$$c := \inf_{\gamma \in \Gamma} \sup_{x \in Q} f(\gamma(x)), \tag{1}$$

where Γ is the set

$$\Gamma = \{\gamma \in C(Q, X) : \ \gamma|_{\partial Q} = \mathrm{id}_{\partial Q}\}. \tag{2}$$

Denoting

$$a := \sup_{\partial Q} f, \tag{3}$$

it is seen from relations (1)-(3) that the inequality below holds

$$c \geq a.$$

Two situations arise: $c > a$ and $c = a$. Corresponding to these cases we have distinct results.

We start by treating the case $c > a$.

Theorem 1. *Let the functional $f : X \to (-\infty, +\infty]$ on the Banach space X satisfy assumption* (H_f). *Let Q be a compact topological submanifold of X with nonempty boundary ∂Q (in the sense of manifolds with boundary) such that*

$$\sup_{Q} f \in \mathbb{R}. \tag{4}$$

If the numbers a and c as defined in (1)-(3) *are such that*

$$c > a \tag{5}$$

and condition $(\mathrm{PS})_{f,c}$ *is satisfied, then c in* (1) *is a critical value of f, i.e. $K_c(f)$ is nonempty.*

The proof is given in Section 2.

The strict inequality (5) is difficult to be checked. Practically, this can be achieved by means of the linking property. To this end, we recall the definition of linking in Motreanu and Panagiotopoulos [6], Chapter 3.

Definition 3. Let Q be a compact topological submanifold of X with nonempty boundary ∂Q (in the sense of manifolds with boundary) and let S be a nonempty closed subset of X. We say that Q links with S provided $\partial Q \cap S = \emptyset$ and $\gamma(Q) \cap S \neq \emptyset$ for every $\gamma \in \Gamma$, where Γ is given by (2).

We deduce from Theorem 1 the following result.

Corollary 1. (Motreanu and Panagiotopoulos [6], Theorem 3.2). *Let the functional $f : X \to (-\infty, +\infty]$ satisfy assumption* (H_f). *Let S and Q link in the sense of Definition* 3. *Assume that*

$$\sup_{Q} f \in \mathbb{R}, \quad b := \inf_{S} f \in \mathbb{R}, \;\; a := \sup_{\partial Q} f < b \tag{6}$$

and condition $(\mathrm{PS})_{f,c}$ *for the number* c *introduced in* (1) (*with* Γ *given by* (2)). *Then* c *in* (1) *is a critical value of* f *with*

$$c \geq b\,. \tag{7}$$

This means in particular that $K_c(f)$ *is nonempty.*

Proof. According to the linking property of Definition 3 we know that $\gamma(Q) \cap S \neq \emptyset$ for all $\gamma \in \Gamma$. By (6) and (1) this ensures that (7) is true. Combining (7) and the last inequality in (6) one obtains that relation (5) is true. We are now in the setting of Theorem 1. The application of Theorem 1 yields the result. ∎

The remaining case $c = a$ is treated in the theorem below.

Theorem 2. (Marano and Motreanu [5]). *Suppose that* Q *and* S *link in the sense of Definition 3 and the function* f *satisfies the following assumptions in addition to* (H_f):

(f_1) $\sup_{x\in Q} f(x) < +\infty$.

(f_2) $\partial Q \subset f_a$ *and* $S \subset f^a$ *for some* $a \in I\!R$.

(f_3)

$$a = \inf_{\gamma\in\Gamma} \sup_{z\in\gamma(Q)} f(z)$$

and condition $(\mathrm{PS})_{f,S,a}$ *holds. Further, there exists* $\varepsilon_0 > 0$ *such that* $N_{\varepsilon_0}(S) \subset D_{\partial\alpha}$ *and the set* $N_\varepsilon(S) \cap f^{a-\varepsilon} \cap f_{a+\varepsilon}$, $\varepsilon \in]0, \varepsilon_0[$, *is closed.*

Then one has $K_a(f) \cap S \neq \emptyset$.

The proof of Theorem 2 is given in Section 2.

Theorems 1 and 2 lead to extensions of celebrated minimax principles in case of nonsmooth functionals satisfying the structure hypothesis (H_f). The classic smooth formulations of these results can be found, for instance, in Rabinowitz [9].

Corollary 2 (Mountain Pass Theorem). *Suppose that the functional* $f : X \to]-\infty, +\infty]$ *satisfy assumption* (H_f) *and*

(f_4) *there exist* $x_1 \in X$, $r > 0$, *and* $a \in I\!R$ *satisfying* $\|x_1\| > r$ *in addition to*

$$\max\{f(0), f(x_1)\} \leq a \leq f(x) \quad \forall x \in \partial B_r\,,$$

(f$_5$) *assumption* (f$_3$) *of Theorem 2 holds with* $Q = [0, x_1]$ (*the segment joining* 0 *and* x_1), $S = \partial B_r$.

Then for Q *and* S *as in* (f$_5$) *the number* c *described in* (1), (2) *is a critical value of* f *and* $c \geq a$.

Proof. Corollary 1 and Theorem 2 can be applied for the sets $Q = [0, x_1]$ and $S = \partial B_r$ observing that we are in the situation of linking as described in Definition 3. ■

Corollary 3 (Saddle Pointy Theorem). *Let* $X = V \oplus E$, *where* $V \neq \{0\}$ *is finite dimensional. If*

(f$_6$) *there are two numbers* $r > 0$ *and* $a \in I\!R$ *such that*

$$\sup_{x \in V \cap B_r} f(x) < +\infty, \quad \partial(V \cap B_r) \subset f_a, \quad E \subset f^a,$$

(f$_7$) *hypothesis* (f$_3$) *of Theorem* 2 *holds for* $S = E$,

then, for $Q = V \cap B_r$ *and* $S = E$, *the number* c *described in* (1), (2) *is a critical value of* f *and* $c \geq a$.

Proof. Corollary 1 and Theorem 2 can be applied for $Q = V \cap B_r$ and $S = E$. We note that the sets Q and S satisfy the linking property in Definition 3. ■

It is important to construct a duality theory in the minimax framework for nonsmooth functionals. We illustrate this topic with the following result.

Theorem 3 (Barletta and Marano [1]). *Suppose that* Q *and* S *link in the sense of Definition* 3. *Let a locally Lipschitz function* $f : X \to I\!R$ *satisfy the Palais-Smale condition, i.e.,* $\alpha = 0$ *in Definition* 2 (*see also Chang* [2]), *together with*

(f$_8$) $\sup_{x \in Q} f(x) < +\infty$,

(f$_9$) $\partial Q \subset f_a$ *and* $S \subset f^a$ *for some* $a \in I\!R$.

Then, setting

$$b := \sup_{\gamma^* \in \Gamma^*} \inf_{x \in S} f(\gamma^*(x)),$$

where

$$\Gamma^* = \{\gamma^* \in C(X, X) : \ \gamma^* \text{ is a homeomorphism}, \ \gamma^*|_{\partial Q} = \mathrm{id}_{\partial Q}\},$$

one has

(i) $a \leq b \leq c$, *with* c *given by* (1), (2),

(ii) $K_b(f) \setminus \partial Q \neq \emptyset$,

(iii) $K_b(f) \cap S \neq \emptyset$ *if* $a = b$.

The proof can be found in Section 3.

The rest of the paper is organized as follows. Section 2 contains the proofs of Theorems 1, 2 and 3. In Section 3 we illustrate the abstract results with applications to nonlinear elliptic boundary value problems with discontinuities and constraints as well as in the unilateral nonsmooth mechanics.

2. Proofs of Abstract Results

Proof of Theorem 1. Relation (4) yields $c < \infty$ by putting $\gamma = \mathrm{id}_Q$ in (1), (2). We have to show that c is a critical value of f. Arguing by contradiction we assume that $K_c(f) = \emptyset$. Then we can take $U = \emptyset$ as the neighborhood of $K_c(f)$ in the deformation result stated in Theorem 3.1 in Motreanu and Panagiotopoulos [6], p. 69. Applying this result with

$$\bar{\varepsilon} = c - a, \tag{8}$$

which is a positive number by (5), we obtain an $\varepsilon \in (0, \bar{\varepsilon})$ as stated in Theorem 3.1 in [6]. By (5) and (8) it follows that

$$\partial Q \subset f_{c-\frac{\varepsilon}{2}} \,. \tag{9}$$

We enlarge the class Γ in (2) by defining

$$\begin{aligned} \Gamma_1 := \{\varphi \in C(Q, X) :\ & \varphi|_{\partial Q} \text{ and } \mathrm{id}_{\partial Q} \text{ are homotopic as} \\ & \text{maps from } \partial Q \text{ into } f_{c-\frac{\varepsilon}{4}} \text{ and } \varphi(\partial Q) \subset f_{c-\frac{\varepsilon}{2}}\}, \end{aligned} \tag{10}$$

which is well defined and nonempty containing at least id_Q (see (9)). Corresponding to the family Γ_1 in (10) we introduce the minimax value

$$c_1 = \inf_{\varphi \in \Gamma_1} \sup_{x \in Q} f(\varphi(x)) \,. \tag{11}$$

We claim that

$$c = c_1. \tag{12}$$

Since $\Gamma \subset \Gamma_1$, from (1), (11) it is clear that $c_1 \leq c$. Assuming by contradiction that $c_1 < c$, there would exist $\varphi \in \Gamma_1$ such that

$$\sup_{x \in Q} f(\varphi(x)) < c.$$

Recalling that $\varphi|_{\partial Q}$ and $\mathrm{id}_{\partial Q}$ are homotopic in $f_{c-\frac{\varepsilon}{4}}$, the (absolute) Homotopy Extension Property guarantees that $\mathrm{id}_{\partial Q}$ can be extended to some $\psi \in C(Q, X)$ such that

$$\sup_{x \in Q} f(\psi(x)) < c \, .$$

This is true because $(Q, \partial Q)$ is a (finitely) triangulable pair in view of the fact that Q is a compact topological manifold with boundary ∂Q. It is thus achieved a contradiction with the definition of c in (1). Hence the claim in (12) is verified.

A direct verification shows that Γ_1 in (10) is a closed subset of the Banach space $C(Q, X)$ with respect to the uniform norm (recall that Q is compact in X)

$$\|\varphi\| = \sup_{x \in Q} \|\varphi(x)\|, \quad \forall\, \varphi \in C(Q, X).$$

Consequently, Γ_1 is a complete metric space.

Let us now introduce the functional $\Pi : C(Q, X) \to (-\infty, +\infty]$ by

$$\Pi(\varphi) = \sup_{x \in Q} f(\varphi(x)), \quad \forall\, \varphi \in C(Q, X) \, . \tag{13}$$

The lower semicontinuity of f implies that $\Pi : \Gamma_1 \to (-\infty, +\infty]$ is lower semicontinuous. Applying Ekeland's variational principle (see [4]) to $\Pi : \Gamma_1 \to (-\infty, +\infty]$ one obtains some $\varphi \in \Gamma_1$ satisfying

$$c \le \Pi(\varphi) \le c + \varepsilon \tag{14}$$

and

$$\Pi(\psi) - \Pi(\varphi) \ge -\varepsilon \|\psi - \varphi\|, \quad \forall\, \psi \in \Gamma_1 \, . \tag{15}$$

The deformation result in Theorem 3.1 in [6] provides a deformation $h_\varphi : W \times [0, \bar{s}] \to X$ corresponding to the compact set $A = \varphi(Q)$, where W is a closed neighborhood of A in X and $\bar{s}$ is a positive number. Following the argument given in the proof of Theorem 3.2 in [6], we can check that for $\bar{s} > 0$ small enough we have

$$h_\varphi(\varphi(\cdot), s) \in \Gamma_1, \quad \forall\, s \in [0, \bar{s}] \, . \tag{16}$$

On the other hand the following properties of the deformation h_φ are supplied by Theorem 3.1 in [6]:

$$\|u - h_\varphi(u, s)\| \le s, \quad \forall\, u \in W, \quad \forall\, s \in [0, \bar{s}]; \tag{17}$$

$$\sup_{u \in A} f(h_\varphi(u, s)) - \sup_{u \in A} f(u) \le -2\varepsilon s, \quad \forall\, s \in [0, \bar{s}]. \tag{18}$$

Combining relations (13)-(18) yields

$$\begin{aligned}-2\varepsilon s &\geq \Pi(h_\varphi(\varphi(\cdot),s)) - \Pi(\varphi)\\ &\geq -\varepsilon\|h_\varphi(\varphi(\cdot),s)-\varphi\| \geq -\varepsilon s, \quad \forall\, 0 \leq s \leq \bar{s}.\end{aligned}$$

This contradiction completes the proof. ∎

Proof of Theorem 2. Suppose on the contrary that $K_a(f) \cap S = \emptyset$. Define $A = \partial Q$, $B = S$, $g = -f$. Then the function g fulfils condition $(\mathrm{PS})_{g,B,-a}$ in Marano and Motreanu [5] while

$$A \cap B = \emptyset\,, \quad A \subset g^{-a}\,, \quad B \subset g_{-a}\,, \quad K_{-a}(g) \cap B = \emptyset\,,$$

where we used the notation

$$K_{-a}(g) = \{u \in X : g(u) = -a \text{ and}$$

$$(-\Phi)^0(u; u-x) - \alpha(u) + \alpha(x) \geq 0 \;\; \forall x \in X\}.$$

By (f_3), the data g and $-a$ satisfy the hypotheses of the deformation result given in Theorem 2.1 of Marano and Motreanu [5]. Thus, there exist $\varepsilon > 0$ and a homeomorphism $\eta : X \to X$ such that

$$\eta(x) = x \;\; \forall x \in \partial Q\,, \quad a + \varepsilon \leq f(\eta(x)) \;\; \forall x \in S\,. \tag{19}$$

According to the formula of a in (f_3) one finds some $\gamma_\varepsilon \in \Gamma$ such that

$$f(\gamma_\varepsilon(x)) < a + \varepsilon\,, \quad \forall x \in Q\,. \tag{20}$$

Since Q links with S and $\eta^{-1} \circ \gamma_\varepsilon \in \Gamma$ there is a point x_ε in Q fulfilling $\eta^{-1}(\gamma_\varepsilon(x_\varepsilon)) \in S$. So, due to (19) and (20), we arrive at the contradiction

$$a + \varepsilon \leq f(\eta(\eta^{-1}(\gamma_\varepsilon(x_\varepsilon)))) = f(\gamma_\varepsilon(x_\varepsilon)) < a + \varepsilon.$$

This completes the proof. ∎

Proof of Theorem 3. Since id $\in \Gamma^*$, by (f_9) one obtains $a \leq b$. In order to show that $b \leq c$, let $\gamma \in \Gamma$ and $\gamma^* \in \Gamma^*$. Taking into account that $(\gamma^*)^{-1} \circ \gamma \in \Gamma$, Definition 3 implies that there exists $z \in Q$ such that $(\gamma^*)^{-1}(\gamma(z)) \in S$. It follows that

$$\inf_{x \in S} f(\gamma^*(x)) \leq f(\gamma^*((\gamma^*)^{-1}(\gamma(z)))) = f(\gamma(z)) \leq \sup_{x \in Q} f(\gamma(x)),$$

so $b \leq c$. Thus assertion (i) is true.

Consider the situation $a < b$. Notice that due to the first inclusion in (f_9) we have $K_b(f) \setminus \partial Q = K_b(f)$. In order to show (ii) suppose on the contrary that $K_b(f) = \emptyset$. Using the deformation result in Chang [2, Theorem 4.1], for the functional $-f$ on X we find a positive number ε, with $\varepsilon < (b-a)/2$, and a homeomorphism $\eta : X \to X$ such that

$$\eta(x) = x, \quad \forall x \in X \setminus f^{-1}([(a+b)/2, (-a+3b)/2[), \tag{21}$$

$$b + \varepsilon \leq f(\eta(x)), \quad \forall x \in f^{b-\varepsilon}. \tag{22}$$

By the definition of b and relation (22) we get $\gamma_\varepsilon^* \in \Gamma^*$ such that

$$b + \varepsilon \leq \inf_{x \in S} f(\eta(\gamma_\varepsilon^*(x))).$$

On the other hand, from the first inclusion in (f_9) and property (21) we see that $\eta \circ \gamma_\varepsilon^* \in \Gamma^*$. The preceding inequality yields the contradiction $b + \varepsilon \leq b$, which proves assertion (ii).

Suppose now that $a = b$. Clearly, in this case it is sufficient to prove only assertion (iii). Arguing by contradiction let us admit $K_b(f) \cap S = \emptyset$. Setting $g = -f$, $d = -b$, $A = \partial Q$, $B = S$, it turns out that $A \subset g^d$, $B \subset g_d$, $K_d(g) \cap B = \emptyset$. In addition, the Palais–Smale condition (in the sense of Chang [2]) for the locally Lipschitz functional g on X is verified. Then the deformation result in Motreanu and Varga [7, Theorem 1.1] implies that there exist a number $\varepsilon > 0$ and a homeomorphism $\eta : X \to X$ such that

$$\eta(x) = x \quad \forall x \in \partial Q, \tag{23}$$

$$b + \varepsilon \leq f(\eta(x)) \quad \forall x \in S. \tag{24}$$

Relation (23) ensures that $\eta \in \Gamma^*$. Then, due to (24), we have

$$b + \varepsilon \leq \inf_{x \in S} f(\eta(x)) \leq b.$$

This contradiction completes the proof. ■

3. Applications

In this Section we exploit Theorem 1 to solve an elliptic boundary value problem expressed as a variational-hemivariational inequality in the sense of Panagiotopoulos [8].

Let Ω be a bounded domain of the real Euclidean space $(I\!R^N, |\cdot|)$, $N \geq 3$, having a smooth boundary $\partial\Omega$. Consider the Sobolev space $H_0^1(\Omega)$ which is the closure of $C_0^\infty(\Omega)$ with respect to the norm

$$\|u\| := \left(\int_\Omega |\nabla u(x)|^2 dx \right)^{1/2}.$$

As usually, denote by 2^* the critical exponent for the Sobolev embedding $H_0^1(\Omega) \subset L^p(\Omega)$, i.e.

$$2^* = \frac{2N}{N-2}.$$

If $p \in [1, 2^*]$ then there exists a constant $c_p > 0$ fulfilling

$$\|u\|_{L^p(\Omega)} \leq c_p \|u\| \quad \forall u \in H_0^1(\Omega), \tag{25}$$

and the embedding is compact whenever $p \in [1, 2^*[$. Let $\{\lambda_n\}$ be the sequence of eigenvalues of the operator $-\Delta$ on $H_0^1(\Omega)$, with

$$0 < \lambda_1 < \lambda_2 \leq \ldots \leq \lambda_n \leq \ldots$$

and let $\{\phi_n\}$ be a corresponding sequence of eigenfunctions normalized as follows:

$$\|\phi_n\|^2 = 1 = \lambda_n \|\phi_n\|^2_{L^2(\Omega)}, \quad n \in \mathbb{N}; \tag{26}$$

$$\int_\Omega \nabla\phi_m(x) \cdot \nabla\phi_n(x)\, dx = \int_\Omega \phi_m(x)\phi_n(x)\, dx = 0 \quad \text{provided } m \neq n.$$

Consider a function $j : \Omega \times \mathbb{R} \to \mathbb{R}$ satisfying the conditions

(j_1) *j is measurable with respect to each variable separately,*

(j_2) *there exist $a_1 > 0$, $p \in]2, 2^*[$ such that*

$$|j(x,t)| \leq a_1 \left(1 + |t|^{p-1}\right) \quad \forall (x,t) \in \Omega \times \mathbb{R}.$$

Then the function $J : \Omega \times \mathbb{R} \to \mathbb{R}$ given by

$$J(x,\xi) = \int_0^\xi -j(x,t)\, dt\,, \quad (x,\xi) \in \Omega \times \mathbb{R},$$

is well defined, $J(\cdot,\xi)$ is measurable, while $J(x,\cdot)$ is locally Lipschitz continuous. So it makes sense to consider its generalized directional derivative with respect to the variable ξ that will be denoted by J^0.

Let us further assume that

(j_3) *$\lambda_k < \lambda_{k+1}$ for some $k \in \mathbb{N}$, $\lambda \in [\lambda_k, \lambda_{k+1}[$, besides*

$$J(x,\xi) \leq \min\{0, a_2(1 - |\xi|^q)\} \quad \text{in } \Omega \times \mathbb{R},$$

where $a_2 > 0$, $q \in]2, p]$;

(j_4) *there exists $r > 0$ fulfilling*

$$\inf_{\|u\|=r} \frac{1}{r^2} \int_\Omega J(x, u(x))\, dx \geq -\frac{1}{2}\left(1 - \frac{\lambda}{\lambda_{k+1}}\right).$$

Given $R \in]0, +\infty[$, denote by (P_R) the following elliptic variational-hemivariational inequality problem:

Find $u \in \overline{B}_R \subset H_0^1(\Omega)$ such that

$$-\int_\Omega \nabla u(x) \cdot \nabla(v-u)(x)\,dx + \lambda \int_\Omega u(x)(v(x)-u(x))\,dx$$

$$\leq \int_\Omega J^0(x, u(x); v(x)-u(x))\,dx$$

for all $v \in \overline{B}_R$.

We state the following result for solving problem (P_R).

Theorem 4. *Suppose that conditions* (j_1)–(j_4) *hold. Then there exists $R_0 \in]0, +\infty[$ such that problem* (P_R) *possesses a solution for every real number $R > R_0$.*

Proof. Set $X = H_0^1(\Omega)$, $V = \mathrm{span}\{\phi_1, \ldots, \phi_k\}$ and $E = V^\perp$. One has $X = V \oplus E$ as well as $\dim(V) = k < +\infty$. Moreover, introduce

$$\Phi(u) = \frac{1}{2}\left(\|u\|^2 - \lambda \|u\|^2_{L^2(\Omega)}\right) + \int_\Omega J(x, u(x))\,dx\,, \quad u \in X\,.$$

By (j_2) the function Φ turns out locally Lipschitz continuous. Setting $W = \mathrm{span}\{\phi_1, \ldots, \phi_{k+1}\}$ it results, for a suitable $\rho > r$,

$$\Phi(u) \leq 0 \quad \forall u \in W \setminus B_\rho. \tag{27}$$

To see this we first note that W is a (finite dimensional) subspace of $L^q(\Omega)$ because $q \leq p < 2^*$. Hence, there exists $a_3 > 0$ fulfilling

$$\|u\|_{L^q(\Omega)} \geq a_3 \|u\|\,, \quad u \in W\,.$$

Denote by $|\Omega|$ the Lebesgue measure of Ω. Since each $u \in W$ can be written as $u = \sum_{i=1}^{k+1} t_i \phi_i$, where $t_1, \ldots, t_{k+1} \in I\!R$, through (26) and (j_3) we obtain

$$\begin{aligned}\Phi(u) &= \frac{1}{2}\sum_{i=1}^{k+1} t_i^2\left(1 - \frac{\lambda}{\lambda_i}\right) + \int_\Omega J(x, u(x))\,dx \\ &\leq \frac{1}{2}t_{k+1}^2\left(1 - \frac{\lambda}{\lambda_{k+1}}\right) + a_2\left(|\Omega| - \|u\|^q_{L^q(\Omega)}\right) \\ &\leq \frac{1}{2}\|u\|^2\left(1 - \frac{\lambda}{\lambda_{k+1}}\right) + a_2\,|\Omega| - a_2 a_3^q \|u\|^q,\end{aligned}$$

which, by assumption $q > 2$, leads to (27).

Now, choose

$$Q = \left(V \cap \overline{B}_\rho\right) \oplus [0, \rho\, \phi_{k+1}], \quad S = \partial B_r \cap E, \quad R_0 = 2\rho.$$

It is known that the compact topological manifold Q links with the closed set S (see Rabinowitz [9]). Further, one has

$$Q \subset B_R, \quad S \subset B_{R/2} \;\; \forall R > R_0. \tag{28}$$

Fix a real number $R > R_0$. If for every $u \in X$ we define

$$\alpha(u) = 0 when$$

u$\in$ $\overline{B}_R$$+\infty otherwise$, $\quad f(u) = \Phi(u) + \alpha(u)$ $,thenhypothesis$(H$_f$) is satisfied, while the first inclusion in (28) yields $f|_Q = \Phi|_Q$. This evidently implies (f_1).

Let us next verify assumption (f_2) with $a = 0$. From (28) and (27) it follows $f(u) \le 0$, $u \in W \cap B_R \setminus B_\rho$. Since each $u \in V \cap B_R$ can be written as $u = \sum_{i=1}^{k} t_i \phi_i$, where $t_1, \ldots, t_k \in I\!R$, by ($j_3$) we get

$$f(u) = \Phi(u) \le \frac{1}{2} \sum_{i=1}^{k} t_i^2 \left(1 - \frac{\lambda}{\lambda_i}\right) \le 0\,.$$

This implies $\partial Q \subset f_0$. Taking now $u \in S$ it results $u = \sum_{i=k+1}^{+\infty} t_i \phi_i$ for suitable $t_i \in I\!R$ in addition to $\|u\| = r$. Through (26) and (j_4) we derive

$$\begin{aligned} f(u) = \Phi(u) &= \frac{1}{2} \sum_{i=k+1}^{+\infty} t_i^2 \left(1 - \frac{\lambda}{\lambda_i}\right) + \int_{\Omega} J(x, u(x))\, dx \\ &\ge \frac{1}{2}\left(1 - \frac{\lambda}{\lambda_{k+1}}\right) \|u\|^2 + \int_{\Omega} J(x, u(x))\, dx \\ &\ge r^2 \left[\frac{1}{2}\left(1 - \frac{\lambda}{\lambda_{k+1}}\right) + \inf_{\|u\|=r} \frac{1}{r^2} \int_{\Omega} J(x, u(x))\, dx\right] \ge 0. \end{aligned}$$

Consequently $S \subset f^0$, which represents the desired conclusion.

Let us finally show (f_3). To this end, let a sequence $\{u_n\} \subset X$ be such that $f(u_n) \to c$ and

$$\Phi^0(u_n; v - u_n) + \alpha(v) - \alpha(u_n) \ge -\varepsilon_n \|v - u_n\| \;\; \forall n \in I\!N,\; v \in X, \tag{29}$$

where $\varepsilon_n \to 0^+$. One has

$$\{u_n\} \subset D_\alpha = \overline{B}_R.$$

So, passing to a subsequence if necessary, we may suppose $u_n \rightharpoonup u$ in X as well as $u_n \to u$ in $L^2(\Omega)$. By (29) with $v = u$ and taking account formula (2) at p. 77 in Clarke [3] we infer that

$$\int_\Omega \nabla u_n(x) \cdot \nabla u(x)\, dx - \lambda \int_\Omega u_n(x)\,(u(x) - u_n(x))\, dx$$

$$+ \int_\Omega J^0(x, u_n(x); u(x) - u_n(x))\, dx + \varepsilon_n \|u_n - u\| \geq \int_\Omega |\nabla u_n(x)|^2\, dx.$$

The upper semicontinuity of J^0 yields

$$\limsup_{n\to+\infty} \|u_n\|^2 \leq \limsup_{n\to+\infty} \int_\Omega J^0(x, u_n(x); u(x) - u_n(x))\, dx + \|u\|^2 = \|u\|^2,$$

thus $u_n \to u$ in X. Hence, either $(\mathrm{PS})_{f,c}$ or $(\mathrm{PS})_{f,S,c}$ holds according to whether $c > 0$ or $c = 0$. In view of the second inclusion in (28) there exists $\varepsilon_0 > 0$ fulfilling

$$N_{\varepsilon_0}(S) \subset B_R = \mathrm{int}(D_{\partial\alpha}).$$

Finally, since for every $\delta \in]0, \varepsilon_0[$ it is true that

$$N_\delta(S) \cap f^{c-\delta} \cap f_{c+\delta} = N_\delta(S) \cap \left\{x \in \overline{B}_R : c - \delta \leq \Phi(x) \leq c + \delta\right\},$$

we see that condition (f_3) holds, too. Now, Theorem 2 can be applied. We obtain a point $u \in X$ such that

$$\Phi^0(u; v - u) + \alpha(v) - \alpha(u) \geq 0 \quad \forall v \in X.$$

The choice of α ensures $u \in \overline{B}_R$ as well as $\Phi^0(u; v - u) \geq 0$ whenever $v \in \overline{B}_R$. Using formula (2) at p. 77 in [3], we thus have

$$\int_\Omega \nabla u(x) \cdot \nabla (v - u)(x)\, dx - \lambda \int_\Omega u(x)(v(x) - u(x))\, dx$$

$$+ \int_\Omega J^0(x, u(x); v(x) - u(v))\, dx \geq 0 \quad \forall v \in \overline{B}_R,$$

i.e. the function u is a solution to problem (P_R). ∎

Remark. The proof of Theorem 4 yields the following value for the constant R_0 entering the statement of Theorem 4:

$$R_0 = 2 \inf \left\{\rho > 0 : \frac{1}{2}\left(1 - \frac{\lambda}{\lambda_{k+1}}\right)\rho^2 + a_4|\Omega| - a_4 a_5^q \rho^q \leq 0\right\}.$$

Finally, we present a formulation in terms of variational–hemivariational inequalities for a model arising in the unilateral mechanics.

Let Ω be a linear elastic body identified with a bounded, open and connected subset $\Omega \subset \mathbb{R}^3$ with a $C^{1,1}$–boundary Γ. In the framework of small deformation theory the displacement field $u : \Omega \to \mathbb{R}^3$ is described by the relations

$$\sigma_{ij,j}(u) + f_i = 0 \text{ and } \sigma_{ij}(u) = C_{ijkl}\varepsilon_{kl}(u) \text{ on } \Omega,\ S_i(u) = \sigma_{ij}(u)n_j \text{ on } \Gamma,$$

with $f = (f_i) \in L^2(\Omega; \mathbb{R}^3)$ and where the elasticity tensor $C_{ijkl} \in L^\infty(\Omega)$ is supposed to satisfy the usual symmetry and ellipticity properties. Assuming that the involved data are sufficiently smooth, the Green formula yields the equality

$$\int_\Omega C_{ijkl}\varepsilon_{ij}(u)\varepsilon_{kl}(v-u)dx = \int_\Omega f\cdot(v-u)dx + \int_\Gamma S(u)\cdot(v-u)ds \quad (30)$$

for all admissible displacements v. One supposes the decomposition $\Gamma = \overline{\Gamma_1} \cup \overline{\Gamma_2} \cup \overline{\Gamma_3}$ with Γ_1, Γ_2, Γ_3 nonempty, mutually disjoint and open (in the induced topology on Γ). We introduce the boundary conditions as follows:

$u(x) \in Q(x) \subset \mathbb{R}^3,\ \forall x \in \Gamma_1,$
with the normal reaction law $-r(x) \in N_{Q(x)}(u(x)),\ \forall x \in \Gamma_1;$
$u(x) = 0,\ \forall x \in \Gamma_2;$
$u_T = C_T$ (given force in the tangential direction) on Γ_3,
$-S_N(x) \in \partial j(x, u_N(x)),\ \forall x \in \Gamma_3,$

where the function $y \in \mathbb{R}^3 \mapsto j(x,y) \in \mathbb{R}$ is locally Lipschitz for all $x \in \Omega$. Using (30), the definition of generalized gradient and the imposed boundary conditions one obtains

$$\begin{array}{l}\int_\Omega C_{ijkl}\varepsilon_{ij}(u)\varepsilon_{kl}(v-u)dx + \int_{\Gamma_3} j^\circ(x,u_N;v_N-u_N)ds \\ -\int_\Omega f\cdot(v-u)dx - \int_{\Gamma_3} C_T\cdot(v_T-u_T)ds \geq 0,\end{array} \quad (31)$$

subject to $v_T = 0$ on Γ_2 and $v(x) \in Q(x)$ for a.e. $x \in \Gamma_1$. Problem (31) fits in the framework of Definition 1 and represents a variational–hemivariational inequality in the sense of P. D. Panagiotopoulos [8], [6] under the functional setting: $X = \{u \in H^1(\Omega;\mathbb{R}^3) : u(x) = 0 \text{ a.e. } x \in \Gamma_2\}$ and α is the indicator function of the set $C = \{u \in X : u(x) \in Q(x) \text{ for a.e. } x \in \Gamma_1\}$, i.e. $\alpha = 0$ on C and $\alpha = +\infty$ otherwise. The set C is supposed to be convex and closed. Then problem (31) reads as finding a critical point (in the sense of Definition 1) $u \in X$ of the nonsmooth functional $f : X \to]-\infty, +\infty]$ of form required in (H_f), where Φ is the potential defined by the left-hand side of (31).

Various applications in Mechanics and Engineering of variational–hemivariational inequalities can be found in Panagiotopoulos [8].

References

[1] G. Barletta and S. A. Marano, Some remarks on critical point theory for locally Lipschitz functions, *Glasgow Math. J.* **45** (2003), 131–141.

[2] K.-C. Chang, Variational methods for non-differentiable functionals and their applications to partial differential equations, *J. Math. Anal. Appl.* **80** (1981), 102–129.

[3] F. H. Clarke, *Optimization and Nonsmooth Analysis*, John Wiley, New York, 1983.

[4] I. Ekeland, Nonconvex Minimization Problems, *Bull. (New Series) Amer. Math.* **1** (1979), 443–474.

[5] S. Marano and D. Motreanu, A deformation theorem and some critical point results for non-differentiable functions, Topol. Methods Nonlinear Anal., in print.

[6] D. Motreanu and P. D. Panagiotopoulos, *Minimax Theorems and Qualitative Properties of the Solutions of Hemivariational Inequalities and Applications*, Kluwer Academic Publishers, Dordrecht, Boston, London, 1999.

[7] D. Motreanu and C. Varga, Some critical point results for locally Lipschitz functionals, *Comm. Appl. Nonlinear Anal.* **4** (1997), 17–33.

[8] P. D. Panagiotopoulos, *Hemivariational inequalities. Applications to mechanics and engineering*, Springer-Verlag, New York, 1993.

[9] P. H. Rabinowitz, *Minimax methods in critical point theory with applications to differential equations*, CBMS Reg. Conf. Ser. in Math. **65**, Amer. Math. Soc., Providence, 1986.

[10] A. Szulkin, Minimax principles for lower semicontinuous functions and applications to nonlinear boundary value problems, *Ann. Inst. H. Poincaré. Analyse nonlinéaire* **3** (1986), 77–109.

Chapter 13

MIN-MAX DUALITY AND SHAKEDOWN THEOREMS IN HARDENING PLASTICITY

Quoc Son Nguyen
Laboratoire de Mécanique des Solides, CNRS-UMR7649
Ecole Polytechnique, Paris, France
son@lms.polytechnique.fr

Abstract This paper gives a comprehensive presentation on shakedown theorems in perfect and in hardening plasticity. The extension of classical shakedown theorems for hardening materials is an interesting subject in plasticity. In particular, the model of generalized standard materials gives a convenient framework to derive appropriate results for common models. The starting point is an extended static shakedown theorem for hardening plasticity. It leads by min-max duality to dual static and kinematic approaches to compute the safety coefficient with respect to shakedown. These approaches are discussed for common models of isotropic and of kinematic hardening. In particular, the kinematic approach leads to new results on the expressions of the safety coefficient.

Keywords: Plasticity, strain-hardening, generalized standard models, shakedown, min-max duality, variational principles, static and kinematic approaches.

Introduction

The elastic shakedown phenomenon is related to the long-term behaviour of a solid under variable loads and expresses the fact that the mechanical response of solids become purely elastic if the load amplitude is small enough or if the hardening effect is strong enough, immaterial of the initial state of the evolution. Shakedown conditions have been discussed in a large number of papers of the literature. Classical shakedown theorems in a quasi-static deformation takes its definitive form from the pioneering works of Bleich (1932), Melan (1936) and Koiter (1960). Its generalization to dynamics has been discussed, cf. Corradi and Maier, 1974. Further extensions to hardening plasticity, non-standard plasticity, to visco-plasticity or to damage mechanics and poroplasticity can be found in a large number of references, e.g. (Maier, 1972), (Mandel, 1976), (Konig, 1987), (Polizzotto et al., 1991), (Debordes, 1976), (Weichert and Maier, 2000), (Maier, 2001), (Bodovillé and de Saxcé, 2001), (Fuschi, 1999). In particular, the reader can refer to (Martin, 1975), or to (Polizzotto, 1982) and (Corigliano et al., 1995) for a rather complete presentation of the theory and historical survey, to (Maier, 2001), (Ponter and Chen, 2001) and (Weichert and Maier, 2000) for new directions on the related subjects.

The objective of this paper is to give a presentation of shakedown theorems in hardening plasticity, available for common models of strain hardening. Our attention is focussed on the framework of generalized standard models of plasticity. This framework is a straightforward extension of perfect plasticity, with the same ingredients of convexity and normality and has been shown to be large enough to cover most common models of hardening plasticity, cf. (Halphen and Nguyen, 1975), (Nguyen, 2000). As in perfect plasticity, e.g. (Koiter, 1960), (Debordes, 1976), (Pycko and Mroz, 1992), the method of min-max duality can be applied to obtain shakedown theorems. The starting point is a static shakedown theorem, given previously by the author in (Nguyen, 1976). This theorem leads to the definition of the safety coefficient with respect to shakedown and, by a min-max duality, to dual expressions of the safety coefficient obtained respectively from static and kinematic approaches. This method is then applied to the particular cases of strain hardening materials for which hardening parameters are the plastic strain or equivalent plastic strain. Our principal goal is to obtain appropriate results for common models of isotropic or kinematic hardening.

1. Static shakedown theorems

The quasi-static evolution of an elastic plastic solid under variable loads on the interval $[0, +\infty[$ is considered here. For dynamic conditions, it is well known that additional terms due to the inertial forces can be taken into account and the same conclusion remains valid as it has been shown in the literature. Let $(u(t), \sigma(t), \epsilon^p(t))$ be the elastic-plastic response of the solid starting from a given initial state on the interval $t \geq 0$. By definition, this response will shake down if the existence of the limit

$$\lim_{t \to \infty} \quad \epsilon^p(t) \tag{1.1}$$

is ensured. It can be noted that this property also ensures that

$$\lim_{t\to\infty} \; u(t) - u_{el}(t) \quad \text{and} \quad \lim_{t\to\infty} \; \sigma(t) - \sigma_{el}(t)$$

exist, under certain additional assumptions, e.g. (Debordes and Nayroles, 1976), (Nayroles, 1977), where $\sigma_{el}(t)$, $u_{el}(t)$) denote the fictitious response of the same solid, assumed to be purely elastic, to the same loading. The proof of this statement is clear for discrete systems while for continua, some difficulties remain concerning the choice of a relevent functional space in the case of perfect plasticity, cf. (Debordes, 1976) and (Suquet, 1981).

1.1 Perfect plasticity

The classical static theorem of shakedown in perfect plasticity, also known as Melan's theorem, states that:

Theorem 13.1 (**Static shakedown theorem**): *If there exists a self-stress field $s^*(x)$, a safety coefficient $m > 1$ and a time τ such that the stress field $m(s^* + \sigma_{el}(t))$ satisfies everywhere and for all $t \geq \tau$ the plastic criterion*

$$f(\; m(s^*(x) + \sigma_{el}(x,t)) \;) \leq 0 \qquad \forall \; x, \quad \forall \;\; t > \tau$$

then there is shakedown, immaterial of the initial conditions.

The proof of Melan's theorem can be obtained in two steps. In the first step, it is shown that under the assumptions of the theorem, the dissipated energy W^d is necessarily bounded. In the second step, this property ensures the existence of $\lim_{t\to\infty} \;\; \epsilon^p(t)$.

To prove the first step, it is useful to note that for all plastically admissible stress fields $\tilde{\sigma}$, the following inequality holds

$$\int_\Omega (\sigma - \tilde{\sigma}) : \dot{\epsilon}^p \; d\Omega \;\geq\; 0.$$

By taking $\tilde{\sigma} = m(s^* + \sigma_{el}(t))$ which is plastically admissible by assumption, it follows that

$$\int_\Omega (\sigma - \sigma^*) : \dot{\epsilon}^p \;\geq\; \frac{m-1}{m} \int_\Omega d_{in} \; d\Omega$$

with $\sigma^* = s^* + \sigma_{el}$. Since $\sigma - \sigma^*$ is a self-stress field and since $\dot{u} - \dot{u}_{el} = 0$ on S_u, one obtains

$$0 = \int_\Omega (\sigma - \sigma^*) : (\dot{\epsilon} - \dot{\epsilon}_{el}) \; d\Omega \;\; \Rightarrow$$

$$\int_\Omega (\sigma - \sigma^*) : \dot{\epsilon}^p + \int_\Omega (\sigma - \sigma^*) : L^{-1} : (\dot{\sigma} - \dot{\sigma}*) \, d\Omega = 0.$$

It follows that

$$-\int_\Omega (\sigma - \sigma^*) : L^{-1} : (\dot{\sigma} - \dot{\sigma}^*) \, d\Omega \; \geq \; \frac{m-1}{m} \int_\Omega d_{in} \, d\Omega,$$

which leads, after a time integration on the interval $[\tau, t]$, to

$$I(\tau) - I(t) \; \geq \; \frac{m-1}{m} W^d \quad \text{with} \tag{1.2}$$

$$I(t) = \int_\Omega \frac{1}{2} (\sigma - \sigma^*) : L^{-1} : (\sigma - \sigma^*) \, d\Omega.$$

The dissipated energy $W^d(t) = \int_\tau^t \int_\Omega \sigma : \dot{\epsilon}^p \, d\Omega d\theta$ thus remains bounded for all initial conditions.

The second step consists of proving that $\epsilon^p(t)$ tends to a limit. The fact that the dissipated energy remains bounded already ensures the existence of a limit of $\epsilon^p(t)$ for any appropriate functional space which is complete with respect to the norm associated with the dissipation, since there exists a constant $c > 0$ such that $\sigma : \dot{\epsilon}^p \geq c\|\dot{\epsilon}^p\|$. This inequality follows from the fact that the origin of stress is strictly inside the elastic domain.

1.2 Hardening plasticity

This discussion is limited to the case of generalized standard models of hardening plasticity. A generalized standard material is defined by the following conditions:

- State variables are (ϵ, α), which represents respectively the strain tensor and a set of internal parameters. Internal parameters α include the plastic strain and other hardening variables. There exists an energy potential $W(\epsilon, \alpha)$ which leads to associated forces

$$\sigma = W_{,\epsilon}\,, \quad A = -W_{,\alpha} \tag{1.3}$$

and to a dissipation per unit volume d_{in}

$$d_{in} = A \cdot \dot{\alpha}. \tag{1.4}$$

- The force A must be plastically admissible, this means that physically admissible forces A^* must remain inside a convex domain C, called the elastic domain and defined by the plastic criterion $f(A^*) \leq 0$.

- Normality law is satisfied for α:

$$\dot{\alpha} = \mu \frac{\partial f}{\partial A}, \quad \mu \geq 0, \quad \mu f = 0. \tag{1.5}$$

Thus, the following maximum dissipation principle is satisfied:

$$A^* \cdot \dot{\alpha} \leq A \cdot \dot{\alpha} = D(\dot{\alpha}) = \max_{A^* \in C} \quad A^* \cdot \dot{\alpha} \tag{1.6}$$

in the spirit of Hill's maximum principle in perfect plasticity. Finally, as in perfect plasticity, the notions of convexity, normality, energy and generalized forces are four principal ingredients of the generalized standard models. The dissipation potential $D(\dot{\alpha})$ is convex, positively homogeneous of degree 1. This function is state-independent, i.e. independent of the present value of state variables (ϵ, α), if the plastic criterion is state-independent. The following result has been obtained under some additional assumptions, cf. (Nguyen, 1976):

Theorem 13.2 (**static shakedown theorem in hardening plasticity**): *It is assumed that the plastic criterion is state-independent and that*

$$W = W_1(\epsilon, \alpha) + W_2(\alpha), \tag{1.7}$$

where W_1 is quadratic and non-negative with respect to (ϵ, α) and $W_2(\alpha)$ is an arbitrary convex and differentiable function. Then there is shakedown whatever the initial state if there exists a time τ, a constant internal parameter field α^* and a safety coefficient $m > 1$ such that the force field $mA^*(t)$ is plastically admissible for all $t > \tau$, where $A^* = -W_{,\alpha}(\epsilon^*, \alpha^*)$ denotes the force defined from the associated response $u^*(t)$.

The associated response u^* denotes the response of a corresponding elastic solid admitting the linear elastic relationship

$$\sigma = W_{,\epsilon} = W_{1,\epsilon} = L : \epsilon + Y \cdot \alpha = L : (\epsilon - \epsilon^{p*}) \quad with \quad \epsilon^{p*} = -L^{-1} : Y \cdot \alpha^*.$$

under the same loading (prescribed forces and displacements) and under the inherent plastic strain ϵ^{p*}. In particular, the associated stress is $\sigma^*(t) = \sigma_{el}(t) + s^*$ where s^* denotes the self-stress field induced by the plastic strain field ϵ^{p*}.

The proof of this proposition is relatively straightforward. Indeed, let $(u(t), \alpha(t))$ be a solution of the evolution problem starting from a given initial condition. Since under the assumptions introduced, mA^* is plastically admissible for all $t \geq \tau$, the following inequality holds

$$\int_\Omega (A - A^*) \cdot \dot{\alpha} \, d\Omega \geq \frac{m-1}{m} \int_\Omega A \cdot \dot{\alpha} \, d\Omega.$$

From the fact that

$$\left\{ \begin{array}{l} (\sigma - \sigma^*) : (\dot{\epsilon} - \dot{\epsilon}^*) - (A - A^*) \cdot (\dot{\alpha} - \dot{\alpha}^*) = \\ \frac{d}{dt}\{W_1(\epsilon - \epsilon^*, \alpha - \alpha^*) + W_2(\alpha) - W_2(\alpha^*) - W_{2,\alpha}(\alpha^*) \cdot (\alpha - \alpha^*)\} \end{array} \right.$$

since $\dot{\alpha}^* = 0$, and that $\int_\Omega \ (\sigma - \sigma^*) : (\dot{\epsilon} - \dot{\epsilon}^*) \, d\Omega = 0$, the following equation results

$$\left\{ \begin{array}{l} -\frac{d}{dt} \int_\Omega \{W_1(\epsilon - \epsilon^*, \alpha - \alpha^*) + W_2(\alpha) - W_2(\alpha^*) \\ -W_{2,\alpha}(\alpha^*) \cdot (\alpha - \alpha^*) \ \} \, d\Omega \ \geq \ \frac{m-1}{m} \int_\Omega d_{in} \, d\Omega. \end{array} \right.$$

Thus the estimate (1.2) is again obtained with

$$\left\{ \begin{array}{l} I(t) = \int_\Omega \{W_1(\epsilon(t) - \epsilon^*(t), \alpha(t) - \alpha^*) + \\ W_2(\alpha(t)) - W_2(\alpha^*) - W_{2,\alpha}(\alpha^*) \cdot (\alpha(t) - \alpha^*) \ \} \, d\Omega. \end{array} \right.$$

Since $I(t)$ is non-negative as a sum of two non-negative terms, the dissipated energy $W^d(t)$ is bounded by $I(\tau)$ for all $t > \tau$. Repeating the same arguments, the existence of the limit $\lim_{t \to \infty} \alpha(t) - \alpha^*$ is ensured again from this estimate on the dissipated energy. In the same spirit as Melan's statement, Theorem 2 gives a sufficient condition ensuring the shakedown of a solid subjected to a given loading path.

2. Safety coefficient and min-max duality

Min-max duality is a well known method of Convex Analysis, cf. for example (Rockafellar, 1970). It has been discussed in perfect plasticity by (Debordes and Nayroles, 1976) and by (Pycko and Mroz, 1992) for shakedown analysis. This method is applied here following the same ideas. For this, the definition of a safety coefficient with respect to shakedown is first considered and dual static and kinematic approches are introduced.

2.1 Safety coefficient

A safety coefficient with respect to shakedown can be introduced from the previous results. For a given loading history on the time interval $[0. + \infty[$, the following result holds

Theorem 13.3 *Let m_s be the safety coefficient defined by the maximum*

$$m_s(\tau) = \max_{\alpha^*} \quad m \tag{2.1}$$

among all constant fields α^* such that

$$\left\{ \begin{array}{l} \forall \ t \geq \tau, \ \ u^*(t) \ \ CA, \ \ \sigma^*(t) \ \ SA, \ \ mA^* \ \ PA, \\ \sigma^* = W_{,\epsilon}\,(\epsilon(u^*), \alpha^*), \ \ A^* = -W_{,\alpha}\,(\epsilon(u^*), \alpha^*), \end{array} \right. \tag{2.2}$$

where CA, SA, PA respectively stand for kinematically, statically and plasticlly admissible fields. Then there is shakedown if there exists $\tau \geq 0$ *such that* $m_s(\tau) > 1$ *while no conclusion is available if* $m_s(\tau) < 1$ *for all* $\tau \geq 0$.

Indeed, if $m_s > 1$ then there exists a constant field α^* in the spirit of Theorem 2 and the conclusion holds. In perfect plasticity, this statement reduces to Melan's theorem.

2.2 Min-max duality

The minimization (2.1) is a convex optimization problem. The dual approach consists of considering dual problems obtained by relaxing some constraints. For this, the initial problem (2.1) is first written as the search of maximum of m in the set of constant fields α^* and time-dependent fields $\tilde{A}(t)$ and of coefficients m such that

$$\begin{cases} \forall \ t \geq \tau, \ \ u^*(t) \ \ CA, \ \ \sigma^*(t) \ \ SA, \ \ m\tilde{A}(t) \ \ PA, \\ \sigma^* = W_{,\epsilon}(\epsilon(u^*), \alpha^*), \ \ A^* = -W_{,\alpha}(\epsilon(u^*), \alpha^*), \ \ \tilde{A}(t) = A^*(t). \end{cases} \tag{2.3}$$

The last constraint is relaxed by the introduction of Lagrange multipliers $\beta(t)$ associated with the constraint $m(\tilde{A}(t) - A^*(t)) = 0$ and the Lagrangian

$$\mathbf{\Lambda}(\beta, m, \alpha^*, \tilde{A}) = m + \int_\tau^\infty \int_\Omega m(\tilde{A} - A^*) \cdot \beta \ \ d\Omega dt. \tag{2.4}$$

The max-min problem

$$\max_{\alpha^*, \tilde{A}, m} \quad \min_{\beta} \quad \mathbf{\Lambda} \tag{2.5}$$

in the set of arbitrary fields $\beta(t)$ and α^*, $\tilde{A}(t)$, m such that

$$\begin{cases} \forall \ t \geq \tau, \ \ u^*(t) \ \ CA, \ \ \sigma^*(t) \ \ SA, \ \ m\tilde{A}(t) \ \ PA, \\ \sigma^* = W_{,\epsilon}(\epsilon(u^*), \alpha^*), \ \ A^* = -W_{,\alpha}(\epsilon(u^*), \alpha^*), \end{cases} \tag{2.6}$$

leads to the initial problem since the result of the minimization with respect to β then gives $\min_\beta \ \mathbf{\Lambda} = m$ if $\tilde{A}(t) = A^*(t)$ for all $t \geq \tau$ and $\min_\beta \ \mathbf{\Lambda} = -\infty$ otherwise.

Theorem 13.4 *The dual problem*

$$m_k = \min_{\beta} \quad \max_{\alpha^*, \ \tilde{A}, \ m} \quad \mathbf{\Lambda} \tag{2.7}$$

in the set of arbitrary fields $\beta(t)$ *and* $\alpha^*, \tilde{A}(t)$ *such that*

$$\begin{cases} \forall \ t \geq \tau, \ \ u^*(t) \ \ CA, \ \ \sigma^*(t) \ \ SA, \ \ m\tilde{A}(t) \ \ PA, \\ \sigma^* = W_{,\epsilon}(\epsilon(u^*), \alpha^*), \ \ A^* = -W_{,\alpha}(\epsilon(u^*), \alpha^*), \end{cases} \tag{2.8}$$

defines the dual approach to compute the safety coefficient. It is clear that

$$m_s(\tau) \leq m_k(\tau) \tag{2.9}$$

according to general results of min-max duality.

In particular, if $m_k < 1$, then it is not possible to find a self-stress satisfying the condition of the previous propositions.

3. Kinematic approach

In the following sections, the dual kinematic approach will be considered for elastic plastic models admitting as internal parameters α the plastic strains or the equivalent plastic strain in view of common models in hardening plasticity.

3.1 Perfect plasticity

For example, the elastic perfectly plastic model admits as internal parameter the plastic strain ϵ^p and the energy is $W = W_1 = 1/2(\epsilon - \epsilon^p) : L : (\epsilon - \epsilon^p)$. The force associated to ϵ^p is the stress tensor σ. Equation (2.4) leads to the strain rate history $\beta = d^p$ and to the Lagrangian

$$\mathbf{\Lambda} = m + \int_\tau^\infty \int_\Omega m(\tilde{\sigma} - \sigma^*) : d^p \, d\Omega dt \tag{3.1}$$

where $m\tilde{\sigma}$ must be plastically admissible and $\sigma^* = s^* + \sigma_{el}$. The operation $\max_{s^*} \mathbf{\Lambda}$ leads to the calculation of

$$\max_{s^*} \int_\tau^\infty \int_\Omega -s^* : d^p \, d\Omega dt \tag{3.2}$$

and of

$$\max_{m\tilde{\sigma} \ PA} \int_\tau^\infty \int_\Omega m\tilde{\sigma} : d^p \, d\Omega dt. \tag{3.3}$$

The last problem is trivial

$$\max_{m\tilde{\sigma} \ PA} \int_\tau^\infty \int_\Omega m\tilde{\sigma} : d^p \, d\Omega dt = \int_\tau^\infty \int_\Omega D(d^p) \, d\Omega dt.$$

with $D(d^p) = \max_{m\tilde{\sigma} \ P.A.} \ m\tilde{\sigma} : d^p$. The first problem leads to

$$\max_{s^* \ self-stress} \int_\Omega -s^* : \int_\tau^\infty d^p dt \, d\Omega$$
$$= \begin{cases} 0 \ if \ \ E^p = \int_\tau^\infty d^p dt \ \ is \ a \ compatible \ field, \\ +\infty \ \ if \ \ E^p \ \ is \ not \ a \ compatible \ field \end{cases}$$

A compatible field means that there exists a displacement field u^p with $u^p = 0$ on S_u such that $E^p = (\nabla u^p)$. It follows that

$$\max_{\epsilon^{p*},\ \tilde{\sigma}} \quad \Lambda = m(1 - \int_0^\infty \int_\Omega \sigma_{el} : d^p \ d\Omega dt) + \int_\mathcal{T}^\infty \int_\Omega D(d^p) \ d\Omega dt.$$

when d^p is admissible and $E^p = \int_\mathcal{T}^\infty d^p dt$ is compatible. Admissible rates d^p must be considered in order to ensure a finite value of $D(d^p)$. For example, if Mises criterion is satisfied, the plastic rate is admissible if $d^p_{kk} = 0$. It is concluded after the maximization with respect to m that the following well known result holds, cf. (Koiter, 1960), (Debordes, 1976), (Pycko and Mroz, 1992):

Theorem 13.5 *In perfect plasticity, the dual kinematic approach leads to a coefficient $m_k \geq m_s$, defined as the minimum*

$$m_k = \min_{d^p} \quad \int_\mathcal{T}^\infty \int_\Omega D(d^p) \ \ d\Omega dt \tag{3.4}$$

among the plastic rates d^p admissible and satisfying

$$E^p = \int_\mathcal{T}^\infty d^p dt \ \textit{compatible}, \qquad \int_\mathcal{T}^\infty \int_\Omega \sigma_{el} : d^p \ d\Omega dt = 1. \tag{3.5}$$

It is useful to note that $m_k \geq 1$ if the following inequality holds

$$\int_\mathcal{T}^\infty \int_\Omega D(d^p) \ \ d\Omega dt \geq \int_\mathcal{T}^\infty \int_\Omega \sigma_{el} : d^p \ d\Omega dt \tag{3.6}$$

for all plastic strain rate d^p admissible and satisfying the condition $E^p = \int_\mathcal{T}^\infty d^p dt$ compatible. This result is classically known as Koiter's shakedown theorem, (Koiter, 1960).

As usually established in min-max duality, equality $m_s = m_k$ is generically satisfied. In particular, this equality holds if the plastic domain is bounded in the stress space as it has been shown by Debordes and Nayroles, 1976. It is expected that this equality always holds although a rigorous proof is lacking for continuum solids.

3.2 Isotropic hardening

The non trivial example of Cam-clay model is considered here in order to illustrate the given method. This model, proposed by the Cambridge school (Burland and Roscoe, 1968), (Schofield and Wroth, 1968), is very popular in soil mechanics since it describes correctly the plastic dilatation or contraction observed experimentally.

If p denotes the mean pressure, $p = \sigma_{kk}/3$ and $q = \|\sigma'\|$, the plastic criterion is given by the yield function

$$((p+R)^2 + bq^2)^{1/2} - R \leq 0$$

where $R \geq 0$ is a critical pressure denoted as the consolidation pressure and $b > 0$ is a material constant.

The elastic domain is thus a family of growing ellipses as shown in Fig. 13.1. This family is homothetical with respect to the origin and there is isotropic hardening. Hardening effects are described by the variation of the consolidation pressure R. For example, the following relation holds

$$R = R_o \exp{(y)},$$

where y is the consolidation parameter, a non-decreasing variable with the plastic deformation to ensure that the consolidation pressure is non-decreasing. It can be defined by the difference between an equivalent plastic strain γ and the plastic dilatation ϵ^p_{kk}

$$y = \gamma - \epsilon^p_{kk}.$$

This choice will be shown to be appropriate. It also ensures that the consolidation is possible in compression and not in tension. This is a generalized standard model with internal parameters ϵ^p, γ and energy

$$W = \frac{1}{2}(\epsilon - \epsilon^p) : L : (\epsilon - \epsilon^p) + V(y) \;\; ; \;\; V'(y) = R \;\; \Rightarrow \;\; V(y) = R_o \exp(y).$$

The generalized forces associated with the plastic dilatation and plastic deviator are $A_o = p + R$ and $A' = \sigma'$ respectively. The generalized force associated with γ is $A_\gamma = -R$. The elastic domain is state-independent since it is defined by the convex cone

$$\sqrt{A_o^2 + bA' : A'} + A_\gamma \leq 0$$

Normality law gives

$$\begin{cases} \dot{\epsilon}^p_{kk} = \mu \frac{A_o}{\sqrt{A_o^2 + bA':A'}} = \mu \frac{(p+R)}{R} \\ \dot{\epsilon}^{p'} = \mu b \frac{A'}{\sqrt{A_o^2 + bA':A'}} = \mu b \frac{\sigma'}{R} \\ \dot{\gamma} = \mu \geq 0. \end{cases}$$

It follows that the plastic volume rate is dilatative if $p > -R$ and contractive if $p > -R$ and that

$$\dot{\gamma} = \{|\dot{\epsilon}^p_{kk}|^2 + \frac{1}{b}\|\dot{\epsilon}^{p'}\|^2\}^{\frac{1}{2}}.$$

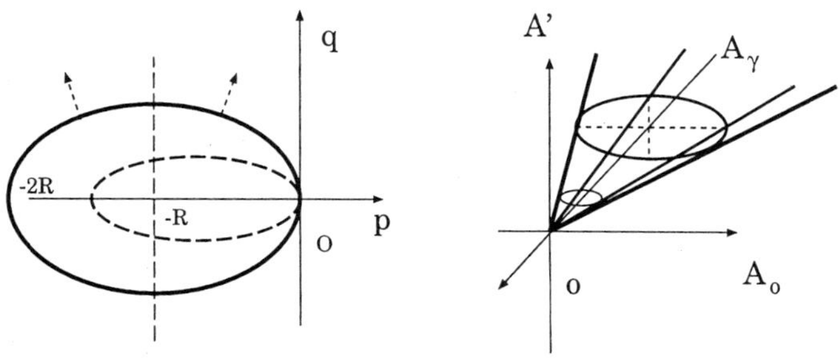

Figure 13.1. Cam-clay model of geomaterials: the elastic domain is limited by a family of ellipses in the stress space $(p \times q)$ and represented by a cone in the force space $(A_o \times A' \times A_\gamma)$.

This relation gives the physical interpretation of γ as a measure of equivalent plastic strain. In particular, the consolidation parameter and the consolidation pressure are non-decreasing since $\dot{y} \geq 0$.

The consolidation pressure R is not limited in the previous description since $R(y)$ varies from 0 to $+\infty$ when y varies from $-\infty$ to $+\infty$. Its expression can be in fact replaced and defined from any suitable monotone increasing curve such that $\lim_{y\to-\infty} \quad R(y) = 0$ and $lim_{y\to+\infty} \quad R(y) = R_{max} < +\infty$. In this case, a model of limited consolidation is obtained.

For such a model, equation (2.4) is now considered with $\tilde{A}_\gamma = A^*_\gamma$. It is clear that

$$\max_{\epsilon^{p*}} \quad -\int_\Omega (s^* + R^* I) : \int_\tau^\infty d^p \; dt d\Omega =$$

$$\begin{cases} -\int_\Omega R^* E^p_{kk} \; d\Omega \;\; if \;\; E^p \;\; is\ compatible, \\ +\infty \;\; if\ not. \end{cases}$$

Finally, R^* must maximise the quantity

$$\int_\tau^\infty \int_\Omega R^* \{(d^p_{kk})^2 + \frac{1}{b}\|d^{p'}\|^2\}^{\frac{1}{2}} \; d\Omega dt d\Omega dt - \int_\Omega R^* E^p_{kk} \; d\Omega \geq 0.$$

Thus the following result holds from the min-max duality,

Theorem 13.6 *For a Cam-clay model with limited consolidation, the kinematic approach leads to a coefficient $m_k \geq m_s$, defined as the minimum*

$$m_k = \min_{d^p} \quad R_{max} \{ \int_\tau^\infty \int_\Omega \{(d^p_{kk})^2 + \frac{1}{b}\|d^{p'}\|^2\}^{\frac{1}{2}} \; d\Omega dt d\Omega dt - \int_\Omega E^p_{kk} \; d\Omega \}$$

$$= \int_{\mathcal{T}}^{\infty} \int_{\Omega} D_m(d^p) \; d\Omega dt \quad , \quad with \quad D_m(d^p) = \max_{\sigma \in C_{max}} \sigma : d^p \qquad (3.7)$$

among the plastic rates d^p satisfying (3.5).

As expected, the result is the same as an elastic perfectly-plastic material with a elastic domain C_{max} associated with the extreme ellipse.

In order to take account of a small resistance to traction, a slight modification of the criterion can also be introduced, cf. (Lade, 1977),

$$((p + R)^2 + bq^2)^{1/2} - R - k \leq 0$$

where $k \geq 0$ is the traction limit. The plastic criterion is thus

$$\sqrt{A_o^2 + bA' : A'} + A_\gamma - k \leq 0$$

for a Cam-clay model with traction resistance. In this case, if $R_{max} < +\infty$, the result remains the same as with an elastic perfectly plastic of plastic criterion associated with R_{max}. If $R_{max} = +\infty$, the method of min-max duality leads to:

- For any rate path (d^p, $\delta\gamma$) such that d^p satisfies (3.5) and the conditions

$$d^{p'} = 0, \qquad d^p_{kk} \geq 0 \quad \forall \; x, t \qquad (3.8)$$

and that $\delta\gamma = d^p_{kk}$, it is concluded that

$$\min_{d^p, \delta\gamma} \quad \max_{\epsilon^{p*}, \gamma^*, m} \quad \mathbf{\Lambda} = \int_{\mathcal{T}}^{+\infty} \int_{\Omega} k d^p_{kk} \; d\Omega dt$$

- If no such plastic rate exists, it is concluded that

$$\min_{d^p, \delta\gamma} \quad \max_{\epsilon^{p*}, \gamma^*, m} \quad \mathbf{\Lambda} = +\infty.$$

As expected, the following result is finally obtained,

Theorem 13.7 *Let Φ be the set of plastic rates satisfying (3.5) and (3.8). For a Cam-clay model with traction resistance and unlimited consolidation, the kinematic approach leads to a coefficient $m_k \geq m_s$, defined as the minimum*

$$m_k = \min_{d^p \in \Phi} \quad \int_{\mathcal{T}}^{\infty} \int_{\Omega} k d^p_{kk} \; d\Omega dt$$

if Φ is not void and $m_k = +\infty$ if Φ is a void set.

3.3 Kinematic hardening

The model of nonlinear and limited kinematic hardening of Fig. 13.2 is considered. For this model, the state variables are ϵ, ϵ^p with $\epsilon^p_{kk} = 0$. The energy potential is $W = \frac{1}{2}(\epsilon - \epsilon^p) : L : (\epsilon - \epsilon^p) + V(\|\epsilon^p\|)$ where $R(y) = V'(y)$ is a monotone increasing function varying in $[0, R_{max}$ when $y \in R^+$. Thus the generalized force is $A = -W,_{\epsilon^p} = \sigma' - C$ with $C = R(\|\epsilon^p\|)\epsilon^p/\|\epsilon^p\|$. With Mises criterion $\|A\| - k \leq 0$, the elastic domain is a sphere of radius k and with center C in the force space. This center C remains near the origin since $\|C\| \leq R_{max}$. It is not difficult to establish that

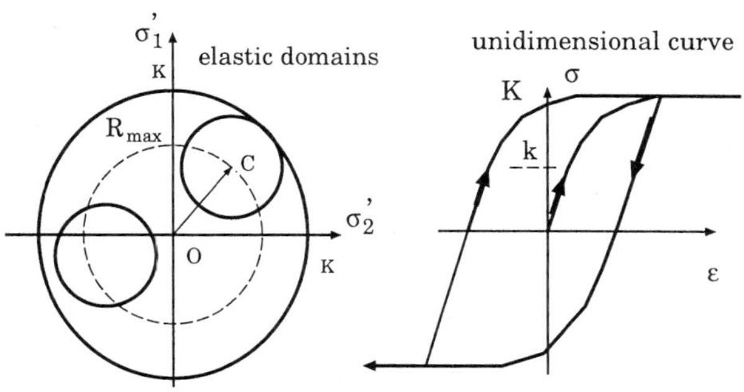

Figure 13.2. A model for limited kinematic hardening

$$\max_{\epsilon^{p*}} \quad - \int_\Omega (s^* - R(\|\epsilon^{p*}\|) \frac{\epsilon^{p*}}{\|\epsilon^{p*}\|}) : \int_\tau^\infty d^p \ \ dt d\Omega =$$
$$\begin{cases} \int_\Omega R_{max}\|E^p\| & if \ \ E^p \ \ iscompatible, \\ +\infty \ \ if \ E^p \ is \ not \ compatible. \end{cases}$$

The same method then leads to,

Theorem 13.8 *For this limited kinematic hardening model, the dual kinematic approach leads to a coefficient $m_k \geq m_s$, defined as the minimum*

$$m_k = \min_{d^p} \quad \int_\Omega R_{max}\|E^p\| \ d\Omega + \int_\tau^\infty \int_\Omega k\|d^p\| \ d\Omega dt \tag{3.9}$$

among plastic rates d^p satisfying $d^p_{kk} = 0$ and (3.5). Again, compatible plastic cycles must be considered as in perfect plasticity or in isotropic hardening.

If Zieger-Prager's model of linear kinematic hardening is considered, then $V = 1/2h\|\epsilon^p\|^2$ and the kinematic hardening is unlimited. It is concluded that

$$\max_{\epsilon^{p*}} \quad -\int_\Omega (s^* - h:\epsilon^{p*}) : \int_\tau^\infty d^p \ \ dtd\Omega = \begin{cases} 0 \ \ if \ \ E^p = \int_\tau^\infty d^p \ dt = 0, \\ +\infty \ \ if \ \ E^p \neq 0. \end{cases}$$

The same method then leads to

Theorem 13.9 *In linear kinematic hardening with Mises criterion, the dual kinematic approach leads to a coefficient $m_k \geq m_s$, defined as the minimum*

$$m_k = \min_{d^p} \quad \int_\tau^\infty \int_\Omega D(d^p) \ \ d\Omega dt \tag{3.10}$$

among the plastic rates d^p satisfying

$$d^p_{kk} = 0, \quad E^p = \int_\tau^\infty d^p \ dt = 0, \quad \int_\tau^\infty \int_\Omega \sigma_{el} : d^p \ d\Omega dt = 1. \tag{3.11}$$

Thus, closed cycles of plastic rates must be considered instead of compatible plastic cycles as in perfect plasticity or in isotropic hardening.

It is clear from this proposition that the result does not depend on the hardening tensor h, the only restriction is its positiveness. The fact that closed cycles must be considered leads to the consideration of the amplitude of the elastic stress and to the well known result that there is shakedown if this amplitude is smaller than the diameter of the yield surface.

In particular, if m_k^0, m_k^{lkh} and m_k^{nkh} denote respectively the kinematic safety coefficients in perfect plasticity, in linear kinematic hardening and in nonlinear kinematic hardening with the same yield stress k, it follows from their definition that

$$m_k^0(\tau) \leq m_k^{nkh}(\tau) \leq m_k^{lkh}(\tau) \ \ for \ all \ \ \tau. \tag{3.12}$$

The case of limited hardening (Theorem 8) appears as the penalization of the unlimited case (Theorem 9), R_{max} is the penalty parameter associated with the constraint $\|E^p\| = 0$.

4. Conclusion

If the extension of shakedown theorems to hardening plasticity has been much discussed in the past, most existing results of the literature are principally concerned with the static approach. In this paper, new results concerning the expressions of kinematic safety coefficients, available for common models of isotropic and kinematic hardening, are given.

The expressions obtained from the min-max duality such as (2.7), (3.9), (3.10), provide a useful complement to the static approach since they deal principally with the kinematic approach. Theorems 4,8,9 are new compared to the existing results of the literature. They are particularly simple and can be easily exploited to approximate by upper bounds the theoretical values in hardening plasticity as it is usually done for shakedown and limit analyses in perfect plasticity. In particular, the amplitude of the plastic rate path, denoted as E^p in this discussion, figures explicitly in the expression of the safety coefficient.

References

Bodovillé, G. and de Saxcé, G. (2001). Plasticity with nonlinear kinematic hardening: modelling and shakedown analysis by the bipotential approach. *Eur. J. Mech. A/Solids*, 20:99–112.

Burland, I. and Roscoe, K. (1968). On the generalized stress/strain behaviour of wet clay. In *Engineering Plasticity, Heyman & Leckie eds.* Cambridge Press.

Corigliano, A., Maier, G., and Pycko, S. (1995). Dynamic shakedown analysis and bounds for elasto-plastic structures with non-associated internal variable constitutive laws. *Int. J. Solids Structures*, 32:3145–3166.

Corradi, L. and Maier, G. (1974). Dynamic non-shakedown theorem for elastic perfectly-plastic continua. *J. Mech. Phys. Solids*, 22:401–413.

Debordes, O. (1976). Dualité des théorèmes statique et cinématique sur la théorie de l'adaptation des milieux continus élasto-plastiques. *C.R. Acad. Sc.*, 282:535–537.

Debordes, O. and Nayroles, B. (1976). Sur la théorie et le calcul à l'adaptation des structures élasto-plastiques. *J. Mécanique*, 20:1–54.

Fuschi, P. (1999). Structural shakedown for elastic-plastic materials with hardening saturation surface. *Int. J. Solids Structures*, 36:219–240.

Halphen, B. and Nguyen, Q. (1975). Sur les matériaux standard généralisés. *J. Mecanique*, 14:1–37.

Koiter, W. (1960). General problems for elastic-plastic solids. In *Progress in Solid Mechanics, Sneddon & Hill eds.*, volume 4, pages 165–221. North Holland.

Konig, J. (1987). *Shakedown of elastic-plastic structures.* Elsevier, Amsterdam.

Lade, P. (1977). Elastoplastic stress-strain theory for cohesionless soil with curved yield surfaces. *Int. J. Solids & Structures*, 13:1019–1035.

Maier, G. (1972). A shake-down matrix theory allowing for workhardening and second order geometric effects. In *Symp. Foundations of Plasticity, Warsaw, Sawzuk ed.*, pages 417–433, Warsaw.

Maier, G. (2001). On some issues in shakedown analysis. *J. Appl. Mech.*, 68:799–808.

Mandel, J. (1976). Adaptation d'une structure plastique écrouissable. *Mech. Res. Com*, 3:251–256.

Martin, J. (1975). *Plasticity: fundamentals and general results.* MIT Press, Cambridge.

Nayroles, B. (1977). Tendances récentes et perspectives à moyen terme en élastoplasticité asymptotique des constructions. In *Proceedings, Congrès Français de Mécanique, Grenoble, France*, Grenoble.

Nguyen, Q. (1976). Extension des théorèmes d'adaptation et d'unicité en écrouissage non linéaire. *C.R. Acad. Sc.*, 282:755–758.

Nguyen, Q. (2000). *Stability and Nonlinear Solid Mechanics.* Wiley, Chichester.

Polizzotto, C. (1982). A unified treatment of shakedown theory and related bounding techniques. *Solid Mech. Arch.*, 7:19–75.

Polizzotto, C., Borino, G., Cademi, S., and Fuschi, P. (1991). Shakedown problems for mechanical models with internal variables. *Eur. J. Mech. A/Solids*, 10:621–639.

Ponter, A. and Chen, H. (2001). A minimum theorem for cyclic load in excess of shakedown with application to the evaluation of a ratchet limit. *J. Mech. Phys. Solids*, 20:539–554.

Pycko, S. and Mroz, Z. (1992). Alternative approach to shakedown as a solution of min-max problem. *Acta Mechanica*, 93:205–222.

Rockafellar, R. (1970). *Convex analysis.* Princeton University Press, Princeton.

Schofield, A. and Wroth, C. (1968). *Critical state soil mechanics.* McGraw-Hill, London.

Suquet, P. (1981). Sur les équations de la plasticité: existence et régularité des solutions. *J. Mecanique*, 20:3–40.

Weichert, D. and Maier, G. (2000). *Inelastic analysis of structures under variable repeated loads.* Kluwer, Dordrecht.

Chapter 14

A FLUID PROBLEM WITH NAVIER-SLIP BOUNDARY CONDITIONS

Adriana Valentina Busuioc

Section de mathématiques, Institut Bernoulli, École Polytechnique Fédérale de Lausanne. CH–1015 Lausanne. Switzerland.

adriana.busuioc@epfl.ch

T. S. Ratiu

Section de mathématiques, Institut Bernoulli, École Polytechnique Fédérale de Lausanne. CH–1015 Lausanne. Switzerland.

tudor.ratiu@epfl.ch

Abstract We study the equations governing the motion of second grade fluids in a bounded domain of $\mathbb{R}^d$, $d = 2, 3$, with Navier slip boundary conditions with and without viscosity (averaged Euler equations). The main results concern the global existence and uniqueness of H^3 solutions in dimension two and, for dimension three, local existence of H^3 solutions for arbitrary initial data and global existence for small initial data and positive viscosity. The last part discusses Liapunov stability conditions for stationary solutions for the averaged Euler equations similar to the Rayleigh-Arnold stability result for the classical Euler equations. This paper presents only the main ideas of the proofs that can be found in [4].

1. Introduction

The second grade fluid equations are a physically accepted model for a solution of polymers in water (see Dunn and Rajagopal [6]). These fluids are a particular class of non-Newtonian fluids of differential type. Rivlin and Ericksen [23] developed in 1955 the theory of isotropic materials for which at time t the stress tensor T depends on the velocity, velocity gradient, and their higher time derivatives (evaluated also at time t)

$$T = -pI + F(A_1, A_2, ..., A_k),$$

where the scalar function p is the pressure, F is an isotropic function, and A_i are the Rivlin-Ericksen tensors defined by

$$A := A_1 = L + L^T, \quad L = \nabla u,$$

$$A_j := \dot{A}_{j-1} + L^T A_{j-1} + A_{j-1} L, \quad j = 2, \ldots, k,$$

where the dot denotes the material derivative. If the function F is a polynomial, the fluid is said to be of grade k. The constitutive law of a fluid of second grade is

$$T = -pI + \nu A_1 + \alpha_1 A_2 + \alpha_2 A_1^2,$$

where α_1 and α_2 are normal stress moduli. Using the conditions

$$\nu \geq 0 \quad \text{and} \quad \alpha_1 + \alpha_2 = 0$$

(see Dunn and Fosdick [10] and Fosdick and Rajagopal [14]), the equation of motion of a second grade fluid becomes:

$$\partial_t v - \nu \Delta u + u \nabla v + \sum_j v_j \nabla u_j = -\nabla p + F, \quad \text{div} u = 0 \tag{1.1}$$

where $v := u - \alpha \Delta u$. In this paper we shall consider the *Navier slip boundary conditions*:

$$u \cdot n = 0, \qquad \Big(D(u) \cdot n\Big)^{tan} = 0 \qquad \text{on} \qquad \partial U, \tag{1.2}$$

where n is the outward unit normal vector field to the boundary ∂U, the symmetric matrix $D(u)$ defined by $D(u) := (\nabla u + (\nabla u)^T)/2$ is the deformation tensor, and the upper index *tan* denotes the tangential component of a vector based at ∂U relative to the orthogonal decomposition given by the tangent and normal spaces to the boundary at that point.

Equations (1.1) with Dirichlet boundary conditions have received a lot of attention. Cioranescu and Ouazar [8] introduced a special Galerkin basis in order to obtain global existence and uniqueness of H^3 solutions in two dimensions and local existence and uniqueness in three dimensions. These authors proved existence of solutions by Galerkin's method in the basis of the eigenfunctions of the operator curlcurl$(u - \alpha\Delta u)$. This choice of basis is optimal because it allows one to prove existence of solutions with minimal restrictions on the data and the domain. The global existence for small initial data in three dimensions is shown in Galdi and Sequeira [15] and Cioranescu and Girault [7].

Recently, equations (1.1) were discovered in a very different and unexpected manner. Camassa and Holm [5] propose the following new model for shallow water:

$$u_t + 2ku_x - u_{xxt} + 3uu_x = 2u_x u_{xx} + u_x u_{xxx}.$$

These equations are the spatial representation of the geodesic spray on the diffeomorphism and the Bott-Virasoro group relative to the right invariant H^1 metric (see [5], [22]). Motivated by these results, Holm, Marsden and Ratiu [17, 18] introduce the H^1 geodesic equations on the group of volume preserving diffeomorphisms of an oriented compact Riemannian manifold relative to the weak right invariant H^1_α Riemannian metric whose value at the identity is given by

$$\|u\|_{H^1_\alpha} = \left(\|u\|^2_{L^2} + \alpha\|\nabla u\|^2_{L^2}\right)^{1/2}.$$

The spatial representation of the resulting geodesic spray is called the α-Euler or averaged Euler equation and it coincides with the second grade fluids equation for vanishing viscosity.

The main results regarding these equations, the second grade fluid and averaged Euler, are:

Theorem 14.1 *Consider equation* (1.1) *in a bounded domain* $U \subset \mathbb{R}^d$, $d = 2, 3$, *with Navier slip boundary conditions* (1.2). *Let* $u_0 \in H^3(U)$ *be a divergence free initial condition verifying the Navier slip boundary conditions and let* $F \in L^1(0, \infty; H^1)$ *denote the external forcing. Then there exists a unique* H^3 *solution* $u \in L^\infty(0, T^*; H^3)$ *where the time of existence* T^* *is estimated as follows:*

1 if $d = 2$ *then* $T^* = +\infty$,

2 if $d = 3$ *and* $\|u_0\|_{H^3(U)} \leq D_1\nu$ *then* $T^* = +\infty$,

3 if $d = 3$ *then* $T^* \geq \dfrac{D_2}{\|u_0\|_{H^3(U)} + \|F\|_{L^1(0,\infty;H^1)}}$,

for some constants $D_1 > 0$ *and* $D_2 > 0$.

Theorem 14.2 *Let* q_e *be the potential vorticity of a stationary solution of the two dimensional averaged Euler equations in a simply connected domain* U *and denote by* ψ_e *the corresponding stream function given as the unique solution of the boundary value problem*

$$q = -(1 - \alpha\Delta)\Delta\psi \quad in \quad U$$

with boundary conditions

$$\psi|_{\partial U} = 0 \quad and \quad 4n_1 n_2 \partial_1 \partial_2 \psi + (\partial_2^2 \psi - \partial_1^2 \psi)(n_2^2 - n_1^2)|_{\partial U} = 0,$$

where (n_1, n_2) *is the outward unit normal. The velocity field* u *is given in terms of the stream function by* $u = (\partial_2 \psi, -\partial_1 \psi)$.

- *If there are constants* $a, A > 0$ *such that*

$$-A \leq \frac{\nabla \psi_e}{\nabla q_e} \leq -a.$$

  then the stationary solution is stable in the H^3*-norm on perturbations of the velocity field.*

- *If there are constants* $a, A > 0$ *such that*

$$a \leq \frac{\nabla \psi_e}{\nabla q_e} \leq A$$

  and $a > k_{min}^{-2}$, *where* k_{min}^2 *is the minimal eigenvalue of* $-(1-\alpha\Delta)\Delta$ *on the space of solutions of the above boundary value problem with* $q \in L^2$, *then the stationary solution is Liapunov stable in the* L^2*-norm on perturbations of the potential vorticity. If the stream function perturbations satisfy in addition the boundary conditions of the elliptic problem above (in particular, the flow rate is preserved), then the Liapunov stability is again in the* H^3 *norm on the perturbations of the velocity.*

- *The eigenfunctions* ψ_λ *of the operator* $-(1-\alpha\Delta)\Delta$ *with the boundary conditions given above are stationary solutions. The ground states* ψ_{λ_0} *are non-isolated global constrained minima of* $\int_U q^2 dxdy$ *on the level sets of the kinetic energy of the fluid; they are Liapunov stable as a family relative to the seminorm on perturbations of the potential vorticity* q *which is the* L^2*-norm on the orthogonal complement of the first eigenfunction.*

2. Preliminary results

We shall present below some key "boundary identities" implied by the Navier slip boundary conditions. We will assume that the exterior normal n, defined a priori only on the boundary of U, is smoothly extended inside U by a vector field again denoted by n.

We start with the two dimensional case. It is easy to check that if the boundary is a straight line and the Navier slip conditions are verified, then curl$u$ vanishes on the boundary. If the boundary is not a straight line, we will prove that the trace of curl$u$ on the boundary can be expressed in terms of u only. More precisely, the following holds.

Proposition 1 *Let u be a vector field verifying the Navier slip boundary conditions (1.2) in a bounded domain $U \subset \mathbb{R}^2$. The following identity holds:*

$$\text{curl} u = 2u \cdot \frac{\partial n}{\partial \tau} \qquad \text{on} \qquad \partial U,$$

where $\frac{\partial}{\partial \tau}$ denotes the tangential derivative given by $\frac{\partial}{\partial \tau} = n_1 \partial_2 - n_2 \partial_1$.

We now study the three dimensional case. We note that if the boundary is planar, then the vector field v is tangent to the boundary. If the boundary is not planar, we prove that the trace of $v \cdot n$ on the boundary can be expressed in terms of u and ∇u only.

Proposition 2 *Let u be a divergence free vector field verifying the Navier slip boundary conditions in a bounded domain $U \subset \mathbb{R}^3$ and let $v = u - \alpha \Delta u$. The following identity holds:*

$$v \cdot n = G(u, \nabla u, n) \quad \text{on} \quad \partial U, \tag{2.1}$$

where $G(u, \nabla u, n)$ is a linear combination of u and ∇u with coefficients depending only on n and its derivatives.

Of special interest in the proof is the following modified Stokes problem

$$\begin{cases} u - \alpha \Delta u = f + \nabla p, \quad \text{div} u = 0, \\ u \text{ verifies the Navier slip boundary conditions on } \partial U. \end{cases} \tag{2.2}$$

Concerning the existence and the regularity of the solution (u, p) to this problem we refer to [24] for the following result:

Theorem 14.3 *Suppose that $f \in H^m(U)^3$, $m \in \mathbb{N}$. There exists a unique (up to a constant for p) solution $(u, p) \in \left(H^{m+2}(U)^3, H^{m+1}(U)\right)$ and there is a constant $K(m)$ such that*

$$\|u\|_{H^{m+2}} + \|p\|_{H^{m+1}} \leq K(m) \|f\|_{H^m}.$$

As an immediate consequence of Theorem 14.3 we obtain the following proposition.

Proposition 3 *Let U be a bounded domain and $m \geq 2$ be a positive integer. Then there is a constant $K > 0$ such that*

$$\|u\|_{H^m(U)} \leq K \|u - \alpha \Delta u\|_{H^{m-2}(U)}$$

for all divergence free H^m vector fields u verifying the Navier slip boundary conditions (1.2).

In connection with the Navier slip boundary conditions, we recall some regularity results.

Proposition 4 *Let U be a bounded domain, $m \geq 1$ and g be a function such that $g - \alpha \Delta g \in H^{m-2}(U)$ and $g\big|_{\partial U} \in H^{m-\frac{1}{2}}(\partial U)$. Then $g \in H^m(U)$ and there is a constant $K > 0$ such that*

$$\|g\|_{H^m(U)} \leq K\|g - \alpha\Delta g\|_{H^{m-2}(U)} + K\|g\|_{H^{m-\frac{1}{2}}(\partial U)}.$$

The next result is proved in Foias and Temam [12].

Proposition 5 *Let U be a bounded domain and m be a positive integer. If u is a L^2 vector field such that $curlu \in H^{m-1}(U)$, $divu \in H^{m-1}(U)$ and $u \cdot n\big|_{\partial U} \in H^{m-\frac{1}{2}}(\partial U)$ then $u \in H^m(U)$ and there is a constant $K > 0$ such that*

$$\begin{aligned} \|u\|_{H^m(U)} \quad \leq \quad & K\|u\|_{L^2(U)} + K\|curlu\|_{H^{m-1}(U)}. \\ & + K\|divu\|_{H^{m-1}(U)} + K\|u \cdot n\|_{H^{m-1/2}(\partial U)} \end{aligned}$$

for all such vector fields u.

We close this section with a regularity result that will be essential in later proofs.

Proposition 6 *There is a constant $K > 0$ such that for every divergence free vector field $u \in H^3(U)$ which verifies the Navier slip boundary conditions* (1.2) *on ∂U the following inequality holds:*

$$\|u\|_{H^3(U)} \leq K\|curl(u - \alpha\Delta u)\|_{L^2(U)} + K\|u\|_{H^1(U)}. \tag{2.3}$$

3. A priori estimates

The aim of this section is to prove some H^3 a priori estimates. As a first step, we deduce H^1 estimates. These estimates are well-known in the case of Dirichlet boundary conditions but the extension to the case of Navier slip boundary conditions must be carefully checked.

After some standard computations we obtain

$$\|u(t)\|_{H^1} \leq C\|u_0\|_{H^1} + C\int_0^t \|F\|_{L^2}. \tag{3.1}$$

We now go to the estimate of the H^3 norm of u. In view of the above H^1 estimates and of Proposition 6, it sufficient to estimate the L^2 norm of curl$v$. We will use the following notations:

$$\text{curl}u = \omega_u, \quad \text{curl}v = \omega_v.$$

Applying the curl to (1.1), it is easy to see that ω_v verifies the following relation:

$$\partial_t \omega_v - \nu \Delta \omega_u + u \cdot \nabla \omega_v - (d-2)\omega_v \cdot \nabla u = \text{curl}F \qquad (d = 2, 3).$$

We will now estimate the L^2 norm of curl$v$. Multiplying the above relation by ω_v yields the following relation:

$$\frac{1}{2}\frac{\mathrm{d}}{\mathrm{d}t}\int_U |\omega_v|^2 \underbrace{-\nu \int_U \Delta\omega_u \cdot \omega_v}_{I_1} + \underbrace{\int_U u \cdot \nabla\omega_v \cdot \omega_v}_{I_2} - \underbrace{(d-2)\int_U \omega_v \cdot \nabla u \cdot \omega_v}_{I_3}$$
$$= \underbrace{\int_U \text{curl}F \cdot \omega_v}_{I_4}. \tag{3.2}$$

Estimating each of these terms we get

$$\frac{\mathrm{d}}{\mathrm{d}t}\|\omega_v\|^2_{L^2} + \frac{\nu}{\alpha}\|\omega_v\|^2_{L^2}$$
$$\leq \frac{\nu}{\alpha}\|u\|^2_{H^1} + 2K_1(d-2)\|\omega_v\|^2_{L^2}\|u\|_{H^3} + 2\|\text{curl}F\|_{L^2}\|\omega_v\|_{L^2}. \tag{3.3}$$

4. The Existence Theorem

4.1 Global estimates, d = 2, 3.

We will now deduce some global a priori estimates. We assume that

$$(d-2)\|u\|_{H^3} \leq \frac{\nu}{4K_1\alpha}. \tag{4.1}$$

Under this hypothesis, we deduce from (3.2) that

$$\frac{\mathrm{d}}{\mathrm{d}t}\|\omega_v\|^2_{L^2} + \frac{\nu}{2\alpha}\|\omega_v\|^2_{L^2} \leq \frac{\nu}{\alpha}\|u\|^2_{H^1} + 2\|\text{curl}F\|_{L^2}\|\omega_v\|_{L^2}. \tag{4.2}$$

If $\nu = 0$ we obtain

$$\|\omega_v(t)\|_{L^2} \leq \|\omega_v(0)\|_{L^2} + \int_0^t \|\text{curl}F\|_{L^2}. \tag{4.3}$$

If $\nu > 0$ we apply Gronwall's lemma in (4.2) to infer that

$$\begin{aligned}\|\omega_v(t)\|_{L^2}^2 &\leq \|\omega_v(0)\|_{L^2}^2 + C\|u_0\|_{H^1}^2 + C\Big(\int_0^t \|F\|_{L^2}\Big)^2 \\ &+ 2\sup_{[0,t]}\|\omega_v\|_{L^2}\int_0^t \|\text{curl}F\|_{L^2}.\end{aligned}$$

We now take the supremum on $[0,T]$ to obtain

$$\begin{aligned}\sup_{[0,T]}\|\omega_v\|_{L^2}^2 &\leq C\|u_0\|_{H^3}^2 + C\Big(\int_0^T \|F\|_{L^2}\Big)^2 + \frac{1}{2}\sup_{[0,T]}\|\omega_v\|_{L^2}^2 \\ &+ 2\Big(\int_0^T \|\text{curl}F\|_{L^2}\Big)^2.\end{aligned}$$

Finally we get

$$\sup_{[0,T]}\|\omega_v\|_{L^2}^2 = f(u_0, F) = \text{ bounded}$$

and hence

$$\sup_{[0,t]}\|u\|_{H^3} \leq C\sup_{[0,t]}\|u\|_{H^1} + C\sup_{[0,t]}\|\text{curl}v\|_{L^2} \leq K_2\|u_0\|_{H^3} + K_2\int_0^t \|F\|_{H^1} \tag{4.4}$$

for some constant $K_2 > 1$.

If $d = 2$ inequality (4.1) is always verified. Therefore (4.4) implies

$$\|u\|_{L^\infty(0,\infty;H^3)} \leq K_2\|u_0\|_{H^3} + K_2\|F\|_{L^1(0,\infty;H^1)} \tag{4.5}$$

which proves the first part of Theorem 14.1.

If $d = 3$ and $\nu > 0$ we have the additional term $C\|\text{curl}v\|_{L^2}^2\|u\|_{H^3}$, hence one needs to assume that u is small, namely we suppose $\|u\|_{H^3} \leq \nu/2\alpha C$. This term in included in $(\nu/\alpha)\|\text{curl}v\|_{L^2}^2$, hence one obtains the same estimates as in dimension two.

Remark. One shows first that the hypothesis $\|u\|_{H^3} \leq \nu/2\alpha C$ holds for u_0 and F small enough.

4.2 Local estimates, d = 3.

In dimension three one can obtain a local estimate which follows from a nonlinear differential inequality. This implies that the following local a priori estimates holds

$$\|u\|_{L^\infty(0,\frac{1}{4K_3M};H^3)} \leq M,$$

where M is given by

$$M = \max\Big(2\sqrt{K_3}\|u_0\|_{H^3}\,,\ 2\sqrt{K_3}\int_0^\infty \|F\|_{L^2}\,,\ 4K_3\int_0^\infty \|\mathrm{curl} F\|_{L^2}\Big). \tag{4.6}$$

This proves the third part of Theorem 14.1.

4.3 Passing to the limit.

We now show how to use the previous estimates to get the existence of the solution. We will use the Galerkin method with a special basis, following the lines of the proof of [7, 8]. Set

$$X_3 = \big\{u \in H^3(U)^d \quad ; \quad \mathrm{div} u = 0, u \text{ verifies the Navier slip boundary conditions on } \partial U\big\},$$

$$X_1 = \text{ closure of } X_3 \text{ for the } H^1 \text{ norm}$$

endowed with the following scalar products

$$\langle u, \phi\rangle_{X_1} = \langle u, \phi\rangle_{L^2} + 2\alpha\langle D(u), D(\phi)\rangle_{L^2}$$

and

$$\langle u, \phi\rangle_{X_3} = \langle \mathrm{curl}(u - \alpha\Delta u), \mathrm{curl}(\phi - \alpha\Delta\phi)\rangle_{L^2} + \langle u, \phi\rangle_{X_1}.$$

The Korn Lemma insures that the norm of X_1 is equivalent to the standard H^1 norm while Proposition 6 implies that the norm of X_3 is equivalent to the H^3 norm.

Remark. It a standard fact that there are eigenvalues λ_i and eigenvectors $e_i \in X_3$, which are the solutions of the problem

$$\langle u, e_i\rangle_{X_3} = \lambda_i\langle u, e_i\rangle_{X_1}$$

for all $u \in X_3$.

One considers

$$u^\ell(t, x) = \sum_{i=1}^{\ell} \alpha_i(t) e_i(x)$$

which is the solution of the problem

$$\langle \partial_t v^\ell, e_i\rangle_{L^2} + \langle -\nu\Delta u^\ell + u^\ell\cdot\nabla v^\ell + \sum_j v_j^\ell \nabla u_j^\ell - F, e_i\rangle_{L^2} = 0 \quad \forall i \in \{1, \ldots, \ell\}. \tag{4.7}$$

Finally we obtain that u^ℓ is bounded in $L^\infty(0, T; H^3)$. A classical passage to the limit yields the result.

5. The Uniqueness Theorem

Let u and $\widetilde{u}$ be two solutions with the same initial data and belonging to $L^\infty(0,T;H^3(U))$. Set $w=\widetilde{u}-u$. Subtracting the equations for u and $\widetilde{u}$ yields:

$$\partial_t(w-\alpha\Delta w)-\nu\Delta w+\widetilde{u}\cdot\nabla\widetilde{v}-u\cdot\nabla v+\sum_j \widetilde{v}_j\nabla\widetilde{u}_j-v_j\nabla u_j=\nabla(\widetilde{p}-p). \quad (5.1)$$

We now multiply by w and integrate in space and time. According to the calculations of the H^1 *a priori* estimates and taking into account that $w(0)=0$, we obtain that

$$\begin{aligned}
&\frac{1}{2}\|w(t)\|_{L^2}^2+\alpha\|D(w(t))\|_{L^2}^2+2\nu\int_0^t\|D(w)\|_{L^2}^2\\
&=-\int_0^t\Big(\underbrace{\int_U w\cdot\nabla v\cdot w}_{I_1}+\underbrace{\int_U u\cdot\nabla(\widetilde{v}-v)\cdot w}_{I_2}\\
&+\int_U w\cdot\nabla(\widetilde{v}-v)\cdot w+\underbrace{\sum_j\int_U v_j\nabla w_j\cdot w}_{I_3}\\
&+\underbrace{\int_U(\widetilde{v}_j-v_j)\nabla u_j\cdot w}_{I_4}+\int_U(\widetilde{v}_j-v_j)\nabla w_j\cdot w\Big).
\end{aligned} \quad (5.2)$$

The same computations as for the H^1 *a priori* estimates show that:

$$\int_U\Big(w\cdot\nabla(\widetilde{v}-v)\cdot w+(\widetilde{v}_j-v_j)\nabla w_j\cdot w\Big)=0.$$

The next step in the proof of uniqueness is to shown that each of the integrals I_1, I_2, I_3 and I_4 on the right-hand side of (5.2) can be written as the sum of terms of one of the following two forms:

$$\int_U \mathcal{D}(w)\mathcal{D}(w)\mathcal{D}(u)g(n) \qquad \textit{form 1}$$

or

$$\int_U w_i\mathcal{D}(w)\mathcal{D}^2(u)g(n), \qquad \textit{form 2}$$

where $g(n)$ denotes some polynomial in n and its derivatives. For a vector field f the notation $\mathcal{D}^m(f)$ denotes a derivative of order $\leq m$ of some component of f.

By the Hölder inequality and the Sobolev embedding $H^2 \hookrightarrow L^\infty$ which hold in both two and three dimensional cases, we deduce that a term of *form 1* can be bounded as follows:

$$\left|\int_U \mathcal{D}(w)\mathcal{D}(w)\mathcal{D}(u)g(n)\right| \leq C\|w\|_{H^1}^2\|u\|_{H^3}.$$

To estimate a term of the type in *form 2*, we will use the Sobolev embedding

$$H^1(U) \hookrightarrow L^p(U) \qquad \text{for all} \qquad p \in [2,6]$$

again true in both the two and three dimensional cases. We obtain:

$$\left|\int_U w_i\mathcal{D}(w)\mathcal{D}^2(u)g(n)\right| \leq C\|w\|_{H^1}^2\|u\|_{H^3}.$$

Inserting these bounds in (5.2), neglecting the viscosity term, and using the Korn Lemma yields

$$\|w(t)\|_{H^1}^2 \leq C\int_0^t \|w\|_{H^1}^2\|u\|_{H^3}.$$

Gronwall's lemma now implies that $w = 0$.

6. Liapunov Stability in the Two Dimensional Inviscid Case

In this section we shall only outline some of the main ideas in the proof of the stability theorem. The result holds only for the averaged Euler equations in a simply connected two dimensional domain. Liapunov stability of the stationary solution u_e will follow from the inequality

$$\sup_{\mathbb{R}^+}\|u - u_e\|_{H^3} \leq C\|u_0 - u_e\|_{H^3}.$$

Since the domain is simply connected and $\operatorname{div} u = 0$, there is a stream function ψ given by $u = (\partial_2\psi, -\partial_1\psi)$ and $\psi = 0$ on the boundary. Let $q := \operatorname{curl} v = -(1-\alpha\triangle)\triangle\psi$ be the potential vorticity.

The second Navier boundary condition $(D(u)\cdot n)^{tan} = 0$ is equivalent to

$$\psi = 4n_1n_2\partial_1\partial_2\psi + (n_2^2 - n_1^2)(\partial_2^2 - \partial_1^2)\psi = 0.$$

In addition, ψ satisfies the following problem

$$-\triangle(\psi - \alpha\triangle\psi) = q$$

in the domain U with the above boundary condition. The equations of motion are

$$\dot{q} = \{\psi, q\} = \partial_1\psi\partial_2 q - \partial_2\psi\partial_1 q.$$

From here it follows that (ψ_e, q_e) is a stationary solution if ∇q_e and $\nabla\psi_e$ are proportional. The equation of motion are Hamiltonian relative to the Lie-Poisson bracket and the Hamiltonian function

$$H(q) := \frac{1}{2}\int_u q\psi.$$

The Casimir function is

$$C(q) := \int_U \Phi(q)$$

where $\Phi : \mathbb{R} \to \mathbb{R}$ is a smooth function. The stationary solutions are found as critical points of the functional $H+C$. At this point one applies the standard energy-Casimir method to obtain the stated result.

Acknowledgment.
T.S.R. was partially supported by the European Commission and the Swiss Federal Government through funding for the Research Training Network *Mechanics and Symmetry in Europe* (MASIE) as well as the Swiss National Science Foundation.

References

[1] S. Agmon, A. Douglis and L. Nirenberg. *Estimates near the boundary for solutions of elliptic partial differential equations satisfying general boundary conditions II.* Comm. Pure Appl. Math., **XVII**, 1964, pp. 35–92.

[2] V.I. Arnold. *Conditions for nonlinear stability of the stationary plane curvilinear flows of an ideal fluid.* Dokl. Akad. Nauk **162**, n° 5, 1965, 773-777.

[3] V.I. Arnold. *An a priori estimate in the theory of hydrodynamic stability.* [English translation] Amer. Math. Soc. Transl. **19**, 1969, 267–269.

[4] A.V. Busuioc and T.S. Ratiu *The second grade fluid and averaged Euler equations with Navier-slip boundary conditions*, submitted.

[5] R. Camassa and D.D. Holm. *An integrable shallow water equation with peaked solitons.* Phys. Rev. Lett., **71**, n° 11, 1993, pp. 1661–1664.

[6] S. Chen, C. Foias, D.D. Holm, E. Olson, E.S. Titi, S. Wynne. *The Camassa-Holm equations as a closure model for turbulent channel and pipe flow.* Phys. rev. Lett., **81**, 1998, pp. 5338-5341.

[7] D. Cioranescu and V. Girault. *Weak and classical solutions of a family of second grade fluids.* Internat. J. Non-Linear Mech., **32**, n° 2, 1997, pp. 317–335.

[8] D. Cioranescu and E. H. Ouazar. *Existence and uniqueness for fluids of second grade.* Nonlinear partial differential equations and their applications. Collège de France Seminar, Vol. VI (Paris, 1982/1983), pp. 178–197. Boston, MA, Pitman, 1984.

[9] G. Derks and T.S. Ratiu *Attracting curves on families of stationary solutions.* Proc. R. Soc. Lond. A, **454**, 1998, pp. 1407–1444.

[10] J. E. Dunn and R. L. Fosdick. *Thermodynamics, stability, and boundedness of fluids of complexity* 2 *and fluids of second grade.* Arch. Rational Mech. Anal., **56**, 1974, pp. 191–252.

[11] J. E. Dunn and K. R. Rajagopal. *Fluids of differential type: critical review and thermodynamic analysis.* Internat. J. Engrg. Sci., **33**, 1995, pp. 689–729.

[12] C. Foias and R. Temam. *Remarques sur les équations de Navier-Stokes stationnaires et les phénomènes successifs de bifurcation.* Ann. Sc. Norm. Pisa, **V**, 1978, pp. 29–63.

[13] C. Foias, D.D. Holm, E.S.Titi. *The three dimensional viscous Camassa-Holm Equations and their relation to the Navier-Stokes equation and turbulence theory.* To appear in J. Dyn. Diff. Eq.

[14] R. L. Fosdick and K. R. Rajagopal, *Anomalous features in the model of "second order fluids"*, Arch. Rational Mech. Anal. **70** , n° 2, 1979, pp.145–152.

[15] G. P. Galdi and A. Sequeira. *Further existence results for classical solutions of the equations of a second-grade fluid.* Arch. Rational Mech. Anal., **128**, n° 4, 1994, pp. 297–312.

[16] A.A. Himonas and G. Misiolek. *The Cauchy Problem for an integrable shallow-water Equation.* Differential and Integral Equations, Vol. **14**, No.7, 2001, pp. 821–831.

[17] D.D. Holm, J.E. Marsden, and T.S. Ratiu. *The Euler-Poincaré equations and semidirect products withx1 applications to continuum theories.* Adv. Math., **137**, n° 1, 1998, pp. 1–81.

[18] D.D. Holm, J.E. Marsden, and T.S. Ratiu. *Euler-Poincaré models of ideal fluids with nonlinear dispersion.* Phys. Rev. Lett., **349**, 1998, pp. 4173–4177.

[19] D.D. Holm, J.E. Marsden, T.S. Ratiu, and A.D. Weinstein. *Nonlinear stability of fluid and plasma equilibria.* Phys. Rep. **123**, 1985, 1–116.

[20] T. Kato. *On classical solutions of the two-dimensional non–stationary Euler equations.* Arch. Rat. Mech. Anal. **24** n° 3, 1967, 302–324.

[21] J.E. Marsden, T.S. Ratiu, and S. Shkoller. *The geometry and analysis of the averaged Euler equations and a new diffeomorphism group.* Geom. and Func. Anal. **10**, 2000, pp.582–599.

[22] G. Misiolek. *A shallow water equation as a geodesic flow on the Bott-Virasoro group.* J. Geom. Phys. **24**, n° 3, 1998, pp. 203–208.

[23] R. S. Rivlin and J. L. Ericksen. *Stress-deformation relations for isotropic materials.* J. Rational Mech. Anal., **4**, 1955, pp. 323–425.

[24] V. E. Ščadilov and V. A. Solonnikov. *On a boundary value problem for a stationary system of Navier-Stokes equations* Proc. Steklov Inst. Math., **125**, 1973, pp. 186–199.

[25] V. A. Solonnikov. *General boundary value problems for Douglis-Nirenberg elliptic systems.* Proc. Steklov Inst. Math. **92**, 1966, pp. 269–339.

[26] M. Vishik. *Incompressible flows of an ideal fluid with vorticity in borderline spaces of Besov type.* Ann. Sci. Ecole Norm. Sup. (4) **32**, 1999, no. 6, pp. 769–812.

[27] W. Wolibner. *Un théorème sur l'existence du mouvement plan d'un fluid parfait homogène, incompressible, pendant un temps infiniment longue.* Math. Zeitschrift **37**, 1933, 698–726.

[28] V.I. Yudovich. *Non–stationary flow of an ideal incompressible liquid.* Zh. Vych. Mat. **3** n° 6, 1963, 1032–1066.

[29] V.I. Yudovich. *Uniqueness theorem for the basic nonstationary problem in dynamics of ideal incompressible fluid.* Math. Res. Lett. **2**, 1995, 27–38.

Chapter 15

AN EXTENSION OF LIMIT ANALYSIS THEOREMS TO INCOMPRESSIBLE MATERIAL WITH A NON-ASSOCIATED FLOW RULE

J. Joachim Telega
Institute of Fundamental Technological Research,
Polish Academy of Sciences, Świętokrzyska 21, 00-049 Warsaw, Poland
jtelega@ippt.gov.pl

Mohammed Hjiaj
Geotechnical Research Group, Dept of Civil Engineering
The University of Newcastle, Callaghan NWS 2308, Australia
mohammed.hjiaj@newcastle.edu.au

Scott W. Sloan
Geotechnical Research Group, Dept of Civil Engineering
The University of Newcastle, Callaghan NWS 2308, Australia
scott.sloan@newcastle.edu.au

Abstract The aim of the paper is to propose an extension of limit analysis theorems for a non-associated flow rule. The approach is to construct, as a first step, a variational formulation for boundary value problems involving rigid plastic materials with a non-associated flow rule. The method used here, proposed by the first author, is based on an extension of Auchmuty's work on variational principles for non-potential operator equations. Having at our disposal the variational formulation, it is a simple matter to derive the bound theorems of limit analysis. It is found that for a non-associated flow rule the lower and the upper bound are coupled.

Keywords: Limit analysis, Non-associated plasticity, Incompressible material, Variational principles, Non-potential operator

1. Introduction

The question of extending limit analysis to materials with a non-associated flow rule is an old but important one. Many frictional soils, such as sands and drained clays, do not display the large volume change upon shearing that is predicted by an associated flow rule, thus calling into question the collapse loads that are found from the classical limit theorems. This question is important in practice, as limit analysis is often used to predict the ultimate (or limit) loads for a wide range of geotechnical structures such as slopes, tunnels, foundations and retaining walls. In 1953, Drucker [6] proved that the limit load for any plastic material with a non-associated flow rule is less than that for the corresponding material with an associated flow rule. After Drucker, Radenkovic [9] proposed the first extension of the limit theorems to a particular class of non-associated plastic materials. These materials must satisfy two assumptions: (a) the plastic potential and the yield function are convex, and (b) the additive constant for the potential can be chosen such that the surface defined by the plastic potential lies inside the elastic domain. He proved that the true limit load lies between the values of the limit loads for two associated materials. The first one, $\mathcal{M}_1$, is defined by a yield surface which coincides with the real yield surface and the second one, $\mathcal{M}_2$, has a yield surface defined by the plastic potential. The limit load for the first one is an upper bound for the real material, and the limit load for the second one is a lower bound for the real material. This result follows from the fact that a statically admissible field for the real material is also a statically admissible stress field for the material $\mathcal{M}_1$, while a kinematically admissible velocity field for the real material is also a kinematically admissible velocity field for the material $\mathcal{M}_2$. If we consider a material defined by a Mohr-Coulomb criterion, the validity of such an approach will depend on the degree of non-associativity of the material. Unfortunately, it is of little use for the practically important case of a frictional material deforming at constant volume. Following Radenkovic, Palmer [8] suggested another approach involving the assembly of a surface inside the elastic domain. For each point σ on the yield surface $f(\boldsymbol{\sigma}) = 0$, a hyperplane orthogonal to the direction of flow is assembled. If these hyperplanes form an envelope that lies completely within the yield surface, then the envelope is necessarily convex. This new surface serves as a yield surface for the material under consideration and lower and upper bounds can be computed in the usual way. For a cone-like yield criterion and potential satisfying Readenkovic's assumptions, the result gives simply the potential. Again, the incompressible case cannot be treated by this approach.

From a mathematical point of view, the lack of bound theorems is explained by the non-existence of a pseudo-potential such that its subgradients yield back the flow rule. Therefore, the non-linear operator governing the problem is not symmetric with respect to a suitable bilinear form. These problems are called non-potential initial-boundary value problems. To recover a "potential structure" of the flow rule (essential for extremum principles), de Saxcé [5] proposed an implicit normality form of the flow rule. He showed that it is possible to derive the flow rule along with its inverse from a single scalar-valued function called a bi-potential that enjoys convexity properties. Using this formulation, he was able to propose an extension of the limit analysis theorems. Unfortunately, their approach doesn't exclude the possibility of the lower bound load multiplier exceeding the upper bound load multiplier.

It is known that in the absence of symmetry of the operator it is very difficult to construct a relevant variational formulation. Several methods have been proposed in the literature. A significant contribution has been made by Brezis and Ekeland ([3] & [4]), whom proposed a variational principle for some parabolic equations. Their approach exploits fundamental concepts of Convex Analysis. A few years later, Tonti [12] proposed a general technique to obtain variational formulations for a general nonlinear problem. His method is based on the choice of a suitable linear, invertible, symmetric kernel function. The later allows the construction, in a systematic way, of an "extended functional" associated with any nonlinear problem. Tonti's ideas has been further developed by Auchmuty ([1] & [2]) who proposed to use an operator split technique to construct an "extended functional" for the linear case. In a recent work, Telega [11] proposed an extension of Auchmuty's approach to more general differential inclusions. He derived extremum and saddlepoint principles for nonlinear, nonconservative and nonpotential elasticity and the stationary and nonstationary quasi-linear heat equation.

The aim of the paper is to exploit the formulation proposed in [11] to derive extremum principles for initial/boundary value problems involving non-associated rigid plastic materials. Based on these principles, we provide an extension of the classical limit analysis theorems. The paper is organized as follows. The next section defines the problem under consideration, describes the governing equations, and introduces the notation used. In section 3, the incompressible non-associated material is presented and an alternative formulation of the flow rule is discussed. Exploiting the partly associated character of the flow rule, the dissipation function is calculated. As expected, this function depends on both the plastic strain rate and the stress. The method for deriving extremum principles for non-potential operator equations is recalled briefly, with reference to solid mechanics, in section 4. In section 5, we apply the technique to boundary value problems involving plastic material with a non-associated flow rule. Having at hand these new principles, a formulation of the bound theorems of limit analysis is proposed.

2. Boundary value problem for rigid plastic solid

In this section, we recall the boundary value problem associated with a rigid plastic material subjected to a quasi-static loading. To ensure the existence of a solution, we must assume that the initial loads do not cause the collapse of the structure, i.e. the initial loads are below the ultimate loads.

Consider a perfectly plastic body occupying a bounded domain $\Omega \subset \mathbb{R}^3$ of boundary Γ, subjected to imposed body force $\overline{\boldsymbol{b}}$, imposed surface traction $\overline{\boldsymbol{t}}_0$ on the part Γ_t of Γ, and imposed velocity $\overline{\dot{\boldsymbol{u}}}$ on the remaining part $\Gamma_{\dot{u}} = \Gamma - \Gamma_t$ of the boundary. We assume that the body is fixed on $\Gamma_{\dot{u}}$, i.e. $\overline{\dot{\boldsymbol{u}}} = \mathbf{0}$. Velocities and strain rates are assumed to be small, so that the geometry change can be neglected and the analysis is performed on the reference configuration. We also suppose that, in the absence of surface tractions, the body can support its own weight. Let $\mathcal{V}$ be the general admissible velocity space defined on Ω. The element $\dot{\boldsymbol{u}} \in \mathcal{V}$ is a vector-valued function $\dot{\boldsymbol{u}} : \Omega \rightarrow \mathbb{R}^3$ satisfying certain differentiability and integrability conditions. The dual space $\mathcal{F} = \mathcal{V}^*$ of $\mathcal{V}$ is the admissible force space on $\overline{\Omega}$ (the bar over a set denotes its closure) whose elements are specified by the body force $\overline{\boldsymbol{b}}$ on Ω and the surface traction $\overline{\boldsymbol{t}}_0$ on Γ_t. Let the geometrical operator (deformation operator) $\mathcal{D}$ be a continuous linear transformation from $\mathcal{V}$ to the space of plastic strain rates fields $\mathcal{E}$:

$$\mathcal{D} : \mathcal{V} \rightarrow \mathcal{E} : \dot{\boldsymbol{u}} \in \mathcal{V} \rightarrow \dot{\varepsilon}^p = \mathcal{D}\dot{\boldsymbol{u}} = \nabla_S \dot{\boldsymbol{u}} \tag{2.1}$$

where ∇_S is the symmetric part of the gradient operator associated with a vector. The dual space $\mathcal{S} = \mathcal{E}^*$ of $\mathcal{E}$ is the admissible stress space on Ω. The element $\boldsymbol{\sigma}(\boldsymbol{x}) \in \mathcal{S}$ is a tensor-valued function $\boldsymbol{\sigma} : \Omega \rightarrow \mathbb{E}_s^3$ satisfying certain differentiability and integrability conditions, say $\boldsymbol{\sigma}(\boldsymbol{x}) \in \mathrm{L}^2(\Omega)$ and $\mathbb{E}_s^3$ stands for the space of symmetric 3×3 matrices. The behavior of the material is specified by its flow rule along with the complementary relations:

$$\dot{\varepsilon}^p = \dot{\lambda}\, \frac{\partial g}{\partial \boldsymbol{\sigma}} \tag{2.2}$$

$$f \leq 0\,, \quad \dot{\lambda} \geq 0\,, \quad \dot{\lambda} f \tag{2.3}$$

where f is the yield surface and g the plastic potential. If the yield surface coincides with the potential ($f \equiv g$) the flow is *associated* and the Drucker inequality holds:

$$\left\{ \begin{array}{l} \forall \boldsymbol{\sigma}' \in K : \\ \dot{\varepsilon}^p \cdot (\boldsymbol{\sigma} - \boldsymbol{\sigma}') \geq 0 \,, \quad \boldsymbol{\sigma} \in K \end{array} \right. \tag{2.4}$$

where $\boldsymbol{\sigma}'$ is any arbitrary stress compelled to belong to K which is the convex set of plastically admissible stresses:

$$K = \{\boldsymbol{\sigma} \in \mathcal{S} \mid f(\boldsymbol{\sigma}) \leq 0\}$$

Using Convex Analysis tools, Moreau [7] further transformed Drucker's inequality and proposed an alternative formulation of the flow rule. The key-idea is to replace the "element of" ($\in$), which constrains the stresses in Drucker's inequality, by the indicator function of the set K, defined by

$$I_K(\boldsymbol{\sigma}) = \left\{ \begin{array}{ll} 0 & \text{if} \quad \boldsymbol{\sigma} \in K, \\ +\infty & \textit{otherwise} \end{array} \right.$$

which is convex because of the convexity of the elastic domain but non-differentiable with respect to $\boldsymbol{\sigma}$. Accordingly, the inequality becomes:

$$I_K(\boldsymbol{\sigma}') + \dot{\varepsilon}^p \cdot (\boldsymbol{\sigma} - \boldsymbol{\sigma}') \geq I_K(\boldsymbol{\sigma})$$

which means the the plastic strain rate belongs to the subdifferential of the indicator function $I_K(\boldsymbol{\sigma})$:

$$\dot{\varepsilon}^p \in \partial I_K(\boldsymbol{\sigma}) \tag{2.5}$$

The subdifferential $\partial I_K(\boldsymbol{\sigma})$ is the set of all subgradients of $I_K(\boldsymbol{\sigma})$ at $\boldsymbol{\sigma}$. It defines a multivalued mapping from $\mathcal{S}$ into $\mathcal{E}$:

$$\partial I_K(\boldsymbol{\sigma}) : \mathcal{S} \mapsto \mathcal{E} : \boldsymbol{\sigma} \mapsto \partial I_K(\boldsymbol{\sigma})$$

In particular, for a differentiable function, the subdifferential is reduced to a singleton which corresponds to the classical gradient. The previous formulation is particularly helpful in deriving extremum principle. However, this formulation exists only for an associated flow rule since Drucker's inequality does not hold for a non-associated one. By applying the Fenchel transform:

$$\sup_{\boldsymbol{\sigma} \in \mathcal{S}} [\boldsymbol{\sigma} \cdot \dot{\varepsilon}^p - I_K(\boldsymbol{\sigma})] = \psi(\dot{\varepsilon}^p) \tag{2.6}$$

the inverse of the flow rule, *i.e.* $\boldsymbol{\sigma}(\dot{\varepsilon}^p)$, can be obtained:

$$\boldsymbol{\sigma} \in \partial \psi(\dot{\varepsilon}^p) \,, \tag{2.7}$$

where $\psi(\dot{\varepsilon}^p)$ is the so-called pseudo-potential of dissipation, which is convex by construction and is a positively homogenous function of order one. Further, the pseudo-potential of dissipation gives the rate of dissipation. In terms of Convex Analysis, the pseudo-potential is said to be the conjugate function of the indicator function and conversely:

$$\psi(\dot{\varepsilon}^p) = (I_K(\boldsymbol{\sigma}))^* \quad \text{and} \quad I_K(\boldsymbol{\sigma}) = (\psi(\dot{\varepsilon}^p))^* .$$

The definition (2.7) of $\psi(\dot{\varepsilon}^p)$ implies that

$$\psi(\dot{\varepsilon}'^p) + I_K(\boldsymbol{\sigma}') \geq \boldsymbol{\sigma}' \cdot \dot{\varepsilon}'^p \tag{2.8}$$

for each $\dot{\varepsilon}'^p \in \mathcal{E}$ and each $\boldsymbol{\sigma} \in \mathcal{S}$. This inequality will play an essential role in the derivation of minimum principles in plasticity. In (2.8) the equality holds if and only if $\dot{\varepsilon}^p \in \partial I_K(\boldsymbol{\sigma})$, or equivalently if and only if $\boldsymbol{\sigma} \in \partial\psi(\dot{\varepsilon}^p)$. Now, we need to define some sets used in the next sections.

2.1 Admissible fields

Static admissibility. In the context of limit analysis, a stress field is said to be statically admissible if it satisfies the local equilibrium equations within the domain and on its boundary and does not violate the plasticity criterion:

$$\begin{array}{rcll} \mathrm{div}\boldsymbol{\sigma} + \overline{\boldsymbol{b}} & = & \mathbf{0} & \text{in } \Omega \\ \boldsymbol{n}\,\boldsymbol{\sigma} & = & \mu_l\,\overline{\boldsymbol{t}}_0 & \text{on } \Gamma_t \\ \boldsymbol{\sigma}(\boldsymbol{x}) & \in & K & \forall \boldsymbol{x} \in \overline{\Omega} \end{array}$$

where μ_l is the maximum admissible load multiplier. The space $\mathcal{S}_{ad}$ defined by

$$\mathcal{S}_{ad} = \left\{\boldsymbol{\sigma} \in \mathcal{S} \mid \mathrm{div}\boldsymbol{\sigma} + \overline{\boldsymbol{b}} = \mathbf{0} \text{ in } \Omega\,,\ \boldsymbol{n}\,\boldsymbol{\sigma} = \mu_s\,\overline{\boldsymbol{t}}_0 \text{ on } \Gamma_t\right\} \tag{2.9}$$

encompasses all stress fields that satisfy the strong equilibrium equations. The space $\mathcal{S}_{sa}$ of statically admissible stresses can be defined as follows:

$$\mathcal{S}_{sa} = \{\boldsymbol{\sigma} \in \mathcal{S} \mid \boldsymbol{\sigma}(\boldsymbol{x}) \in \mathcal{S}_{ad} \cap K\} \tag{2.10}$$

Kinematic admissibility. A velocity field is said to be kinematically admissible if it satisfies the local compatibility equations within the domain and on its boundary and complies with the flow rule:

$$\begin{array}{rcll} \dot{\varepsilon}^p & = & (\nabla_S \dot{\boldsymbol{u}}) \neq \mathbf{0} & \text{in } \Omega \\ \dot{\boldsymbol{u}} & = & \mathbf{0} & \text{on } \Gamma_{\dot{u}} \\ \dot{\varepsilon}^p(\dot{\boldsymbol{u}}(\boldsymbol{x})) & \in & \partial I_K(\boldsymbol{\sigma}) & \text{in } \overline{\Omega} \end{array}$$

The condition $(\nabla_S \dot{\boldsymbol{u}}) \neq \mathbf{0}$ means that we are looking for a velocity field in which the external loads produce non-negative work:

$$\int_{\Gamma_t} \mathbf{t} \cdot \dot{\boldsymbol{u}} \, \mathrm{d}\Gamma > 0$$

The space $\mathcal{V}_{ad}$ defined by

$$\mathcal{V}_{ad} = \{\dot{\boldsymbol{u}} \in \mathcal{V} \mid \dot{\varepsilon}^p = (\nabla_S \dot{\boldsymbol{u}}) \ \text{in}\, \Omega \ , \ \dot{\boldsymbol{u}} = \mathbf{0} \ \text{on}\, \Gamma_{\dot{u}}\} \tag{2.11}$$

contains all velocity fields that satisfy the local compatibility equations. The kinematically admissible velocity space $\mathcal{V}_{ka}$ is defined by

$$\mathcal{V}_{ka} = \left\{ \dot{\boldsymbol{u}} \in \mathcal{V}_{ad} \mid \dot{\varepsilon}^p(\dot{\boldsymbol{u}}(\boldsymbol{x})) \in \partial I_K(\boldsymbol{\sigma}) \ , \ \int_{\Gamma_t} \mathbf{t} \cdot \dot{\boldsymbol{u}} \, \mathrm{d}\Gamma > 0 \right\} \tag{2.12}$$

Here, the velocity may eventually be discontinuous at the crossing of certain surfaces, but this discontinuity is not arbitrary, it must be compatible with the flow rule. Finally, we denote the space of admissible plastic strain rates by

$$\mathcal{E}_{ad} = \{\dot{\varepsilon}^p \in \mathcal{E} \mid \exists \dot{\boldsymbol{u}} \in \mathcal{V}_{ad} \quad \text{such that} \quad \dot{\varepsilon}^p = \nabla_S \dot{\boldsymbol{u}}\} \tag{2.13}$$

and the space of kinematically admissible plastic strain rates by

$$\mathcal{E}_{ka} = \left\{ \dot{\varepsilon}^p \in \mathcal{E}_{ad} \mid \dot{\varepsilon}^p = \dot{\lambda} \, \frac{\partial g}{\partial \boldsymbol{\sigma}} \right\} \tag{2.14}$$

The boundary value problem The boundary value problem of limit analysis involves finding a triple $(\mu_l, \boldsymbol{\sigma}, \dot{\boldsymbol{u}})$ such that the stress field $\boldsymbol{\sigma}$ is statically admissible, the velocity field $\dot{\boldsymbol{u}}$ is kinematically admissible, and μ_l is the associated limit load multiplier. We can write this problem formally by:

$$\begin{cases} \textit{Find} \quad (\mu_l, \boldsymbol{\sigma}, \dot{\boldsymbol{u}}) \\ \textit{such that} \quad \boldsymbol{\sigma} \in \mathcal{S}_{sa} \ , \ \dot{\boldsymbol{u}} \in \mathcal{V}_{ka} \end{cases} \tag{2.15}$$

3. The incompressible non-associated plastic model

It is well known that soil materials have a very complicated behavior. Idealizations are, therefore, often necessary in order to develop simple mathematical constitutive laws for practical applications. Several models can be found in the literature, most of them are complex and require many parameters, some of which are physically meaningless. The relative simplicity of the Drucker-Prager model, which simulates several key characteristics of soil behavior with only three parameters, explains why it is widely used. Here, the governing constitutive relations are developed with reference to the classical elastic perfectly plastic Drucker-Prager model. Hence, the model is described by the dual variables $\boldsymbol{\sigma} = (\boldsymbol{\sigma}_d, s_m)$ and $\dot{\varepsilon}^p = (\dot{\mathbf{e}}^p_d, \dot{e}^p_m)$ where

$$\begin{array}{rcl} s_m & = & \frac{1}{3}\mathrm{Tr}(\boldsymbol{\sigma}) \\ \boldsymbol{\sigma}_d & = & \boldsymbol{\sigma} - s_m\, \mathbb{I} \\ \dot{e}^p_m & = & \mathrm{Tr}(\dot{\varepsilon}^p) \\ \dot{\mathbf{e}}^p_d & = & \dot{\varepsilon}^p - \frac{1}{3}\dot{e}^p_m\, \mathbb{I} \end{array}$$

where $\mathbb{I}$ is the 3×3 unit matrix. The convex set of plastically admissible stresses (Figure 3.1) is given by

$$K_\phi = \left\{ \boldsymbol{\sigma} \in \mathcal{S} \mid \frac{1}{k_d}\|\boldsymbol{\sigma}_d\| + s_m \tan\phi - c \leq 0 \right\} \tag{3.1}$$

where where c is the cohesion, ϕ the friction angle, k_d a constant and $\|\bullet\|$ denotes the classical Euclidean norm. Its boundary defines the yield function:

$$f = \frac{1}{k_d}\|\boldsymbol{\sigma}_d\| + s_m \tan\phi - c = 0\,. \tag{3.2}$$

The subscript ϕ attached to K in (3.1) is used to distinguish the admissible domain of a frictional material with the one, denoted by K_c, of a purely cohesive material ($\phi = 0$). The flow rule is given by (2.2) where g is defined by

$$g = \frac{1}{k_d}\|\boldsymbol{\sigma}_d\| - \xi \tag{3.4}$$

in which ξ is a positive constant. By adopting the plastic potential (3.3) the plastic flow is incompressible

$$\mathrm{Tr}(\dot{\varepsilon}^p) = 0$$

In the expression (3.4) of the potential g, the magnitude of the constant is arbitrary. Therefore, we can associate to the potential the family of the convex sets

$$\widetilde{K}_c = \left\{ \boldsymbol{\sigma} \in \mathcal{S} \mid \frac{1}{k_d}\|\boldsymbol{\sigma}_d\| - \xi \leq 0 \right\} \tag{3.5}$$

By setting $\xi = c$, we recover the convex set K_c.

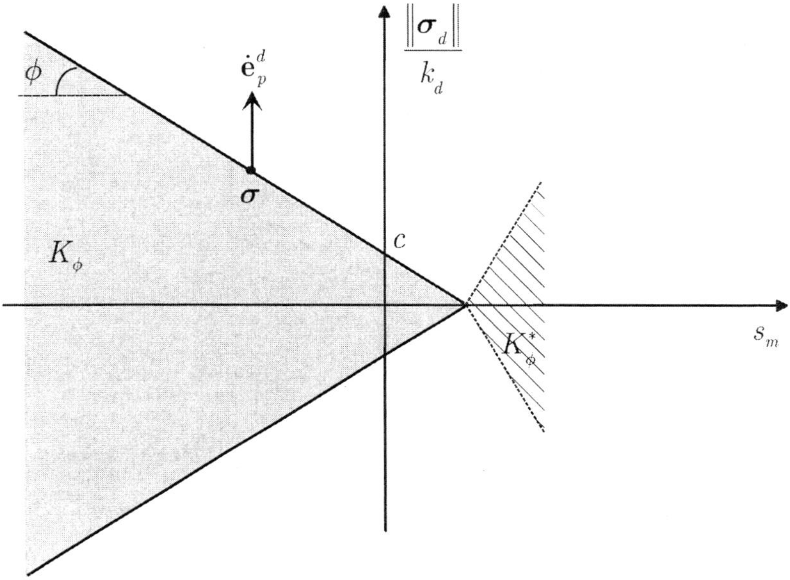

Figure 15.1. The Drucker-Prager cone

Alternative formulation of the flow rule The rigid plastic model presented above is non-associated and we can not derive a convex pseudo-potential that describes it. This fact can be easily proved using the cyclic monotonicity condition. In fact, the law being non-monotonic, it is not cyclically monotonic and no convex pseudo-potential can be derived. However, if we consider it admissible to include a dependence of the pseudo-potential on the state, then normality holds and we obtain a family of convex pseudo-potentials. The state-dependent pseudo-potential is called a quasi-pseudo-potential. The term "quasi" is used to emphasize its dependence on the state. The existence of a quasi-pseudo-potential results from the partly associated character of the law (see [10]). In fact, for the present model the deviator part of the plastic strain rate is associated. By "freezing" the mean stress s_m, the convex cone K becomes a function of $\boldsymbol{\sigma}_d$ only and the flow rule can be expressed under the form of a quasi-variational inequality:

$$(3.6) \qquad \begin{cases} \forall \boldsymbol{\sigma}'_d \in K(s_m) : \\ \dot{\mathbf{e}}_d^p \cdot (\boldsymbol{\sigma}_d - \boldsymbol{\sigma}'_d) \geq 0\,, \quad \boldsymbol{\sigma}_d \in K(s_m) \end{cases}$$

where $K(s_m)$ is defined by

$$(3.7) \qquad K(s_m) = \left\{ \boldsymbol{\sigma}_d \in \mathcal{S} \mid \frac{1}{k_d} \|\boldsymbol{\sigma}_d\| \leq \bar{c} \right\} \quad \text{with} \quad \bar{c} = c - s_m \tan\phi$$

The previous inequality suggests that the flow rule can be formulated in the form of the principle of maximim dissipation: the actual deviatoric stress is an optimal solution for the maximization problem in the variable $\boldsymbol{\sigma}_d$:

$$\begin{cases} \textit{maximize} & \dot{\mathbf{e}}_d^p \cdot \boldsymbol{\sigma}_d \\ \textit{subject to} & \boldsymbol{\sigma}_d \in K(s_m) \end{cases}$$

To further transform inequality (3.5) in the same way as for the perfect plastic model with an associated flow rule, we introduce the indicator function of the closed convex set $K(s_m)$:

$$I_{K(s_m)}(\boldsymbol{\sigma}_d) = \begin{cases} 0 & \text{if} \quad \boldsymbol{\sigma}_d \in K(s_m), \\ +\infty & \textit{otherwise} \end{cases}$$

This function is subsequently used to enforce the constraints on $\boldsymbol{\sigma}_d$:

$$(3.8) \qquad I_{K(s_m)}(\boldsymbol{\sigma}'_d) + \dot{\mathbf{e}}^p \cdot (\boldsymbol{\sigma}_d - \boldsymbol{\sigma}'_d) \geq I_{K(s_m)}(\boldsymbol{\sigma}_d)$$

The inequality (3.8) corresponds to the convexity condition for a non-differentiable function applied to $I_{K(s_m)}(\boldsymbol{\sigma}_d)$:

$$(3.9) \qquad \dot{\mathbf{e}}^p \in \partial I_{K(s_m)}(\boldsymbol{\sigma}_d)$$

Applying the Fenchel transform,

$$(3.10) \quad \psi_{s_m}(\dot{\mathbf{e}}^p) = \sup_{\boldsymbol{\sigma}_d \in \mathcal{S}} \left[\boldsymbol{\sigma}_d \cdot \dot{\mathbf{e}}^p - I_{K(s_m)}(\boldsymbol{\sigma}_d) \right] = k_d \,(c - s_m \tan\phi)\, \|\dot{\mathbf{e}}^p\|$$

the inversion of the flow rule is achieved

$$(3.11) \qquad \boldsymbol{\sigma}_d \in \partial \psi_{s_m}(\dot{\mathbf{e}}^p)$$

The two quasi-pseudo-potentials are related by the Fenchel inequality:

$$(3.12) \qquad \psi_{s_m}(\dot{\mathbf{e}}'^p) + I_{K(s_m)}(\boldsymbol{\sigma}'_d) \geq \boldsymbol{\sigma}'_d \cdot \dot{\mathbf{e}}'^p$$

The equality is reached in (3.12) if and only if (3.9) is satisfied. In particular, if ϕ is equal to zero, we recover the expression of the pseudo-potential of dissipation for a purely cohesive material with an associated flow rule

$$(3.13) \qquad \psi_c(\dot{\mathbf{e}}^p) = k_d\, c\, \|\dot{\mathbf{e}}^p\|$$

4. Minimum principle for non-potential operator

In this section, we recall briefly some fundamental results introduced in [11]. It is found that the term variational may be understood in different ways. This term can mean: weak formulation (*i.e.* principle of virtual work), stationary principle or extremum principle. Although a weak formulation can be associated with most of the physical problems, a few of them admit a stationary principle and lesser number admit an extremum principle. Extremum principles are particularly attractive from both a mathematical and a computational point of view. In fact, mathematicians used them to prove existence and eventually uniqueness of the solution to the corresponding boundary value problem. The possibility of searching for the solution of a physical problem as a minimum point of a convex functional on a convex set is especially relevant from the computational point view. Consider an operator equation, not necessarily linear,

$$N(\boldsymbol{u}) = \boldsymbol{f}$$

where $\boldsymbol{u}$ belongs to a properly chosen function space V and $\boldsymbol{f}$ to its dual V^*.

(i) The weak formulation, being a rather general form of the virtual work principle, is given by

$$\boldsymbol{f} \in V^* \qquad < N(\boldsymbol{u}), \boldsymbol{v} > = < \boldsymbol{f}, \boldsymbol{v} > \qquad \forall \boldsymbol{v} \in V$$

Here $< \cdot, \cdot >$ stands for the duality pairing on $V^* \times V$.

(ii) Stronger is the stationary principle

$$\mathcal{J}(\boldsymbol{u}) \rightarrow \text{stationary} \quad \text{over} \quad V$$

Then, an arbitrary variation of the functional yields back the strong form of the equation

$$\delta \mathcal{J} = < N(\boldsymbol{u}) - \boldsymbol{f}, \delta \boldsymbol{u} > = 0 \quad \Rightarrow \quad N(\boldsymbol{u}) = \boldsymbol{f}$$

(iii) The strongest are extremum principles (minimum or maximum principles and min-max principles or saddle-point principles).

For instance,

$$\mathcal{J}(\boldsymbol{u}) \longrightarrow \min \quad \text{over} \quad V$$

where

$$\mathcal{J} = \mathcal{J}_1 + I_C$$

Here C is a set of constraints and I_C denotes its indicator function [16,32,33]:

$$I_C(\boldsymbol{u}) = \begin{cases} 0 & \text{if} \quad \boldsymbol{u} \in C, \\ +\infty & \textit{otherwise} \end{cases}$$

In the next subsection, we recall the method, proposed in [11], to derive extremum principles for a non-potential operator equation. The relations are developed for the general case and specifications are made with respect to our problem.

4.1 Abstract framework

Consider the problem of solving the following *differential inclusion*

$$F(N(\boldsymbol{u})) + \boldsymbol{f} \in \partial_{\boldsymbol{u}} \Phi(\Lambda \boldsymbol{u}) \tag{4.1}$$

where
- $\Lambda : V \to Y$ is a continous linear operator,
- Y is another space whose elements are denoted by $\boldsymbol{q}$,
- Y^* is the dual space of Y,
- $N : V \to V_1$ is a continuous operator, not necessarily linear,
- $F : V_1 \to V^*$ is a continuous mapping.

For solid mechanics applications, the operator Λ is the deformation operator assumed here to be linear. Accordingly, we have that $Y \equiv \mathcal{E}$ and $Y^* \equiv \mathcal{S}$. The adjoint operator of Λ, denoted by Λ^*, is the equilibrium operator:

$$\Lambda \boldsymbol{u} = \begin{bmatrix} \nabla_S \boldsymbol{u} \\ \boldsymbol{u} \end{bmatrix} \begin{matrix} \text{in } \Omega \\ \text{on } \Gamma_u \end{matrix} \quad \text{and} \quad \Lambda^* \boldsymbol{q}^* = \begin{bmatrix} -\text{div} \boldsymbol{q}^* \\ \boldsymbol{n}\, \boldsymbol{q}^* \end{bmatrix} \begin{matrix} \text{in } \Omega \\ \text{on } \Gamma_t \end{matrix}$$

Associated with the problem, consider the lagrangian $\mathcal{L} : V \times V \to \mathbb{R}$ defined by :

$$\mathcal{L}(\boldsymbol{u}, \boldsymbol{v}) = \Phi(\Lambda \boldsymbol{u}) - \Phi(\Lambda \boldsymbol{v}) - \langle F(N(\boldsymbol{u})) + \boldsymbol{f}, \boldsymbol{v} - \boldsymbol{u} \rangle. \tag{4.2}$$

By setting:

$$\boldsymbol{q} = \Lambda \boldsymbol{v} \quad \text{and} \quad \Lambda^* \boldsymbol{q}^* = F(N(\boldsymbol{u})) + \boldsymbol{f}$$

the lagrangian (4.2) becomes

$$\mathcal{L}(\boldsymbol{u}, \boldsymbol{q}^*; \boldsymbol{q}) = \Phi(\Lambda \boldsymbol{u}) - \Phi(\boldsymbol{q}) + \langle \boldsymbol{q}^*, \boldsymbol{q} \rangle - \langle F(N(\boldsymbol{u})) + \boldsymbol{f}, \boldsymbol{u} \rangle. \tag{4.3}$$

Now, we define the functional $\mathcal{G}(\boldsymbol{u}, \boldsymbol{q}^*)$ as:

$$\begin{aligned} \mathcal{G}(\boldsymbol{u}, \boldsymbol{q}^*) &= \sup_{q \in Y^*} \mathcal{L}(\boldsymbol{u}, \boldsymbol{q}^*; \boldsymbol{q}) \\ &= \sup_{q \in Y^*} [\langle \boldsymbol{q}^*, \boldsymbol{q} \rangle - \Phi(\boldsymbol{q})] - \langle F(N(\boldsymbol{u})) + \boldsymbol{f}, \boldsymbol{u} \rangle + \Phi(\Lambda \boldsymbol{u}) \\ &= \Phi(\Lambda \boldsymbol{u}) + \Phi^*(\boldsymbol{q}^*) - \langle F(N(\boldsymbol{u})) + \boldsymbol{f}, \boldsymbol{u} \rangle \end{aligned} \tag{4.4}$$

The minimum principle associated with (4.1) means evaluating

$$\left| \begin{array}{l} \textit{Find} \\ \boldsymbol{\alpha} = \inf\limits_{\boldsymbol{u} \in V,\, \boldsymbol{q}^* \in Y^*} \{\Phi(\Lambda \boldsymbol{u}) + \Phi^*(\boldsymbol{q}^*) - \langle F(N(\boldsymbol{u})) + \boldsymbol{f}, \boldsymbol{u} \rangle\} \qquad (\mathcal{P}) \\ \textit{subject to} \quad \Lambda^* \boldsymbol{q}^* = F(N(\boldsymbol{u})) + \boldsymbol{f} \end{array} \right.$$

Remark 4.1 We observe that the last principle involves the adjoint operator Λ^* which usually can be calculated easily. Furthermore, we have that

$$\mathcal{G}(\boldsymbol{u}, \boldsymbol{q}^*) \geq 0 \qquad \text{and} \quad \mathcal{G}(\hat{\boldsymbol{u}}, \hat{\boldsymbol{q}}^*) = 0 \tag{4.5}$$

if only if $(\hat{\boldsymbol{u}}, \hat{\boldsymbol{q}}^*)$ is a solution to (4.1)
The dual problem means evaluating

$$\boldsymbol{\alpha}^* = \sup_{q \in Y} \mathcal{H}(\boldsymbol{q}) \quad (\mathcal{P}^*). \tag{4.6}$$

where

$$\mathcal{H}(\boldsymbol{q}) = \inf_{\boldsymbol{u} \in V,\ \boldsymbol{q}^* \in Y^*} \{\mathcal{L}(\boldsymbol{u}, \boldsymbol{q}^*; \boldsymbol{q}) \quad , \quad \Lambda^* \boldsymbol{q}^* = F(N(\boldsymbol{u})) + \boldsymbol{f}\}\,. \tag{4.7}$$

A point $(\hat{\boldsymbol{u}}, \hat{\boldsymbol{q}}^*; \hat{\boldsymbol{q}}) \in V \times Y^* \times Y$ is a saddle point of $\mathcal{L}$ if,

$$\mathcal{L}(\hat{\boldsymbol{u}}, \hat{\boldsymbol{q}}^*; \boldsymbol{q}) \leq \mathcal{L}(\hat{\boldsymbol{u}}, \hat{\boldsymbol{q}}^*; \hat{\boldsymbol{q}}) \leq \mathcal{L}(\boldsymbol{u}, \boldsymbol{q}^*; \hat{\boldsymbol{q}}) \ \ \forall\, \boldsymbol{u} \in V,\ \boldsymbol{q} \in Y,\ \boldsymbol{q}^* \in Y^*$$

satisfying the constraint $\boldsymbol{q}^* = F(N(\boldsymbol{u})) + \boldsymbol{f}$

Properties:

1 If (P) and $(\mathcal{P}^*)$ are nontrivial, then $\boldsymbol{\alpha}$ and $\boldsymbol{\alpha}^*$ are finite with $\boldsymbol{\alpha}^* \leq \boldsymbol{\alpha}$.

2 If $\boldsymbol{\alpha}^* > 0$, then (4.1) does not posses a solution.

3 If there exists a solution $\hat{\boldsymbol{q}}$ in Y of $(\mathcal{P}^*)$ with $\mathcal{H}(\hat{\boldsymbol{q}}) = 0$ and a $(\hat{\boldsymbol{u}}, \hat{\boldsymbol{q}}^*)$ in $V \times Y^*$ such that $(\hat{\boldsymbol{u}}, \hat{\boldsymbol{q}}^*; \hat{\boldsymbol{q}})$ is a saddle point of $\mathcal{L}$, then $(\hat{\boldsymbol{u}}, \hat{\boldsymbol{q}}^*)$ will be a solution of (4.1).

4 If $(\hat{\boldsymbol{u}}, \hat{\boldsymbol{q}}^*; \hat{\boldsymbol{q}})$ is a saddle point of L, then

$$0 = \mathcal{G}(\hat{\boldsymbol{u}}, \hat{\boldsymbol{q}}^*) = \mathcal{L}(\hat{\boldsymbol{u}}, \hat{\boldsymbol{q}}^*; \hat{\boldsymbol{q}}) = \mathcal{H}(\hat{\boldsymbol{q}}).$$

4.2 Applications

4.2.1 Example 1. Let us study the problem $(\mathcal{P})$ more closely and consider the particular case where

$$F(N(\mathbf{u})) \equiv \mathbf{0}.$$

Then, we have

$$\mathcal{G}(\boldsymbol{u}, \boldsymbol{q}^*) = \Phi(\Lambda \boldsymbol{u}) + \Phi^*(\boldsymbol{q}^*) - \langle \boldsymbol{f}, \boldsymbol{u} \rangle. \tag{4.9}$$

and the problem $\mathcal{P}$ can be written as follows

$$\inf\{\Phi(\Lambda \boldsymbol{u}) - \langle \boldsymbol{f}, \boldsymbol{u} \rangle \mid \boldsymbol{u} \in V\} \; + \; \inf\{\Phi^*(\boldsymbol{q}^*) \mid \boldsymbol{q}^* \in Y^*, \Lambda^* \boldsymbol{q}^* = \boldsymbol{f}\}$$

Since $\mathcal{G}(\boldsymbol{u}, \boldsymbol{q}^*) \geq \mathbf{0}$, we have

$$\Phi(\Lambda \boldsymbol{u}) - \langle \boldsymbol{f}, \boldsymbol{u} \rangle \geq -\Phi^*(\boldsymbol{q}^*) \quad \forall \boldsymbol{u} \in V \;, \forall \boldsymbol{q}^* \in Y^* \tag{4.10}$$

Thus we recover, for instance, the principle of the total potential energy of linear elasticity:

$$\Phi(\Lambda \boldsymbol{u}) \longrightarrow \text{minimum}$$

and its dual (the principle of the total complementary energy):

$$-\Phi^*(\boldsymbol{q}^*) \longrightarrow \sup$$

It suffices to set $\boldsymbol{\varepsilon}(\boldsymbol{u}) = \Lambda \boldsymbol{u}$ and $\boldsymbol{q}^* = \boldsymbol{\sigma}$.

4.2.2 Example 2. Consider the boundary value problem associated with an elastic solid described by a non-symmetric elastic operator:

$$\boldsymbol{\sigma} = \mathbb{A}\, \boldsymbol{\varepsilon} \qquad \text{with } \mathbb{A} \neq \mathbb{A}^{\mathrm{T}} \tag{4.11}$$

In this case N is linear. In term of displacement, the strong form of the governing equations are:

$$\begin{aligned} \operatorname{div}(\mathbb{A}(\nabla_S \boldsymbol{u})) + \overline{\boldsymbol{b}} &= \mathbf{0} \quad \text{in } \Omega \\ \mathbf{n}\,(\mathbb{A}(\nabla_S \boldsymbol{u})) &= \overline{\boldsymbol{t}} \quad \text{on } \Gamma_t \\ \boldsymbol{u} &= \mathbf{0} \quad \text{on } \Gamma_u \end{aligned}$$

Let us make the following decomposition:

$$\mathbb{A} = \mathbb{A}_1 + \mathbb{A}_2 \tag{4.12}$$

where $\mathbb{A}_1$ is such that its elements satisfy $a^1_{ijkl} = a^1_{klij}$. Taking into account the decomposition (4.12), the problem can be written as (4.1) where

$$F(\boldsymbol{u}) = \begin{bmatrix} \operatorname{div}(\mathbb{A}_2(\nabla_S \boldsymbol{u}) \\ -\boldsymbol{n}\,(\mathbb{A}_2(\nabla_S \boldsymbol{u}) \end{bmatrix} \begin{matrix} \text{in } \Omega \\ \text{on } \Gamma_t \end{matrix} \quad , \quad \boldsymbol{f} = \begin{bmatrix} \overline{\boldsymbol{b}} \\ \overline{\boldsymbol{t}} \end{bmatrix} \begin{matrix} \text{in } \Omega \\ \text{on } \Gamma_t \end{matrix} \tag{4.13}$$

and

$$\begin{aligned} \Phi(\Lambda \boldsymbol{u}) &= \langle -\operatorname{div}\left(\mathbb{A}_1(\nabla_S \boldsymbol{u})\right), \boldsymbol{u})\rangle + \langle (\boldsymbol{n}\left(\mathbb{A}_1(\nabla_S \boldsymbol{u})\right), \boldsymbol{u})\rangle \\ &= \int_\Omega \mathcal{U}(\boldsymbol{\varepsilon}_1)\,\mathrm{d}\Omega \end{aligned}$$

where $\boldsymbol{\varepsilon}_1 = \mathbb{A}_1^{-1}\boldsymbol{\sigma}$ and $\mathcal{U}$ is the potential energy functional. Furthermore, it is clear that

$$\boldsymbol{q}^* \equiv \boldsymbol{\sigma}_1 = \mathbb{A}_1\,\boldsymbol{\varepsilon}.$$

The conjugate functional Φ^* is given by

$$\Phi^*(\boldsymbol{q}^*) = \int_\Omega \mathcal{W}(\boldsymbol{\sigma}_1)\,\mathrm{d}\Omega$$

where $\mathcal{W}$ is the complementary potential energy functional. Now, we are able to give the expression of the functional $\mathcal{H}(\boldsymbol{u}, \boldsymbol{q}^*)$

$$\begin{aligned} \inf_{\boldsymbol{u},\, \boldsymbol{q}^*} \Bigg\{ &\int_\Omega \left(\mathcal{U}(\boldsymbol{\varepsilon}_1(\boldsymbol{u})) + \mathcal{W}(\boldsymbol{q}^*)\right)\,\mathrm{d}\Omega - \int_\Omega \overline{\boldsymbol{b}} \cdot \boldsymbol{u}\,\mathrm{d}\Omega - \int_{\Gamma_t} \overline{\boldsymbol{t}} \cdot \boldsymbol{u}\,\mathrm{d}\Gamma \\ &- \int_\Omega \operatorname{div} \boldsymbol{\sigma}_2 \cdot \boldsymbol{u}\,\mathrm{d}\Omega + \int_{\Gamma_t} (\boldsymbol{n}\,\boldsymbol{\sigma}_2) \cdot \boldsymbol{u}\,\mathrm{d}\Gamma \Bigg\} \end{aligned} \tag{4.14}$$

where $\boldsymbol{\sigma}_2 \equiv \mathbb{A}_2(\nabla_S \boldsymbol{u})$.

The last two terms in (4.14) can be rearranged and we obtain

$$\begin{aligned} \inf_{\boldsymbol{u},\, \boldsymbol{q}^*} \Bigg\{ &\int_\Omega \left(\mathcal{U}(\boldsymbol{\varepsilon}_1(\boldsymbol{u})) + \mathcal{W}(\boldsymbol{q}^*)\right)\,\mathrm{d}\Omega - \int_\Omega \overline{\boldsymbol{b}} \cdot \boldsymbol{u}\,\mathrm{d}\Omega - \int_{\Gamma_t} \overline{\boldsymbol{t}} \cdot \boldsymbol{u}\,\mathrm{d}\Gamma \\ &+ \int_\Omega \boldsymbol{\sigma}_2(\boldsymbol{u}) \cdot \boldsymbol{\varepsilon}(\boldsymbol{u})\,\mathrm{d}\Omega \Bigg\} \end{aligned} \tag{4.15}$$

In general, this functional is not convex because of the last term. However, the linearity of the last term ensures that the functional is bi-convex with respect to $(\boldsymbol{u}, \boldsymbol{q}^*)$. As a result, we can not prove that the solution of the problem is unique.

5. The limit analysis problem

5.1 The associated case

In plasticity, the constitutive operator is non-linear and multivalued. In what follows, we will denote this operator by $\mathcal{A}$ and its inverse by $\mathcal{A}^{-1}$ such that we can write the flow rule and its inverse respectively as

$$\dot{\varepsilon}^p \in \mathcal{A}^{-1}(\boldsymbol{\sigma}) \quad \text{and} \quad \boldsymbol{\sigma} \in \mathcal{A}(\dot{\varepsilon}^p) \tag{5.1}$$

If the flow rule is associated, the operator $\mathcal{A}$ corresponds to the subdifferential of the pseudo-potential of dissipation. We consider first an incompressible rigid plastic material with an associated flow rule. We denote the velocity field by $\dot{\boldsymbol{u}}$, which belongs to the divergence-free space $\mathcal{V}_{ad}$.:

$$\mathcal{V}_{ad} = \left\{ \dot{\boldsymbol{u}} \in [\mathrm{H}^1(\Omega)]^3 \mid \dot{\boldsymbol{u}} = \mathbf{0} \text{ on } \Gamma_u \, , \ \operatorname{div} \dot{\boldsymbol{u}} = \mathbf{0} \text{ in } \Omega \right\}$$

The linear operator Λ is the deformation operator $\mathcal{D}$ consisting of a transformation from $\mathcal{V}_{ad}$ to the space of plastic strain rate $\mathcal{E}$:

$$\dot{\mathbf{e}}^p_d(\dot{\boldsymbol{u}}) = \nabla_S(\dot{\boldsymbol{u}})$$

Let μ be any load multiplier such that $\mu \leq \mu_l$. The boundary value problem can be stated as

$$\begin{array}{rcll} -\operatorname{div}\left(\mathcal{A}(\nabla_S \dot{\boldsymbol{u}})\right) & = & \overline{\boldsymbol{b}} & \text{in } \Omega \\ \boldsymbol{n}\left(\mathcal{A}(\nabla_S \boldsymbol{u})\right) & = & \mu\, \overline{\boldsymbol{t}}_0 & \text{on } \Gamma_t \\ \dot{\boldsymbol{u}} & = & \mathbf{0} & \text{on } \Gamma_u \\ \mathcal{A}(\dot{\varepsilon}^p) & \in & K & \text{in } \Omega \end{array}$$

or equivalently

$$\boldsymbol{f} \in \partial\Phi(\Lambda \dot{\boldsymbol{u}}) \tag{5.2}$$

where $\boldsymbol{f}$ is given by (4.13b) and $\Phi(\Lambda u)$ by

$$\int_\Omega \psi_c(\dot{\mathbf{e}}^p_d(\dot{\boldsymbol{u}}))\, \mathrm{d}\Omega \tag{5.3}$$

where $\psi_c(\dot{\mathbf{e}}^p_d)$ is given by (3.13). We note that the integral (5.3) contains also the dissipation terms on the surfaces of discontinuities. The above expression is simplified by the fact that $F \equiv 0$. It is quite straightforward to calculate the dual functional $\Phi^*(\boldsymbol{q}^*)$ where $\boldsymbol{q}^* \equiv \boldsymbol{\sigma}$:

$$\Phi^*(\boldsymbol{\sigma}) = I_{\mathcal{C}_c}(\boldsymbol{\sigma}) \tag{5.4}$$

where $\mathcal{C}_c$ is the set of stress fields that do not violate the yield criterion:

$$\mathcal{C}_c = \{\boldsymbol{\sigma} = (\boldsymbol{\sigma}_d, s_m)\,, \boldsymbol{\sigma} \in \mathrm{L}^2(\Omega, \mathbb{E}_s^3) \mid \boldsymbol{\sigma}(\boldsymbol{x}) \in K_c(\boldsymbol{x})\, \forall \boldsymbol{x} \in \overline{\Omega}\}$$

The functional $\mathcal{G}(\dot{\boldsymbol{u}}, \boldsymbol{\sigma})$ is now given by

$$\mathcal{G}(\dot{\boldsymbol{u}}, \boldsymbol{\sigma}) = \int_{\Omega} \psi_c(\dot{\mathrm{e}}_d^p(\dot{\boldsymbol{u}}))\,\mathrm{d}\Omega + I_{\mathcal{C}_c}(\boldsymbol{\sigma}) - \int_{\Omega} \overline{\boldsymbol{b}} \cdot \dot{\boldsymbol{u}}\,\mathrm{d}\Omega - \mu \int_{\Gamma_t} \overline{\boldsymbol{t}}_0 \cdot \dot{\boldsymbol{u}}\,\mathrm{d}\Gamma \tag{5.5}$$

which is separable into a function of $\dot{\boldsymbol{u}}$ and a function of $\boldsymbol{\sigma}$. Hence, we get the extremum principle:

$$\inf_{\dot{\boldsymbol{u}} \in \mathcal{V}_{ka}} \left\{ \int_{\Omega} \psi_c(\dot{\mathrm{e}}_d^p(\dot{\boldsymbol{u}}))\,\mathrm{d}\Omega - \int_{\Omega} \overline{\boldsymbol{b}} \cdot \dot{\boldsymbol{u}}\,\mathrm{d}\Omega - \mu \int_{\Gamma_t} \overline{\boldsymbol{t}}_0 \cdot \dot{\boldsymbol{u}}\,\mathrm{d}\Gamma \right\} \tag{5.6}$$

and the dual one

$$\sup_{\boldsymbol{\sigma} \in \mathcal{S}_{sa}} \{-I_{\mathcal{C}_c}(\boldsymbol{\sigma})\}\,. \tag{5.7}$$

Introducing the kinematic load multiplier

$$\mu_k(\dot{\boldsymbol{u}}) = \frac{\displaystyle\int_{\Omega} \psi_c(\dot{\mathrm{e}}_d^p(\dot{\boldsymbol{u}}))\,\mathrm{d}\Omega - \int_{\Omega} \overline{\boldsymbol{b}} \cdot \dot{\boldsymbol{u}}\,\mathrm{d}\Omega}{\displaystyle\int_{\Gamma_t} \overline{\boldsymbol{t}}_0 \cdot \dot{\boldsymbol{u}}\,\mathrm{d}\Gamma}, \tag{5.8}$$

the minimum principle (5.6) yields back the upper bound theorem:

$$\mu_l = \inf_{\dot{\boldsymbol{u}} \in \mathcal{V}_{ka}} \mu_k(\dot{\boldsymbol{u}}) \tag{5.9}$$

The maximum principle (5.7) gives the lower bound theorem:

$$\mu_l = \sup_{\boldsymbol{\sigma} \in \mathcal{S}_{sa} \cap K} \mu_s(\boldsymbol{\sigma}). \tag{5.10}$$

Since $\mathcal{G}(\dot{\boldsymbol{u}}, \boldsymbol{\sigma}) \geq 0$, therefore

$$\mu_s \leq \mu_l \leq \mu_k. \tag{5.12}$$

5.2 The non-associated case

Now we assume a rigid plastic material with a non-associated flow rule (2.2) and (2.3). Let us define two fictitious materials with associated flow rules. The first one, $\mathcal{M}_1$, is defined by a yield surface (3.2) which coincides with the real yield surface and the second one, $\mathcal{M}_2$, has a yield surface defined by the plastic potential. The dissipation function for $\mathcal{M}_1$ is:

$$\psi_\phi(\dot{\varepsilon}^p) = \sup_{\sigma\in\mathcal{S}}\left[\boldsymbol{\sigma}\cdot\dot{\varepsilon}^p - I_{K_\phi}(\boldsymbol{\sigma})\right] = \frac{c\,\dot{\mathrm{e}}^p_m}{\tan\phi} + I_{K^*_\phi}(\dot{\varepsilon}^p), \tag{5.13}$$

where the convex set K^*_ϕ (see Figure 3.1) defined by

$$K^*_\phi(\dot{\varepsilon}^p) = \left\{\ \dot{\varepsilon}^p \in \mathcal{E} \mid \dot{\mathrm{e}}^p_m \geq k_d \tan\phi\,\|\dot{\mathbf{e}}^p_d\|\right\}.$$

is the dual of K_ϕ (3.1). The pseudo-potential of dissipation for $\mathcal{M}_2$ is denoted by $\widetilde{\psi}_c(\dot{\mathbf{e}}^p)$ and its expression is obtained from (3.13) by replacing c with *cte*. Referring to (4.1), now we see that $F(N(\dot{\boldsymbol{u}}))) \neq \mathbf{0}$. Different choices of $F(N(\dot{\boldsymbol{u}}))$ are possible. Let us consider two such choices.

5.2.1 First choice. One can consider the dissipation function of $\mathcal{M}_1$ to construct the functional Φ:

$$\Phi(\Lambda\boldsymbol{u}) = \int_\Omega \psi_\phi(\dot{\varepsilon}^p(\dot{\boldsymbol{u}}))\,\mathrm{d}\Omega. \tag{5.14}$$

Let us denote by $\boldsymbol{\sigma}_1$ ($\boldsymbol{\sigma}_1 \equiv \boldsymbol{q}^*$) the stress satisfying the inverse flow rule for the solid $\mathcal{M}_1$

$$\boldsymbol{\sigma}_1 \in \partial\psi_\phi(\dot{\varepsilon}^p) \tag{5.15}$$

Then $N(\dot{\boldsymbol{u}}) \equiv \boldsymbol{\sigma}_2(\dot{\boldsymbol{u}})$ is necessarily given by

$$N(\dot{\boldsymbol{u}}) = \partial_{\dot{\mathbf{e}}^p_d}\psi_{s_m}(s_m(\dot{\boldsymbol{u}}), \dot{\mathbf{e}}^p_d(\dot{\boldsymbol{u}})) - \partial\psi_\phi(\dot{\varepsilon}^p(\dot{\boldsymbol{u}})) \tag{5.16}$$

where ψ_{s_m} is given by (3.9). The total stress is

$$\boldsymbol{\sigma} = \boldsymbol{\sigma}_1 + \boldsymbol{\sigma}_2 \tag{5.17}$$

where $\boldsymbol{\sigma}_1$ is given by (5.15). The functional dual of Φ, denoted by Φ^*, is given by:

$$\Phi^*(\boldsymbol{\sigma}_1) = I_{\mathcal{C}_\phi}(\boldsymbol{\sigma}_1) \tag{5.18}$$

where

$$\mathcal{C}_\phi = \{\boldsymbol{\sigma} = (\boldsymbol{\sigma}_d, s_m)\ , \boldsymbol{\sigma} \in \mathrm{L}^2(\Omega, \mathbb{E}^3_s) \mid \boldsymbol{\sigma}(\boldsymbol{x}) \in K_\phi(\boldsymbol{x})\ \forall \boldsymbol{x} \in \overline{\Omega}\}.$$

Using the decomposition (5.17), we can specify the operator F as

$$F(N(\dot{\boldsymbol{u}})) = \begin{bmatrix} \operatorname{div} \boldsymbol{\sigma}_2(\dot{\boldsymbol{u}}) \\ -\boldsymbol{n}\, \boldsymbol{\sigma}_2(\dot{\boldsymbol{u}}) \end{bmatrix} \begin{matrix} \text{in } \Omega \\ \text{on } \Gamma_t \end{matrix},$$

and formulate the minimum principle associated with the boundary value problem for non-associated rigid plastic solids:

$$\beta = \inf_{\dot{\boldsymbol{u}}, \boldsymbol{\sigma}_1} \left\{ \int_\Omega \psi_\phi(\dot{\mathrm{e}}_d^p(\dot{\boldsymbol{u}}))\, \mathrm{d}\Omega + I_{C_\phi}(\boldsymbol{\sigma}_1) - \int_\Omega \overline{\boldsymbol{b}} \cdot \dot{\boldsymbol{u}}\, \mathrm{d}\Omega - \mu_l \int_{\Gamma_t} \overline{\boldsymbol{t}}_0 \cdot \dot{\boldsymbol{u}}\, \mathrm{d}\Gamma + \int_\Omega \boldsymbol{\sigma}_2(\dot{\boldsymbol{u}}) \cdot \dot{\varepsilon}^p(\dot{\boldsymbol{u}})\, \mathrm{d}\Gamma \right\} \tag{5.19}$$

Accordingly, the upper theorem of limit analysis is:

$$\mu_l = \inf_{\dot{\boldsymbol{u}} \in \mathcal{V}_{ka}} \mu_k(\dot{\boldsymbol{u}}, \boldsymbol{\sigma}_2(\dot{\boldsymbol{u}})) \tag{5.20}$$

where

$$\mu_k(\dot{\boldsymbol{u}}, \boldsymbol{\sigma}_2(\dot{\boldsymbol{u}})) = \frac{\int_\Omega \psi_\phi(\dot{\mathrm{e}}_d^p(\dot{\boldsymbol{u}}))\, \mathrm{d}\Omega + \int_\Omega \boldsymbol{\sigma}_2(\dot{\boldsymbol{u}}) \cdot \dot{\varepsilon}^p(\dot{\boldsymbol{u}})\, \mathrm{d}\Gamma - \int_\Omega \overline{\boldsymbol{b}} \cdot \dot{\boldsymbol{u}}\, \mathrm{d}\Omega}{\int_{\Gamma_t} \overline{\boldsymbol{t}}_0 \cdot \dot{\boldsymbol{u}}\, \mathrm{d}\Gamma}.$$

The lower bound theorem is:

$$\mu_l(\boldsymbol{\sigma}_1, \boldsymbol{\sigma}_2(\dot{\boldsymbol{u}})) = \sup_{\boldsymbol{\sigma}(\dot{\boldsymbol{u}}) \in \mathcal{S}_{sa} \cap \widetilde{\mathcal{C}}} \mu_s(\boldsymbol{\sigma}_1 + \boldsymbol{\sigma}_2(\dot{\boldsymbol{u}})) \tag{5.21}$$

Remark 5.1 In the lower bound theorem $\boldsymbol{\sigma}_2(\dot{\boldsymbol{u}})$ is taken from the upper bound theorem (5.20). Consequently, the kinematic and the static approach are still coupled through the term $\boldsymbol{\sigma}_2(\dot{\boldsymbol{u}})$.

5.2.2 Second choice. Now, can consider the dissipation function of $\mathcal{M}_2$ to construct Φ:

$$\Phi(\Lambda \boldsymbol{u}) = \int_\Omega \widetilde{\psi}_c(\dot{\mathrm{e}}_d^p(\dot{\boldsymbol{u}}))\, \mathrm{d}\Omega \tag{5.22}$$

Then we follow the same steps as for the first choice except that

$$N(\dot{\boldsymbol{u}}) = \boldsymbol{\sigma}_2(\dot{\boldsymbol{u}}) = \partial_{\dot{\mathrm{e}}_d^p}\psi_{s_m}(s_m(\dot{\boldsymbol{u}}), \dot{\mathrm{e}}_d^p(\dot{\boldsymbol{u}})) - \partial\widetilde{\psi}_c(\dot{\mathrm{e}}_d^p(\dot{\boldsymbol{u}})), \qquad (5.24)$$

$$\Phi^*(\boldsymbol{\sigma}_1) = I_{\widetilde{\mathcal{C}}_c}(\boldsymbol{\sigma}_1), \qquad (5.23)$$

where

$$\widetilde{\mathcal{C}}_c = \{\boldsymbol{\sigma} = (\boldsymbol{\sigma}_d, s_m)\ , \boldsymbol{\sigma} \in \mathrm{L}^2(\Omega, \mathbb{E}_s^3) \mid \boldsymbol{\sigma}(\boldsymbol{x}) \in \widetilde{K}_c(\boldsymbol{x})\, \forall \boldsymbol{x} \in \overline{\Omega}\}.$$

Accordingly, the lower and the upper bound are obtained by replacing Φ, Φ^* and $\boldsymbol{\sigma}_2(\dot{\boldsymbol{u}})$ in (5.19) by their expressions (5.22), (5.23) and (5.24).

Remark 5.2 In the last formulation, the dissipation function depends on the constant ξ which is arbitrary. However, this does not affect the functional since this constant can be eliminated by rearranging the first and the last terms in (5.19).

6. Summary

In this paper, a new formulation of the theorems of limit analysis applicable to non-associated plastic material has been given. The study deals with the non-associated incompressible plastic material but can be extended to more general cases where the dilatancy angle is non-zero. The derivation is based on a method, proposed by Telega, for constructing extremum principles for a non-potential operator equation. The functional enjoys convexity properties that can be exploited in the future for devising appropriate numerical algorithms.

Acknowledgments

The last two authors wish to thank the Australian Research Council for their financial support.

References

[1] Auchmuty, Giles, Variational principles for operator equations and initial value problems, *Nonlinear Anal., Theory, Methods & Applications*, **12** (1988), 531-564.

[2] Auchmuty, Giles, Min-max problems for non-potential operator equations, *Contemporary Mathematics*, **209** (1997), 19-28.

[3] Brezis, H. and I. Ekeland, Un principle variationnel associé à certaines équations paraboliques. Le cas indépendant du temps, *C. R. Acad. Sci. Paris*, Sér. A, **282** (1976), 971-974.

[4] Brezis, H. and I. Ekeland, Un principle variationnel associé à certaines équations paraboliques. Le cas indépendant du temps, *C. R. Acad. Sci. Paris*, Sér. A, **282** (1976), 1197-1198.

[5] de Saxcé, G. and L. Bousshine, Limit analysis theorems for implicit standard materials : application to the unilateral contact with dry friction and the non-associated flow rule in soils and rocks, *International Journal of Mechanical Sciences*, **40** (1998), 387-398.

[6] Drucker, D.C. , Coulomb friction, plasticity and limit loads, *Journal of Applied Mechanics Trans. ASME*, **21** (1954), 71-74.

[7] Moreau, J.J. , Sur les lois de frottement, de plasticité et de viscosité, *C. R. Acad. Sci. Paris*, Sér. A, **21** (1970), 608-611.

[8] Palmer, A.C. , A limit theorem for materials with non-associated flow laws, *Journal de Mécanique*, **5** (1966), 217-222.

[9] Radenkovic, D. , Limit theorems for a Coulomb material with a non standard dilatation. *C. R. Acad. Sci. Paris*, **252** (1961), 4103-4104.

[10] Telega, J.J. , Limit analysis theorems in the case of Signiroini's boundary conditions and friction, *Arch. Mech.*, **37** (2002), 549-562.

[11] Telega, J.J. , Extremum principles for nonpotential and initial value problems, *Arch. Mech.*, **54** (2002), 565-592.

[12] Tonti, E. , Variational formulations for every nonlinear problem, *Int. J. Engng. Sci.*, **22** (1984), 1343–1371.

Chapter 16

PERIODIC SOLITON RESONANCES

Masayoshi Tajiri

Graduate School of Engineering, Osaka Prefecture University, Sakai, Osaka 599-8531, Japan

tajiri@ms.osakafu-u.ac.jp

Abstract The soliton solutions to the Davey-Stewartson equation are analyzed to show the existence of periodic soliton resonances. The periodic soliton resonances are irrelevant to the divergence of the soliton solution. This is crucially different from the soliton resonance between relatively inclined two line solitons.

Keywords: Soliton, periodic soliton, periodic soliton resonances

1. Introduction

The Kadomtsev-Petviashvili (KP) equation with positive dispersion has three kinds of soliton solutions; conventional line soliton, algebraic soliton and periodic soliton solutions. We have investigated the interactions between two periodic solitons, between line soliton and periodic soliton and algebraic soliton and periodic soliton to the KP equation with positive dispersion and found the periodic soliton resonance in each case [8, 9]. The periodic soliton resonance is irrelevant to the divergence of the soliton solutions. This is crucially different from the resonant interaction between two line solitons to the KP equation with negative dispersion which is associated with the parametric end points of the singular regime for interactions between two line solitons [5]. There are two types of peculiar interactions, one is the resonant interaction and the other is a long-range interaction. Typical resonant interaction between line soliton and y-periodic soliton is shown in Figure 16.1 [9]. As the y-periodic soliton is approaching to the line soliton, the hump of periodic soliton is decreasing, which looks as if the line soliton swallowed the hump. The purpose of this paper is review our recent work about periodic soliton reonances which follow the Davey-Stewartson(DS) equation [10, 12, 14].

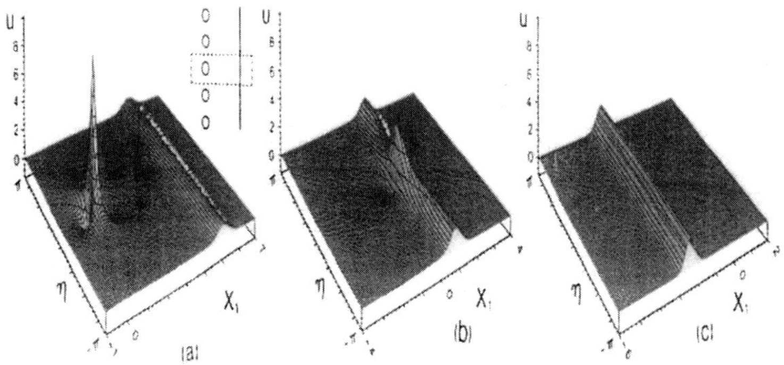

Figure 16.1. The sequence of snapshots of the resonant interaction between y-periodic soliton and line soliton with the parameters close to the resonant condition, which follows the KP equation with positive dispersion. The area inside the dotted lines in (a) is shown in figures [9].

2. Periodic Soliton Resonances: Solutions to the DS Equation

The DS equation may be written as [4]

$$\begin{cases} iu_t + pu_{xx} + u_{yy} + r|u|^2u - 2uv = 0\,, \\ v_{xx} - pv_{yy} - r\left(|u|^2\right)_{xx} = 0\,, \end{cases} \tag{2.1}$$

where $p = \pm 1$ and r is constant. Equation (2.1) with $p = 1$ and $p = -1$ are called the DS I and DS II equations, respectively. The DS equation has also the algebraic soliton and periodic soliton solutions in addition to the line soliton solution. Recently, we have investigated the interactions between two periodic solitons, between line soliton and periodic soliton and algebraic soliton and periodic soliton to the DS equation [2, 10, 12, 14]. It was shown that the periodic soliton resonance exists in each case.

At first, let us consider the resonant interaction between line soliton and y-periodic soliton mutually parallel propagation to the $x-$direction of the DSI (p=+1) equation as an example [10]. The solution described the interaction can be constructed from the three soliton solution of Satsuma and Ablowitz, which is written as [7, 10],

$$u = u_0 e^{i\zeta}\frac{g}{f}\ ,\quad v = \ -2\left(\ln f\right)_{xx} \tag{2.2}$$

with

$$\begin{aligned} f &= 1 + e^{\xi_P} \cos\eta + \frac{M}{4} e^{2\xi_P} \\ &+ e^{\xi_L}\{1 + N e^{\xi_P} \cos\eta + \frac{MN^2}{4} e^{2\xi_P}\}, \end{aligned} \tag{2.3}$$

$$\begin{aligned} g &= 1 + e^{\xi_P + i\phi_P} \cos\eta + \frac{M}{4} e^{2(\xi_P + i\phi_P)} \\ &+ e^{\xi_L + i\phi_L}\{1 + N e^{\xi_P + i\phi_P} \cos\eta + \frac{MN^2}{4} e^{2(\xi_P + i\phi_P)}\}, \end{aligned} \tag{2.4}$$

where

$$\begin{aligned} &\zeta = kx + ly - \omega t \ , \ \xi_P = \alpha x - \Omega_P t + \sigma_P \, , \\ &\xi_L = Px - \Omega_L t + \sigma_L \ , \ \eta = \delta y - \gamma_P t + \theta_P \, , \\ &\sin^2 \frac{\phi_P}{2} = \frac{\alpha^2 + \delta^2}{2ru_0^2} \ , \ \sin^2 \frac{\phi_L}{2} = \frac{P^2}{2ru_0^2} \, , \\ &\Omega_P = 2k\alpha - (\alpha^2 - \delta^2) \cot \frac{\phi_P}{2} \ , \ \gamma_P = 2l\delta \, , \\ &\Omega_L = P(2k - P \cot \frac{\phi_L}{2}) \ , \ M = \{1 - \frac{(\alpha^2 + \delta^2)^2}{2\delta^2 r u_0^2}\}^{-1} \, , \\ &N = \frac{2ru_0^2 \sin(\phi_P/2)\sin(\psi_L/2)\cos\{(\phi_P - \psi_L)/2\} - \alpha P}{2ru_0^2 \sin(\phi_P/2)\sin(\psi_L/2)\cos\{(\phi_P + \psi_L)/2\} - \alpha P} \, , \end{aligned}$$

When we assume that α and P are positive and $(\Omega_P/\alpha) > (\Omega_L/P)$, two separated y-periodic soliton and line soliton long before and long after the interaction are given by

$$f_P = 1 + e^{\xi_P} \cos(\eta) + \frac{M}{4} e^{2\xi_P} \ , \ f_L = \frac{M}{4} e^{2\xi_P}\{1 + N^2 e^{\xi_L}\} \, , \tag{2.5}$$

and

$$f_P = e^{\xi_L}\{1 + N e^{\xi_P} \cos\eta + \frac{MN^2}{4} e^{2\xi_P}\} \ , \ f_L = 1 + e^{\xi_L} \, , \tag{2.6}$$

respectively. Comparing eq.(2.5) and eq.(2.6), we find that N is related to the phase shift after the collision. The condition $|N| = \infty$ corresponds to the phase shift in the propagation direction becomes infinity. In the case $\alpha P > 0$, this means that the period of intermediate state, where the periodic soliton propagates together with the line soliton, persists infinitly. This is thought as a resonance between y-periodic soliton and line soliton. By equating the denominator of N to zero, we have the resonant condition,

$$2ru_0^2 \sin \frac{\phi_P}{2} \sin \frac{\phi_L}{2} \cos \frac{\phi_P + \phi_L}{2} - \alpha P = 0 \, . \tag{2.7}$$

When the resonant condition (2.7) is satisfied, the y-periodic soliton and line soliton interact so as to form a new resonant line soliton which is given by

$$f = 1+\frac{MN^2}{4}e^{2\xi_P+\xi_L} = 1+\frac{M}{4}e^{\widetilde{\xi}}, \tag{2.8}$$

$$g = 1+\frac{MN^2}{4}e^{2\xi_P+\xi_L+i(2\phi_P+\phi_L)} = 1+\frac{M}{4}e^{\widetilde{\xi}+i(2\phi_P+\phi_L)}. \tag{2.9}$$

where $\widetilde{\xi} = 2\xi_P + \xi_L + 2\ln|N|$. Using the dispersion relations of the line soliton and the y-periodic soliton and eq.(2.7), we can obtain the following relation

$$\sin^2\frac{2\phi_P+\phi_L}{2} = \frac{(2\alpha+P)^2}{2ru_0^2},$$

$$2\Omega_P+\Omega_L = (2\alpha+P)\{2k-(2\alpha+P)\cot\frac{2\phi_P+\phi_L}{2}\},$$

which are the dispersion relation for the resonant line soliton. The typical resonant interaction between y-periodic soliton and line soliton is shown in Figure 16.2 [10]. As the y-periodic soliton is approaching to the line soliton, the humps of the periodic soliton is decreasing, which looks as if the line soliton swallowed the humps. Figure 16.2(c) shows that the interaction yields a new resonant line soliton as predicated by eqs.(2.8) and (2.9).

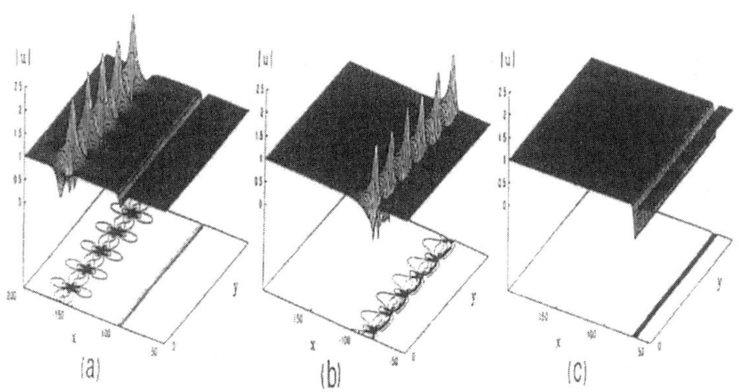

Figure 16.2. The sequence of snapshots of the resonant interaction between y-periodic soliton and line soliton with the parameters close to the resonant condition, which follows the DSI equation [10].

Here, It should be noted that this periodic soliton resonance is irrelevant to the divergence of the solution, because the coefficient of

$\exp(2\xi_P + \xi_L)$, which is the main term expressing the intermediate line soliton, includes N in the square N^2. This is crucially different from the resonant interaction between two line solitons which is associated with the parametric end points of the regime where the solution becomes singular [1].

The condition $N = 0$ corresponds to the phase shift in the propagating direction becomes negative infinite for $\alpha P > 0$. This means that two solitons can interact infinitely apart each other, which is thought as long-range interaction between line soliton and y-periodic soliton.

Next, we consider the resonant interaction between growing-and-decaying mode and line soliton as an another interesting example [11, 12]. The growing-and-decaying mode solution to the DS equation describes the nonlinear evolution of monocromatic perturbation with the Benjamin-Feir unstable condition in two-dimension.

The growing-and-decaying mode solution is given by [11]

$$u = u_0 e^{i\zeta} \frac{g}{f} \ , \quad v = \ -2\,(\ln f)_{xx} \tag{2.10}$$

with

$$f = 1 + e^{-\Omega_g t + \sigma_g} \cos \eta_g + \frac{M}{4} e^{2(-\Omega_g t + \sigma_g)},$$

$$g = 1 + e^{-\Omega_g t + \sigma_g + i\phi_g} \cos \eta_g + \frac{M}{4} e^{2(-\Omega_g t + \sigma_g + i\phi_g)}$$

where

$$\eta_g = \beta x + \delta y - \gamma_g t + \theta_g \ , \quad \Omega_g = (\beta^2 + \delta^2) \cot \frac{\phi_g}{2} \ ,$$

$$\gamma_g = 2k\beta + 2l\delta \ , \quad \sin^2 \frac{\phi_g}{2} = \frac{\delta^2 - \beta^2}{2ru_0^2} \ , \quad M = \frac{2}{1 + \cos \phi_g} > 1 \, ,$$

σ_g and θ_g are arbitrary phase constants. The existence condition for the nonsingular solution is given by $M > 1$ for real ϕ_g, which is satisfied for

$$0 < (\delta^2 - \beta^2) < 2ru_0^2 \tag{2.11}$$

which is in agreement with the condition which the plane wave,

$$u^0 = u_0 \exp\{i(kx + ly - \omega t)\} \ , \quad v^0 = 0 \, , \tag{2.12}$$

is linearly unstable to modulational perturbation and Ω_g is in agreement with the growth rate of the Benjamin-Feir instability [3]. This solution grows exponentially at an initial stage, reaches a state of maximum modulation, demodulates and finally returns to an unmodulated initial state. The solution decribing the interaction between line soliton and growing-and-decaying mode to the DS I equation is given by

$$\begin{aligned} f &= 1+e^{-\Omega_g t+\sigma_g}\cos\eta_g+\frac{M}{4}e^{2(-\Omega_g t+\sigma_g)}+e^{\xi}\{1+(N_r\cos\eta_g \\ &\qquad - N_i\sin\eta_g)e^{-\Omega_g t+\sigma_g}+\frac{M|\widetilde{N}|^2}{4}e^{2(-\Omega_g t+\sigma_g)}\}, \end{aligned} \tag{2.13}$$

$$\begin{aligned} g &= 1+e^{-\Omega_g t+\sigma_g+i\phi_g}\cos\eta_g+\frac{M}{4}e^{2(-\Omega_g t+\sigma_g+i\phi_g)} \\ &\qquad + e^{\xi+i\phi_L}\{1+(N_r\cos\eta_g - N_i\sin\eta_g)e^{-\Omega_g t+\sigma_g+i\phi_g} \\ &\qquad + \frac{M|\widetilde{N}|^2}{4}e^{2(-\Omega_g t+\sigma_g+i\phi_g)}\}, \end{aligned} \tag{2.14}$$

where

$$\begin{aligned} &\xi = Px+Qy-\Omega_L t+\sigma_L\ ,\ \ \eta_g=\beta x+\delta y-\gamma_g t+\theta_g\,, \\ &\sin^2\frac{\phi_g}{2}=\frac{\delta^2-\beta^2}{2ru_0^2}\ ,\ \ \sin^2\frac{\phi_L}{2}=\frac{P^2-Q^2}{2ru_0^2}\,, \\ &\Omega_g=(\beta^2+\delta^2)\cot\frac{\phi_g}{2}\ ,\ \ \gamma_g=2k\beta+2l\delta\,, \\ &\Omega_L=2kP+2lQ-(P^2+Q^2)\cot\frac{\phi_L}{2}\ ,\ \ M=\frac{2}{1+\cos\phi_g}\,, \\ &\widetilde{N}=N_r+iN_i \\ &=\frac{2ru_0^2\sin(\phi_g/2)\sin(\phi_L/2)\cos[(\phi_g-\phi_L)/2]-i(\beta P-\delta Q)}{2ru_0^2\sin(\phi_g/2)\sin(\phi_L/2)\cos[(\phi_g+\phi_L)/2]-i(\beta P-\delta Q)}\,, \end{aligned}$$

where σ_g, σ_L and θ_g are arbitrary constants. When we consider the case: $0 < \Omega_g$ the solutions of line soliton long before and after the growth of growing-and-decaying mode are given by

$$f=\frac{M}{4}e^{2(-\Omega_g t+\sigma_g)}\{1+e^{\xi+\sigma_3}\}\,, \tag{2.15}$$

and

$$f=1+e^{\xi}\,, \tag{2.16}$$

respectively, where $\sigma_3 = 2\ln|N|$. Comparing these equations, we see that the phase of the line soliton shifts by the amount $2\ln|N|$ due to the growing-and-decaying mode. $\ln|N| = \infty$ and $-\infty$ may be regarded as the conditions for resonance between line soliton and growing-and-decaying mode which are obtained by equating the denominator and numerator to zero, respectively:

$$\cos\frac{\phi_g+\phi_L}{2}=0\ ,\ \ \frac{Q}{P}=\frac{\beta}{\delta}\,, \tag{2.17}$$

and

$$\cos\frac{\phi_g - \phi_L}{2} = 0 \ , \ \frac{Q}{P} = \frac{\beta}{\delta} \ . \tag{2.18}$$

The existence of resonance between line soliton and growing-and-decaying mode was first pointed out by Pelinovsky [6]. Although the growing-and-decaying mode exists substantially during only a finite time, an infinite phase shift happens to the line soliton. The time evolution of wave field was investigated under a quasi-resonant condition [12]. The interaction between line soliton and growing-and-decaying mode having the paremeters near resonant condition ($1 \ll |N|$; quasi-resonance) is shown in Figure 16.3.

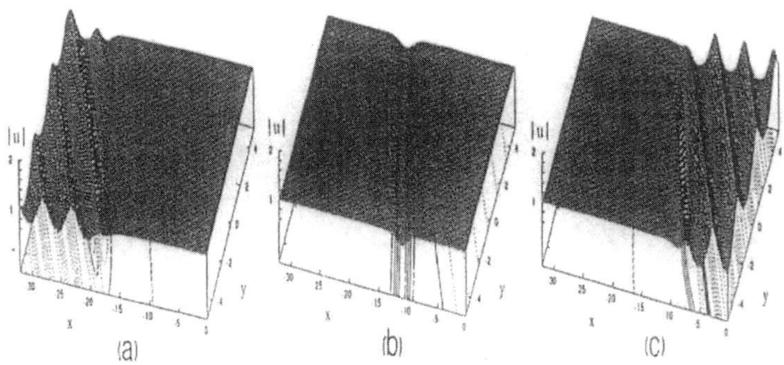

Figure 16.3. The sequence of snapshots of the resonant interaction between line soliton and growing-and-decaying mode which follows the DSI equation [12].

In the period, $t \sim \frac{\sigma_g}{\Omega_g}$ (namely $e^{-\Omega_g t+\sigma_g} \sim O(1)$) the solutions in the backward region and forward region of line soliton are approximated as follows,

$$f \simeq 1 + e^{-\Omega_g t+\sigma_g}\cos\eta_g + \frac{M}{4}e^{2(-\Omega_g t+\sigma_g)} \ , \tag{2.19}$$

and

$$f \simeq \frac{M}{4}e^{\xi+\sigma_3+2(-\Omega_g t+\sigma_g)} \ , \tag{2.20}$$

respectively. Equation (2.20) denotes a uniform state, in this period, the mode is growing only in the backward region of the line soliton as shown in Figure 16.3(a) [12]. In the period, $t \sim \frac{\sigma_g+\frac{1}{4}\sigma_3}{\Omega_g}$; $(\sqrt{|N|}e^{-\Omega_g t+\sigma_g} \sim O(1))$, the mode in the backward region ended the sequence of evolution and the wave field shifts to intermediate state, where only the line soliton in the resonant state exists in the wave field as shown in Figure 16.3(b). This intermediate state persists over compratively long time interval for $|N| \gg 1$. In the period, $t \sim \frac{\sigma_g + \frac{1}{2}\sigma_3}{\Omega_g}$; $(|N|e^{-\Omega_g t + \sigma_g} \sim O(1))$, the solution in the backward region and forward region of the line soliton are given by

$$f \simeq 1\,, \tag{2.21}$$

and

$$f \simeq e^{\xi}\{1 + (N_r \cos\eta_g - N_i \sin\eta_g)e^{-\Omega_g t + \sigma_g} + \frac{M|\widetilde{N}|^2}{4} e^{2(-\Omega_g t + \sigma_g)}\}\,, \tag{2.22}$$

respectively. In this period, the mode is developed only in the forward region of the line soliton as shown in Figure 16.3(c) [12]. If resonant condition is satisfied exactly, the growing-and-decaying mode exists only in the backward region of the line soliton.

3. Soliton Stability by the Hirota Method

In this section, we discuss the stability problem of the line soliton to the DSI equation with the use of the Hirota Method [13]. The resonant line soliton produced by the interaction between line soliton and y-periodic soliton maintains its resonant state all along the time after the collision, if resonant condition were satisfied exactly. But the solution with parameters slightly different from the resonant condition, consists of the resonant line soliton and small disturbance. The solution describing the interaction between line soliton and y-periodic soliton is given by eq.(2.2) . Figure 16.4 is the schematic diagram of the world line of the soliton humps in the (x, t) plane for the quasi-resonant interaction. Line **AB** is the world line of the resonant line soliton. The quasi-resonant state reaches the most resonant state at the midpoint **C** of the world line **AB** , where the disturbance becomes minimum. After the most resonant state, the state begins to deresonate and eventually the resonant line soliton decays into the lower amplitude line soliton and y-periodic soliton which are the same ones as before collision. The solution after reaching the most resonat state describes the time evolution of the line soliton undergoing the instability to the small transverse disturbance. The line **Cb** in Figure 16.4 corresponds to the period of linear development of the disturbance. When we replace $\xi_P + \frac{1}{2}\ln|N|$ and $\xi_L + \ln|N|$ with $\widetilde{\xi_P}$ and $\widetilde{\xi_L}$, respectively, so as to $\widetilde{\xi_P} \sim \widetilde{\xi_L} \sim O(1)$ at the most resonant point **C**, eqs.(2.3) and (2.4) are rewritten as follows [13]:

$$\begin{cases} f = f^{(0)} + \frac{1}{\sqrt{|N|}} f^{(1)} \cos\eta + \frac{1}{|N|} f^{(2)}\,, \\ g = g^{(0)} + \frac{1}{\sqrt{|N|}} g^{(1)} \cos\eta + \frac{1}{|N|} g^{(2)}\,, \end{cases} \tag{3.1}$$

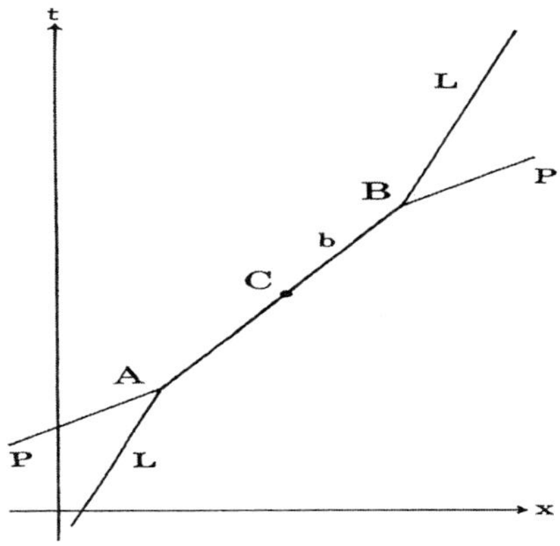

Figure 16.4. The schematic diagram of the world lines of the soliton humps in the x-t plane.

where

$$f^{(0)} = 1 + \frac{M}{4}\exp(\widetilde{\xi_0}), \tag{3.2}$$

$$g^{(0)} = 1 + \frac{M}{4}\exp(\widetilde{\xi_0} + \mathrm{i}\Phi), \tag{3.3}$$

$$f^{(1)}(\widetilde{\xi_P}, \widetilde{\xi_L}) = \exp(\widetilde{\xi_P})\left\{1 + \varepsilon\exp(\widetilde{\xi_L})\right\}, \tag{3.4}$$

$$f^{(2)}(\widetilde{\xi_P}, \widetilde{\xi_L}) = \exp(\widetilde{\xi_L}) + \frac{M}{4}\exp(2\widetilde{\xi_P}), \tag{3.5}$$

$$g^{(j)} = f^{(j)}(\widetilde{\xi_P} + \mathrm{i}\phi_P, \widetilde{\xi_L} + \mathrm{i}\phi_L) \qquad (j = 1, 2), \tag{3.6}$$

where $\Phi = 2\phi_P + \phi_L$, $P_0 = 2\alpha + P$, $\Omega_0 = 2\Omega_P + \Omega_L$, $\widetilde{\xi_0} = 2\widetilde{\xi_P} + \widetilde{\xi_L} = P_0 x - \Omega_0 t + 2\widetilde{\sigma_P} + \widetilde{\sigma_L}$ and $\varepsilon = 1$ for $0 < N$, $\varepsilon = -1$ for $N < 0$. Taking the limit $|N| \to \infty$, we have the resonant line soliton solution with wave number P_0 and phase Φ as follows,

$$u_0 = u_0 \mathrm{e}^{\mathrm{i}\zeta}\frac{1 + \mathrm{e}^{\xi_0 + \mathrm{i}\Phi}}{1 + \mathrm{e}^{\xi_0}}, \quad v_0 = -\frac{P_0^2}{2}\operatorname{sech}^2\frac{1}{2}\xi_0, \tag{3.7}$$

where $\xi_0 = \widetilde{\xi_0} + (1/2)\log(M/4)$. Using the resonant condition, and dispersion relations of y-periodic soliton and line soliton, we obtain

$$P^2 = P_0^2 - 4\delta^2 \pm 4\left(\sqrt{2ru_0^2 - P_0^2}\right)\delta\,, \tag{3.8}$$

Taking account of the relations, $\alpha = (P_0 - P)/2, \sin^2(\phi_L/2) = P^2/2ru_0^2$ and $\phi_P = (\Phi - \phi_L)/2$, we see that $P, \alpha, \phi_L, \phi_P, \Omega_P$ and Ω_L are expressed by P_0, Φ and δ, Therefore, the quasi-resonant solution is expressed by parameters (P_0, Φ, δ), which are the wave number and the phase of resonant line soliton and the wave number of transverse disturbance on the resonant line soliton, respectively. Neglecting the terms of $O(1/N)$, we have [13].

$$u \;=\; u_0 \mathrm{e}^{\mathrm{i}\zeta}\left[\frac{g^{(0)}}{f^{(0)}} + \frac{1}{\sqrt{|N|}}\frac{g^{(0)}}{f^{(0)}}\left(\frac{g^{(1)}}{g^{(0)}} - \frac{f^{(1)}}{f^{(0)}}\right)\cos\eta\right]. \tag{3.9}$$

It should be noted that the maginitude of the initial disturbance (the magnitude of the disturbance on the resonant line soliton at the most resonant state) is of the order $(1/\sqrt{|N|})$. The second term of eq.(3.9) describes the linear development of the disturbance on the resonant line soliton. Regarding the quasi-resonant soliton solution at the most resonant state as the initial state, we can size the full-time evolution up to the nonlinear stage of the soliton with a small disturbance undergoing the instability by using the quasi-resonant solution.

4. Summary

We showed the existence of the periodic soliton resonances for the DSI equation. The priodic soliton resonance is irrelevant to the divergence of the soliton solution. We expect that the periodic soliton resonances play fundamental role in the nonlinear development of the higher-dimensional wave field as the existence of the periodic soliton resonance may be related to the instability of solitons, accompanied by their decay and merging.

References

[1] D. Anker and N. C. Freeman, Proc. R. Soc. London A **360**, 529 (1978).

[2] T. Arai and Y. Takeuchi and M. Tajiri, J. Phys. Soc. Jpn. **70**, 55 (2001).

[3] T. B. Benjamin and J. E. Feir, J. Fluid Mech. **27**, 417 (1967), T. B. Benjamin, Proc. R. Soc. London, Ser. A **299**, 59 (1967), J. E. Feir, Proc. R. Soc. London, Ser. A **299**, 54 (1967).

[4] A. Davey and K. Stewartson, Proc. R. Soc. London, Ser. A **338**, 101 (1974).

[5] J. W. Miles, J. Fluid Mech. **79**, 157, 171 (1977).

[6] D. Pelinovsky, Phys. Lett. **196A**, 181 (1994).

[7] J. Satsuma and M. J. Ablowitz, J. Math. Phys. **20**, 1496 (1979).

[8] M. Tajiri and Y. Murakami, Phys. Lett. A **143**, 217 (1990), Y. Murakami and M. Tajiri, Wave Motion **14**, 169 (1991).

[9] M. Tajiri and Y. Fujimura and Y. Murakami, J. Phys. Soc. Jpn. **61**, 783 (1992), Y. Murakami and M. Tajiri, J. Phys. Soc. Jpn. **61**, 791 (1992).

[10] M. Tajiri and T. Arai and Y. Watanabe, J. Phys. Soc. Jpn. **67**, 4051 (1998).

[11] M. Tajiri and T. Arai, Phys. Rev. E **60**, 2297 (1999).

[12] M. Tajiri and K. Takeuchi and T. Arai, Phys. Rev. E **64**, 056622-1 (2001).

[13] M. Tajiri and K. Takeuchi and T. Arai, J. Phys. Soc. Jpn. **70**, 1505 (2001).

[14] Y. Watanabe and M. Tajiri, J. Phys. Soc. Jpn. **67**, 705 (1998).

Chapter 17

GENERALIZED LEGENDRE-FENCHEL TRANSFORMATION

Claude Vallée
L3MA, SP2MI,Boulevard Pierre et Marie Curie B.P. 30179
86962 Futuroscope Chasseneuil Cedex, France
claude.vallee@l3ma.univ-poitiers.fr

Mohammed Hjiaj
Geotechnical Research Group, Dept of Civil Engineering
The University of Newcastle, Callaghan NWS 2308, Australia
mohammed.hjiaj@newcastle.edu.au

Danièle Fortuné
L3MA, SP2MI,Boulevard Pierre et Marie Curie B.P. 30179
86962 Futuroscope Chasseneuil Cedex, France
daniele.fortune@l3ma.univ-poitiers.fr

Géry de Saxcé
Laboratoire de Mécanique de Lille URA CNRS 1441
Cité Scientifique, 59655 Villeneuve d'Ascq Cedex, France
gery.desaxce@univ-lille1.fr

Abstract The aim of the paper is to characterize transformations that preserve the potential structure of a relationship between dual variables. The first step consists in "geometrizing" the condition for the existence of a potential. Having at hand this formulation, it becomes clear that the canonical similitudes represents the class of transformations that preserves the potential form of a relationship. Next, we derive the conditions under which canonical similitudes preserve the convexity of the potential or change it into concavity. This new class of transformations provides a generalization of the Legendre-Fenchel transformation.

Keywords: Potential, Differential geometry, Lagrangian submanifolds, Symplectic linear space, Canonical similitudes, Convexity

1. Introduction

Often in physics, it is of particular interest to express the relation existing between the dual variables of the problem under consideration in term of a scalar-valued function, called potential. For instance, in solid mechanics, a potential form of the relationship between the generalized stress and the generalized strain is particularly attractive. The condition of potentiality of an operator is closely related to the symmetry of its derivative or equivalently to the conservativity of the operator that is the vanishing of the related circuital integral along any closed curve in the domain of the operator. A direct advantage of the potential form of the constitutive relation is the existence of stationary principles that become extremum principles when the potential is convex or concave. Accordingly, existence of solutions to boundary value problems can be investigated in an elegant way. Furthermore, such principles are a natural basis for applying discretization methods such as the finite element method. In some circumstances, it may be advantageous to apply a change of variables that yields to a more convenient expression of the potential regarding, for instance, to numerical solution. In this paper, we discuss the conditions under which a change of variables preserves the potential form of the relationship. Convexity or concavity being an important feature, we also characterize transformations that preserve the convexity of the potential or change it into concavity. To do so, an essential step consists in "geometrizing" the condition for the existence of potential. Having at our disposal this "geometrical" interpretation, it becomes more easy to study the conditions under which a change of variables maintain the potential form of the law and the convexity or the concavity of the potential. In particular, it is seen that canonical transformations preserve the potential structure after transformation. The rest of the paper is organized as follows. The next section discusses different formulations of the condition ensuring the existence of a potential. A "geometrization" of this condition is introduced, that is, the Lagrange 2-form must be null. The section 3 deals with the canonical similitudes and their properties since they preserve the Lagrange form. Two characterizations of the infinitesimal canonical similitudes are proposed. In section 4, a relationship between the second derivative of the original potential and the second derivative of the potential obtained after transformation is established. This relationship is used in Section 5 to derive the conditions under which the convexity is maintained or changed into concavity.

2. Condition for the existence of a potential: formulations

2.1 Symmetry of the second derivative

To model a physical phenomenon, two fundamental steps need to be performed. In the first stage, pertinent variables have to be identified before relationships between them are set up in the second stage. To check the validity of the model, experimental observations are compared with the predictions of the model. In solid mechanics, two physically different groups of variables can be distinguished: the geometrical variables and the dynamical variables. The first one comprises the generalized displacements or velocities that allow characterizing, at each time, the current configuration of the solid. These variables are often called primal variables and the space to which they belong, the space of configurations. The generalized deformations are related to the generalized velocities through a geometrical operator that can be nonlinear. These variables reveal the deformation state of the solid. The dynamical variables comprise also two sub-groups: the generalized forces that represent a part of the external actions and the generalized stresses, where a degree of freedom exists in their choice. However, the generalized stresses need to be in equilibrium with the external forces that define the equilibrium operator. To complete the definition of the problem, we need to specify the relation between the generalized stresses and generalized deformations. To fix the ideas, a variable $\mathbf{x}$ (for example the strain or strain rate tensor) belonging to an n-dimensional linear space $\mathcal{E}$ is considered. The conjugate variable, for instance the stress or stress rate tensor, denoted $\mathbf{y}$, belongs to the dual n-dimensional linear space $\mathcal{E}^*$. The relationship between $\mathbf{x}$ and $\mathbf{y}$ can be expressed in the following explicit form:

$$\mathbf{y} = \mathcal{C}(\mathbf{x}) \tag{2.1}$$

Operator $\mathcal{C}(\cdot)$ transforms $\mathbf{x}$ defined in the domain $\mathcal{D}(\mathcal{C}) \subset \mathcal{E}$ into $\mathbf{y}$ defined in the range of the operator $\mathcal{R}(\mathcal{C}) \subset \mathcal{E}^*$. Note that, in what follows, we will always assume that the domain of the operator $\mathcal{C}(\cdot)$ coincides with the space $\mathcal{E}$ and the range with its dual. In the most general case, the constitutive mapping $\mathcal{C}$ could be a nonlinear and non-smooth operator:

$$\mathcal{C} : \mathcal{E} \to \mathcal{E}^* : \mathbf{x} \mapsto \mathbf{y} = \mathcal{C}(\mathbf{x})$$

The operator $\mathcal{C}$ is said single-valued or multivalued at $\mathbf{x}$ according to whether $\mathcal{C}(\mathbf{x})$ is a singleton or a set containing more than one element. The non-smoothness is a desirable feature for multivalued law like the flow rule in plasticity. In what follows, only invertible single-valued operator are considered. If the map $\mathcal{C}$ is invertible, the inverse law has the form:

$$\mathcal{C}^{-1} : \mathcal{E}^* \to \mathcal{E} : \mathbf{y} \mapsto \mathbf{x} = \mathcal{C}^{-1}(\mathbf{y})$$

For physical reasons closely related to conservativity, a potential form of the constitutive relation is often preferred. In that case, the constitutive operator can be obtained as the gradient of a scalar-valued function ϕ for all its elements of its domain:

$$\mathbf{y} = D\phi(\mathbf{x}) \tag{2.2}$$

Properties of the potential such as convexity allow us to investigate upon the existence and the properties of solutions of boundary value problems and to devise suitable numerical procedures for their computation. For a single-valued operator, the condition ensuring that such function does, in fact, exist, is the conservativity of the operator, i.e. the vanishing of the related circuital integral along every closed curve in the domain of the operator. If the constitutive operator is differentiable, this condition is ensured by the symmetry of its first Fréchet derivative

$$[D\mathcal{C}(\mathbf{x})\mathrm{d}\mathbf{x}]\delta\mathbf{x} = [D\mathcal{C}(\mathbf{x})\delta\mathbf{x}]\mathrm{d}\mathbf{x} \tag{2.3}$$

for any vectors $\mathrm{d}\mathbf{x}$ and $\delta\mathbf{x}$ in $\mathcal{E}$. If the Elie Cartan notations are adopted, the previous condition becomes: the exterior derivative $\mathrm{d}\mathcal{C}$ of the 1-form $\mathcal{C}$ is the null 2-form.

2.2 The Souriau and Leibniz notations

It is worth investigating how Lagrange would have written the condition (2.3)? Or, how the work of Souriau [8] let suppose he would have written it? We have denoted $\mathrm{d}\mathbf{x}$ and $\delta\mathbf{x}$ the current vectors because, following Leibniz, and after him Souriau [8], we shall use the short notations $\mathrm{d}\mathbf{y}$ for $D\mathcal{C}(\mathbf{x})\mathrm{d}\mathbf{x}$ and $\delta\mathbf{y}$ for $D\mathcal{C}(\mathbf{x})\delta\mathbf{x}$ whenever $\mathbf{y}$ is related to $\mathbf{x}$ through $\mathcal{C}$. By doing so, the condition insuring that a relationship between $\mathbf{x}$ and $\mathbf{y}$ derive from a potential can be simply written as

$$\mathrm{d}\mathbf{y}\,\delta\mathbf{x} = \delta\mathbf{y}\,\mathrm{d}\mathbf{x} \tag{2.4}$$

2.3 Nullity of the Lagrange 2-form

The previous consideration is among the reasons that led Joseph Louis Lagrange [3], in his pioneering book “Mécanique Analytique”, to introduce the 2-form

$$\mathcal{S}\left(\begin{bmatrix} \mathrm{d}\mathbf{x} \\ \mathrm{d}\mathbf{y} \end{bmatrix}, \begin{bmatrix} \delta\mathbf{x} \\ \delta\mathbf{y} \end{bmatrix}\right) = \delta\mathbf{y}\mathrm{d}\mathbf{x} - \mathrm{d}\mathbf{y}\delta\mathbf{x} \tag{2.5}$$

defined on the linear product space $\mathcal{E} \times \mathcal{E}^*$. Nowadays, this 2-form is called the *Lagrange 2-form* and the linear space $\mathcal{E} \times \mathcal{E}^*$ is said to be *symplectic*. With the definition (2.5), a constitutive relation derive from a potential if

$$\mathcal{S}\left(\begin{bmatrix} \mathrm{dx} \\ \mathrm{dy} \end{bmatrix}, \begin{bmatrix} \delta\mathrm{x} \\ \delta\mathrm{y} \end{bmatrix}\right) = 0 \tag{2.6}$$

2.4 The Lagrange 2-form in continuum mechanics

The previous condition has been defined for vectors and an extension to tensors is sought. In continuum mechanics, the space $\mathcal{E}$ is the space of the symmetric second-order tensors, classically identified with its dual. Then, the dimension of the space $\mathcal{E}$ is six. By taking advantage of the linear operator trace, which we shall denote by "tr", a relevant Lagrange 2-form can be defined on $\mathcal{E} \times \mathcal{E}$:

$$\mathcal{S}\left(\begin{bmatrix} \mathrm{dx} \\ \mathrm{dy} \end{bmatrix}, \begin{bmatrix} \delta\mathrm{x} \\ \delta\mathrm{y} \end{bmatrix}\right) = \mathrm{tr}\,(\delta\mathbf{y}\mathrm{d}\mathbf{x}) - \mathrm{tr}\,(\mathrm{d}\mathbf{y}\delta\mathbf{x})$$

As we can see, vectors and tensors can be used interchangeably. The distinction between tensors and vectors is formal since a tensor can be represented by a six-dimensional vector. For the sake of exemplifying the symplectic notion in a simple generic model, let us suppose that $\mathcal{E}$ and $\mathcal{E}^*$ are the ordinary n-dimensional linear space . Then $\mathcal{E} \times \mathcal{E}^*$ is $\mathbb{R}^{2n}$ made symplectic by the Lagrange form

$$\mathcal{S}\left(\begin{bmatrix} \mathrm{dx} \\ \mathrm{dy} \end{bmatrix}, \begin{bmatrix} \delta\mathrm{x} \\ \delta\mathrm{y} \end{bmatrix}\right) = \langle \delta\mathbf{y}, \mathrm{d}\mathbf{x}\rangle - \langle \mathrm{d}\mathbf{y}, \delta\mathbf{x}\rangle$$

where the bracket denotes the "dot" product in $\mathbb{R}^n$. In what follows, we shall use matrix notations in both $\mathbb{R}^n$ and $\mathbb{R}^{2n}$ where the variables $\mathbf{y}$ and $\mathbf{x}$ are represented by n-dimensional vectors. With these notations, the Lagrange form becomes

$$\mathcal{S}\left(\begin{bmatrix} \mathrm{dx} \\ \mathrm{dy} \end{bmatrix}, \begin{bmatrix} \delta\mathrm{x} \\ \delta\mathrm{y} \end{bmatrix}\right) = \delta\mathbf{y}^{\mathrm{T}}\mathrm{d}\mathbf{x} - \mathrm{d}\mathbf{y}^{\mathrm{T}}\delta\mathbf{x}$$

where T in the exponent stands for the usual transposition. By denoting the $n \times n$ and $2n \times 2n$ unit matrices, by $\mathbf{I}_n$ and $\mathbf{I}_{2n}$, respectively, the previous relation can be written as:

$$\mathcal{S}\left(\begin{bmatrix} \mathrm{dx} \\ \mathrm{dy} \end{bmatrix}, \begin{bmatrix} \delta\mathrm{x} \\ \delta\mathrm{y} \end{bmatrix}\right) = \begin{bmatrix} \mathrm{d}\mathbf{x}^{\mathrm{T}} & \mathrm{d}\mathbf{y}^{\mathrm{T}} \end{bmatrix} \mathbf{J} \begin{bmatrix} \delta\mathbf{x} \\ \delta\mathbf{y} \end{bmatrix} \tag{2.7}$$

where $\mathbf{J}$ defined by

$$\mathbf{J} = \begin{bmatrix} \mathbf{0} & \mathbf{I}_n \\ -\mathbf{I}_n & \mathbf{0} \end{bmatrix} \tag{2.8}$$

is a skew-symmetric $2n \times 2n$ matrix satisfying

$$\mathbf{J}^2 = -\mathbf{I}_{2n}. \tag{2.9}$$

From the relation (2.9), it is clear that $\mathbf{J}$ is invertible, its inverse is $-\mathbf{J}$ and its determinant is equal to 1.

2.5 Lagrangian submanifolds

Let us denote by n the common dimension of $\mathcal{E}$ and $\mathcal{E}^*$. If $\mathbf{x}$ and $\mathbf{y}$ are related by a constitutive law deriving from a potential ϕ then the pair $(\mathbf{x}, \mathbf{y})$ describes a non-degenerate submanifold in $\mathcal{E} \times \mathcal{E}^*$. The *dimension* of this submanifold is equal to the dimension of $\mathcal{E}$ since $2n$ variables related by n independent equations represent merely n independent variables. Further, two vectors $\begin{bmatrix} \mathrm{d}\mathbf{x} \\ \mathrm{d}\mathbf{y} \end{bmatrix}$ and $\begin{bmatrix} \delta\mathbf{x} \\ \delta\mathbf{y} \end{bmatrix}$ tangent to the submanifold at $(\mathbf{x}, \mathbf{y})$ satisfy

$$\mathcal{S}\left(\begin{bmatrix} \mathrm{d}\mathbf{x} \\ \mathrm{d}\mathbf{y} \end{bmatrix}, \begin{bmatrix} \delta\mathbf{x} \\ \delta\mathbf{y} \end{bmatrix}\right) = 0$$

Let us proof this assertion and by the way illustrate the interest of the Souriau-Leibniz notations. If $\begin{bmatrix} \mathrm{d}\mathbf{x} \\ \mathrm{d}\mathbf{y} \end{bmatrix}$ and $\begin{bmatrix} \delta\mathbf{x} \\ \delta\mathbf{y} \end{bmatrix}$ are tangent at $(\mathbf{x}, \mathbf{y})$ to the submanifold described by the equation

$$\mathbf{y} = D\phi(\mathbf{x}),$$

then $\mathrm{d}\mathbf{x}$ and $\delta\mathbf{x}$ can be arbitrary but $\mathrm{d}\mathbf{y}$ and $\delta\mathbf{y}$ have to satisfy

$$\mathrm{d}\mathbf{y} = D(D\phi)(\mathbf{x})\mathrm{d}\mathbf{x} \quad \text{and} \quad \delta\mathbf{y} = D(D\phi)(\mathbf{x})\delta\mathbf{x}$$

Then, we can see that the Lagrange 2-form equal to

$$\mathcal{S}\left(\begin{bmatrix} \mathrm{d}\mathbf{x} \\ \mathrm{d}\mathbf{y} \end{bmatrix}, \begin{bmatrix} \delta\mathbf{x} \\ \delta\mathbf{y} \end{bmatrix}\right) = \big(D(D\phi)(\mathbf{x})\delta\mathbf{x}\big)\mathrm{d}\mathbf{x} - \big(D(D\phi)(\mathbf{x})\mathrm{d}\mathbf{x}\big)\delta\mathbf{x}$$

is null. The nullity of the Lagrange two-form results from the symmetry of the second derivative of ϕ. The reciprocal property is also true. Consider in $\mathcal{E} \times \mathcal{E}^*$ a submanifold such that its dimension is equal to half of the dimension of $\mathcal{E} \times \mathcal{E}^*$. If the Lagrange form is null on the tangent space of this submanifold, then it is described by the equation:

$$\mathbf{y} = D\phi(\mathbf{x})$$

for a certain potential ϕ. Such a submanifold is said to be *lagrangian*. If the Lagrange form applied to two arbitrary vectors of the tangent space of a submanifold is null, then this submanifold is said to be *isotropic*. The maximal dimension of an isotropic submanifold corresponds to half of the dimension of the space $\mathcal{E} \times \mathcal{E}^*$.

2.6 Geometrization of the condition ensuring the existence of a potential

From the above considerations, it appears that defining a one-to-one relationship between $\mathbf{x}$ and $\mathbf{y}$ is equivalent to give a submanifold in the linear space $\mathcal{E} \times \mathcal{E}^*$ whose dimension corresponds to half of the dimension of the linear space. Claiming the existence of a potential is requiring this submanifold to be lagrangian. In that way, the condition ensuring the existence of a potential is geometrized. For instance, when $\mathcal{E}$ is the space of symmetric second-order tensors identified with its dual, requiring a potential form of the relationship between strain and stress tensors is equivalent to select a lagrangian 6-dimensional submanifold of the 12-dimensional space $\mathcal{E} \times \mathcal{E}$. It is worth to mention that the present geometrical interpretation provides a natural framework for solving continuum mechanics problems since classical results from symplectic geometry can be used. For example, some properties Maslow's [5] index have been applied by Cardin & Spera [1] to solid mechanics.

3. Canonical similitudes and generalized Legendre transformation

Within the framework of the previously quoted symplectic linear space model $\mathbb{R}^{2n}$, we begin this section by examining what happen when the variables $(\mathbf{x}, \mathbf{y})$ are replaced by other variables $(\mathbf{X}, \mathbf{Y})$, trough a transformation $\mathcal{F}$. More precisely we seek to characterize transformations $\mathcal{F}$ that do not destroy the symplectic structure.

3.1 Canonical and anti-canonical transformations

A transformation is said to be *canonical* if the Lagrange form is preserved. It is said to be *anti-canonical* if the Lagrange form is preserved up to the sign.

Examples.

- The transformation defined by

$$\begin{cases} \mathbf{X} = \mathbf{y} \\ \mathbf{Y} = -\mathbf{x} \end{cases}$$

is canonical as one can easily check that

$$\langle \mathrm{d}\mathbf{X}, \delta\mathbf{Y}\rangle - \langle \delta\mathbf{X}, \mathrm{d}\mathbf{Y}\rangle = -\langle \mathrm{d}\mathbf{y}, \delta\mathbf{x}\rangle + \langle \delta\mathbf{y}, \mathrm{d}\mathbf{x}\rangle.$$

- But, just swapping the variables

$$\begin{cases} \mathbf{X} = \mathbf{y} \\ \mathbf{Y} = \mathbf{x} \end{cases}$$

is anti-canonical:

$$\langle \mathrm{d}\mathbf{X}, \delta\mathbf{Y}\rangle - \langle \delta\mathbf{X}, \mathrm{d}\mathbf{Y}\rangle = \langle \mathrm{d}\mathbf{y}, \delta\mathbf{x}\rangle - \langle \delta\mathbf{y}, \mathrm{d}\mathbf{x}\rangle$$

3.2 Canonical similitudes

A canonical transformation is said to be a *canonical similitude* if the lagrangian form is preserved up to a scalar multiplicative factor (a priori, not necessarily constant). It follows that a canonical transformation is a canonical similitude for which the value of the ratio between the new Lagrange form and the initial one is 1. An *anti-canonical* transformation is a canonical similitude where the value of the ratio is -1. Let us denote the ratio by $\rho(\mathbf{x}, \mathbf{y})$. Then, a change of variables is a canonical similitude if

$$\langle \mathrm{d}\mathbf{X}, \delta\mathbf{Y}\rangle - \langle \delta\mathbf{X}, \mathrm{d}\mathbf{Y}\rangle = \rho(\mathbf{x}, \mathbf{y})[\langle \delta\mathbf{y}, \mathrm{d}\mathbf{x}\rangle - \langle \mathrm{d}\mathbf{y}, \delta\mathbf{x}\rangle] \tag{3.1}$$

This formula, expressed in the short Souriau-Leibniz notations, means that a change of variables

$$\begin{bmatrix} \mathbf{X} \\ \mathbf{Y} \end{bmatrix} = \mathcal{F}\left(\begin{bmatrix} \mathbf{x} \\ \mathbf{y} \end{bmatrix}\right) := \begin{bmatrix} \mathcal{X}(\mathbf{x}, \mathbf{y}) \\ \mathcal{Y}(\mathbf{x}, \mathbf{y}) \end{bmatrix}$$

is a canonical similitude if the Fréchet derivative $D\mathcal{F}\left(\begin{bmatrix} \mathbf{x} \\ \mathbf{y} \end{bmatrix}\right)$ of the transformation $\mathcal{F}$ satisfies

$$\left[D\mathcal{F}\left(\begin{bmatrix} \mathbf{x} \\ \mathbf{y} \end{bmatrix}\right)\right]^{\mathrm{T}} \mathbf{J} \left[D\mathcal{F}\left(\begin{bmatrix} \mathbf{x} \\ \mathbf{y} \end{bmatrix}\right)\right] = \rho(\mathbf{x}, \mathbf{y})\mathbf{J} \tag{3.2}$$

Let χ denotes the matrix composed of the partial derivatives of the transformation components $\mathcal{X}$ and $\mathcal{Y}$

$$\begin{bmatrix} D_x\mathcal{X} & D_y\mathcal{X} \\ D_x\mathcal{Y} & D_y\mathcal{Y} \end{bmatrix}$$

Then, condition (3.2) reads

$$\chi^{\mathrm{T}}\mathbf{J}\chi = \rho(\mathbf{x},\mathbf{y})\mathbf{J} \tag{3.3}$$

3.2.1 Ratio of a canonical similitude. We recall that if the dimension of a symplectic linear space is greater than 2, the ratio of every canonical similitude is *constant.*

3.3 Characterizations of a canonical similitude

This subsection deals with the Fréchet derivative of the canonical similitudes; we shall call them *infinitesimal canonical similitudes.* They are $2n \times 2n$ matrices, which can be detailed in four $n \times n$ block matrices. If $\mathcal{F}$ denotes a canonical similitude then the matrix χ is an infinitesimal canonical similitude.

3.3.1 First characterization. By adopting the following notations

$$D_x\mathcal{X} := \mathbf{A}\,, \quad D_y\mathcal{X} := \mathbf{B}\,, \quad D_x\mathcal{Y} := \mathbf{C}\,, \quad D_y\mathcal{Y} := \mathbf{D}\,,$$

the $2n \times 2n$ matrix

$$\chi := \begin{bmatrix} \mathbf{A} & \mathbf{B} \\ \mathbf{C} & \mathbf{D} \end{bmatrix}$$

is the derivative of a canonical similitude of ratio ρ if and only if

$$\begin{cases} \mathbf{A}^{\mathrm{T}}\mathbf{C} - \mathbf{C}^{\mathrm{T}}\mathbf{A} = \mathbf{0} \\ \mathbf{A}^{\mathrm{T}}\mathbf{D} - \mathbf{C}^{\mathrm{T}}\mathbf{B} = \rho\mathbf{I}_n \\ \mathbf{B}^{\mathrm{T}}\mathbf{D} - \mathbf{D}^{\mathrm{T}}\mathbf{B} = \mathbf{0} \end{cases} \tag{3.4}$$

In the most general case, the sub-matrices **A**, **B**, **C** and **D** are functions of the variables **x** and **y**.

Proof: By applying straightforwardly the condition (3.3), we obtain that the matrix χ is an infinitesimal canonical similitude if

$$\begin{bmatrix} \mathbf{A}^{\mathrm{T}} & \mathbf{C}^{\mathrm{T}} \\ \mathbf{B}^{\mathrm{T}} & \mathbf{D}^{\mathrm{T}} \end{bmatrix} \begin{bmatrix} \mathbf{0} & \mathbf{I}_n \\ -\mathbf{I}_n & \mathbf{0} \end{bmatrix} \begin{bmatrix} \mathbf{A} & \mathbf{B} \\ \mathbf{C} & \mathbf{D} \end{bmatrix} = \rho \begin{bmatrix} \mathbf{0} & \mathbf{I}_n \\ -\mathbf{I}_n & \mathbf{0} \end{bmatrix}$$

that is

$$\begin{bmatrix} -\mathbf{C}^{\mathrm{T}} & \mathbf{A}^{\mathrm{T}} \\ -\mathbf{D}^{\mathrm{T}} & \mathbf{B}^{\mathrm{T}} \end{bmatrix} \begin{bmatrix} \mathbf{A} & \mathbf{B} \\ \mathbf{C} & \mathbf{D} \end{bmatrix} = \begin{bmatrix} \mathbf{0} & \rho\mathbf{I}_n \\ -\rho\mathbf{I}_n & \mathbf{0} \end{bmatrix}$$

or

$$\mathbf{A}^{\mathrm{T}}\mathbf{C} - \mathbf{C}^{\mathrm{T}}\mathbf{A} = \mathbf{0} \qquad \mathbf{B}^{\mathrm{T}}\mathbf{C} - \mathbf{D}^{\mathrm{T}}\mathbf{A} = -\rho\mathbf{I}_n$$

$$\mathbf{A}^{\mathrm{T}}\mathbf{D} - \mathbf{C}^{\mathrm{T}}\mathbf{B} = \rho\mathbf{I}_n \qquad \mathbf{B}^{\mathrm{T}}\mathbf{D} - \mathbf{D}^{\mathrm{T}}\mathbf{B} = \mathbf{0}$$

It can be shown that if χ is an infinitesimal canonical similitude then χ^{T} is also an infinitesimal canonical similitude providing a second characterization, which leads to equivalent relations.

3.3.2 Second characterization. The matrix

$$\chi^{\mathrm{T}} = \begin{bmatrix} \mathbf{A}^{\mathrm{T}} & \mathbf{C}^{\mathrm{T}} \\ \mathbf{B}^{\mathrm{T}} & \mathbf{D}^{\mathrm{T}} \end{bmatrix}$$

is also an infinitesimal canonical similitude of same ratio ρ.
Proof: By pre-multiplying each member of (3.3) by $-\mathbf{J}$ and taking into account (2.9), we obtain

$$\left(-\mathbf{J}\chi^{\mathrm{T}}\mathbf{J}\right)\chi = \rho\,\mathbf{I}_{2n}$$

The previous relation means that the matrix $\left(-\mathbf{J}\chi^{\mathrm{T}}\mathbf{J}\right)$ is the inverse of the matrix $\dfrac{1}{\rho}\chi$. Taking into account that the inverse is unique, one has also

$$\chi\left(-\mathbf{J}\chi^{\mathrm{T}}\mathbf{J}\right) = -\rho\,\mathbf{J}^2$$

or after post-multiplying each member of the previous equation by $-\mathbf{J}^{-1}$

$$(\chi^{\mathrm{T}})^{\mathrm{T}}\mathbf{J}\chi^{\mathrm{T}} = \rho\,\mathbf{J}$$

showing that χ^{T} is an infinitesimal canonical similitude. Accordingly, by replacing $\mathbf{A}$ by $\mathbf{A}^{\mathrm{T}}$, $\mathbf{B}$ by $\mathbf{C}^{\mathrm{T}}$, $\mathbf{C}$ by $\mathbf{B}^{\mathrm{T}}$ and $\mathbf{D}$ by $\mathbf{D}^{\mathrm{T}}$ in (3.4), a second set of conditions is obtained:

$$\left\{\begin{array}{l} \mathbf{A}\mathbf{B}^{\mathrm{T}} - \mathbf{B}\mathbf{A}^{\mathrm{T}} = \mathbf{0} \\ \mathbf{A}\mathbf{D}^{\mathrm{T}} - \mathbf{B}\mathbf{C}^{\mathrm{T}} = \rho\,\mathbf{I}_n \\ \mathbf{C}\mathbf{D}^{\mathrm{T}} - \mathbf{D}\mathbf{C}^{\mathrm{T}} = \mathbf{0} \end{array}\right. \tag{3.5}$$

This set of conditions turns to be equivalent to the first one.

3.4 The symplectic group

If we deal only with linear canonical similitudes (i.e. constant infinitesimal canonical similitude), characterized by a ratio equal to one ($\rho = 1$), we obtain the symplectic group $Sp(n)$. This group is composed of the $2n \times 2n$ matrices satisfying one of the two sets of conditions (3.4) or (3.5). This group acts linearly on $\mathbb{R}^{2n}$ through the following relation:

$$\begin{bmatrix} \mathbf{A} & \mathbf{B} \\ \mathbf{C} & \mathbf{D} \end{bmatrix}\begin{bmatrix} \mathbf{x} \\ \mathbf{y} \end{bmatrix} = \begin{bmatrix} \mathbf{A}\mathbf{x} + \mathbf{B}\mathbf{y} \\ \mathbf{C}\mathbf{x} + \mathbf{D}\mathbf{y} \end{bmatrix}$$

To check that $Sp(n)$ is a group, we need to prove the three following properties.

• The composition of two constant infinitesimal canonical similitudes χ and χ' is a constant infinitesimal canonical similitude:

$$(\chi\chi')^{\mathrm{T}}\mathbf{J}(\chi\chi') = \chi'^{\mathrm{T}}(\chi^{\mathrm{T}}\mathbf{J}\chi)\chi' = \chi'^{\mathrm{T}}\mathbf{J}\chi' = \mathbf{J}$$

• The neutral element is a constant infinitesimal canonical similitude: obviously $\mathbf{I}_{2n}$ is a canonical transformation.
• The inverse of an infinitesimal canonical similitude is a infinitesimal canonical similitude.
By taking the determinant of both sides of the equality $\chi^{\mathrm{T}}\mathbf{J}\chi = \mathbf{J}$, we obtain

$$\det(\chi^{\mathrm{T}})\det(\mathbf{J})\det(\chi) = \det(\mathbf{J})$$

or

$$(\det(\chi))^2 = 1$$

showing that the determinant of χ is non-null and therefore χ is invertible. We already know that the transpose of a canonical similitude is a canonical similitude of same ratio, i.e.

$$\chi^{\mathrm{T}}\mathbf{J}\chi = \mathbf{J} \Rightarrow \chi\mathbf{J}\chi^{\mathrm{T}} = \mathbf{J}$$

By taking the inverse of the last equality, we obtain

$$(\chi^{-1})^{\mathrm{T}}\mathbf{J}^{-1}\chi^{-1} = \mathbf{J}^{-1} \quad \text{or} \quad (\chi^{\mathrm{T}})^{-1}\mathbf{J}\chi^{-1} = \mathbf{J}$$

proving that χ^{-1} is also a canonical similitude. Consequently, $Sp(n)$ is a group.

3.5 Properties of canonical similitudes

3.5.1 Transport of Lagrangian submanifolds. The vanishing of the Lagrange form is obviously preserved after a canonical similitude. If $\mathbf{x}$ and $\mathbf{y}$ describe a lagrangian submanifold, the new variables $\mathbf{X}$ and $\mathbf{Y}$, resulting from $\mathbf{x}$ and $\mathbf{y}$ after a canonical similitude, also describe a lagrangian submanifold.

3.5.2 Transport of the potentiality. If variables $\mathbf{x}$ and $\mathbf{y}$ are related by a constitutive law deriving from a potential ϕ, they describe the lagrangian submanifold defined by the equation

$$\mathbf{y} = grad\,\phi(\mathbf{x}) \tag{3.6}$$

After a canonical similitude, the new variables $\mathbf{X}$ and $\mathbf{Y}$ describe, in turn, a lagrangian submanifold, which can be represented by the equation

$$\mathbf{Y} = grad\,\Psi(\mathbf{X}) \tag{3.7}$$

where Ψ is a new potential. Without the geometric interpretation of the condition ensuring the existence of a potential, long developments are needed to obtain this result.

3.6 Particular canonical similitudes

3.6.1 The Legendre transformation. The Legendre transformation [7] allows inverting the relationship between $\mathbf{x}$ and $\mathbf{y}$ and can be viewed as a change of variable:

$$\begin{cases} \mathbf{X} = \mathbf{y} \\ \mathbf{Y} = \mathbf{x} \end{cases}$$

As we already noticed, the swap of the 2 variables is an anti-canonical transformation of ratio equal to -1. If $\mathbf{y}$ is related to $\mathbf{x}$ by means of a potential ϕ

$$\mathbf{y} = grad\, \phi(\mathbf{x}),$$

Legendre showed that $\mathbf{x}$ is, in turn, related to $\mathbf{y}$ through a potential ϕ^* such that

$$\mathbf{x} = grad\, \phi^*(\mathbf{y})$$

The potential $\phi^*(\mathbf{y})$ is equal to

$$\langle \mathbf{x}, grad\, \phi(\mathbf{x}) \rangle - \phi(\mathbf{x}),$$

and its expression in terms of $\mathbf{y}$ is

$$\langle \mathcal{C}^{-1}\mathbf{y}, grad\, \phi(\mathcal{C}^{-1}\mathbf{y}) \rangle - \phi(\mathcal{C}^{-1}\mathbf{y})$$

3.6.2 The Legendre-Fenchel transformation. Another transformation that swap the variables is given by Fenchel [2, 6]. However, this transformation is going a step further by characterizing the derivative of the potential obtained after transformation. It is known that:
- when ϕ is convex, ϕ^* is also convex and can be obtained by a maximization procedure using the closed formula

$$\phi^*(\mathbf{y}) = \sup_{\mathbf{x}} \left[\langle \mathbf{x}, \mathbf{y} \rangle - \phi(\mathbf{x}) \right]$$

- when ϕ is concave, ϕ^* is also concave and can be obtained by a minimization procedure using the closed formula

$$\phi^*(\mathbf{y}) = \inf_{\mathbf{x}} \left[\langle \mathbf{x}, \mathbf{y} \rangle - \phi(\mathbf{x}) \right]$$

In other words, convexity and concavity are transported by these Legendre-Fenchel transformations. We can remark that, with these definitions, there is no need for the potential to be differentiable.

3.7 Generalized Legendre-Fenchel transformation

To every canonical similitude,

$$\left[\begin{array}{c}\mathbf{X}\\ \mathbf{Y}\end{array}\right]=\mathcal{F}\left(\left[\begin{array}{c}\mathbf{x}\\ \mathbf{y}\end{array}\right]\right)$$

thanks to the equivalence

$$\mathbf{y}=grad\,\phi(\mathbf{x})\Leftrightarrow\mathbf{Y}=grad\,\Psi(\mathbf{X})$$

we can associate a generalized Legendre transformation which transforms a potential ϕ into a new potential Ψ. The question is now what canonical similitudes are associated with a generalized Legendre transformation, which transport the convexity/concavity or swap the convexity with the concavity and how to characterize them. It is the goal of this paper to answer this question.

4. Transport of the second derivative of a potential

The aim of this section is to establish a relationship between the second derivative of the original potential and the second derivative of the potential obtained after transformation.

4.1 Transport of vectorial tangent spaces

If $\left[\begin{array}{c}\mathrm{d}\mathbf{x}\\ \mathrm{d}\mathbf{y}\end{array}\right]$ is tangent at $\left[\begin{array}{c}\mathbf{x}\\ \mathbf{y}\end{array}\right]$ to the lagrangian submanifold defined by the equation

$$\mathbf{y}=grad\,\phi(\mathbf{x})$$

and if $\left[\begin{array}{c}\mathbf{X}\\ \mathbf{Y}\end{array}\right]=\mathcal{F}\left(\left[\begin{array}{c}\mathbf{x}\\ \mathbf{y}\end{array}\right]\right)$ is a canonical similitude, then $\left[\begin{array}{c}\mathrm{d}\mathbf{X}\\ \mathrm{d}\mathbf{Y}\end{array}\right]$ is tangent at $\left[\begin{array}{c}\mathbf{X}\\ \mathbf{Y}\end{array}\right]$ to the submanifold defined by the equation

$$\mathbf{Y}=grad\,\Psi(\mathbf{X})$$

which is also lagrangian. In other words,

$$\mathrm{d}\mathbf{y}=D(grad\,\phi)(\mathbf{x})\mathrm{d}\mathbf{x}\Rightarrow\mathrm{d}\mathbf{Y}=D(grad\,\Psi)(\mathbf{X})\mathrm{d}\mathbf{X}$$

which can be written as

$$\left[D(grad\,\phi)(\mathbf{x})\,\vdots\,-\mathbf{I}_n\right]\left[\begin{array}{c}\mathrm{d}\mathbf{x}\\ \mathrm{d}\mathbf{y}\end{array}\right]=0\Rightarrow\left[D(grad\,\Psi)(\mathbf{X})\,\vdots\,-\mathbf{I}_n\right]\left[\begin{array}{c}\mathrm{d}\mathbf{X}\\ \mathrm{d}\mathbf{Y}\end{array}\right]=0$$

The previous condition can be stated in another way: any vector lying in the nullspace of the kernel of the $n \times 2n$ matrix

$$\left[D(grad\,\phi)(\mathbf{x}) \vdots -\mathbf{I}_n\right]$$

belongs to the kernel of the $n \times 2n$ matrix

$$\left[D(grad\,\Psi)(\mathbf{X}) \vdots -\mathbf{I}_n\right]\left[\begin{array}{cc} D_x\mathcal{X} & D_y\mathcal{X} \\ D_x\mathcal{Y} & D_y\mathcal{Y} \end{array}\right]$$

Let us denote respectively by $\boldsymbol{s}$ and $\boldsymbol{S}$ the second derivatives of the functions ϕ and Ψ, respectively. The Lagrange multipliers technique claims the existence of a $n \times n$ matrix Λ such that

$$[\boldsymbol{S} - \mathbf{I}_n]\left[\begin{array}{cc} \mathbf{A} & \mathbf{B} \\ \mathbf{C} & \mathbf{D} \end{array}\right] = \Lambda\,[\boldsymbol{s} - \mathbf{I}_n]\,,$$

that is to say

$$\left\{\begin{array}{l} \boldsymbol{S}\mathbf{A} - \mathbf{C} = \Lambda \boldsymbol{s} \\ \boldsymbol{S}\mathbf{B} - \mathbf{D} = -\Lambda \end{array}\right. \tag{4.1}$$

4.2 Relationship between the second derivatives of the potentials

Theorem 1. The second derivative $\boldsymbol{S}$ of the transformed potential can be expressed in terms of the second derivative $\boldsymbol{s}$ of the original potential through

$$\boldsymbol{S} = \left(\boldsymbol{s}\mathbf{B}^{\mathrm{T}} + \mathbf{A}^{\mathrm{T}}\right)^{-1}\left(\boldsymbol{s}\mathbf{D}^{\mathrm{T}} + \mathbf{C}^{\mathrm{T}}\right) \tag{4.2}$$

Proof: Let us consider the system (4.1) and post-multiply each member of the first equation by $\mathbf{D}^{\mathrm{T}}$ and each member of the second equation by $\mathbf{C}^{\mathrm{T}}$. After that we substract both equations and obtain

$$\boldsymbol{S}\left(\mathbf{A}\mathbf{D}^{\mathrm{T}} - \mathbf{B}\mathbf{C}^{\mathrm{T}}\right) + \left(\mathbf{D}\mathbf{C}^{\mathrm{T}} - \mathbf{C}\mathbf{D}^{\mathrm{T}}\right) = \Lambda\left(\boldsymbol{s}\mathbf{D}^{\mathrm{T}} + \mathbf{C}^{\mathrm{T}}\right)$$

Using the second set of conditions (3.5), we obtain

$$\rho \boldsymbol{S} = \Lambda\left(\boldsymbol{s}\mathbf{D}^{\mathrm{T}} + \mathbf{C}^{\mathrm{T}}\right) \tag{4.3}$$

Let us post-multiply each member of the first equation in (4.1) by $\mathbf{B}^{\mathrm{T}}$, each member of the second by $\mathbf{A}^{\mathrm{T}}$ and subtract both equations:

$$\boldsymbol{S}\left(\mathbf{A}\mathbf{B}^{\mathrm{T}} - \mathbf{B}\mathbf{A}^{\mathrm{T}}\right) + \left(\mathbf{D}\mathbf{A}^{\mathrm{T}} - \mathbf{C}\mathbf{B}^{\mathrm{T}}\right) = \Lambda\left(\boldsymbol{s}\mathbf{B}^{\mathrm{T}} + \mathbf{A}^{\mathrm{T}}\right)$$

Using again the second set of conditions (3.5), we have

$$\rho \mathbf{I}_n = \Lambda \left(s\,\mathbf{B}^{\mathrm{T}} + \mathbf{A}^{\mathrm{T}} \right),$$

which proves both Λ and $\left(s\,\mathbf{B}^{\mathrm{T}} + \mathbf{A}^{\mathrm{T}} \right)$ are invertible. Accordingly,

$$\Lambda = \rho \left(s\,\mathbf{B}^{\mathrm{T}} + \mathbf{A}^{\mathrm{T}} \right)^{-1} \tag{4.4}$$

that achieves the proof by putting together both conditions (4.3) and (4.4).

4.3 Action of the $Sp(n)$ group on $n \times n$ matrix

We can view the formula (4.2) as the action $\underline{\chi}$ of the element χ on the symmetric matrix s:

$$\underline{\chi}(s) = \left(s\,\mathbf{B}^{\mathrm{T}} + \mathbf{A}^{\mathrm{T}} \right)^{-1} \left(s\,\mathbf{D}^{\mathrm{T}} + \mathbf{C}^{\mathrm{T}} \right)$$

This action is well-known [4] when χ is an element of $Sp(n)$. However, it is not so straightforward to prove that for two elements χ and χ' of $Sp(n)$, we have

$$\underline{\chi'} \left(\underline{\chi}(s) \right) = \underline{\chi' \chi}(s)$$

In our presentation, this result is established by observing that if we compose two canonical similitudes $\mathcal{F}$ and $\mathcal{F}'$ of ratio equal to 1, then a potential ϕ is transformed into Ψ by $\mathcal{F}$ which is transformed to θ by $\mathcal{F}'$. It is clear that ϕ is transformed into θ directly by $\mathcal{F}' \circ \mathcal{F}$. Therefore the second derivative of θ is related to the second derivative of ϕ by $\underline{\chi' \chi}$ or by $\underline{\chi'} \circ \underline{\chi}$.

4.4 Examples

4.4.1. First example. Swapping $\mathbf{x}$ by $\mathbf{y}$ defines an anti-canonical transformation, the derivative of which is

$$\begin{bmatrix} \mathbf{A} & \mathbf{B} \\ \mathbf{C} & \mathbf{D} \end{bmatrix} = \begin{bmatrix} \mathbf{0} & \mathbf{I}_n \\ \mathbf{I}_n & \mathbf{0} \end{bmatrix}$$

As a result, we have $S = s^{-1}$. This result is well known by users of the Legendre-Fenchel transformation. It can be seen that the convexity is maintained

$$S \geq \mathbf{0} \Leftrightarrow s^{-1} \geq \mathbf{0} \Leftrightarrow s \geq \mathbf{0}$$

The second derivative $D(grad\,\Psi)(\mathbf{X})$ is positive if and only if the second derivative $D(grad\,\phi)(\mathbf{x})$ is positive.

4.4.2. Second example. If $\mathbf{B}$ and $\mathbf{C}$ are null matrices, the condition satisfied by $\mathbf{A}$ and $\mathbf{D}$ such that $\begin{bmatrix} \mathrm{A} & \mathbf{0} \\ \mathbf{0} & \mathbf{D} \end{bmatrix}$ is an infinitesimal canonical similitude is

$$\mathbf{A}\,\mathbf{D}^{\mathrm{T}} = \rho\,\mathbf{I}_n \qquad \text{or} \qquad \mathbf{D}\,\mathbf{A}^{\mathrm{T}} = \rho\,\mathbf{I}_n$$

Under these conditions,

$$\boldsymbol{S} = (\mathbf{A}^{\mathrm{T}})^{-1}\,\boldsymbol{s}\,\mathbf{D}^{\mathrm{T}}$$

Since the inverse of $\mathbf{A}^{\mathrm{T}}$ is $\dfrac{1}{\rho}\mathbf{D}$,we have

$$\boldsymbol{S} = \frac{1}{\rho}\,\mathbf{D}\,\boldsymbol{s}\,\mathbf{D}^{\mathrm{T}}$$

The convexity and the concavity are conserved after transformation if ρ is positive while they are exchanged if ρ is negative.

4.4.3. Third example. If $\mathbf{A}$ and $\mathbf{D}$ are null matrices, the condition satisfied by $\mathbf{B}$ and $\mathbf{C}$ such that $\begin{bmatrix} \mathbf{0} & \mathrm{B} \\ \mathrm{C} & \mathbf{0} \end{bmatrix}$ is an infinitesimal canonical similitude is

$$\mathbf{B}\,\mathbf{C}^{\mathrm{T}} = -\rho\,\mathbf{I}_n \quad \text{or} \quad \mathbf{C}\,\mathbf{B}^{\mathrm{T}} = -\rho\,\mathbf{I}_n$$

Under these conditions,

$$\boldsymbol{S} = (\mathbf{B}^{\mathrm{T}})^{-1}\,\boldsymbol{s}^{-1}\,\mathbf{C}^{\mathrm{T}}$$

Since the inverse of $\mathbf{B}^{\mathrm{T}}$ is $-\dfrac{1}{\rho}\mathbf{C}$,we have

$$\boldsymbol{S} = \frac{1}{\rho}\,\mathbf{C}\,\boldsymbol{s}^{-1}\,\mathbf{C}^{\mathrm{T}}$$

The Fenchel-Legendre-Moreau transformation corresponds to

$$\rho = -1, \quad \mathbf{B} = \mathbf{I}_n, \quad \mathbf{C} = \mathbf{I}_n$$

The convexity and the concavity are conserved after transformation if ρ is positive while they are exchanged if ρ is negative.

4.5 Self-consistence of the characterization

The relation (4.2) provides the expression of the second derivative of the new potential in terms of the second derivative of the initial potential and the transformation. The aim of this subsection is to show that the formula (4.2) is self-consistent. In others words, we shall prove that in the relation (4.2) $\boldsymbol{S}$ and $\boldsymbol{s}$ are simultaneously symmetric if and only if the transformation χ is an infinitesimal canonical similitude. To prove that, let us forget that $\boldsymbol{s}$ and $\boldsymbol{S}$ are symmetric. The matrix $\boldsymbol{S}$ will be symmetric if

$$\boldsymbol{S} = \left(\boldsymbol{s}\mathbf{B}^{\mathrm{T}} + \mathbf{A}^{\mathrm{T}}\right)^{-1}\left(\boldsymbol{s}\mathbf{D}^{\mathrm{T}} + \mathbf{C}^{\mathrm{T}}\right) = \left(\mathbf{D}\boldsymbol{s}^{\mathrm{T}} + \mathbf{C}\right)\left(\mathbf{B}\boldsymbol{s}^{\mathrm{T}} + \mathbf{A}\right)^{-1} = \boldsymbol{S}^{\mathrm{T}}$$

After eliminating the inverse matrices by pre- and post-porducts, we obtain

$$\left(\boldsymbol{s}\mathbf{D}^{\mathrm{T}} + \mathbf{C}^{\mathrm{T}}\right)\left(\mathbf{B}\boldsymbol{s}^{\mathrm{T}} + \mathbf{A}\right) = \left(\boldsymbol{s}\mathbf{B}^{\mathrm{T}} + \mathbf{A}^{\mathrm{T}}\right)\left(\mathbf{D}\boldsymbol{s}^{\mathrm{T}} + \mathbf{C}\right)$$

or, equivalently

$$\boldsymbol{s}\left(\mathbf{D}^{\mathrm{T}}\mathbf{B} - \mathbf{B}^{\mathrm{T}}\mathbf{D}\right)\boldsymbol{s}^{\mathrm{T}} + \boldsymbol{s}\left(\mathbf{D}^{\mathrm{T}}\mathbf{A} - \mathbf{B}^{\mathrm{T}}\mathbf{C}\right) + \left(\mathbf{C}^{\mathrm{T}}\mathbf{B} - \mathbf{A}^{\mathrm{T}}\mathbf{D}\right)\boldsymbol{s}^{\mathrm{T}} + \left(\mathbf{C}^{\mathrm{T}}\mathbf{A} - \mathbf{A}^{\mathrm{T}}\mathbf{C}\right) = \mathbf{0}$$

Now, we will show that this equality holds for every symmetric matrix $\boldsymbol{s}$ if and only if the transformation χ is an infinitesimal canonical similitude.

Proposition 1. The operation defines by (4.2) transforms a symmetric $n \times n$ matrix $\boldsymbol{s}$ into a symmetric matrix $\boldsymbol{S}$ if and only if χ is an infinitesimal canonical similitude.

Proof : Let us change $\boldsymbol{s}$ by $\mu\boldsymbol{s}$ for some scalar μ, then the above equality becomes

$$\mu^2\,\boldsymbol{s}\left(\mathbf{D}^{\mathrm{T}}\mathbf{B} - \mathbf{B}^{\mathrm{T}}\mathbf{D}\right)\boldsymbol{s} + \mu\,\boldsymbol{s}\left(\mathbf{D}^{\mathrm{T}}\mathbf{A} - \mathbf{B}^{\mathrm{T}}\mathbf{C}\right) + \mu\left(\mathbf{C}^{\mathrm{T}}\mathbf{B} - \mathbf{A}^{\mathrm{T}}\mathbf{D}\right)\boldsymbol{s} + \left(\mathbf{C}^{\mathrm{T}}\mathbf{A} - \mathbf{A}^{\mathrm{T}}\mathbf{C}\right) = \mathbf{0}$$

The last equality holds for every μ only if

(*i*) $\mathbf{C}^{\mathrm{T}}\mathbf{A} - \mathbf{A}^{\mathrm{T}}\mathbf{C} = \mathbf{0}$

(*ii*) $\forall \boldsymbol{s},\ \boldsymbol{s}\left(\mathbf{D}^{\mathrm{T}}\mathbf{B} - \mathbf{B}^{\mathrm{T}}\mathbf{D}\right)\boldsymbol{s} = \mathbf{0}$

(*iii*) $\forall \boldsymbol{s},\ \boldsymbol{s}\left(\mathbf{D}^{\mathrm{T}}\mathbf{A} - \mathbf{B}^{\mathrm{T}}\mathbf{C}\right) = \left(\mathbf{D}^{\mathrm{T}}\mathbf{A} - \mathbf{B}^{\mathrm{T}}\mathbf{C}\right)^{\mathrm{T}}\boldsymbol{s}$

These three conditions must be fulfilled. The condition (*ii*) leads, by choosing $\boldsymbol{s} = \mathbf{I}_n$, to

$$\mathbf{D}^{\mathrm{T}}\mathbf{B} - \mathbf{B}^{\mathrm{T}}\mathbf{D} = \mathbf{0}$$

By choosing also $\boldsymbol{s} = \mathbf{I}_n$, the condition (*iii*) shows that the $n \times n$ matrix $\mathbf{E}$, defined by

$$\mathbf{E} = \mathbf{D}^{\mathrm{T}}\mathbf{A} - \mathbf{B}^{\mathrm{T}}\mathbf{C}\,,$$

is symmetric. Moreover, for every symmetric matrix $\boldsymbol{s}$

$$\boldsymbol{s}\,\mathbf{E} = \mathbf{E}\,\boldsymbol{s}$$

Let us choose $\boldsymbol{s} = \mathbf{u}\,\mathbf{u}^{\mathrm{T}}$ where $\mathbf{u}$ is a unit vector, then

$$\mathbf{u}\,(\mathbf{E}\,\mathbf{u})^{\mathrm{T}} = (\mathbf{E}\,\mathbf{u})\,\mathbf{u}^{\mathrm{T}}$$

Let us apply this identity to the vector $\mathbf{u}$ itself, then

$$\langle\, \mathbf{E}\,\mathbf{u}, \mathbf{u}\rangle\, \mathbf{u} = \mathbf{u}\left[(\mathbf{E}\,\mathbf{u})^{\mathrm{T}}\,\mathbf{u}\right] = \mathbf{E}\,\mathbf{u} \tag{4.5}$$

The dot product between brackets is a scalar and therefore the relation (4.5) shows that every vector $\mathbf{u}$ is an eigenvector of $\mathbf{E}$. Consequently, there exists a scalar ρ such that

$$\mathbf{E} = \rho\,\mathbf{I}_n$$

In others words,

$$\mathbf{A}^{\mathrm{T}}\mathbf{D} - \mathbf{C}^{\mathrm{T}}\mathbf{B} = \rho\,\mathbf{I}_n$$

Finally, we recover the first set of conditions and χ is an infinitesimal canonical similitude.

5. Characterization of canonical similitudes transporting convexity

Let us suppose that $\boldsymbol{s}$ is positive, we are looking for the conditions on $\mathbf{A}$, $\mathbf{D}$, $\mathbf{C}$ and $\mathbf{D}$ ensuring that the matrix $\boldsymbol{S}$ defined by (4.2) is also positive. To find these conditions, let us examine the sign of

$$\langle \mathbf{V}, \boldsymbol{S}\mathbf{V}\rangle = \langle (\mathbf{B}\,\boldsymbol{s} + \mathbf{A})^{-1}\,\mathbf{V}, \left(\boldsymbol{s}\,\mathbf{D}^{\mathrm{T}} + \mathbf{C}^{\mathrm{T}}\right)\mathbf{V}\rangle$$

For convenience, the following change of variables is made:

$$\mathbf{W} = (\mathbf{B}\,\boldsymbol{s} + \mathbf{A})^{-1}\,\mathbf{V}$$

It is now equivalent to study the sign of

$$\langle \mathbf{W}, \left(\boldsymbol{s}\,\mathbf{D}^{\mathrm{T}} + \mathbf{C}^{\mathrm{T}}\right)(\mathbf{B}\,\boldsymbol{s} + \mathbf{A})\,\mathbf{W}\rangle$$

By developing the previous relation, the problem reduces to study the positiveness of the matrix $\mathbf{Q}$

$$\mathbf{Q} = \boldsymbol{s}\,\mathbf{D}^{\mathrm{T}}\mathbf{B}\,\boldsymbol{s} + \mathbf{C}^{\mathrm{T}}\mathbf{B}\,\boldsymbol{s} + \boldsymbol{s}\,\mathbf{D}^{\mathrm{T}}\mathbf{A} + \mathbf{C}^{\mathrm{T}}\mathbf{A} \tag{5.1}$$

which is symmetric

$$\mathbf{Q}^{\mathrm{T}} = \boldsymbol{s}\,\mathbf{B}^{\mathrm{T}}\mathbf{D}\,\boldsymbol{s} + \boldsymbol{s}\,\mathbf{B}^{\mathrm{T}}\mathbf{C} + \mathbf{A}^{\mathrm{T}}\mathbf{D}\,\boldsymbol{s} + \mathbf{A}^{\mathrm{T}}\mathbf{C}$$

thanks to the first characterization. In fact, considering the properties of the canonical similitudes, we obtain

$$\mathbf{Q}^{\mathrm{T}} = \boldsymbol{s}\,\mathbf{D}^{\mathrm{T}}\mathbf{B}\,\boldsymbol{s} + \boldsymbol{s}\left(\mathbf{D}^{\mathrm{T}}\mathbf{A} - \rho\,\mathbf{I}_n\right) + \left(\mathbf{C}^{\mathrm{T}}\mathbf{B} + \rho\,\mathbf{I}_n\right)\boldsymbol{s} + \mathbf{C}^{\mathrm{T}}\mathbf{A} = \mathbf{Q}$$

Let us introduce the following matrices

$$\boldsymbol{\alpha} = \mathbf{D}^{\mathrm{T}}\mathbf{B}, \quad \boldsymbol{\gamma} = \mathbf{C}^{\mathrm{T}}\mathbf{A}, \quad \boldsymbol{\beta} = \mathbf{C}^{\mathrm{T}}\mathbf{B} + \frac{\rho}{2}\,\mathbf{I}_n \tag{5.2}$$

where $\boldsymbol{\alpha}$ and $\boldsymbol{\gamma}$ are symmetric. With the notations (5.2), the matrix $\mathbf{Q}$ becomes

$$\mathbf{Q} = \boldsymbol{s}\,\boldsymbol{\alpha}\,\boldsymbol{s} + \boldsymbol{\beta}\,\boldsymbol{s} + \boldsymbol{s}\,\boldsymbol{\beta}^{\mathrm{T}} + \boldsymbol{\gamma} \tag{5.3}$$

To derive the conditions ensuring the positiveness of the matrix $\mathbf{Q}$, we need to study the sign of this second order trinomial matrix equation (5.3) for positive matrix $\boldsymbol{s}$. Let us start with some preliminary lemmas.

Lemma 1. The following identities hold:

$$\boldsymbol{\gamma}\,\boldsymbol{\alpha} = \boldsymbol{\beta}^2 - \frac{\rho^2}{4}\mathbf{I}_n \quad \text{and} \quad \boldsymbol{\alpha}\,\boldsymbol{\gamma} = \left(\boldsymbol{\beta}^{\mathrm{T}}\right)^2 - \frac{\rho^2}{4}\mathbf{I}_n \tag{5.4}$$

Proof: By developing the matrix product of $\boldsymbol{\gamma}$ by $\boldsymbol{\alpha}$, we obtain

$$\boldsymbol{\gamma}\,\boldsymbol{\alpha} = \mathbf{C}^{\mathrm{T}}\left(\mathbf{A}\mathbf{D}^{\mathrm{T}}\right)\mathbf{B}$$

Using the first characterization (3.4), we have

$$\mathbf{C}^{\mathrm{T}}\left(\mathbf{A}\mathbf{D}^{\mathrm{T}}\right)\mathbf{B} = \left(\mathbf{C}^{\mathrm{T}}\mathbf{B}\right)\left(\mathbf{C}^{\mathrm{T}}\mathbf{B}\right) + \rho\left(\mathbf{C}^{\mathrm{T}}\mathbf{B}\right) = \left(\mathbf{C}^{\mathrm{T}}\mathbf{B} + \frac{\rho}{2}\mathbf{I}_n\right)^2 - \frac{\rho^2}{4}\mathbf{I}_n$$

Using the definition of $\boldsymbol{\beta}$ (5.2b), we obtain

$$\boldsymbol{\gamma}\,\boldsymbol{\alpha} = \boldsymbol{\beta}^2 - \frac{\rho^2}{4}\mathbf{I}_n$$

and after transposition

$$\boldsymbol{\alpha}\,\boldsymbol{\gamma} = \left(\boldsymbol{\beta}^{\mathrm{T}}\right)^2 - \frac{\rho^2}{4}\mathbf{I}_n$$

Lemma 2. $\boldsymbol{\alpha}$ and $\boldsymbol{\beta}$ are related by

$$\boldsymbol{\alpha}\,\boldsymbol{\beta} = \boldsymbol{\beta}^{\mathrm{T}}\boldsymbol{\alpha} \tag{5.5}$$

Proof: Let us develop the matrix product $\boldsymbol{\alpha}\,\boldsymbol{\beta}$ using definitions (5.2):

$$\boldsymbol{\alpha}\,\boldsymbol{\beta} = \mathbf{D}^{\mathrm{T}}\left(\mathbf{B}\mathbf{C}^{\mathrm{T}}\right)\mathbf{B} + \frac{\rho}{2}\,\mathbf{D}^{\mathrm{T}}\mathbf{B} \tag{5.6}$$

Using the second characterization (3.5)

$$\mathbf{B}\mathbf{C}^{\mathrm{T}} = \mathbf{A}\mathbf{D}^{\mathrm{T}} - \rho\,\mathbf{I}_n,$$

the second member of (5.6) can be transformed in

$$(\mathbf{D}^{\mathrm{T}}\mathbf{A})\,(\mathbf{D}^{\mathrm{T}}\mathbf{B}) - \frac{\rho}{2}\,\mathbf{D}^{\mathrm{T}}\mathbf{B} \tag{5.7}$$

¿From the first characterization (3.4), we know that

$$\mathbf{D}^{\mathrm{T}}\mathbf{A} = \mathbf{B}^{\mathrm{T}}\mathbf{C} + \rho\,\mathbf{I}_n \tag{5.8}$$

combining (5.7) and (5.8), we find that

$$\left(\mathbf{B}^{\mathrm{T}}\mathbf{C} + \frac{\rho}{2}\,\mathbf{I}_n\right)\mathbf{D}^{\mathrm{T}}\mathbf{B}$$

which is equal to $\boldsymbol{\beta}^{\mathrm{T}}\boldsymbol{\alpha}$.

Lemma 3. $\boldsymbol{\gamma}$ and $\boldsymbol{\beta}$ are related by

$$\boldsymbol{\gamma}\,\boldsymbol{\beta}^{\mathrm{T}} = \boldsymbol{\beta}\,\boldsymbol{\gamma}$$

Proof: By developing the matrix product $\boldsymbol{\gamma}\,\boldsymbol{\beta}^{\mathrm{T}}$ using definitions (5.2), we get

$$\boldsymbol{\gamma}\,\boldsymbol{\beta}^{\mathrm{T}} = \mathbf{A}^{\mathrm{T}}\,(\mathbf{C}\mathbf{B}^{\mathrm{T}})\,\mathbf{C} + \frac{\rho}{2}\,\mathbf{A}^{\mathrm{T}}\mathbf{C} \tag{5.9}$$

The second characterization gives

$$\mathbf{C}\mathbf{B}^{\mathrm{T}} = \mathbf{D}\mathbf{A}^{\mathrm{T}} - \rho\,\mathbf{I}_n,$$

which, combined with (5.9), leads to

$$\boldsymbol{\gamma}\,\boldsymbol{\beta}^{\mathrm{T}} = \mathbf{A}^{\mathrm{T}}\left(\mathbf{D}\mathbf{A}^{\mathrm{T}} - \frac{\rho}{2}\,\mathbf{I}_n\right)\mathbf{C}$$

By developing the matrix product and replacing $\mathbf{A}^{\mathrm{T}}\mathbf{D}$ by $\mathbf{C}^{\mathrm{T}}\mathbf{B} + \rho\,\mathbf{I}_n$ according to the first characterization, we get

$$\boldsymbol{\gamma}\,\boldsymbol{\beta}^{\mathrm{T}} = \left(\mathbf{C}^{\mathrm{T}}\mathbf{B} + \frac{\rho}{2}\,\mathbf{I}_n\right)\mathbf{A}^{\mathrm{T}}\mathbf{C}$$

which is by definition equal to $\boldsymbol{\beta}\,\boldsymbol{\gamma}$.

After proving these lemmas, let us come back to the relation (5.3). By simply considering $\boldsymbol{s}$ and $\boldsymbol{s}^{-1}$ close to the null matrix $\mathbf{0}$, we discover *necessary* conditions ensuring that $\mathbf{Q}$ is definite-positive: the symmetric matrice $\boldsymbol{\alpha}$ and γ should be positive. Let assume that $\boldsymbol{\alpha}$ is definite-positive and let $\boldsymbol{\alpha}^{\frac{1}{2}}$ be its square root. If we pre-multiply and post-multiply both member of (5.3) by $\boldsymbol{\alpha}^{\frac{1}{2}}$, it is equivalent to study the positivitness of

$$\widetilde{\mathbf{Q}} = \widetilde{s}^2 + \widetilde{\boldsymbol{\beta}}\,\widetilde{s} + \widetilde{s}\,\widetilde{\boldsymbol{\beta}}^{\mathrm{T}} + \widetilde{\gamma}$$

where the following change of variables has been made

$$\widetilde{\mathbf{Q}} = \boldsymbol{\alpha}^{\frac{1}{2}}\,\mathbf{Q}\,\boldsymbol{\alpha}^{\frac{1}{2}}, \quad \widetilde{s} = \boldsymbol{\alpha}^{\frac{1}{2}}\,s\,\boldsymbol{\alpha}^{\frac{1}{2}}, \quad \widetilde{\boldsymbol{\beta}} = \boldsymbol{\alpha}^{\frac{1}{2}}\,\boldsymbol{\beta}\,\boldsymbol{\alpha}^{-\frac{1}{2}}, \quad \widetilde{\gamma} = \boldsymbol{\alpha}^{\frac{1}{2}}\,\gamma\,\boldsymbol{\alpha}^{\frac{1}{2}}$$

When the matrices $\boldsymbol{\alpha}$ and γ are symmetric and positive, the matrix $\widetilde{\gamma}$ is also symmetric and positive. Let us recall the results of the lemmas with the new variables:

- **Lemma 1**:

$$\widetilde{\gamma} = \widetilde{\boldsymbol{\beta}}^2 - \frac{\rho^2}{4}\mathbf{I}_n$$

- **Lemma 2**:

$$\widetilde{\boldsymbol{\beta}} = \widetilde{\boldsymbol{\beta}}^{\mathrm{T}}$$

- **Lemma 3**:

$$\widetilde{\gamma}\,\widetilde{\boldsymbol{\beta}} = \widetilde{\boldsymbol{\beta}}\,\widetilde{\gamma}$$

Using the lemmas expressed in the new quantities, the matrix $\widetilde{\mathbf{Q}}$ becomes

$$\widetilde{\mathbf{Q}} = \left(\widetilde{s} + \widetilde{\boldsymbol{\beta}}\right)^2 - \left(\frac{\rho}{2}\right)^2 \mathbf{I}_n$$

The question turns now into: is the matrix $\widetilde{\mathbf{Q}}$ defined by

$$\widetilde{\mathbf{Q}} = \left(\widetilde{s} + \widetilde{\boldsymbol{\beta}} + \frac{\rho}{2}\mathbf{I}_n\right)\left(\widetilde{s} + \widetilde{\boldsymbol{\beta}} - \frac{\rho}{2}\mathbf{I}_n\right) \tag{5.10}$$

positive whenever $\widetilde{s}$ is positive? The answer is affirmative if

$$\widetilde{\boldsymbol{\beta}} \geq \frac{|\rho|}{2}\mathbf{I}_n \tag{5.11}$$

Proof: Assume, as a particular case, that the symmetric matrix $\widetilde{s}$ has the same eigenvectors as the symmetric matrix $\widetilde{\boldsymbol{\beta}}$ and let $\lambda^i_{\widetilde{s}}$ and $\lambda^i_{\widetilde{\beta}}$ be their i^{th} eigenvalues. Then, the condition (5.10) can be written in the eigenvectors base:

$$\forall\, i\ (i = 1, \dots n), \qquad \left(\lambda^i_{\widetilde{s}} + \lambda^i_{\widetilde{\beta}} + \frac{\rho}{2}\right)\left(\lambda^i_{\widetilde{s}} + \lambda^i_{\widetilde{\beta}} - \frac{\rho}{2}\right)$$

For every i the ordinary trinomial must be positive for every $\lambda^i_{\widetilde{s}} > 0$. This is possible only if both roots $-\lambda^i_{\widetilde{\beta}} - \dfrac{\rho}{2}$ and $-\lambda^i_{\widetilde{\beta}} + \dfrac{\rho}{2}$ are negative, which means

$$\lambda^i_{\widetilde{\beta}} \geq \frac{|\rho|}{2} \qquad \forall\, i\ (i = 1, \dots n),$$

or in matrix form,

$$\widetilde{\boldsymbol{\beta}} \geq \frac{|\rho|}{2}\mathbf{I}_n$$

The condition (5.11) has been proven for a particular case and is so far a necessary condition. We need now to show that this condition is also sufficient. The product of the symmetric and positive matrices

$$\left(\widetilde{\boldsymbol{s}} + \widetilde{\boldsymbol{\beta}} + \frac{\rho}{2}\mathbf{I}_n\right) \quad \text{and} \quad \left(\widetilde{\boldsymbol{s}} + \widetilde{\boldsymbol{\beta}} - \frac{\rho}{2}\mathbf{I}_n\right)$$

is a matrix that has all its eigenvalues positives. Therefore, the matrix

$$\widetilde{\mathbf{Q}} = \widetilde{\boldsymbol{s}}^2 + \widetilde{\boldsymbol{\beta}}\,\widetilde{\boldsymbol{s}} + \widetilde{\boldsymbol{s}}\,\widetilde{\boldsymbol{\beta}}^{\mathrm{T}} + \widetilde{\boldsymbol{\gamma}}$$

has positive eigenvalues. Since we already know that the matrix $\mathbf{Q}$ is symmetric, it is now positive. So in the case $\boldsymbol{\alpha} = \mathbf{D}^{\mathrm{T}}\mathbf{B}$ definite-positive, the condition (5.11) is necessary and sufficient.

Conclusion: A canonical similitude $\mathcal{F}$ of ration ρ will preserve the convexity if its derivative $\begin{bmatrix} \mathbf{A} & \mathbf{B} \\ \mathbf{C} & \mathbf{D} \end{bmatrix}$ satisfy

1) $\mathbf{D}^{\mathrm{T}}\mathbf{B} \geq 0$ and $\mathbf{C}^{\mathrm{T}}\mathbf{A} \geq 0$

2) $\mathbf{B}^{\mathrm{T}}(\mathbf{D}\mathbf{C}^{\mathrm{T}})\,\mathbf{B} \geq 0$ if $\rho > 0$ or $\mathbf{D}^{\mathrm{T}}(\mathbf{B}\mathbf{A}^{\mathrm{T}})\,\mathbf{D} \geq 0$ if $\rho < 0$

6. Summary

In this paper, we have shown that canonical similitudes are transformations that preserve the potential structure of a relationship between dual variables. This can be observed after an appropriate formulation of the condition for the existence of a potential using geometrical arguments. Within this particular class of transformation, conditions under which the transformation will preserve the convexity have been derived. These conditions take the form of positiveness conditions on symmetric matrices corresponding to the components of the transformation derivative.

References

[1] Cardin, F. and Spera, M. , On the Internal Work in Generalized Hyperelastic Materials, *Meccanica*, **30** (1995), 727-734.

[2] Fenchel, W. , On conjugate convex functions, Canadian Journal of Mathematics, **1** (1949), 73-77.

[3] Lagrange, Joseph-Louis (1965). *Mécanique Analytique.* Librairie scientifique et technique Albert Blanchard, Paris. (reedition of the original book edited in 1888 by Gaston Darboux)

[4] Mneimné, R. and Testard, F. (1986). *Introduction à la théorie des groupes de Lie Classiques*, Hermann, Paris.

[5] Maslov, V.P. and Samborski, S.N. (1992). *Idempotent Analysis*, American Mathematical Society, Providence, Rhode Island.

[6] Moreau Jean-Jacques (1967). *Séminaire sur les opérations aux dérivées partielles II - Fonctionnelles Convexes*, Collège de France, Paris.

[7] Rockafellar, R.T. and Wets R.J.B. (1998). *Variational Analysis.* Springer-Verlag, Berlin.

[8] Souriau, Jean-Marie (1970). *Structure des Systèmes Dynamiques.* Dunod, Paris.

Chapter 18

A ROBUST VARIATIONAL FORMULATION FOR A ROD SUBJECT TO INEQUALITY CONSTRAINTS

G.H.M. van der Heijden
Centre for Nonlinear Dynamics
University College London
Gower Street
London WC1E 6BT, UK
g.heijden@ucl.ac.uk

Abstract We present a variational formulation of a constrained rod using soft wall contact. Care is taken to formulate a computationally robust method. The method is applied to a simple model for a drill string. Special boundary conditions are considered which allow the results to be compared with known (planar) results in the literature. The agreement is found to be good. An interesting dependence of the post-buckling behaviour on the boundary conditions is found. Finally, some 3D post-buckling solutions are computed as well.

Keywords: Rod mechanics, Cosserat theory, constrained equilibria, quaternions, contact, penalty method, constrained (torsional) buckling, drill string

Introduction

There is significant recent interest in contact problems involving elastic rods undergoing large deflections, both from a theoretical [16] and an applied point of view. Applications include models for supercoiled DNA [5, 17] and twisted textile yarn [9], to name a few. In [11] a general formulation for a rod constrained to lie on a cylinder was presented which underlies the mechanics of plied structures such as DNA molecules and spun textile yarn in the case of two strands winding about a straight line of mutual contact. Localised solutions were studied through dynamical systems techniques by taking advantage of the fact that these localised solutions are well approximated by infinite-length solutions which are homoclinic orbits in a (reduced) dynamical system. It is an instance of the so-called Kirchhoff kinetic analogy [2, 15]: arclength along the rod can be treated as time in a dynamical system. This approach has been successful for a range of studies in structural localisation (see [4] and references therein). The dynamical system may be integrable (typically if some symmetry properties hold), or non-integrable. In the latter case spatial chaos is often found with an infinity of multi-pulse homoclinic solutions (post-buckling solutions), as for instance in [12]. However, real problems in structural mechanics typically deal with a boundary-value problem posed on a finite-length structure, rather than the initial-value formulation of dynamical systems theory.

The work in [11] can be considered as a simple model for a drill string deforming inside a narrow borehole. Such a drill string is known to be in contact with the borehole wall for long stretches of pipe, but occasionally may leave the wall. Indeed, it was shown in [11] that for localised solutions the reaction force changes sign across the localised deformation. At points of such a sign change the force starts to point in the wrong direction and contact can no longer be maintained: the rod lifts off. This lift off is particularly likely to occur at critical loads where the string changes configuration. For instance, a known (but still poorly understood) transition is from a 'snaking' solution (with the string oscillating about the straight configuration) to a helical solution [1, 14]. This transitional behaviour will depend on the precise form of the applied boundary conditions. Indeed, the influence of different boundary conditions on the configuration of the rod is another poorly understood phenomenon of significant interest to the drilling industry.

In this paper we investigate the problem of inequality constraints for finite-length rods using a numerical approach based on penalties. We apply it to a simple idealised drill string model with as little distraction by non-essential details as possible. (More inclusive models have been studied by Tucker & Wang [18].) This approach allows us to use powerful numerical continuation and bifurcation techniques from dynamical systems theory. Since for this simplified problem the buckling loads can be obtained exactly, we have a check on the performance of the proposed computational scheme. Later, more complicated models will be considered.

The method is more robust than methods based on (multiple) shooting (e.g., [8]), a feature which is particularly felt for 3D deformations. Shooting methods have the disadvantage that one has to make assumptions on what is going on inside the constraining surface: one has to assume the number of contact points. Therefore it is not straightforward to accommodate the creation or destruction of point contacts. Our method can also easily be generalised to more general types of rod (extensible, shearable, etc.). One interesting extension would be to different wall geometries. This would provide a model for drill strings in curved boreholes (as are often used these days in strongly inclined, sometimes even near-horizontal, drilling) or perhaps surgical wires in human arteries or veins during angioplasty.

1. A soft-contact formulation

1.1 Equilibrium equations

In [11, 12] an equality cylindrical constraint was considered. Choosing special (cylindrical) co-ordinates adapted to the surface in fact allowed the constraint to be eliminated entirely. The resulting model gives an idealised description of long stretches of drill string inside a narrow borehole (especially for sections of a helically deformed structure). For longer lengths of pipe the drill string may have segments that are in contact and segments that are free. For each of these segments individually a reduced description can be used but one would have to patch together solutions at points of contact. This is rather tedious and also requires knowledge of the number of contact points. Therefore, rather than seeking a reduced description by using integrals of motion and a local co-ordinate system adapted to the type of surface considered, we consider a penalty (or soft-contact) formulation to model contact. The advantages of this wider approach are its generality (we do not rely on special simple geometries) and its robustness in numerical computations. The cylinder will be imagined vertical, so that to a good approximation gravity can be ignored.

We follow Maddocks and co-workers [6] and employ a constrained variational approach based on Cosserat theory [2]. This means that we view a rod as a mathematical line endowed with elastic properties. The full configuration of the rod is described by a smooth vector function $\boldsymbol{r}(s)$ representing the rod's centreline, and a local orthonormal frame $\{\boldsymbol{d}_1(s), \boldsymbol{d}_2(s), \boldsymbol{d}_3(s)\}$ of so-called directors serving to locate the varying orientation in space of the rod's cross-section. Here s measures arclength in the undeformed rod. For simplicity we shall consider an inextensible and unshearable rod so that for $\boldsymbol{d}_3$ we can take the local tangent to the rod, i.e.,

$$\boldsymbol{r}' = \boldsymbol{d}_3, \tag{1.1}$$

while $\boldsymbol{d}_1$ and $\boldsymbol{d}_2$ are two vectors in the normal cross-section, taken coincident with the principal bending axes.

Next we make the common assumption that the elastic energies depend quadratically on the strains (this amounts to linear constitutive behaviour), so that we can write for the total potential energy of the free rod

$$W(u_i) = \frac{1}{2}\sum_{i=1}^{3} K_i u_i^2. \tag{1.2}$$

Here the u_i $(i = 1, 2, 3)$ are the generalised strains of the theory, which satisfy

$$\boldsymbol{d}_i' = \boldsymbol{u} \times \boldsymbol{d}_i, \qquad u_i = \boldsymbol{u} \cdot \boldsymbol{d}_i, \tag{1.3}$$

and thus can be expressed as

$$u_k = \frac{1}{2}\sum_{i,j=1}^{3} \epsilon_{ijk} \boldsymbol{d}_i' \cdot \boldsymbol{d}_j, \tag{1.4}$$

where ϵ_{ijk} is the completely anti-symmetric symbol. u_1 and u_2 are the curvatures about $\boldsymbol{d}_1$ and $\boldsymbol{d}_2$, respectively, while u_3 is the twist. The K_i are the associated elastic stiffnesses. In writing down (1.2) we have assumed the rod to be straight and prismatic in the stress-free state.

We parametrise the directors $\boldsymbol{d}_i$ (or, equivalently, the group $SO(3)$ of 3×3 proper rotation matrices) relative to a fixed co-ordinate system $\{\boldsymbol{e}_1, \boldsymbol{e}_2, \boldsymbol{e}_3\}$ by means of so-called Euler parameters, or quaternions, i.e., by a quadruple of real numbers $\boldsymbol{q} = (q_1, q_2, q_3, q_4)$ satisfying the norm condition

$$\boldsymbol{q} \cdot \boldsymbol{q} = 1. \tag{1.5}$$

This parametrisation in terms of the (slightly redundant) *four* Euler parameters avoids the polar singularity plaguing parametrisations based on the more common *three* Euler angles. Also, the relevant expressions in terms of Euler parameters become polynomial while Euler angles lead to awkward trigonometric expressions. These two advantages make the Euler parameter formulation numerically much better behaved than one based on Euler angles.

Thus we obtain a mechanical system with seven generalised co-ordinates, $(\boldsymbol{r}, \boldsymbol{q})$, in terms of which we shall now express the potential energy. First we define the components of the rod's centreline relative to the fixed frame as $\boldsymbol{r} = (x, y, z)$, and parametrise the directors in terms of the

Euler parameters as

$$d_1 = \begin{pmatrix} q_1^2 - q_2^2 - q_3^2 + q_4^2 \\ 2(q_1q_2 + q_3q_4) \\ 2(q_1q_3 - q_2q_4) \end{pmatrix}, \tag{1.6}$$

$$d_2 = \begin{pmatrix} 2(q_1q_2 - q_3q_4) \\ -q_1^2 + q_2^2 - q_3^2 + q_4^2 \\ 2(q_1q_4 + q_2q_3) \end{pmatrix}, \tag{1.7}$$

$$d_3 = \begin{pmatrix} 2(q_1q_3 + q_2q_4) \\ 2(q_2q_3 - q_1q_4) \\ -q_1^2 - q_2^2 + q_3^2 + q_4^2 \end{pmatrix}. \tag{1.8}$$

From (1.4) the strains can then be written as

$$u_i = \frac{2\boldsymbol{B}_i\boldsymbol{q}\cdot\boldsymbol{q}'}{|\boldsymbol{q}|^2} \qquad (i = 1, 2, 3), \tag{1.9}$$

where the $\boldsymbol{B}_i$ are 4×4 skew-symmetric matrices satisfying the relations

$$\boldsymbol{B}_1\boldsymbol{q} = (q_4, q_3, -q_2, -q_1), \tag{1.10}$$

$$\boldsymbol{B}_2\boldsymbol{q} = (-q_3, q_4, q_1, -q_2), \tag{1.11}$$

$$\boldsymbol{B}_3\boldsymbol{q} = (q_2, -q_1, q_4, -q_3). \tag{1.12}$$

The skewness of the matrices $\boldsymbol{B}_i$ together with the norm condition (1.5) implies that the four vectors $\{\boldsymbol{q}, \boldsymbol{B}_1\boldsymbol{q}, \boldsymbol{B}_2\boldsymbol{q}, \boldsymbol{B}_3\boldsymbol{q}, \boldsymbol{B}_4\boldsymbol{q}\}$ form an orthonormal basis of $\mathbb{R}^4$, a result that will be used repeatedly in the following calculations.

Statical equilibrium solutions are constrained critical points of the energy functional

$$E[\boldsymbol{r}, \boldsymbol{q}] = \int_0^1 \left[\frac{1}{2}\sum_{i=1}^{3} K_i u_i^2(\boldsymbol{q}'(s), \boldsymbol{q}(s)) + V(\boldsymbol{r}(s))\right] ds,$$

subject to the pointwise constraints $\boldsymbol{r}' = \boldsymbol{d}_3$ and $\boldsymbol{q}\cdot\boldsymbol{q}' = 0$, and appropriate boundary conditions, including the requirement that the Euler parameters have unit norm at one end point. Here we have scaled the length of the rod to 1 and included the potential energy V. We let

$$V(\boldsymbol{r}) = -\frac{a}{g(\boldsymbol{r})}\exp\frac{g(\boldsymbol{r})}{b}, \tag{1.13}$$

where $g(\boldsymbol{r}) = x^2 + y^2 - r^2$ and a and b are small positive constants, in order to model a soft confining cylinder of radius r. (This repulsive potential was used in [10] to avoid self-intersection of deforming rods modelling DNA. Bulman & Manning [3] consider the $1/r$ potential $V(\boldsymbol{r}) = \frac{a}{x+r} - \frac{a}{x-r}$ for buckling between two parallel walls a distance $2r$ apart, but we find the advantage of our choice of potential that it is considerably more localised.)

Equivalently, the equilibrium solutions are critical points of the Lagrangian

$$L(\boldsymbol{r}, \boldsymbol{q}, \boldsymbol{r}', \boldsymbol{q}') = W(u_i) + V(\boldsymbol{r}) + \boldsymbol{\lambda} \cdot (\boldsymbol{r}' - \boldsymbol{d}_3) + \nu \boldsymbol{q} \cdot \boldsymbol{q}', \tag{1.14}$$

to be interpreted as a function of the axis $\boldsymbol{r}$, the quaternion $\boldsymbol{q}$ and their derivatives. $\boldsymbol{\lambda}$ and ν are Lagrange multipliers. The conjugate momenta are

$$\boldsymbol{n} = \frac{\partial L}{\partial \boldsymbol{r}'} = \boldsymbol{\lambda}, \tag{1.15}$$

$$\boldsymbol{\mu} = \frac{\partial L}{\partial \boldsymbol{q}'} = \sum_{i=1}^{3} K_i u_i \frac{2\boldsymbol{B}_i \boldsymbol{q}}{|\boldsymbol{q}|^2} + \nu \boldsymbol{q}. \tag{1.16}$$

With the help of (1.9) the second equation can be inverted to give

$$\boldsymbol{q}' = \sum_{i=1}^{3} \frac{\boldsymbol{\mu} \cdot \boldsymbol{B}_i \boldsymbol{q}}{4K_i} \boldsymbol{B}_i \boldsymbol{q}, \tag{1.17}$$

while (1.1) gives an expression for $\boldsymbol{r}'$. The Legendre transform becomes

$$\begin{aligned} H(\boldsymbol{r}, \boldsymbol{q}, \boldsymbol{n}, \boldsymbol{\mu}) &:= \begin{pmatrix} \boldsymbol{n} \\ \boldsymbol{\mu} \end{pmatrix} \cdot \begin{pmatrix} \boldsymbol{r}' \\ \boldsymbol{q}' \end{pmatrix} - L(\boldsymbol{r}, \boldsymbol{q}, \boldsymbol{r}', \boldsymbol{q}') \\ &= \sum_{i=1}^{3} \frac{(\boldsymbol{\mu} \cdot \boldsymbol{B}_i \boldsymbol{q})^2}{8K_i} + \boldsymbol{n} \cdot \boldsymbol{d}_3 - V(\boldsymbol{r}) = \sum_{i=1}^{3} \frac{m_i^2}{2K_i} + \boldsymbol{n} \cdot \boldsymbol{d}_3 - V(\boldsymbol{r}), \end{aligned} \tag{1.18}$$

where $m_i = \boldsymbol{m} \cdot \boldsymbol{d}_i$ are the components of the moment $\boldsymbol{m}$ in the rod with respect to the director frame, for which we can write

$$m_i = \frac{\partial W}{\partial u_i} = K_i u_i = 2K_i \boldsymbol{B}_i \boldsymbol{q} \cdot \boldsymbol{q}' / |\boldsymbol{q}|^2 = \frac{1}{2} \boldsymbol{\mu} \cdot \boldsymbol{B}_i \boldsymbol{q}. \tag{1.19}$$

(Note that it is the special form of constraint (1.1) that allows us to write down the Legendre transform without the Lagrangian (1.14) being strictly convex in $\boldsymbol{r}'$.) Notice that the multiplier ν does not appear in H.

H is the (unconstrained) Hamiltonian for rod statics, with $\boldsymbol{n}$, the variable conjugate to the axis $\boldsymbol{r}$, equal to the force acting across the cross-section of the rod [6]. The canonical Hamiltonian equations then become

$$\boldsymbol{r}' = \frac{\partial H}{\partial \boldsymbol{n}} = \boldsymbol{d}_3, \tag{1.20}$$

$$\boldsymbol{q}' = \frac{\partial H}{\partial \boldsymbol{\mu}} = \sum_{i=1}^{3} \frac{\boldsymbol{\mu}\cdot B_i\boldsymbol{q}}{4K_i} B_i\boldsymbol{q}, \tag{1.21}$$

$$\boldsymbol{n}' = -\frac{\partial H}{\partial \boldsymbol{r}} = \frac{\partial V(\boldsymbol{r})}{\partial \boldsymbol{r}}, \tag{1.22}$$

$$\boldsymbol{\mu}' = -\frac{\partial H}{\partial \boldsymbol{q}} = \sum_{i=1}^{3} \frac{\boldsymbol{\mu}\cdot B_i\boldsymbol{q}}{4K_i} B_i\boldsymbol{\mu} - \left(\frac{\partial \boldsymbol{d}_3}{\partial \boldsymbol{q}}\right)^T \boldsymbol{n}. \tag{1.23}$$

Here we have used that $(\partial/\partial\boldsymbol{q})(\boldsymbol{\mu}\cdot B_i\boldsymbol{q}) = B_i\boldsymbol{\mu}$.

1.2 Boundary conditions

To find rod solutions we solve the first-order system of 14 ordinary differential equations (1.20)–(1.23) subject to appropriate boundary conditions. We consider three cases.

1. Clamped, pure thrust:
Valid boundary conditions are

$$\boldsymbol{r}(0) = (0,0,0),\quad \boldsymbol{r}(1) = (0,0,*),\quad \boldsymbol{q}(0) = (0,0,0,1),\quad \boldsymbol{q}(1) = (0,0,0,1),$$
$$\boldsymbol{n}(1) = (*,*,-P), \tag{1.24}$$

where P is an applied compressive force in the $-\boldsymbol{e}_3$ direction and the $*$ signifies that there is no condition on the corresponding component. The conditions on $\boldsymbol{q}$ specify that the directors at the ends of the rod are aligned with the fixed axes $\{\boldsymbol{e}_1,\boldsymbol{e}_2,\boldsymbol{e}_3\}$ and that no twist angle is imposed.

However, the boundary-value problem (1.20)–(1.24) has non-isolated solutions, which prohibit numerical solution via continuation methods. In particular, the equations are invariant under the infinitesimal transformation $\boldsymbol{\mu} \to \boldsymbol{\mu} + \epsilon\boldsymbol{q}$, the generator of the integral $|\boldsymbol{q}|^2$ of the Hamiltonian system [6]. We can remove the non-isolation by replacing the above boundary conditions by equivalent ones which factor out the symmetry. Thus we replace the condition $q_4(1) = 0$, which is implied by the remaining conditions and the fact that $|\boldsymbol{q}|^2$ is an integral, by the condition $\mu_4(0) = 0$, which selects a single representative of the family of possible $\boldsymbol{\mu}$ values. The full set of boundary conditions then becomes

$$\boldsymbol{r}(0) = (0,0,0),\quad \boldsymbol{r}(1) = (0,0,*),\quad \boldsymbol{q}(0) = (0,0,0,1),\quad \boldsymbol{q}(1) = (0,0,0,*),$$

$$\boldsymbol{n}(1) = (*, *, -P), \quad \boldsymbol{\mu}(0) = (*, *, *, 0). \tag{1.25}$$

2. Clamped, pure torque:
Now the 14 boundary conditions are

$$\boldsymbol{r}(0) = (0,0,0), \quad \boldsymbol{r}(1) = (0,0,*), \quad \boldsymbol{q}(0) = (0,0,0,1), \quad \boldsymbol{q}(1) = (0,0,*,*),$$
$$\boldsymbol{n}(1) = (*,*,0), \quad m_3(1) = \frac{1}{2}\boldsymbol{\mu}(1)\cdot \boldsymbol{B}_3\boldsymbol{q}(1) = M, \quad \boldsymbol{\mu}(0) = (*,*,*,0), \tag{1.26}$$

where M is the applied torque. Note that the conditions on $\boldsymbol{q}(1)$, in view of (1.6)–(1.8), allow the frame of directors $\{\boldsymbol{d}_1, \boldsymbol{d}_2, \boldsymbol{d}_3\}$ to rotate only about $\boldsymbol{d}_3$.

3. Hinged:
A valid set of boundary conditions in this case is

$$\boldsymbol{r}(0) = (0,0,0), \quad \boldsymbol{r}(1) = (0,0,*), \quad \boldsymbol{d}_2(0) = (0,1,0), \quad \boldsymbol{d}_2(1) = (0,*,0),$$
$$\boldsymbol{n}(1) = (*,*,-P), \quad m_2(0) = \frac{1}{2}\boldsymbol{\mu}(0)\cdot \boldsymbol{B}_2\boldsymbol{q}(0) = 0,$$
$$m_2(1) = \frac{1}{2}\boldsymbol{\mu}(1)\cdot \boldsymbol{B}_2\boldsymbol{q}(1) = 0, \quad \boldsymbol{\mu}(0) = (*,*,*,0). \tag{1.27}$$

The ends are hinged about $\boldsymbol{d}_2$ and under the action of the compressive force P the rod will therefore deform in the x-z plane. The conditions on $\boldsymbol{d}_2$ can be written in terms of the q_i by using (1.7). Note that we have again replaced an implied condition ($d_{2y}(1) = 1$, because $\boldsymbol{d}_2$ is a unit vector) by the gauge condition $\mu_4(0) = 0$.

2. Numerical results

We shall restrict ourselves to isotropic rods, i.e., rods for which $K_1 = K_2 =: K$. When such a rod is clamped at both ends and subjected to both compression and twist then it buckles at a critical load given by

$$\cos\pi\sqrt{m^2+4p} - \cos\pi m + \left(2\pi p \sin\pi\sqrt{m^2+4p}\right)/\sqrt{m^2+4p} = 0 \tag{2.1}$$

(see [13]), where p and m are the dimensionless loads given by

$$p = \frac{PL^2}{4\pi^2 K}, \qquad m = \frac{ML}{2\pi K}. \tag{2.2}$$

In the case of pure-thrust buckling we have $m = 0$ and the primary buckling load is $p_c = 1$. In the case of pure torsional buckling (i.e., $p = 0$) (2.1) reduces to $\tan\pi m = \pi m$, giving the first-mode critical load $m_c = 1.4303$.

For a hinged rod primary buckling is governed by $m^2 = 1 - 4p$, giving $p_c = 1/4$ for pure-thrust buckling.

In each of the three cases we start the continuation from the trivial (straight and untwisted) solution placed at the centre of the cylinder,

$$\boldsymbol{r}(0) = (0,0,0), \quad \boldsymbol{r}(1) = (0,0,1), \quad \boldsymbol{q}(0) = (0,0,0,1), \quad \boldsymbol{q}(1) = (0,0,0,1), \tag{2.3}$$

and apply an increasing load so as to buckle the rod. In the hinged case we choose K_1 slightly larger than K_2 in order to force the primary buckling mode to be in the x-z plane. (The actual value of K_1 is then irrelevant as all bending is about $\boldsymbol{d}_2$, and K_2 is to be used for K in (2.2).) We take $a = b = 0.01$ in (1.13). The boundary-value problem was implemented in the software package AUTO [7] for single-parameter continuation of solutions and detection of bifurcations (such as primary buckling loads). This well-tested code uses orthogonal collocation allied to pseudo-arclength continuation.

Since under pure-thrust loading the rod remains planar we can compare our results with those of Domokos et al. [8], who constructed shapes for a hinged planar Euler elastica buckling between two parallel plates by patching together pieces of free rod and pieces of straight rod lying against the wall (hard wall contact). We confirm their buckling scenario (see Fig. 18.1(a)). Upon contact with the cylinder the point contact grows into line contact until, under a further increase of the load, the straight central segment buckles inward (it effectively buckles as a clamped rod), leaving two point contacts. The buckled segment eventually reaches the opposite side of the cylinder. The corresponding load-deflection diagram in Fig. 18.2(a) reveals that the load P reaches two limit points (a local maximum and a local minimum), in agreement with results in [8]. In fact, the curve is a smoothed version of the curve found in [8] with the first limit point (near $p = 3.6$) corresponding to the intersection point of two curves, one with configurations before and one with configurations after the buckling of the central flat segment, while the second limit point (near $p = 2.25$) corresponds to the intersection with a curve of configurations with three point contacts, two on one side of the cylinder and the middle on the other side. Clearly, under controlled load the solution will become unstable at $p = 3.6$ and a jump will take place to a remote stable solution (snap buckling).

For a clamped rod we find an entirely different buckling scenario (see Fig. 18.1(b)): upon contact with the cylinder no line contact develops; instead, the rod bends away near the ends where buckles develop that grow until (point) contact is made at the opposite side of the cylinder. This behaviour can be understood as follows. Line contact would

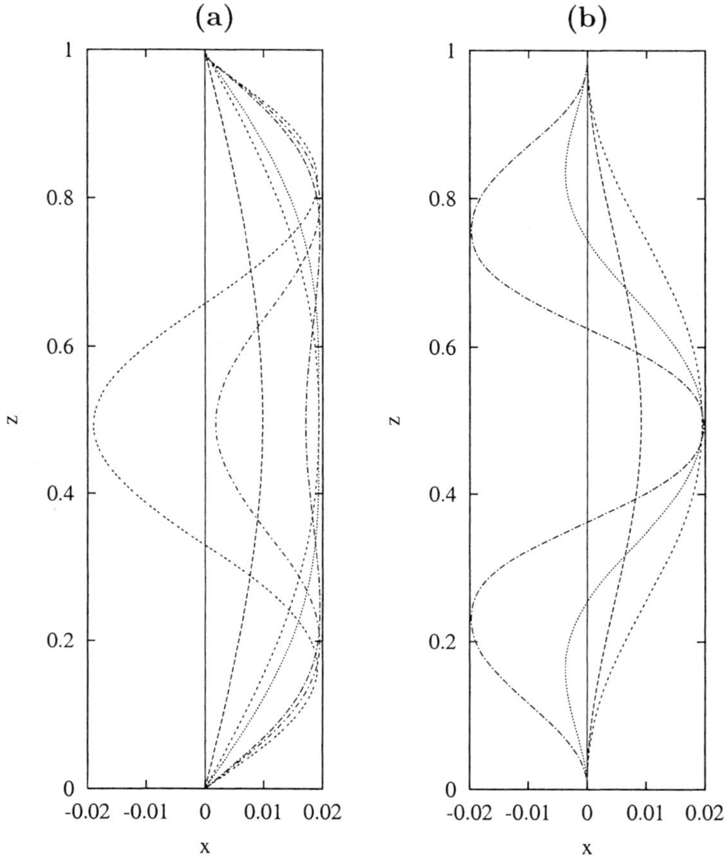

Figure 18.1. Planar rod configurations under (a) hinged and (b) clamped (pure thrust) boundary conditions. ($r/L = 0.02$.)

require the touch-down point of the rod to be moment-free, i.e., an inflection point. By symmetry, then, the end point would also have to be moment-free, which of course it is for a hinged end but in general it is not for a clamped end (the end may be moment-free at isolated points of the loading path, but at fixed boundary conditions (1.25) the end will just sweep through such points and the rod cannot lie down in line contact). Fig. 18.2(b) gives the load-deflection diagram for the clamped case. Rapid changes in the end shortening $1 - z$ take place during the bending phases of the rod's deformation until (new) contact is established. The load increases monotonically: it does not attain local maxima or minima.

Fig. 18.3 shows results for clamped boundary conditions under a pure torque.

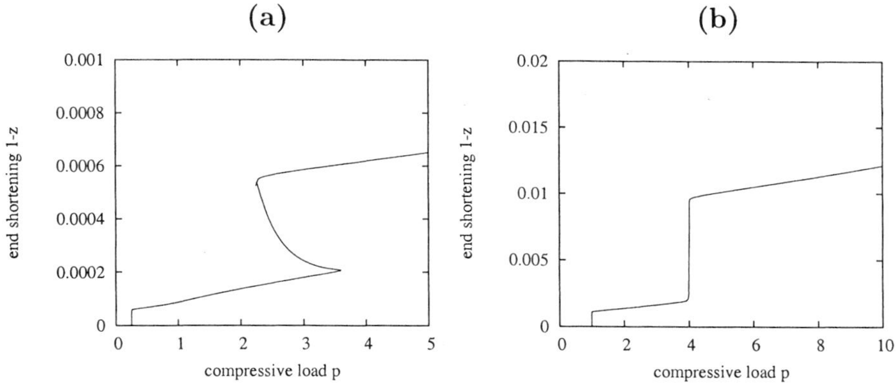

Figure 18.2. End shortening against compressive load for a rod under (a) hinged (numerical buckling load $p_c = 0.250041$), and (b) clamped (numerical buckling load $p_c = 1.000065$) boundary conditions. ($r/L = 0.02$.)

3. Discussion

We have presented a computational approach to dealing with unilateral constraints in contact problems for elastic rods. This approach will be used in future work to study the post-buckling behaviour of confined drill strings. The results of planar test problems agree with 2D elastica results of Domokos et al. [8] for the case of hinged rods. However, clamped rods are more common in applications and we find that constrained buckling in the clamped case proceeds along a different path from that in the hinged case. This behaviour deserves a closer study. For instance, it would be interesting to back these results up by an analytical approach as in [8] on the planar elastica. Finally, we have also presented numerical results for fully 3D torsional buckling inside a cylinder.

Acknowledgments

The author wishes to thank John Maddocks and Gábor Domokos for stimulating discussions. This work was supported by a Royal Society Research Fellowship.

References

[1] A.N. Bedrossian, F. Abbassian, V.A. Dunayevsky & S.A. Hope, Analysis of post-buckling behaviour of drillstrings, *8th UK ABAQUS User Group Conference*, 15 September 1993, Leicester University.

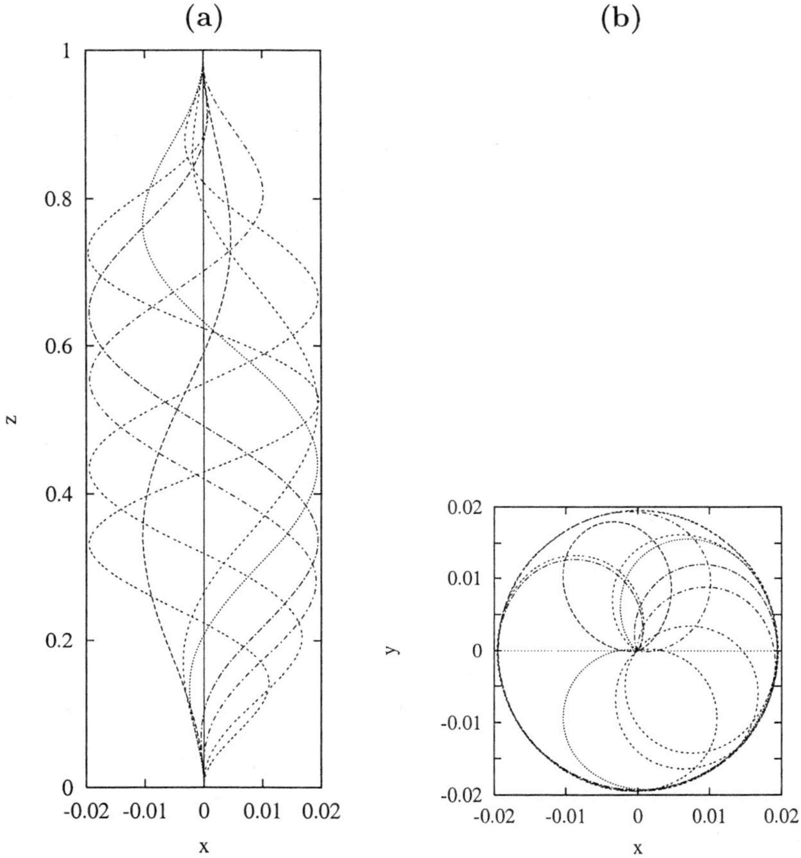

Figure 18.3. Two views of a buckled rod under clamped boundary conditions (pure torque, numerical buckling load $m_c = 1.430747$). ($K_3/K = 3/4$, $r/L = 0.02$.)

[2] S.S. Antman, *Nonlinear Problems of Elasticity* (Springer-Verlag, New York, 1995).

[3] G.B. Bulman & R.S. Manning, Conjugate points for an integrodifferential second variation operator: stability of an elastic rod buckling into a soft wall, preprint.

[4] A.R. Champneys, G.W. Hunt & J.M.T. Thompson (eds), Localization and Solitary Waves in Solid Mechanics, Advanced Series in Nonlinear Dynamics, Vol. 12 (World Scientific, Singapore, 1999).

[5] B.D. Coleman & D. Swigon, Theory of supercoiled elastic rings with self-contact and its application to DNA plasmids, *J. Elasticity* **60**, 173-221 (2000).

[6] D.J. Dichmann, Y.W. Li & J.H. Maddocks, Hamiltonian formulations and symmetries in rod mechanics, in *Mathematical Approaches to Biomolecular Structure and Dynamics*, J. Mesirov, K. Schulten & D. Sumners (eds), The IMA Volumes in Mathematics and its Applications, Vol. 82 (Springer-Verlag, New York, 1996), pp. 71-113.

[7] E.J. Doedel, A.R. Champneys, T.R. Fairgrieve, Yu.A. Kuznetsov, B. Sandstede & X.J. Wang, AUTO97: Continuation and bifurcation software for ordinary differential equations (1997) (available by anonymous ftp from `ftp.cs.concordia.ca/pub/doedel/auto`).

[8] G. Domokos, P. Holmes & B. Royce, Constrained Euler buckling, *J. Nonlinear Sci.* **7**, 281-314 (1997).

[9] W.B. Fraser & D.M. Stump, The equilibrium of the convergence point in two-strand yarn plying, *Int. J. Solids Struct.* **35**, 285-298 (1998).

[10] D. Garrivier & B. Fourcade, Twisting DNA with variable intrinsic curvature, *Europhys. Lett.* **49**, 390-395 (2000).

[11] G.H.M. van der Heijden, The static deformation of a twisted elastic rod constrained to lie on a cylinder, *Proc. R. Soc. Lond. A* **457**, 695-715 (2001).

[12] G.H.M. van der Heijden, A.R. Champneys & J.M.T. Thompson, Spatially complex localisation in twisted elastic rods constrained to a cylinder, *Int. J. Solids Struct.* **39**, 1863-1883 (2002).

[13] G.H.M. van der Heijden, S. Neukirch, V.G.A. Goss & J.M.T. Thompson, Instability and self-contact phenomena in the writhing of clamped rods, *Int. J. Mech. Sci.* **45**, 161-196 (2003).

[14] N.C. Huang & P.D. Pattillo, Helical buckling of a tube in an inclined wellbore, *Int. J. Non-linear Mech.* **35**, 911-923 (2000).

[15] S. Kehrbaum & J.H. Maddocks, Elastic rods, rigid bodies, quaternions, and the last quadrature, *Phil. Trans. R. Soc. Lond. A* **355**, 2117-2136 (1997).

[16] F. Schuricht, A variational approach to obstacle problems for shearable nonlinearly elastic rods, *Arch. Rat. Mech. Anal.* **140**, 103-159 (1997).

[17] J.M.T. Thompson, G.H.M. van der Heijden, S. Neukirch, Supercoiling of DNA plasmids: mechanics of the generalized ply, *Proc. R. Soc. Lond. A* **458**, 959-985 (2002).

[18] R.W. Tucker & C. Wang, An integrated model for drill-string dynamics, *J. Sound Vibr.* **224**, 123-165 (1999).

Chapter 19

COMPUTING FEM SOLUTIONS OF PLASTICITY PROBLEMS VIA NONLINEAR MIXED VARIATIONAL INEQUALITIES

Paolo Venini
Department of Structural Mechanics
University of Pavia, Via Ferrata 1, I27100 Pavia, Italy
paolo.venini@unipv.it

Roberto Nascimbene
Department of Structural Mechanics
University of Pavia, Via Ferrata 1, I27100 Pavia, Italy
roberto.nascimbene@unipv.it

Abstract A fixed–point algorithm for the solution of hardening plasticity problems is proposed. The continuous problem may be classified as a mixed nonlinear non–differentiable variational inequality of the second type that is discretized by means of a truly mixed finite–element scheme. The non–differentiability of the dissipation functional is coped with via regularization whereas the nonlinearity is approached with a fixed–point iterative strategy. Numerical results are proposed to assess the sensitivity of the approach with respect to the mesh size and the regularization parameter. The simplicity of the proposed fixed–point scheme with respect to more classical return mapping approaches seems to represent one of the key features of our algorithm. Eventually an extension is proposed that allows to analyze the propagation of a cohesive crack within an elastoplastic bulk. A suitable modification of the variational formulation is introduced to this scope that takes full advantage of the inherent discontinuity of the displacement field.

Keywords: Variational Inequalities, Plasticity, Finite Elements, Mixed Methods, Cohesive Cracks.

1. Introduction

Hardening plasticity may be considered a mature topic from the theoretical and algorithmic viewpoints. A detailed and comprehensive presentation of return–mapping algorithms is given in Simo and Hughes ([6]) where stability, consistency and convergence of various available algorithms are discussed. Several variants of strain–driven elastic–predictor plastic–corrector algorithms are at disposal of the analyst that rely on the notion of consistent tangent operator that is crucial in order to preserve the quadratic convergence of Newton's method ([7]). Complementary mixed formulations making use of the maximum plastic dissipation principle have conversely received much less attention by the engineering community. One of the first investigations along this mainstream seems to be ([8]) where stresses belongs to L^2, i.e. they are globally discontinuous, whereas displacements are in H^1 and are therefore more regular. Following ([3]), the *truly* mixed approach is considered herein, see e.g. ([1]), according to which displacements are the discontinuous Lagrange multipliers and the stresses are "more regular", i.e. belong to the anisotropic space $H(\text{div})$. One of the peculiarities of our approach is the use of the composite triangular element developed by Johnson and Mercier ([2]) for the approximation of the stress field. The resulting variational inequality is non–differentiable so that a regularization technique is used for this purpose ([4]). The nonlinearity is then approached by means of a fixed–point iterative strategy that applies to the single loading step. An innovative modification is eventually presented that allows to consider the presence of an evolving cohesive crack within the elastoplastic core. The full implementation of the resulting approach is left to a future communication while numerical results are presented herein concerning an existing and fixed crack. The paper outline is as follows. The mathematical formulation of the continuous problem is presented first with attention paid to the mixed variational formulation leading to the FEM discretization. An un–cracked medium is considered initially that is promptly extended to the case of cohesive crack propagation. FEM discretization is then considered that is based on the introduction of proper functional spaces for stresses, plastic strains, displacements and plastic multipliers. The numerical algorithm for the solution of the elastoplastic problem is then introduced. Rather than relying on classical predictor–corrector strategies, our approach takes full advantage of the variational structure of the problem in that the nonlinear non–differentiable variational inequality is regularized to form a differentiable nonlinear problem that is solved by means of a recursive fixed–point algorithm. A few numerical examples are eventually presented to assess the convergence rate of the proposed approach.

2. Mathematical formulation

Within a small displacement–gradients setting, let $\boldsymbol{\sigma}$ be the symmetric stress tensor, $\mathbf{p}$ the trace–free plastic strain tensor, $\mathbf{u}$ the displacement vector and λ a scalar Lagrange multiplier to be used to enforce plastic incompressibility in weak form. We move from the Hellinger–Reissner principle considering the version in which the stresses are "differentiable" and the displacements are discontinuous. Upon denoting with Ω the physical domain of the system, letting Latin indexes run from one to two and after introducing the (auxiliary) stress space

$$\Sigma = \{\boldsymbol{\tau} : \tau_{ij} = \tau_{ji}, \tau_{ij} \in L^2(\Omega)\},$$

that is an Hilbert space with inner product and norm respectively defined as

$$(\boldsymbol{\tau}, \boldsymbol{\sigma})_\Sigma = \int_\Omega \boldsymbol{\tau} : \boldsymbol{\sigma} dx = \int_\Omega \tau_{ij}\sigma_{ij} dx, \quad \| \boldsymbol{\tau} \|_\Sigma = (\boldsymbol{\tau}, \boldsymbol{\tau})^{1/2},$$

we introduce the displacement and stress spaces as

$$W = [L^2(\Omega)]^2, \qquad H = H(\text{div}, \Omega) = \{\boldsymbol{\tau} \in \Sigma,\ \underline{\text{div}}\ \boldsymbol{\tau} \in W\},$$

where W has the standard L^2–product norm while H is endowed with the graph norm

$$\| \boldsymbol{\tau} \|_H = \left(\| \boldsymbol{\tau} \|_\Sigma^2 + \| \underline{\text{div}}\ \boldsymbol{\tau} \|_W^2 \right)^{\frac{1}{2}}, \tag{2.1}$$

that makes $H(\text{div}; \Omega)$ an Hilbert space. The elastoplastic problem we are going to consider reads: find

$$\begin{aligned} \bar{\boldsymbol{\sigma}} &= (\boldsymbol{\sigma}, \mathbf{p}) \in \overline{H} = H \times Q = H(\text{div}, \Omega) \times [L^2(\Omega)]_s^{2\times 2} \text{ and} \\ m &= (\mathbf{u}, \lambda) \in L = W \times \Lambda = [L^2(\Omega)]^2 \times L^2(\Omega) \end{aligned}$$

such that

$$\begin{cases} \langle A_1 \bar{\boldsymbol{\sigma}}, \bar{\boldsymbol{\tau}} - \bar{\boldsymbol{\sigma}} \rangle + j(\bar{\boldsymbol{\tau}}) - j(\bar{\boldsymbol{\sigma}}) + b_3(\bar{\boldsymbol{\tau}} - \bar{\boldsymbol{\sigma}}, m) \geq 0 \quad \forall \bar{\boldsymbol{\tau}} = (\boldsymbol{\tau}, \mathbf{q}) \in \overline{H} \\ \qquad\qquad b_3(\bar{\boldsymbol{\sigma}}, n) = \langle g, n \rangle \quad \forall n = (\mathbf{v}, \mu) \in L, \end{cases} \tag{2.2}$$

where, in the case of isotropic elasticity and nonlinear kinematic hardening, denoting with a superscript D the deviatoric part of a symmetric tensor, one has

$$\langle A_1 \bar{\boldsymbol{\sigma}}, \bar{\boldsymbol{\tau}} \rangle = \int_\Omega \left[\left(\frac{1}{2\mu} \boldsymbol{\sigma}^D + \mathbf{p} \right) : \boldsymbol{\tau}^D + \frac{1}{\lambda + \mu} \text{tr}\boldsymbol{\sigma}\text{tr}\boldsymbol{\tau} + (\boldsymbol{\sigma}_0(\mathbf{p}) - \boldsymbol{\sigma})^D : \mathbf{q} \right] dx, \tag{2.3}$$

where, with abuse of notation, λ indicates a Lamé constant (and not a Lagrange multiplier as e.g. in the definition of m prior to relation (2.2)). The backstress $\boldsymbol{\sigma}_0(\mathbf{p})$ is the center of the elastic domain ([6]) accounting for nonlinear kinematic hardening and evolving according to the law

$$\boldsymbol{\sigma}_0(\mathbf{p}) = h(|\mathbf{p}|)\mathbf{p}, \tag{2.4}$$

where $h(\alpha) = h_0 + h_1 e^{-\eta\alpha}$ is the exponential hardening function and h_0, h_1 and η are material dependent constants. The bilinear form $b_3(\cdot,\cdot)$, used to enforce equilibrium and plastic incompressibility, and the work of the body force $\mathbf{b}$ are respectively computed as

$$b_3(\bar{\boldsymbol{\tau}}, m) = \int_\Omega (\mathbf{u} \cdot \underline{\mathrm{div}}\ \boldsymbol{\tau} - \lambda \mathrm{tr}\mathbf{q})dx, \quad \langle g, n\rangle = \int_\Omega \mathbf{b} \cdot \mathbf{v} dx.$$

Furthermore, the dissipation functional j is defined as

$$j : [L^2(\Omega)]_s^{2\times 2} \to \mathbb{R}, \quad j(\mathbf{q}) = \int_\Omega D(\mathbf{q}(x))dx,$$

in which the Von Mises dissipation function D reads

$$D(\mathbf{q}) = k|\mathbf{q}| = k\sqrt{q_{ij}q_{ij}}, \tag{2.5}$$

k being the material–dependent yielding stress. Our interest is focussed on the regularized version of problem (2.2) where the non–differentiable functional is approximated by a differentiable one so as to switch from a non–differentiable variational inequality of the second kind to an elliptic mixed variational equation. This choice allows an easier implementation and, most important, error estimates for the regularized problem are available that ensure the convergence of the discrete regularized problem to the continuous non–differentiable one ([3]). Let now ε be a positive scalar and consider the following regularized dissipation indexed on ε

$$D_\varepsilon(\mathbf{q}) = k \begin{cases} |\mathbf{q}| - \varepsilon/2 & |\mathbf{q}| > \varepsilon \\ \dfrac{1}{2\varepsilon}\mathbf{q} : \mathbf{q} & |\mathbf{q}| \leq \varepsilon, \end{cases}$$

that turns out to be differentiable. A straightforward calculation in fact allows one to write

$$\underline{\underline{D}}'_\varepsilon(\mathbf{q}) = k \begin{cases} \dfrac{\mathbf{q}}{|\mathbf{q}|} & |\mathbf{q}| > \varepsilon \\ \\ \dfrac{\mathbf{q}}{\varepsilon} & |\mathbf{q}| \leq \varepsilon, \end{cases}$$

giving sense to the duality pairing $\langle j'_\varepsilon(\mathbf{p}), \mathbf{q}\rangle$. The regularized overall mixed nonlinear variational problem may then be written as follows. Find

$$\begin{aligned} \bar{\boldsymbol{\sigma}} &= (\boldsymbol{\sigma}, \mathbf{p}) \in \overline{H} = H \times Q = H(\mathrm{div}, \Omega) \times [L^2(\Omega)]_s^{2\times 2} \text{ and} \\ m &= (\mathbf{u}, \lambda) \in L = W \times \Lambda = [L^2(\Omega)]^2 \times L^2(\Omega), \end{aligned}$$

such that

$$\begin{cases} \langle A_1 \bar{\boldsymbol{\sigma}}, \bar{\boldsymbol{\tau}}\rangle + \langle j'_\varepsilon(\mathbf{p}), \mathbf{q}\rangle + b_3(\bar{\boldsymbol{\tau}}, m) &= 0 \qquad \forall \bar{\boldsymbol{\tau}} = (\boldsymbol{\tau}, \mathbf{q}) \in \overline{H} \\ b_3(\bar{\boldsymbol{\sigma}}, n) &= \langle g, n\rangle \quad \forall n = (\mathbf{v}, \mu) \in L \end{cases} . \tag{2.6}$$

The Johnson–Mercier space, used for the approximation of the stresses, is then introduced as follows:

$$JM(K) = \{\boldsymbol{\sigma} |\ \boldsymbol{\sigma} \in H(\mathrm{div}; K),\ \boldsymbol{\sigma}|_{T_j} \in [P_1(T_j)]_s^{2\times 2},\ j = 1, 2, 3\}, \tag{2.7}$$

where K indicates any triangular element of the mesh taht is itself divided into 3 sub–triangles T_i, $i = 1, 2, 3$ and $P_1(T_j)$ is the space of the polynomials of degree ≤ 1 on T_j. Definition (2.7) calls for $\boldsymbol{\sigma}\cdot\mathbf{n}$ to be continuous across each sub–triangle. An element $\boldsymbol{\sigma}$ of $JM(T)$ is uniquely determined by the following 15 degrees of freedom ([2]):

$$\int_{e_i} (\boldsymbol{\sigma}\cdot\mathbf{n})\cdot\mathbf{w}\,ds, \quad \forall \mathbf{w} \in (P_1(e_i))^2, \quad i = 1, 2, 3,$$

$$\int_T \boldsymbol{\sigma} : \boldsymbol{\omega}\,dx, \qquad \forall \boldsymbol{\omega} \in (P_0(T))_s^{2\times 2},$$

where e_i denote the edges of the element. The degrees–of–freedom of the JM element do not correspond to point values but are flux values $\boldsymbol{\sigma}\cdot\mathbf{n}$ and average stress values. It is however interesting to note that we shall not need to reconstruct point stress values to check the plasticity condition since no yielding function enters our formulation (but its convex dual, i.e. the dissipation function). Stresses are then approximated in the space

$$H_h = \{\boldsymbol{\sigma}_h \in H(\mathrm{div}, \Omega),\ \boldsymbol{\sigma}_h|_K \in JM(K)\}. \tag{2.8}$$

As to displacements, plastic strains and plastic incompressibility multipliers, they are respectively approximated by standard polynomial spaces, i.e.

$$W_h = \left\{\mathbf{v}_h \in W :\ \mathbf{v}_h|_K \in [P_1(K)]^2\right\}, \tag{2.9}$$

$$Q_h = \left\{\mathbf{q} \in Q :\ \mathbf{q}_h|_K \in [P_1(K)]^{2\times 2}\right\}, \Lambda_h = P_1(K), \tag{2.10}$$

FEM spaces in Equations (2.8–2.10) are chosen in accordance with ([3]) so that the relevant abstract convergence analysis holds. The nonlinear equations governing the elastoplastic problem are hereafter derived moving from the regularized problem of Equation (2.6). In the spirit of the principle of virtual work, independent variations are given to the test functions $\boldsymbol{\tau}$, $\mathbf{q}$, $\mathbf{v}$ and μ thus obtaining four sets of equations that may be grouped in the non–linear system

$$\begin{bmatrix} A_{\sigma\sigma} & A_{\sigma p} & B_{\sigma u} & 0 \\ A_{p\sigma} & A_{pp}(p) & 0 & B_{p\lambda} \\ B_{u\sigma} & 0 & 0 & 0 \\ 0 & B_{\lambda p} & 0 & 0 \end{bmatrix} \begin{Bmatrix} \sigma \\ p \\ u \\ \lambda \end{Bmatrix} = \begin{Bmatrix} 0 \\ 0 \\ g \\ 0 \end{Bmatrix} \tag{2.11}$$

which, apart from the nonlinearity due to the dependence of A_{pp} on p, is governed by the classical mixed matrix ([1])

$$\mathcal{M}(p) = \begin{pmatrix} A & B^T \\ B & 0 \end{pmatrix},$$

where

$$A = \begin{bmatrix} A_{\sigma\sigma} & A_{\sigma p} \\ A_{p\sigma} & A_{pp}(p) \end{bmatrix}, \quad B = \begin{bmatrix} B_{u\sigma} & 0 \\ 0 & B_{\lambda p} \end{bmatrix}.$$

Block matrices appearing in Equation (2.11) are defined as

$$A_{\sigma\sigma} = \int_\Omega \left[\frac{1}{2\mu} \boldsymbol{\sigma}^D : \boldsymbol{\tau}^D + \frac{1}{\lambda+\mu} \mathrm{tr}\boldsymbol{\sigma}\mathrm{tr}\boldsymbol{\tau} \right] dx,$$

$$A_{\sigma p} = \int_\Omega \mathbf{p} : \boldsymbol{\tau}^D dx,$$

$$B_{\sigma u} = \int_\Omega \mathbf{u} \cdot \underline{\mathrm{div}}\ \boldsymbol{\tau} dx,$$

$$A_{p\sigma} = \int_\Omega \boldsymbol{\sigma}^D : \mathbf{q} dx,$$

$$A_{pp}(p) = \int_\Omega h(|\mathbf{p}|)\mathbf{p} : q dx + \int_\Omega \underline{\underline{D}}'_\varepsilon(\mathbf{p}) : \mathbf{q} dx,$$

$$B_{p\lambda} = -\int_\Omega \lambda \mathrm{tr}\mathbf{q} dx,$$

$$B_{u\sigma} = \int_\Omega \mathbf{v} \cdot \underline{\mathrm{div}}\ \boldsymbol{\sigma} dx,$$

$$B_{\lambda p} = -\int_\Omega \mu \mathrm{tr}\mathbf{p} dx,$$

$$g = \int_\Omega \mathbf{b} \cdot \mathbf{v} dx.$$

2.1 Extension to cohesive crack propagation

One of the key–features of the approach presented above is the inherent discontinuity of the displacement field. In the mainstream of discontinuous finite element methods that have been developed in the last few years for the analysis of cohesive crack propagation ([?]), ([?]), ([?]), our approach is in a certain sense the most *natural* since displacement discontinuities are naturally present all over the medium and may be activated when a given stress threshold is reached, e.g. when the maximum tensile stress reaches a critical value. Let now Γ denote the interface between two adjacent elements where the cohesive crack is expected to develop. By considering Equation $(2.6)_1$ at the element level, i.e. before assemblage, and denoting with a subscript K the element number, one may write

$$\begin{aligned} \langle A_1\bar{\boldsymbol{\sigma}}, \bar{\boldsymbol{\tau}}\rangle|_K &- \int_{\Gamma_K} \mathbf{u}\cdot(\boldsymbol{\tau}\cdot\mathbf{n})ds \\ &+ \langle j'_\varepsilon(\mathbf{p}), \mathbf{q}\rangle|_K + b_3(\bar{\boldsymbol{\tau}}, m)|_K = 0 \quad \forall\bar{\boldsymbol{\tau}} = (\boldsymbol{\tau}, \mathbf{q}) \in \overline{H}. \end{aligned} \tag{2.12}$$

If now the assemblage process is invoked and by denoting with subscript K_1 and K_2 the adjacent elements sharing the edge Γ_K, the second addendum in Equation (2.12) gives rise to the following assembled integral

$$\int_{\Gamma_K} (\mathbf{u}_{K_1} - \mathbf{u}_{K2})\cdot(\boldsymbol{\tau}\cdot\mathbf{n})\,ds = \int_{\Gamma_K} [\![\mathbf{u}]\!]\cdot(\boldsymbol{\tau}\cdot\mathbf{n})\,ds, \tag{2.13}$$

where $[\![\cdot]\!]$ denotes the jump of the relevant quantity. Within the framework of cohesive crack theory, a traction vs displacement–jump constitutive law is introduced in abstract form as

$$\boldsymbol{\sigma}\cdot\mathbf{n} \equiv \underline{t} = \mathcal{F}([\![\mathbf{u}]\!]), \tag{2.14}$$

that, in view of the application of an iterative approach, may be inverted and consistently linearized (at least in its rate form) as

$$[\![\mathbf{u}]\!] = \mathcal{C}^{-1}(\boldsymbol{\sigma}\cdot\mathbf{n}),$$

where $\mathcal{C}$ is the second–order tensor of linearized material coefficients. Within the framework of the Hellinger–Reissner approach, it is worth comparing the different forms of the energy functionals in the case of elastic, elastic–plastic and elastic–plastic fractured media. One may in fact write:

<u>Elastic medium</u>

$$\langle A_0\boldsymbol{\sigma}, \boldsymbol{\tau}\rangle = \int_K \mathbb{C}^{-1}\boldsymbol{\sigma} : \boldsymbol{\tau}dx,$$

where $\mathbb{C}$ is the fourth–order tensor of elastic coefficients.

Elastic–plastic medium

$$\langle A_1\bar{\boldsymbol{\sigma}}, \bar{\boldsymbol{\tau}}\rangle = \int_\Omega \left[\left(\frac{1}{2\mu}\boldsymbol{\sigma}^D + \mathbf{p}\right) : \boldsymbol{\tau}^D + \frac{1}{\lambda+\mu}\mathrm{tr}\boldsymbol{\sigma}\mathrm{tr}\boldsymbol{\tau} + (\boldsymbol{\sigma}_0(\mathbf{p}) - \boldsymbol{\sigma})^D : \mathbf{q}\right] dx,$$

reported here for completeness from Equation (2.3).

Elastic–plastic medium in the presence of cracks

$$\langle A_2\bar{\boldsymbol{\sigma}}, \bar{\boldsymbol{\tau}}\rangle = \langle A_1\bar{\boldsymbol{\sigma}}, \bar{\boldsymbol{\tau}}\rangle - \int_{\Gamma_K} \mathcal{C}^{-1}\,(\boldsymbol{\sigma}\cdot\mathbf{n})\cdot(\boldsymbol{\tau}\cdot\mathbf{n})\,ds. \tag{2.15}$$

A few key issues are worth being addressed at this stage:

1 thanks to our truly mixed approach, the only ingredient needed to account for cohesive crack propagation is the computation of the extra dissipative term in Equation (2.15).

2 no extra *discontinuous displacement mode* is to be introduced since the displacement field is inherently discontinuous. This is a remarkable advantage over nearly all available approaches in the literature. The minor downside is the necessity to have the crack coincide with an interface between adjacent elements. Such further issue will be tackled in a forthcoming paper;

3 the continuity of the stress fluxes, that is usually imposed as an additional constraint, is already satisfied a–priori and is in fact guaranteed by the condition $\boldsymbol{\sigma} \in H(\mathrm{div}, \Omega)$.

3. Numerical studies

We consider the well–known Cook membrane problem under the action of a uniform load pointing upward (rather than the more classical free–end boundary load). Furthermore, in addition to the analysis of an elastoplastic membrane we study the effect of a crack that is fully developed as in Figure 19.3. As to the numerical parameters adopted in the simulations reference is made to ([2]). Attention is paid to the convergence rate of the proposed approach as well as to its capability to handle the presence of a crack. Another remarkable feature of our approach is its the inherent capability of handling the analysis of incompressible media. Such kind of problems was investigated in ([5]) for linear–elastic isotropic media and will be extended to the incompressible elastic–plastic case in a forthcoming contribution by the authors.

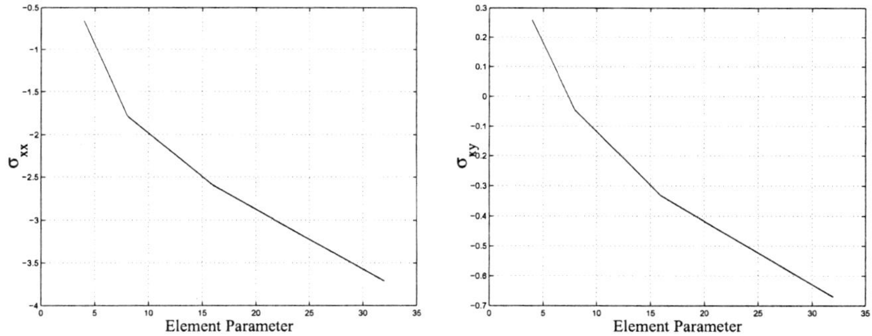

Figure 19.1. σ_{xx} and σ_{xy} convergence patterns

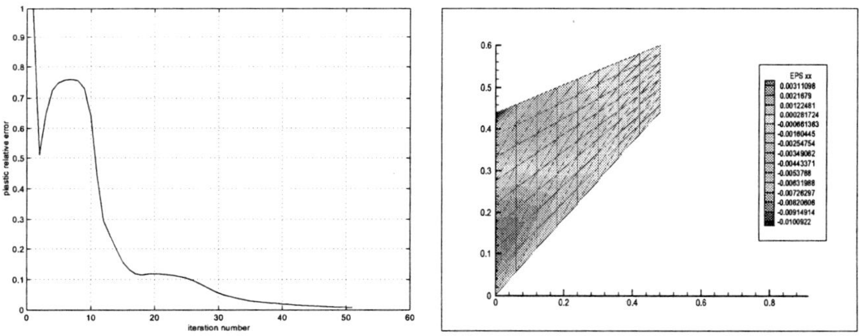

Figure 19.2. Plastic strain relative error and plastic strain p_{xx} pattern

3.1 Un–cracked elastoplastic material

A few representative numerical results are reported. in particular, Figure 19.1 shows the convergence patterns of stress components σ_{xx} and σ_{xy} whereas Figure 19.2 shows the plastic strain relative error versus the iteration number and the plastic strain component p_{xx}.

3.2 Cracked media

The presence of a crack as shown in Figure 19.3 is considered next. Notice that no further discontinuous mode is included in the formulation to address the presence of a crack as is normally done within classical discontinuous finite element approaches. Conversely, the "simple" evaluation of the dissipation energy integral in Equation (2.15) suffices for our scope. Figure 19.4 shows the variation of the stress component σ_{xx} enhancing the presence of a stress concentration at the crack tip.

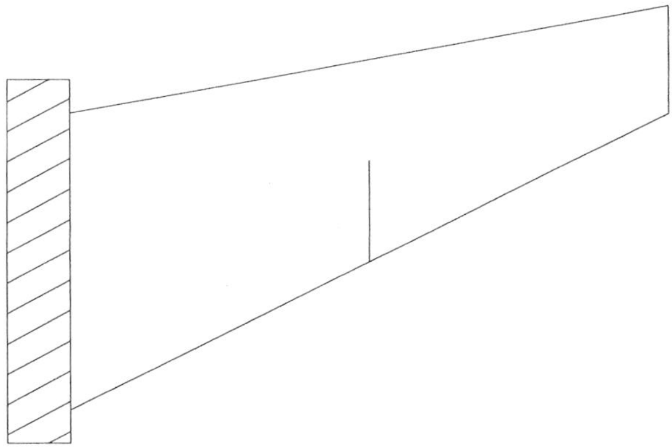

Figure 19.3. Cracked Cook's membrane

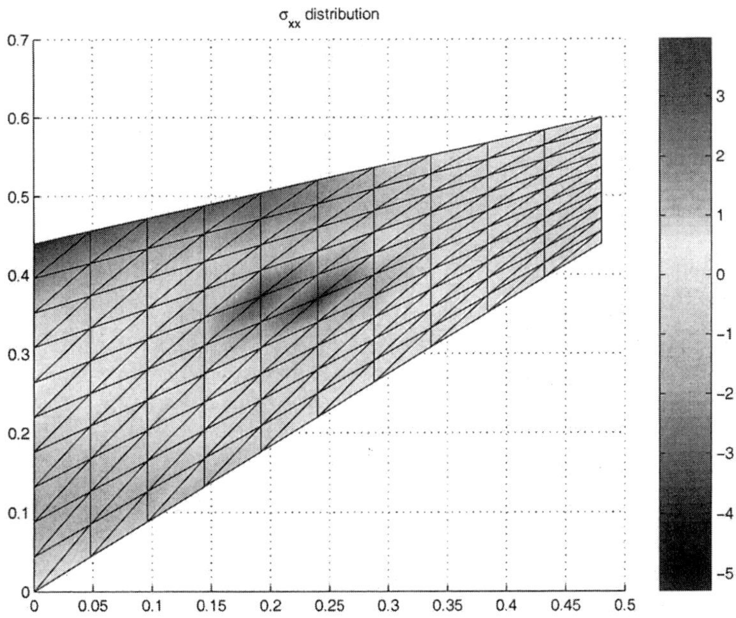

Figure 19.4. σ_{xx} field

Furthermore Figure 19.5 shows the displacement field that is clearly influenced by the presence of the cohesive crack.

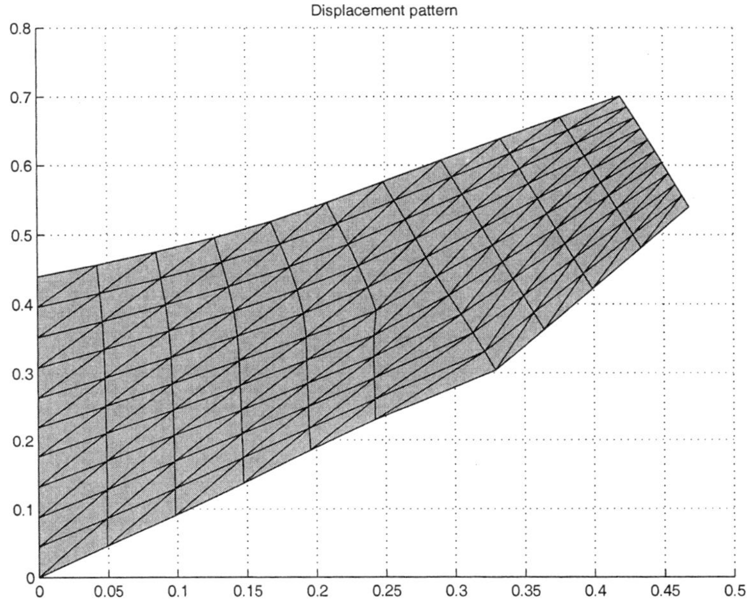

Figure 19.5. Displacement field in the presence of a crack

4. Conclusions

We have presented a novel numerical approach for the solution of elastoplastic problems that extends the approach presented in ([5]). The core of the method is the numerical solution of a regularized mixed variational inequality that is tackled by a two–step algorithm consisting of a fixed–point approach within which an Uzawa–type algorithm is nested as to the solution of the resulting mixed linear system. Convergence rates are numerically assessed and preliminary successful results concerning a cracked medium presented that are based on an extended variational formulation that is developed in the body of the paper.

References

[1] Brezzi, F. and M. Fortin, *Mixed and Hybrid Finite Element Methods*, Springer–Verlag, New York, 1991.

[2] Johnson, C. and B. Mercier, Some equilibrium finite elements methods for two dimensional elasticity problems, *Numer. Math.*, **30** (1978), 103-116.

[3] Han, W. and B.D. Reddy, On the finite element method for mixed variational inequalities arising in elastoplasticity, *SIAM J. Numer.*

Anal., **32** (1995), 1778-1807.

[4] Han, W. and S. Jensen, The Ka**v**canov method for a nonlinear variational inequality of the second kind arising in elastoplasticty, Technical Report, Dept. of Mathematics, University of Iowa, 1994.

[5] Nascimbene, R. and P. Venini, A new locking–free equilibrium mixed element for plane elasticity with continuous displacement interpolation, *Comp. Meth. Appl. Mech. Engrg.*, **191** (2001), 1843-1860.

[6] Simo, J.C. and T.J.R. Hughes, *Computational Inelasticity*, Springer, Interdisciplinary Applied Mathematics 1997.

[7] Simo, J.C. and R.L. Taylor, Consistent tangent operators for rate–independent elastoplasticity, *Comp. Meth. Appl. Mech. Engrg.*, **48** (1985), 101-118.

[8] Simo J.C. R.L. Taylor and K.S. Pister, Complementary mixed finite element formulation for elastoplasticity, *Comp. Meth. Appl. Mech. Engrg.*, **51** (1985), 177-208.

Chapter 20

FINITE ELEMENT DUAL ANALYSIS IN PIEZOELECTRIC CRACK ESTIMATION

Chang-Chun Wu
School of Civil Engineering and Mechanics,
Shanghai Jiao Tong University,
Shanghai, China, 200240
ccwu@sjtu.edu.cn

Zi-Ran Li, Lei Li
Department of Modern Mechanics,
Univ. of Science and Technology of China,
Hefei, China, 230026

G. Yagawa
Department of Quantum Engineering and Systems Science,
University of Tokyo, 7-3-1,
Tokyo 113, Japan

Abstract Owing to the complexity of piezoelectric crack problems, derivation of their closed-form solutions is virtually impossible and numerical solutions are largely resorted to. Hence, the upper and lower bound estimation for piezoelectric fracture parameters are of theoretical and practical values. An alternative assessment approach for electromechanical coupling cracked system is suggested in the paper. The dual path-independent integrals and the related bound theorems are presented. Consequently, the dual piezoelectric finite elements are formulated and implemented in bound analysis.

Keywords: fracture assessment, piezoelectric, bound theorem, dual finite elements

1. Introduction

Piezoelectric ceramics has been widely used as electromechanical transducers, sensors, actuators and adaptive structures as well as various resonators. piezoelectric ceramics are brittle and susceptible to cracking at all scales from domains to devices. When subjected to loading, these materials can fail prematurely due to the propagation of flaws or defects induced during the manufacturing process and by the in-service electromechanical loading. Hence, it is necessary to understand and be able to analyze the fracture characteristics of piezoelectric materials so that reliable service life predictions of the pertinent devices can be conducted.

Electro-mechanical modeling of piezoelectric fracture is complicated by the fact that the piezoelectric materials exhibit electro-elastic coupling behavior as well as anisotropy behavior. Among the theoretical studies of cracked piezoelectric bodies, Permeable and impermeable conditions at the crack face by Parton [1] and Deeg [2] respectively are mostly adopted. Under Parton's permeable condition, the electric potential and the normal component of the electric displacement are assumed to be continuous across the crack. This is considered to be acceptable for slender cracks in which the separation between the faces is negligible. Deeg's impermeable condition is based on the observation that the permitivity of air/vacuum is typically three order of magnitude lower than that of the piezoelectric materials. The crack cavity induces an insulating condition in which the normal component of the electric displacement vanishes. In other words, the crack face is charge-free. Whichever condition at the crack face is adopted, it is hard to get the exact solutions except few cases and the numerical method is often resorted to [3,4].

In the numerical analysis, the upper/lower bound estimation becomes a matter of great significance as it can help obtain an accurate/reliable solution. A serial of dual analysis for path integrals and bounds for ordinary crack parameter is presented by Wu et al [5-7]. In this paper, the dual analysis is extended to the fracture of piezoelectric materials. An alternating path-independent I-integral is presented, which can be considered as the complementary energy release rate of a fracture system of piezoelectric material model.

In the bound analysis, the conventional compatible displacement-electric potential model can be employed to estimate the lower bound of $J-$integral. For I, however, its upper bound should be estimated by stress-electric displacement equilibrium elements. In this paper, a penalty-equilibrium hybrid element is suggested to estimate the upper bound for piezoelectric crack systems.

2. Dual Path-Independent Integrals

Rice's J-integral method is developed to electromechanical fracture analysis, and a path-independent integral for piezoelectric medium is defined as [8,9]

$$J(u_i,\ \varphi) = \int_s [\ Hdx_2 - (\sigma_{ij}n_j\frac{\partial u_i}{\partial x_1} + D_j n_j \frac{\partial \varphi}{\partial x_1})ds\ ] \tag{2.1}$$

where the electric enthalpy

$$H(\varepsilon_{ij}, E_i) = \frac{1}{2}c^E_{ijkl}\varepsilon_{ij}\varepsilon_{kl} - \frac{1}{2} \in^{\varepsilon}_{ij} E_i E_j - e_{ikl}\varepsilon_{kl}E_i \tag{2.2}$$

with the strain $\varepsilon_{ij} = \frac{1}{2}(u_{i,j} + u_{j,i})$ and the electric field strength $E_i = -\varphi_{,i}$. For a given linear elastic crack system with potential energy

$$\Pi_p(u_i, \varphi) = \int_v H(\varepsilon_{ij}, E_i) - \int_{s_\sigma} \overline{T_i}u_i ds + \int_{s_w} \overline{q_s}\varphi ds.. \tag{2.3}$$

As the dual of J, another path-independent integral for piezoelectric medium can be defined as

$$I(\sigma_{ij}, D_i) = \int_s [\ - Gdx_2 + (\sigma_{i2}\frac{\partial u_i}{\partial x_j} + D_2\frac{\partial \varphi}{\partial x_j})dx_j\ ] \tag{2.4}$$

where the complementary electric enthalpy

$$G(\sigma_{ij}, D_i) = \frac{1}{2}s^D_{ijkl}\sigma_{ij}\sigma_{kl} - \frac{1}{2}\beta^{\sigma}_{ij}D_i D_j + g_{ijk}\sigma_{ij}D_k \tag{2.5}$$

with the stress σ_{ij} and the electric displacement D_i. It can be verified that I is identical to J [5]. Let the system complementary energy

$$\Pi_c(\sigma_{ij}, D_i) = \int_v Gdv - \int_{S_u} \overline{u_i}\sigma_{ij}n_j ds - \int_{S_\varphi} \overline{\varphi}D_j n_j ds \tag{2.6}$$

the I-integral can be defined as the complementary energy release rate, i.e. $I = d\Pi_c/da$.

3. Lower Bound Theorem For J

For a given piezoelectric crack system with homogeneous boundary constraints, $\bar{u}_i\,|_{Su} = 0$, $\bar{\varphi}\,|_{\varphi} = 0$, the approximate solution of the J-integral based on potential energy principle is always a lower bound of its actual one:

$$J(\widetilde{u}_i, \widetilde{\varphi})\ \le\ J(u_i, \varphi) \tag{3.1}$$

where (u_i, φ) and $(\widetilde{u}_i, \widetilde{\varphi})$ are respectively the actual solution and that given by the displacement-electric potential compatible finite element method.

Proof. Let $\widetilde{u}_i = u_i + \delta u_i$, $\delta u_i$are compatible virtual displacements; $\varphi = \varphi + \delta\varphi$, $\delta\varphi$ is the compatible virtual electric potential . Then the approximate potential energy can be expressed as

$$\Pi_p(\widetilde{u}_i, \widetilde{\varphi}) = \Pi_p(u_i, \varphi) + \delta\Pi_p + \delta^2\Pi_p \tag{3.2}$$

As a stationary conditions of $\Pi_p(\widetilde{u}_i, \widetilde{\varphi})$, $\delta\Pi_p = 0$. On the other hand, for the given exact displacements u_i and electric potentialφwhich satisfied the boundary conditions $\bar{u}_i\,|_{Su} = 0$, $\bar{\varphi}\,|_{S\varphi} = 0$,

$$\int_{S_\sigma} \overline{T_i} u_i - \int_{S_w} \overline{q_s}\varphi = 2\int_v H(\varepsilon_{ij}, E_i)dv \tag{3.3}$$

should be holding, such that

$$\begin{aligned} \Pi_p(u_i, \varphi) &= \textstyle\int_v H(\varepsilon_{ij}, E_i) - \int_{S_\sigma} \overline{T_i} u_i ds + \int_{S_w} \overline{q_s}\varphi ds \\ &= -\textstyle\int_v H(\varepsilon_{ij}, E_i) = -\int_v H(u_i, \varphi) \end{aligned} . \tag{3.4}$$

Besides it is obvious that

$$\delta^2\Pi_p = \int_v H(\delta u_i, \delta\varphi)$$

In accordance with the definition of J-integral and its positive definite attribute, we have

$$J(\widetilde{u}_i, \widetilde{\varphi}) = -\frac{d}{da}\Pi_p(\widetilde{u}_i, \widetilde{\varphi}) = J(u_i, \varphi) + \delta^2 J \tag{3.5}$$

where

$$J(u_i, \varphi) = -\frac{d}{da}\Pi_p(u_i, \varphi) = \frac{d}{da}\int_v H(\delta u_i, \delta\varphi)dv \geq 0 \tag{3.6}$$

$$\delta^2 J = -\frac{d}{da}\delta^2\Pi_p = -\frac{d}{da}\int_v H(\delta u_i, \delta\varphi)dv \tag{3.7}$$

Observing that $H(u_i, \varphi)$ and $H(\delta u_i, \delta\varphi)$are all the energy functions, they possess the same functions configurations, and must take the same varying regularity when stable crack in the system is developing. Thus, the comparison of eqns (3.6) and (3.7) indicate that $\delta^2 J \leq 0$, and then the inequality (3.1) must be true.

4. Upper Bound Theorem For I

For a given piezoelectric crack system with homogeneous boundary constraints, $\bar{u}_i\,|_{Su} = 0$, $\bar{\varphi}\,|_{S\varphi} = 0$, the approximate solution of the I-integral based on complementary energy principle is always an upper bound of its actual one:

$$I(\widetilde{\sigma}_{ij}, \widetilde{D}_i) \ \geq I(\sigma_{ij}, D_i) \tag{4.1}$$

where $(\widetilde{\sigma}_{ij}, \widetilde{D_i})$ and (σ_{ij}, D_i) are respectively the actual solution and that given by the stress-electric displacement equilibrium finite element method.

Proof. Let $\widetilde{\sigma}_{ij} = \sigma_{ij} + \delta\sigma_{ij}$, $\delta\sigma_{ij}$are virtual stresses; $\widetilde{D}_i = D_i + \delta D_i$,$\delta D_i$ is the virtual electric displacement.

$$\Pi_c(\widetilde{\sigma}_{ij}, \widetilde{D}_i) = \Pi_p(\sigma_{ij}, D_i) + \delta\Pi_c + \delta^2\Pi_c \tag{4.2}$$

As a stationary conditions of $\Pi_c(\widetilde{\sigma}_{ij}, \widetilde{D}_i)$, $\delta\Pi_c = 0$. On the other hand, due to $\bar{u}_i\,|_{Su} = 0$, $\bar{\varphi}\,|_{S\varphi} = 0$, the complimentary energy functional is now

$$\Pi_c(\sigma_{ij}, D_i) = \int_v G(\sigma_{ij}, D_i) dv \tag{4.3}$$

Besides

$$\delta^2\Pi_c = \int_v G(\delta\sigma_{ij,}\delta D_i) dv \tag{4.4}$$

In respect to the exact solutions, $I(\sigma_{ij}, D_i) = J(u_i, \varphi) \geq 0$. Thus, by means of eqns (4.2)-(4.4), we have

$$I(\widetilde{\sigma}_{ij}, \widetilde{D}_i) = I(\sigma_{ij}, D_i) + \delta^2 I \tag{4.5}$$

$$I(\sigma_{ij}, D_i) = \frac{d}{da}\Pi_c(\sigma_{ij}, D_i) = \frac{d}{da}\int_v G(\sigma_{ij,}D_i) dv \geq 0 \tag{4.6}$$

$$\delta^2 I = \frac{d}{da}\delta^2\Pi_c = \frac{d}{da}\int_v G(\delta\sigma_{ij,}\delta D_i) dv \tag{4.7}$$

Observing that $G(\sigma_{ij,}D_i)$ and $G(\delta\sigma_{ij,}\delta D_i)$are all the positive definite complementary energy functions, and they possess the same functions configurations. Therefor, they must take the same varying regularity when the system is suffering from a stable crack development. By comparison of eqns (4.6) and (4.7), we find $\delta^2 I \geq 0$, and then the inequality (4.1) must be true.

5. Dual Finite Element Formulation

In respect to 2D problem, a 4-node piezoelectric isoparametric element can be formulated by using the bilinear interpolation functions for $\widetilde{u}_i$ and $\widetilde{\varphi}$ respectively. The resulting compatible model can be employed to estimate the lower bound of $J-$integral as mentioned above. In this section, we attempt to develop the hybrid element approach into the bound analysis for I-integral. Our goal is to endow the hybrid model with the ability to estimate bounds of I as that the equilibrium model does. In the piezoelectric hybrid element, the nodal d.o.f.s include the two displacement components and the electric potential. They and the co-ordinates are interpolated, i.e.

$$\left\{ \begin{array}{c} u \\ v \\ \varphi \end{array} \right\} = \frac{1}{4}\sum_{i=1}^{4}(1+\xi_i\xi)(1+\eta_i\eta)\left\{ \begin{array}{c} u_i \\ v_i \\ \varphi_i \end{array} \right\} = \mathbf{N}\,\mathbf{q} \tag{5.1}$$

$$\mathbf{q} = \{u_1, v_1, \varphi_1, \ldots, u_4, v_4, \varphi_4\}^T \tag{5.2}$$

The stress field of Pian-Sumihara element is employed [10], i.e.

$$\left\{ \begin{array}{c} \sigma_x \\ \sigma_y \\ \tau_{xy} \end{array} \right\} = \left[\begin{array}{ccccc} 1 & 0 & 0 & a_3^2\xi & a_1^2\eta \\ 0 & 1 & 0 & b_3^2\xi & b_1^2\eta \\ 0 & 0 & 1 & a_3b_3\xi & a_1b_1\eta \end{array} \right] \left\{ \begin{array}{c} \beta_1 \\ \vdots \\ \beta_5 \end{array} \right\} = \mathbf{P}_m\beta_m \tag{5.3}$$

in which the geometric parameters are :

$$\left[\begin{array}{cc} a_1 & b_1 \\ a_2 & b_2 \\ a_3 & b_3 \end{array} \right] = \frac{1}{4}\left[\begin{array}{cccc} -1 & 1 & 1 & -1 \\ 1 & -1 & 1 & -1 \\ -1 & -1 & 1 & 1 \end{array} \right] \left[\begin{array}{cc} x_1 & y_1 \\ x_2 & y_2 \\ x_3 & y_3 \\ x_4 & y_4 \end{array} \right] \tag{5.4}$$

The element stress (5.3) can be degenerated from uncoupled complete linear expansions by the mechanical energy consistency condition [11,12].

For the element electric displacement, it is initially assumed as a linear function:

$$\mathbf{D} = \left\{ \begin{array}{c} D_x \\ D_y \end{array} \right\} = \left[\begin{array}{cccccc} 1 & 0 & \xi & \eta & 0 & 0 \\ 0 & 1 & 0 & 0 & \xi & \eta \end{array} \right] \left\{ \begin{array}{c} \beta_1 \\ \beta_2 \\ \beta_3 \\ \vdots \\ \beta_6 \end{array} \right\} = \boldsymbol{D}_h + \boldsymbol{D}_c \tag{5.5}$$

where $\mathbf{D}_c$ and $\mathbf{D}_h$ are the constant mode and the high order mode respectively. For purpose of improving numerical performance of the piezoelectric hybrid element, the electric energy consistency condition [13] will be introduced into $\mathbf{D}_h$. Such that two β-parameters will be eliminated from Eqn. (5.5), and the desirable electric displacement can be formulated as

$$\left\{ \begin{array}{c} D_x \\ D_y \end{array} \right\} = \left[\begin{array}{cccc} 1 & 0 & a_3\xi & a_1\eta \\ 0 & 1 & b_3\xi & b_1\eta \end{array} \right] \left\{ \begin{array}{c} \beta_6 \\ \vdots \\ \beta_9 \end{array} \right\} = \mathbf{P}_e\beta_e \tag{5.6}$$

For finite element formulation, the Reissner functional can be written as :

$$\Pi_R(\mathbf{u}, \varphi, \sigma, \mathbf{D}) = \sum_{elements} \Pi_R^e(\mathbf{u}^e, \varphi^e, \sigma^e, \mathbf{D}^e) \tag{5.7}$$

is the elementwise functional. The superscript "e" denotes an individual element. After invoking Eqn.(5.1), Eqn.(5.3) and Eqn.(5.6), the elementwise functional becomes :

$$\Pi_R^e(\beta, \mathbf{q}) = \beta^T \mathbf{G}\,\mathbf{q} - \frac{1}{2}\beta^T \mathbf{H}\beta - \text{load terms} \tag{5.8}$$

where

$$\mathbf{G} = \int_{V^e} \mathbf{P}^T \mathbf{D}\mathbf{N} dv, \quad \mathbf{H} = \int_{V^e} \mathbf{P}^T \mathbf{S}\mathbf{P} dv, \quad \mathbf{P} = \left[\begin{array}{cc} \mathbf{P}_m & \mathbf{0}_{3\times4} \\ \mathbf{0}_{2\times5} & \mathbf{P}_e \end{array} \right]^T$$

here $\boldsymbol{D}$ is the strain-displacement matrix and $\boldsymbol{S}$ is the piezoelectricity constitution matrix [4]

As the stationary condition of Π_R, we have

$$\mathbf{H} \left\{ \begin{array}{c} b_m \\ b_e \end{array} \right\} = \mathbf{G}\,\mathbf{q} \text{and thus}$$

$$\Pi_R^e = \frac{1}{2}\mathbf{q}^T (\mathbf{G}^T \mathbf{H}^{-1} \mathbf{G})\,\mathbf{q} - \text{load terms} \tag{5.9}$$

Hence, The element matrix is $\mathbf{G}^T\mathbf{H}^{-1}\mathbf{G}$. The element stress and electric displacement can be retrieved by using Eqns (5.3) and (5.5).

On the other hand, for avoiding the difficulty to find reliable and practical equilibrium elements, a penalty-equilibrium element is developed here. For individual piezoelectric element, Reissner energy functional $\Pi_R^e(u_i, \sigma_{ij}, \varphi, D_i)$ is now generalized as

$$\Pi_{RG}^e = \Pi_R^e - (\ \alpha_1 \int_{V^e} \sigma_{ij,j}\sigma_{ij,j} dv \text{ - } \alpha_2 \int_{V^e} D_{i,j} D_{i,j} dv\ ) \tag{5.10}$$

in which the penalty factor α_1 and α_2 are taken to be a large constant (e.g. 10^4).

$$\Pi^e_{RG}(\beta, \mathbf{q}) = \beta^T \mathbf{G}\,\mathbf{q} - \frac{1}{2}\beta^T \mathbf{H}\beta - \beta^T \mathbf{H}_p \beta \tag{5.11}$$

where

$$\mathbf{H}_p = \iint_{v^e} (\partial \mathbf{P})^T (\partial \mathbf{P}) dv,$$

here ∂ is the differential operator.

In such a way, the equilibrium equations, $\sigma_{ij,j} = 0$ and $D_{i,j} = 0$, will be enforced within element, such that the functional $\Pi^e_R(u_i, \sigma_{ij}, \varphi, D_i) \Rightarrow \Pi^e_c(\sigma_{ij}, D_i)$ being the complementary energy one. Consequently, the hybrid element based on Π^e_R will degenerate into an equilibrium element based on Π^e_c.

Besides the traction free condition $\sigma_{ij}n_j = 0,\ D_i n_i = 0$ on the crack surface should be satisfied exactly because the stresses and strains are singular in the domain near crack tip, and the numerical solutions I, is very sensitive to the free traction constraint. In comparison with the assumed displacement finite element, one of advantages for hybrid stress element would be easily to impose the homogeneous boundary traction condition. Here we take the quadrilateral element in Figure 5.1 as an illustrating example. The element edge $\eta = 1$ is lies in the crack face. The free traction condition can be expressed as

$$\sigma_y = 0,\ \tau_{xy} = 0,\ D_y = 0$$

A smooth function $\Omega(\eta) = \frac{1}{4}(1+\eta)^2$ is introduced to enforce the traction condition. Therefore, the revised stress field and the electric displacement field can be expressed as follows

$$\begin{Bmatrix} \sigma_x \\ \sigma_y \\ \tau_{xy} \\ D_x \\ D_y \end{Bmatrix} = \begin{bmatrix} 1 & & & & \\ & \Omega(\eta) & & & \\ & & \Omega(\eta) & & \\ & & & 1 & \\ & & & & \Omega(\eta) \end{bmatrix}$$
$$\begin{bmatrix} 1 & & & & & a_1^2\eta & a_3^2\xi & 0 & 0 \\ & 1 & & & & b_1^2\eta & b_3^2\xi & 0 & 0 \\ & & 1 & & & a_1b_1\eta & a_3b_3\xi & 0 & 0 \\ & & & 1 & & 0 & 0 & a_1\eta & a_3\xi \\ & & & & 1 & 0 & 0 & b_1\eta & b_3\xi \end{bmatrix} \begin{Bmatrix} \beta_1 \\ \beta_2 \\ \beta_3 \\ \vdots \\ \beta_9 \end{Bmatrix} \tag{5.12}$$

Many numerical results show that the obtained penalty model becomes a wonderful tool to estimate the upper bound of $I-$integral.

6. Numerical Implementation

In numerical calculations, the well known 4-node isoparametric element Q4 and the present penalty-equilibrium element P-S(α) are employed to estimate J and I respectively. The center cracked panel CCP

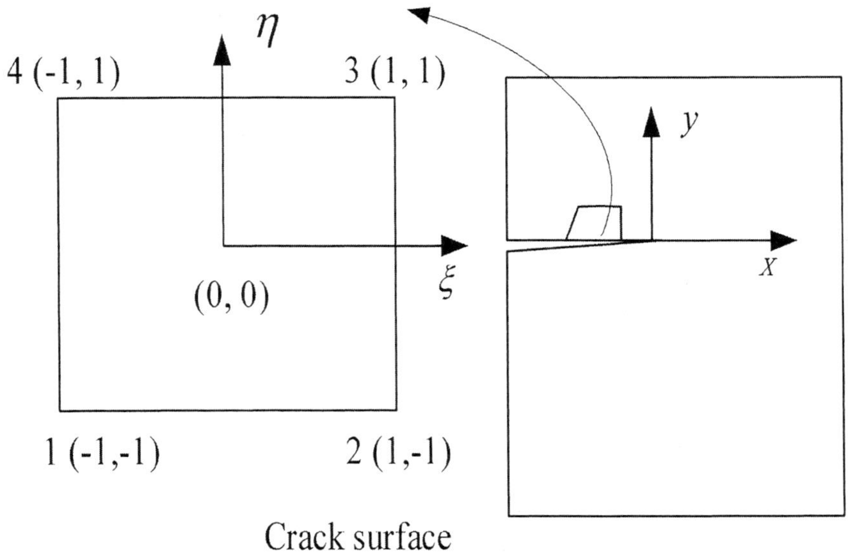

Figure5.1. The crack surface element

(shown in Figure 6.1) with pure mechanical loading and mixed mechanical/ electrical loading is calculated. Only a quarter of the 8×8 specimen needs to be considered due to the symmetry. Three finite element meshes and two integral paths are shown in Figure 6.2-a, b, c. For the analysis, PZT-4 and PZT-5H material properties, as given in Table 6.1, were used.

Table 6.1. Material constants for PZT-4 and PZT-5H where C_{ij}'s are in 10GNm^{-2}, e_{ij}'s are in Cm^{-2} and $\in_{ij}$'s are μCV^{-1}m^{-1}.

	C_{11}	C_{12}	C_{13}	C_{33}	C_{44}	e_{31}	e_{33}	e_{15}	$\in_{11}$	$\in_{33}$
PZT-4	13.9	7.78	7.43	11.3	2.56	6.98	13.84	13.44	6.00	5.47
PZT-5H	12.6	5.5	5.3	11.7	3.53	-6.5	23.3	17.0	15.1	13.0

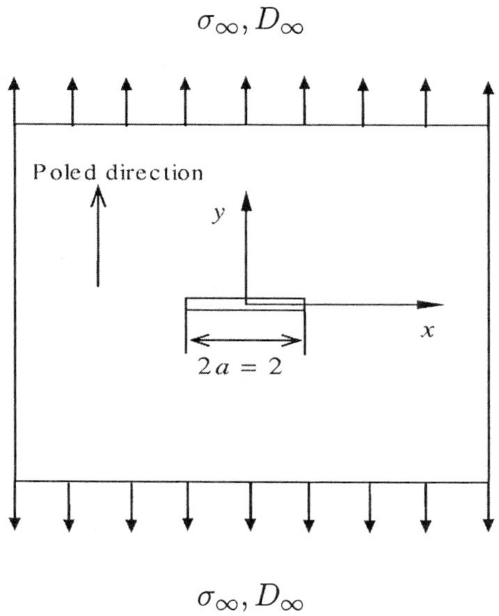

Figure 6.1. The center crack panel specimen (CCP)

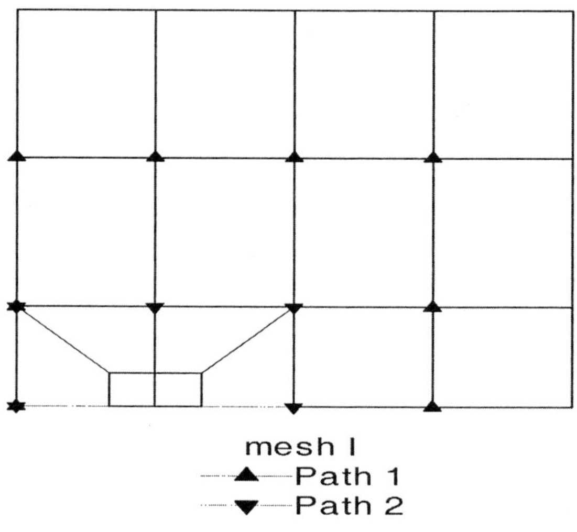

Figure 6.2-a.

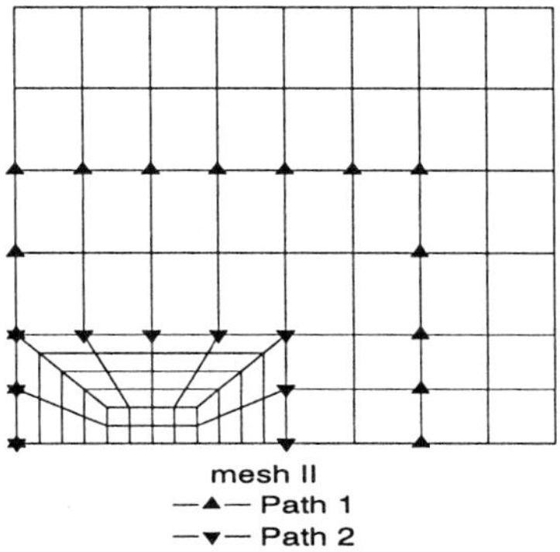

Figure 6.2-b.

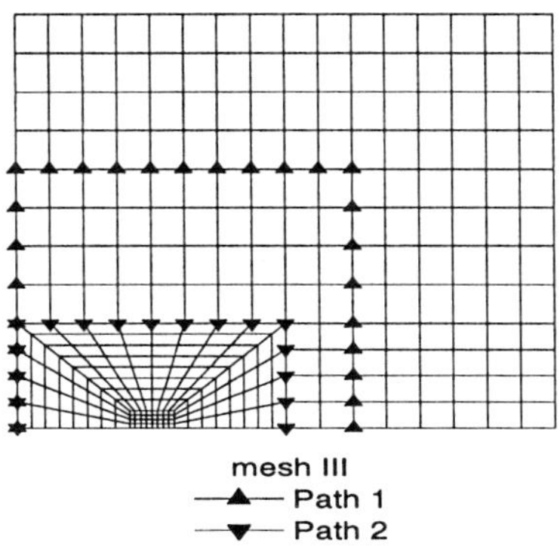

Figure 6.2-c.

Figure 6.2. The employed finite element meshes and the selected integral paths

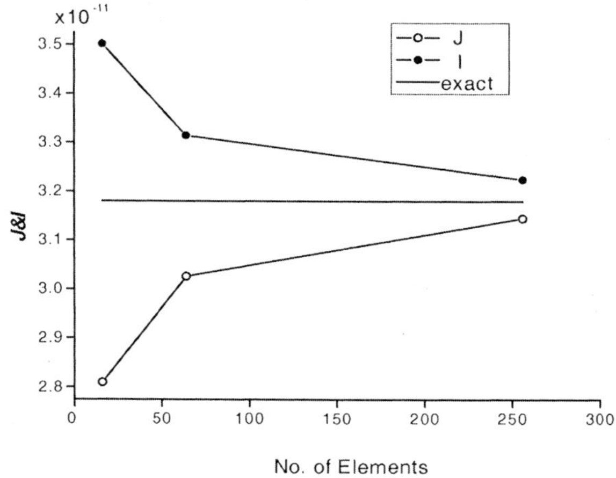

$(\boldsymbol{D}_\infty = 0, \sigma_\infty = 1.0)$

Figure 6.3-a

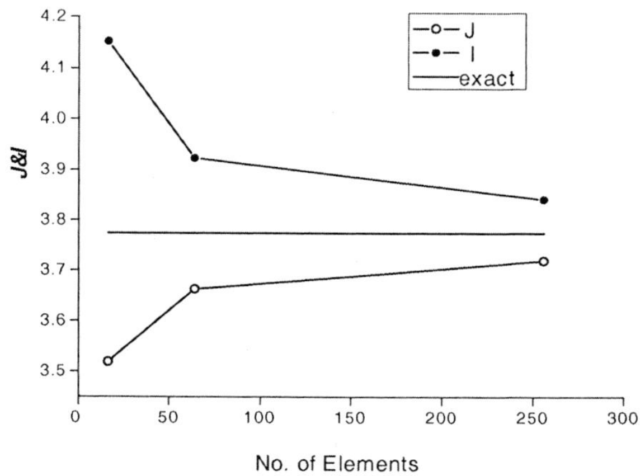

$(\boldsymbol{D}_\infty = 1.0 \times 10^{-10}, \sigma_\infty = 1.0)$

Figure 6.3-b

Figure 6.3. The bound behavior of the solutions of J and I in three meshes with different densities(PZT-4)

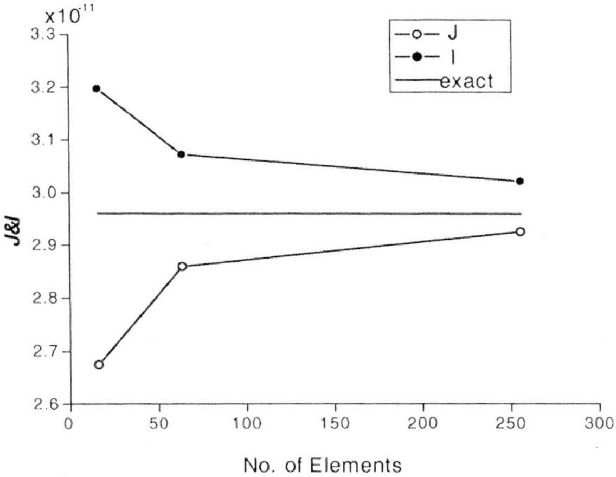

$(\boldsymbol{D}_\infty = 0, \sigma_\infty = 1.0)$

Figure 6.4-a

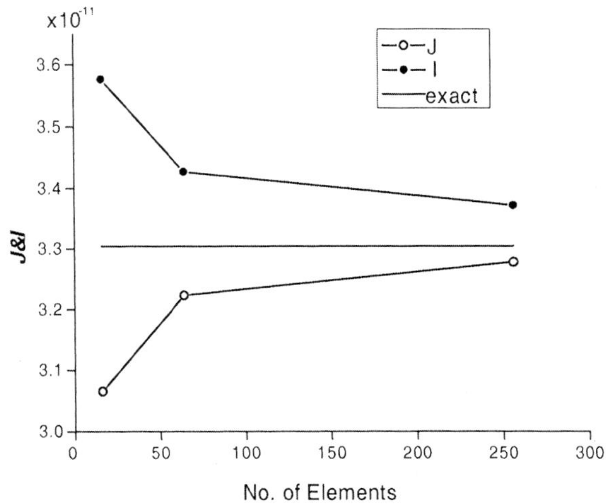

$(\boldsymbol{D}_\infty = 1.0 \times 10^{-10}, \sigma_\infty = 1.0)$

Figure 6.4-b

Figure 6.4. The bound behavior of the solutions of J and I in three meshes with different densities(PZT-5h)

To inspect the convergence behavior of the solutions of J and I, three meshes with different densities and two independent integral paths are simultaneously considered for each specimen. The results are the average of that of the two paths as they are so close. From the results shown in Figure 6.3-a,b and Figure 6.4-a,b it can be seen that for different mechanical and/or electrical loading, the solutions of J by Q4 always converge to the exact one from a certain lower bound. On the contrary, the solutions of I by P-S(α) always converge to the exact one from a certain upper bound. All the numerical solutions demonstrate the bound theorems presented in the paper.

7. Conclusions

The suggested I-integral makes the dual analysis possible for piezoelectric fracture. In general, we have $J(\widetilde{u}_i, \widetilde{\varphi}) \leq J(u_i, \varphi) = I(\sigma_{ij}, D_i) \leq I(\widetilde{\sigma}_{ij}, \widetilde{D}_i)$.

The lower bound of J can be estimated by the present piezoelectric compatible element.

The upper bound of Ican be estimated by the present piezoelectric penalty-equilibrium element.

8. Acknowledgments

The assistance from the *National Natural Science Foundation of China* through project No. 10172078 and 10102018 is gratefully acknowledged. The support from Ministry of ECSST, Japan (under Scientist Exchange Program) is also gratefully acknowledged.

References

[1] Parton, V. Z. (1976) Fracture mechanics of piezoelectric materials. *Acta Astronaut.* **3**, 671-683.

[2] Deeg, W. F. (1980) *The analysis of dislocation, crack and inclusion problems in piezoelectric solids.* Ph.D. dissertation, Stanford University, CA.

[3] Kumar, S. and Singh, R. N. (1996) Crack propagation in piezoelectric materials under combined mechanical and electrical loadings. *Acta Mater.* **44**, 173-200.

[4] Kumar, S. and Singh, R. N. (1997) Energy release rate and crack propagation in piezoelectric materials. Part I: mechanical/electrical load, Part II: combined mechanical and electrical loads. *Acta Mater.*, **45**, 849-868.

[5] Wu C.C., Xiao Q.Z. and Yagawa G.(1998), Dual analysis for path integrals and bounds for crack parameter, *Int.J. Solids and Structures*, 35, 1635-1652.

[6] Wu C.C., Xiao Q.Z. and Li, Z.R. Fracture estimation: bound theorem and numerical strategy, In *Comput. Meth. in Struct. Eng.*, eds F.Y.Cheng and Y.Gu, 349-360, Elsevier, 1999.

[7] Sze, K.Y., Wu, C.C. and Huang Y.Q.(1999)., Upper and lower bounds for evaluation of nonlinear fracture parameters, *Int.J. Engineering Fracture Mechanics,* 62,185-201

[8] Pak, Y. E. (1990) Crack extension force in a piezoelectric material. *J.Appl.Mech.* **57**, 647-653.

[9] Suo Z., Kuo C.-M., Barnett D.M. and Willis J.R.(739-765). Fracture mechanics in piezoelectric material, *J.Mech.Phys.Solids.* 40:4, 739-765

[10] Pian, T. H. H. and Sumihara, K. (1984) Rational approach for assumed stress finite elements. *Inter.J.Numer.Methods Engrg.* **20**, 1685-1695.

[11] Pian, T. H. H. and Wu, C. C. (1988) A rational approach for choosing stress term for hybrid finite element formulations. *Inter.J.Numer.Methods Engrg.* **26**, 2331-2343.

[12] Wu, C. C. and Cheung, Y. K. (1995) On optimization approach of hybrid stress elements, *Finite Elements in Analysis and Design* **21**, 111-128.

[13] Liu, M. and Wu, C. C. (1999) Numerical analysis of crack-tip fields of piezoelectric media by hybrid finite element., *Chinese Science Bulletin* **44**, 995-1000.

Chapter 21

DUALITY AND COMPLEMENTARITY IN CONSTRAINED MECHANICAL SYSTEMS

Hiroaki Yoshimura
Department of Mechanical Engineering, Waseda University
3-4-1, Okubo, Shinjuku, Tokyo 169-8555, JAPAN
yoshimura@waseda.jp

Abstract The duality and complementarity appeared in classical mechanics are investigated, specifically focusing upon mechanical systems with ideal constraints. Structural relations of constrained mechanical systems are defined by employing the nonenergic condition, and a duality principle is illustrated in conjunction with the structural relations and the nonenergic condition. It is demonstrated how dynamical equations of motion can be obtained as complementary and dual relations associated to given kinematical constraints in the context of the variational principle of Hamilton.

Keywords: Constrained mechanical systems, connection matrices, duality principle, complementarity, variational principles.

1. Introduction

In analysis of classical mechanical systems such as space structures comprised of flexible and rigid bodies with joints, vehicles and artificial manipulators, we often meet constrained systems with ideal kinematical constraints, whether holonomic or nonholonomic. The kinematical constraints implicitly induce dynamical relations associated to constraint forces, which are usually represented by employing Lagrange multipliers. From the viewpoint of modeling, as the number of elements or bodies of the systems is increasing, it would be very difficult even to manipulate such kinematical and dynamical constraints as well as to formulate mathematical models (Kane,1965; Roberson and Wittenburg,1966; Schiehlen,1997). This implies that we need some systematic idea to structurally investigate complicated mechanical systems. The geometric mechanics approach is no doubt a powerful mathematical tool, because it enables us intrinsically not only to understand the structure of mechanical systems but also to formulate dynamics without using any specific coordinate systems (Abraham and Marsden, 1978; Arnold, 1984; Marsden and Ratiu,1999). Many researches have been done to geometrically formulate the constrained mechanical systems such as nonholonomic systems (Vershik and Faddeev,1981; Weber,1986; Koiller,1992; Bloch, Krishnaprasad, Marsden and Murray,1996).

The idea of interconnections has been a powerful modeling tool since it enables us to regard a complicated system as an aggregated system of interconnecting subsystems (Kron,1963). The interconnection originally came from electric network theory as a tool to recognize a large scale system as a network throughout interconnections of subsystems, which was also studied from the differential geometric viewpoint (Kron,1934; Kondo,1962). Recently, the interconnection of conservative physical systems was investigated in the context of Hamiltonian systems (van der Schaft and Maschke, 1995). It is noteworthy that there exists a beautiful duality structure called Planck-Okada-Arsove's logic in network theory (Iri, Shirakawa, Kajitani and Shinoda,1983; Arsove,1953).

In classical mechanics, although it has been known that there exist a formal duality and complementarity in theory of elasticity (Gao,1999), the analogue of the duality and complementarity in mechanics of particle and rigid body systems has been rarely discussed. This is because, in mechanics of the particle and rigid body systems, we normally focus upon the primal formulation of dynamical equations of motion based on the variational principle of Hamilton rather than the dual formulation of kinematical constraints, and eventually we have not paid much attention to structural relations existing between kinematical constraints and their dual dynamical equations of motion (Arthurs,1970; Crandall, Karnopp, Kurz, and Pridmore-Brown,1968). In other word, there might be a formal duality and complementarity in mechanics, that is, required dynamical equations of motion might be regarded as complementary and dual relations associated to the given kinematical constraints throughout the variational principle.

In the paper, we study the duality and complementarity appearing in constrained mechanical systems in the context of the variational principle, and we illustrate a systematic treatment of kinematical and dynamical constraints by developing a set of dual connection matrices. First we describe the nonenergic condition coined by Birkhoff (1927) in the context of the Lagrangian mechanics. Second we show how the system structures can be defined by employing the nonenergic condition together with a set of dual connection matrices. Then we demonstrate that the duality principle plays an essential role in understanding structures of the constrained mechanical systems. Third we describe that the set of dual connection matrices provides structural properties of the constrained mechanical systems, by which we develop projections that enable us to formulate kinematical and dynamical relations of constrained mechanical systems. Finally we illustrate how required dynamical equations of motion are to be complementary and dual relations as to original kinematical constraints and also how the duality structure of the constrained mechanical systems is incorporated into the context of variational principle.

2. Fundamentals

2.1 Constrained Lagrangian Mechanics

Consider a conservative mechanical system whose configuration is denoted by n generalized coordinates $x^1, ..., x^n$ and the motion of the mechanical system is given by

$$t \mapsto x(t) = (x^1(t), ..., x^n(t)), \tag{2.1}$$

where t is time defined in some open interval of $\mathbb{R}$ including 0. The time derivative of (2.1) at $t = \tau$ denotes the velocity of the point x as

$$v = \sum_{i=1}^{n} \left.\frac{dx^i}{dt}\right|_{t=\tau} \left(\frac{\partial}{\partial x^i}\right).$$

Because v can be recognized to be a differential operator which is applied to a function $f(x)$ such that

$$vf = \left.\frac{df(x(t))}{dt}\right|_{t=\tau} = \sum_{i=1}^{n} \left.\frac{dx^i}{dt}\right|_{t=\tau} \left(\frac{\partial f}{\partial x^i}\right),$$

differential operators such as v consist of a vector space called the tangent space at x, and hence admissible displacements or velocities are apparently tangent vectors. On the other hand, the differential of f at x can be regarded as a map acting on the tangent vector at $x(\tau)$ such that

$$v \mapsto \langle df, v\rangle = \sum_{i=1}^{n} \left.\frac{dx^i}{dt}\right|_{t=\tau} \left(\frac{\partial f}{\partial x^i}\right) \in \mathbb{R},$$

where

$$df = \sum_{i=1}^{n} \frac{\partial f}{\partial x^i} dx^i.$$

In the above, $\langle , \rangle$ indicates a pairing of dual vectors. The differentials at x of such functions consist of a vector space called the cotangent space at x. For example, since conservative forces are derived from a potential function $f(x)$ such that $F_i = -\partial f / \partial x^i$, the force $F = (F_1, ..., F_n)$ is regarded as a cotangent vector at x.

Variational Principle. Let $L(x, \dot{x})$ be the Lagrangian of the mechanical system. The variational principle of Hamilton is to seek the curve $x(t)$ such that the action functional is stationary as

$$\delta \int_{t_1}^{t_2} L(x, \dot{x}) dt = 0,$$

where the variations of $x^i(t)$ at the fixed endpoints are assumed to be zero, that is, $\delta x^i(t_1) = \delta x^i(t_2) = 0$. Then we have the Lagrange-d'Alembert principle as

$$\sum_{i=1}^{n} \left(\frac{d}{dt} \frac{\partial L}{\partial \dot{x}^i} - \frac{\partial L}{\partial x^i} \right) \delta x^i = 0.$$

If δx^i are linearly independent, we have

$$\frac{d}{dt} \frac{\partial L}{\partial \dot{x}^i} - \frac{\partial L}{\partial x^i} = 0, \qquad i = 1, ..., n.$$

Consider the case that there exist m kinematical constraints as

$$\sum_{i=1}^{n} a_{ri} \, \dot{x}^i = 0, \qquad r = 1, ..., m < n,$$

and we have

$$\frac{d}{dt} \frac{\partial L}{\partial \dot{x}^i} - \frac{\partial L}{\partial x^i} = F_i, \qquad i = 1, ..., n.$$

Here we denote by $F_1, ..., F_n$ constraint forces, which are represented, using Lagrange's multipliers $\lambda_1, ..., \lambda_m$, by

$$F_i = \sum_{r=1}^{m} a_{ri} \lambda_r, \qquad i = 1, ..., n.$$

Thus, as the mathematical model of the constrained mechanical system, we obtain a set of $n + m$ differential algebraic equations, which consists of n dynamical equations of motion and m algebraic equations of the kinematical constraints as follows:

$$\begin{aligned} &\frac{d}{dt}\frac{\partial L}{\partial \dot{x}^i} - \frac{\partial L}{\partial x^i} = \sum_{r=1}^{m} a_{ri}\lambda_r, && i = 1, ..., n, \\ &\sum_{i=1}^{n} a_{ri}\, \dot{x}^i = 0, && r = 1, ..., m. \end{aligned} \tag{2.2}$$

2.2 Nonenergic Condition

Let us consider a coordinate transformation

$$x^i = x^i(q^1, ..., q^n), \quad i = 1, ..., n.$$

The admissible variations of generalized coordinates are

$$\delta x^i = \sum_{j=1}^{n} \frac{\partial x^i}{\partial q^j}\delta q^j = \frac{\partial x^i}{\partial q^1}\delta q^1 + \cdots + \frac{\partial x^i}{\partial q^n}\delta q^n, \quad i = 1, ..., n,$$

which are also denoted, in matrix, by

$$\delta x = A(q)\, \delta q,$$

where $A(q) = \left[\partial x^i / \partial q^j\right]$, $\delta x = (\delta x^1, ..., \delta x^n)$ and $\delta q = (\delta q^1, ..., \delta q^n)$. We denote by $F = (F_1, ..., F_n)$ dual forces as to $\delta x = (\delta x^1, ..., \delta x^n)$ and the virtual work done by the forces is described by

$$\omega = \sum_{i=1}^{n} F_i\, \delta x^i.$$

The above relation is also represented by

$$\omega = \langle F, \delta x \rangle .$$

As we have the invariance of the virtual work as

$$\sum_{i=1}^{n} F_i \delta x^i = \sum_{j=1}^{n} Q_j \delta q^j, \text{ that is, } \langle F, \delta x \rangle = \langle Q, \delta q \rangle , \tag{2.3}$$

we can derive dynamical relations between $F = (F_1, ..., F_n)$ and $Q = (Q_1, ..., Q_n)$ such that

$$Q_j = \sum_{i=1}^{n} F_i \frac{\partial x^i}{\partial q^j} = F_1 \frac{\partial x^1}{\partial q^j} + \cdots + F_n \frac{\partial x^n}{\partial q^j}, \quad j = 1, ..., n.$$

That is, in matrix,

$$Q = A^T(q)\,F,$$

which is so obtained that

$$\langle F,\ A\,\delta q\rangle = \left\langle A^T\,F,\ \delta q\right\rangle. \tag{2.4}$$

It is apparent that equation (2.4) is another saying of the invariance of the virtual work in equation (2.3). Thus, we have the following theorem with respect to transformation rules of admissible displacements and forces under a coordinate transformation.

Theorem 21.1 *In any mechanical system, a coordinate transformation $x^i = x^i(q^1, ..., q^n)$, $i = 1, ..., n$, induces transformation rules for dual variables, i.e., admissible displacements and forces*

$$\delta x = A(q)\,\delta q,$$
$$Q = A^T(q)\,F,$$

such that the invariance of the virtual work holds: $\langle F, A\,\delta q\rangle = \left\langle A^T F,\ \delta q\right\rangle$, that is, $\langle F, \delta x\rangle = \langle Q, \delta q\rangle$, where $A(q) = \left[\partial x^i / \partial q^j\right]$.

The above theorem is usually known in mechanics as the covariancy associated to a coordinate transformation, which implies that physical laws are invariant under any coordinate transformation (Lanczos,1962). Let us consider the following condition for the paring of dual variables F and δx:

$$\langle F, \delta x\rangle = \sum_{i=1}^{n} F_i\ \delta x^i = 0. \tag{2.5}$$

The above relation implies the condition that the force F does not work, that is, the principle of virtual work. Specifically in the case that $\delta x = \dot{x}$, it was called the nonenergicness by Birkhoff (1927). So, in general, let us call the condition in equation (2.5) *a nonenergic condition.* The nonenergic condition typically appears in constrained mechanical systems, by which kinematical constraints induce their dual dynamical constraints, which we shall describe in details in the following section.

3. Duality in Constraints and Connection Matrices

3.1 Kinematical Constraints

Consider m kinematical constraints by the following Pfaffian forms

$$\omega_r(x) = \sum_{i=1}^{n} a_{ri}\, dx^{\,i} = 0, \quad r = 1, ..., m, \tag{3.1}$$

where a_{ri} are functions of x. If the Pfaffian forms in equation (3.1) are completely integrable, the constraints are said to be holonomic, otherwise nonholonomic.

The admissible displacements or tangent vectors at x are denoted by

$$\delta x = \sum_{i=1}^{n} \delta x^i \left(\frac{\partial}{\partial x^i} \right),$$

which satisfies the kinematical constraints as

$$\langle \delta x, \; \omega_r \rangle = 0, \quad r = 1, ..., m.$$

So we obtain kinematical constraints as

$$\sum_{i=1}^{n} a_{ri} \, \delta x^i = 0, \;\; r = 1, ..., m,$$

which is represented, in matrix, by

$$a(x) \, \delta x = o_m.$$

In the above, we employ Kronecker's delta $\langle \frac{\partial}{\partial x^i}, dx^j \rangle = \delta_i^j$, the matrix $a(x)$ is an $m \times n$ matrix whose coefficients a_{ri} are smooth functions of coordinates x^i, and o_m indicates the m-dimensional zero vector.

Let us split $\delta x = (\delta x^1, ..., \delta x^n)$ into $\delta x_1 = (\delta x^1, ..., \delta x^m)$ and $\delta x_2 = (\delta x^{m+1}, ..., \delta x^n)$ such that

$$a(x)\delta x = \left[\; a_1(x) \mid a_2(x) \; \right] \left[\frac{\delta x_1}{\delta x_2} \right] = o_m$$

and the coefficient matrix $a_1(x)$ as to δx_1 has a full rank m. Because δx_1 can be described in terms of δx_2 as $\delta x_1 = J(x)\delta x_2$, we have

$$\sum_{i=1}^{n} N_{ri} \, \delta x^i = 0, \quad r = 1, ..., m,$$

which is also represented, in matrix, by

$$N(x) \, \delta x = \left[\; I_m \mid -J(x) \; \right] \left[\frac{\delta x_1}{\delta x_2} \right] = o_m. \tag{3.2}$$

Here $J(x) \, (= -a_1^{-1} a_2)$ is an $m \times (n - m)$ matrix and we denote by I_m an $m \times m$ unit matrix. We call equation (3.2) *a causal form of kinematical constraints* and we also call $N(x)$ *a connection matrix*, by which an input-output relation between δx_1 and δx_2 can be assigned in an implicit form. From equation (3.2), we obtain

$$\delta x^i = \sum_{\alpha=m+1}^{n} B_{i\alpha} u^\alpha, \quad i = 1, ..., n.$$

In matrix representation,

$$\delta x = B\, u_2, \quad \text{that is,} \quad \left[\frac{\delta x_1}{\delta x_2}\right] = \left[\frac{J}{I_{n-m}}\right] u_2, \tag{3.3}$$

where $u_2 = (u^{m+1}, ..., u^n)$ indicates independent admissible displacements.

3.2 Dynamical Constraints

Let us derive dynamical constraints induced from the original kinematical constraints from a duality viewpoint. To do this, we consider the nonenergic condition $\langle F, \delta x\rangle = 0$. Substituting kinematical constraints in equation (3.3) into the nonenergic condition, we have

$$\langle F, \delta x\rangle = \langle F, Bu_2\rangle = \langle B^T F, u_2\rangle = 0.$$

Since the above relation is satisfied for all $u_2 = (u^{m+1}, ..., u^n)$, we obtain induced dynamical constraints by

$$B^T(x)\, F = o_{n-m}, \quad \text{or} \quad \left[\, J^T(x) \mid I_{n-m} \,\right] \left[\frac{F_1}{F_2}\right] = o_{n-m}, \tag{3.4}$$

where the force $F = (F_1, ..., F_n)$ is also split into $F_1 = (F_1, ..., F_m)$ and $F_2 = (F_{m+1}, ..., F_n)$. We call equation (3.4) *a causal form of dynamical constraints*, by which an input-output relation between F_1 and F_2 is assigned such that $F_1 = -J^T(x)F_2$. Further we also call $B^T(x)$ *a dual connection matrix*. From (3.4), we can derive

$$F = N^T f_1, \quad \text{or} \quad \left[\frac{F_1}{F_2}\right] = \left[\frac{I_m}{-J^T}\right] f_1,$$

where $f_1 = (f_1, ..., f_m)$ may be regarded as Lagrange's multipliers. From Theorem 1, recall that

$$\delta x = B\, u_2,$$
$$f_2 = B^T F = o_{n-m}.$$

Notice that the constraint force F is vanished by imposing the dual connection matrix B^T. Then, we can formally illustrate that the invariance of the nonenergic condition holds

$$\langle F, \delta x\rangle = \langle f_2, u_2\rangle = 0$$

for arbitrary $u_2 = (u^{m+1}, ..., u^n)$.

4. Duality Principle in Constrained Systems

4.1 Structure of Mechanical Systems

We have demonstrated how kinematical constraints and their dual dynamical constraints are represented by employing a set of dual connection matrices N and B^T. Let us introduce a conceptual idea of *structural relations of constrained mechanical systems.* The structure of the constrained system satisfying the nonenergic condition $\langle F, v\rangle = 0$ is defined by

$$\begin{aligned} N(x)\, v &= o_m && (\textit{Kinematical Constraints}),\\ B^T(x)\, F &= o_{n-m} && (\textit{Dynamical Constraints}). \end{aligned} \tag{4.1}$$

In the above, N and B^T are dual connection matrices, which have the orthogonal property as $NB = O$.

4.2 Duality Principle

Let us investigate how kinematical constraints, their dual dynamical constraints and the nonenergic condition are interrelated with each other throughout *a duality principle.* The dynamical constraints are derivable by imposing the nonenergic condition on the kinematical constraints. Conversely, the kinematical constraints can be derived by imposing the nonenergic condition on the dynamical constraints. To show this, we rewrite the nonenergic condition (2.5) as

$$\langle F, v\rangle = \sum_{i=1}^{n} F_i\, v^i = 0, \tag{4.2}$$

where $v = (v^1, ..., v^n)$ denotes a velocity vector and $F = (F_1, ..., F_n)$ indicates the dual vector of v. As we have derived before, the dynamical constraints (3.4) may be represented by

$$F_i = \sum_{r=1}^{m} N_{ri}\, f_r, \quad i = 1, ..., n, \quad \text{that is,} \quad F = N^T f_1, \tag{4.3}$$

where $f_1 = (f_1, ..., f_m)$ are m independent dual vectors. Substituting equation (4.3) into equation (4.2),

$$\langle F, v\rangle = \left\langle N^T f_1,\, v\right\rangle = \langle f_1,\, Nv\rangle = 0,$$

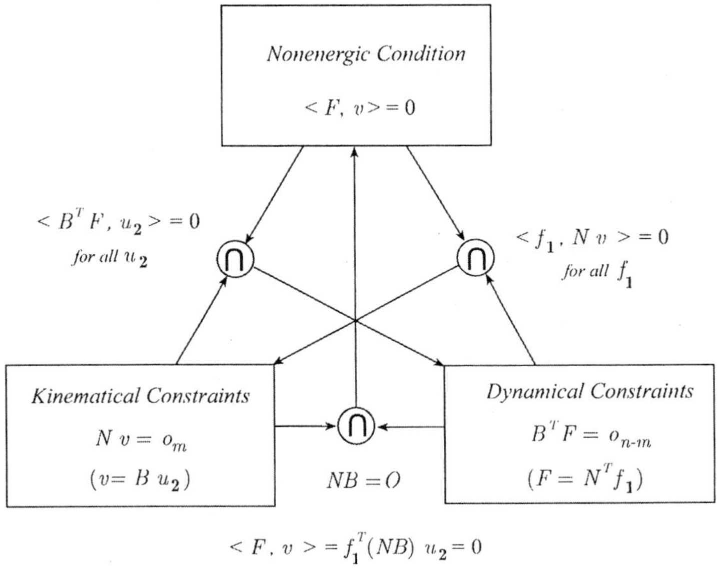

Figure 21.1. Duality Principle and Connection Matrices

which holds for all $f_1 = (f_1, ..., f_m)$, then we have

$$u_1 = N\,v = o_m, \quad \text{that is,} \quad u_1 = \left[\; I_m \;\middle|\; -J(x) \;\right] \left[\frac{v_1}{v_2}\right] = o_m. \quad (4.4)$$

Notice that the invariance of the virtual work principle, that is, the nonenergic condition in the complementary form is denoted by

$$\langle F, v\rangle = \langle f_1, u_1\rangle = 0,$$

which is satisfied for all f_1.
Let us illustrate the structural relations of the constrained mechanical system in Figure 3. It is apparent that the dynamical constraints are induced by imposing the nonenergic condition on the kinematical constraints, while the kinematical constraints can be obtained, conversely, by imposing the nonenergic condition on the dynamical constraints. The kinematical and their dual dynamical constraints satisfy the nonenergic condition. Hence we have the following theorem as to the *duality principle* in mechanics.

Theorem 21.2 *For any mechanical system with ideal constraints, there exists a logical implication between the structural relations and the nonen-*

ergic condition:

1. *Kinematical constraints :* $N\,v = o_m$,
2. *Dynamical constraints :* $B^T\,F = o_{n-m}$,
3. *Nonenergic condition :* $\langle F, v\rangle = 0$.

That is, any one of the three relations in the above can be deduced from the rest of two.

The duality principle is employed in various fields of physical systems such as electrical circuits and elastic bodies (Gao,1999). In the theory of electrical circuits, the analogous theorem is known as a network postulate called *Planck-Okada-Arsove's logic* (Arsove,1953; Takahashi,1974; Iri, Shirakawa, Kajitani and Shinoda,1983); each of kinematical and dynamical constraints corresponds to constraints due to KCL and KVL (Kirchhoff's Current and Voltage Laws) respectively, and the nonenergic condition corresponds to the Tellegen's theorem. The structural relations of mechanical systems are generally nonlinear, which are essentially different from those of the electric circuits, however the analogy of the duality principle is still, locally, satisfied in mechanics (Yoshimura and Kawase, 2001).

4.3 Geometric Properties of Local Structures

We shall see some geometric properties of local structures of constrained mechanical systems. As we have seen, the structural relations can be locally represented by a dual pair of connection matrices N, B^T with the orthogonal property:

$$B^T N^T = O, \quad \text{where } B = \left[\frac{J}{I_{n-m}}\right] \quad \text{and } N^T = \left[\frac{I_m}{-J^T}\right].$$

Let $\{b_{m+1}, ..., b_n\}$ and $\{n^1, ..., n^m\}$ be column vectors of B and N^T respectively, and then we can express

$$v = \sum_{\alpha=m+1}^{n} u^\alpha\, b_\alpha, \qquad F = \sum_{r=1}^{m} f_r\, n^r,$$

In matrix representation,

$$v = B\,u_{\mathbf{2}}, \qquad F = N^T f_{\mathbf{1}}.$$

In the above, v and F indicate a vector on V^{n-m} and its dual vector on $V^{*\,m}$ respectively, where V^{n-m} and $V^{*\,m}$ denote respectively an $n-m$

and m dimensional locally defined admissible (tangent and cotangent) subspaces associated to kinematical constraints and their dual dynamical constraints.

Let us introduce complementary subspaces V^m and $V^{*\,n-m}$ such that base vectors of V^m and $V^{*\,n-m}$ are respectively given by

$$\{b_1, ..., b_m\} = \bar{B}, \quad \{n^{m+1}, ..., n^n\} = \bar{N}^T,$$

where $\bar{B}$ and $\bar{N}^T$ are denoted by

$$\bar{B} = \left[\frac{I_m}{O}\right] \text{ and } \bar{N}^T = \left[\frac{O}{I_{n-m}}\right].$$

Notice that $\bar{B}$ and $\bar{N}^T$ are also orthogonal with each other as

$$\bar{B}^T \bar{N}^T = O.$$

Thus, we may regard v and F respectively as a vector on V^n and its dual vector on $V^{*\,n}$ such that

$$\begin{aligned} v = v_1 + v_2 &= \bar{B}\, u_1 + B\, u_2, \\ F = F_1 + F_2 &= N^T f_1 + \bar{N}^T f_2. \end{aligned} \tag{4.5}$$

In equation (4.5), $v \in V^n$ is split into $v_1 \in V^m$ and $v_2 \in V^{n-m}$, while $F \in V^{*n}$ into $F_1 \in V^{*m}$ and $F_2 \in V^{*n-m}$. It is apparent that $V^n = V^m \oplus V^{n-m}$ and $V^{*n} = V^{*m} \oplus V^{*n-m}$ and also that $u_1 = 0$ and $f_2 = 0$ because of the dual constraints of the structural relations in equation (4.1). Equation (4.5) gives local expressions for the kinematical and their dynamical relations. The base vectors in V^n, V^{*n} are respectively given in local coordinates by

$$[\,\bar{B} \mid B\,] = \left[\begin{array}{c|c} I_m & J \\ O & I_{n-m} \end{array}\right] \text{ and } [N^T \,|\bar{N}^T\,] = \left[\begin{array}{c|c} I_m & O \\ -J^T & I_{n-m} \end{array}\right].$$

Since $\bar{B}^T N^T = I_m$ and $B^T \bar{N}^T = I_{n-m}$ are satisfied, we can define the following projections:

$$\begin{aligned}
P1 &: V^n \to V^{n-m} &&: B\,\bar{N} = \begin{bmatrix} O & J \\ O & I_{n-m} \end{bmatrix}, \\
P2 &: V^n \to V^m &&: \bar{B}\,N = \begin{bmatrix} I_m & -J \\ O & O \end{bmatrix}, \\
P1^* &: V^{*n} \to V^{*\,n-m} &&: \bar{N}^T B^T = \begin{bmatrix} O & O \\ J^T & I_{n-m} \end{bmatrix}, \\
P2^* &: V^{*n} \to V^{*m} &&: N^T \bar{B}^T = \begin{bmatrix} I_m & O \\ -J^T & O \end{bmatrix},
\end{aligned}$$

where $P1 = (P1)^2$, $P2 = I_n - P1$, $P1^* = (P1^*)^2$, and $P2^* = I_n - P1^*$. By the projection $P2 : V^n \to V^m$, we obtain the following relations as to the velocity:

$$\begin{aligned} P2 \circ v_1 &= v_1, \\ P2 \circ v_2 &= 0. \end{aligned}$$

Then

$$\begin{aligned} v_1 &= P2 \circ v, \\ v_2 &= v - P2 \circ v = P1 \circ v. \end{aligned}$$

In matrix, recalling that $N\bar{B} = I_m, \bar{N}B = I_{n-m}, NB = O, \bar{N}\bar{B} = O$, we have

$$\begin{aligned} v_1 &= P2 \circ v = (\bar{B}\ N)\ (v_1 + v_2) \\ &= (\bar{B}\ N)\ (\bar{B}\ u_1 + B\ u_2) \\ &= \bar{B}\ u_1, \\ v_2 &= (I - P2) \circ v = P1 \circ v = (B\ \bar{N})\ (v_1 + v_2) \\ &= (B\ \bar{N})\ (\bar{B}\ u_1 + B\ u_2) \\ &= B\ u_2. \end{aligned}$$

As to the force, similarly, by the projection $P1^* : V^{*n} \to V^{*n-m}$,

$$\begin{aligned} P1^* \circ F_1 &= 0, \\ P1^* \circ F_2 &= F_2, \end{aligned}$$

we have

$$\begin{aligned} F_2 &= P1^* \circ F, \\ F_1 &= F - P1^* \circ F = P2^* \circ F. \end{aligned}$$

In matrix,

$$\begin{aligned} F_2 &= P1^* \circ F = (\bar{N}^T B^T)\ (F_1 + F_2) \\ &= (\bar{N}^T B^T)\ (N^T\ f_1 + \bar{N}^T\ f_2) \\ &= \bar{N}^T f_2, \\ F_1 &= (I_n - P1^*) \circ F = P2^* \circ F \\ &= (N^T \bar{B}^T)\ (F_1 + F_2) \\ &= (N^T \bar{B}^T)\ (N^T\ f_1 + \bar{N}^T\ f_2) \\ &= N^T f_1. \end{aligned}$$

The principle of virtual work, i.e., the nonenergic condition can be described by

$$\begin{aligned}\langle F,\ v\rangle &= \langle F,\ v_2\rangle = \big\langle F,\ (B\bar{N})v\big\rangle \\ &= \langle F,\ B\,u_2\rangle = \big\langle B^T\,F,\ u_2\big\rangle \\ &= \langle f_2,\ u_2\rangle = 0,\end{aligned}$$

which is satisfied for arbitrary u_2, dynamical constraints are derived by

$$f_2 = B^T\ F = o_{n-m}.$$

Similarly, we have a complementary virtual work principle such that

$$\begin{aligned}\langle F,\ v\rangle &= \langle F_1,\ v\rangle = \big\langle (N^T\bar{B}^T)\ F,\ v\big\rangle \\ &= \big\langle N^T f_1,\ v\big\rangle = \langle f_1,\ N\ v\rangle \\ &= \langle f_1,\ u_1\rangle = 0,\end{aligned}$$

which is satisfied for arbitrary f_1, kinematical constraints can be obtained by

$$u_1 = N\ v = o_m.$$

5. Complementarity and Duality in Variational Principles

5.1 Lagrange-d'Alembert Principle

Let $v^i\ (= \delta x^i)$ be admissible variations of x^i, $i = 1, ..., n$, and the Lagrange-d'Alembert principle is given by

$$\sum_{i=1}^{n}\left(\frac{d}{dt}\frac{\partial L}{\partial \dot{x}^i} - \frac{\partial L}{\partial x^i}\right) v^i = 0,$$

which is also denoted, in matrix, by

$$\left\langle \frac{d}{dt}\frac{\partial L}{\partial \dot{x}} - \frac{\partial L}{\partial x},\ v\right\rangle = 0$$

together with the following kinematical constraints:

$$\sum_{j=1}^{n} N_{rj}\, v^j = 0, \quad r = 1, ..., m, \quad \text{that is,} \quad N(x)\, v = o_m. \tag{5.1}$$

Let F be the constraint force associated to the kinematical constraints in equation (5.1). As the constraint force does not work, the nonenergic condition is satisfied as

$$\langle F, v\rangle = \sum_{i=1}^{n} F_i\, v^i = 0.$$

Then, we have

$$\sum_{i=1}^{n}\left(\frac{d}{dt}\frac{\partial L}{\partial \dot{x}^i}-\frac{\partial L}{\partial x^i}-F_i\right)v^i=0.$$

In matrix representation,

$$\left\langle \frac{d}{dt}\frac{\partial L}{\partial \dot{x}}-\frac{\partial L}{\partial x}-F,\ v\right\rangle=0.$$

Because the admissible displacement v and the constraint force F can be expressed by

$$v^i=\sum_{\alpha=m+1}^{n}B_{i\,\alpha}\,u^{\alpha},\quad F_i=\sum_{r=1}^{m}N_{r\,i}\,f_r,$$

i.e., in matrix representation,

$$v=B\,u_{\mathbf{2}},\quad F=N^T f_{\mathbf{1}},$$

we have

$$\sum_{\alpha=m+1}^{n}\sum_{i=1}^{n}B_{i\,\alpha}\left(\frac{d}{dt}\frac{\partial L}{\partial \dot{x}^i}-\frac{\partial L}{\partial x^i}-\sum_{r=1}^{m}N_{r\,i}f_r\right)u^{\alpha}=0,$$

which is also denoted, in matrix, by

$$\left\langle B^T\left(\frac{d}{dt}\frac{\partial L}{\partial \dot{x}}-\frac{\partial L}{\partial x}-N^T f_{\mathbf{1}}\right),\ u_{\mathbf{2}}\right\rangle=0.$$

Since the above relation is satisfied for arbitrary $u_{\mathbf{2}}=(u^{m+1},...,u^n)$ and the constraint forces $F=N^T f_{\mathbf{1}}$ are eliminated such that

$$B^T F=B^T(N^T f_{\mathbf{1}})=(B^T N^T)\,f_{\mathbf{1}}=o_{n-m},$$

we obtain

$$\sum_{i=1}^{n}B_{i\,\alpha}\left(\frac{d}{dt}\frac{\partial L}{\partial \dot{x}^i}-\frac{\partial L}{\partial x^i}\right)=0,\quad \alpha=m+1,...,n,$$

and, thus, in matrix representation,

$$B^T\left(\frac{d}{dt}\frac{\partial L}{\partial \dot{x}}-\frac{\partial L}{\partial x}\right)=o_{n-m}. \tag{5.2}$$

5.2 Complementary Relations

Let us illustrate that the kinematical constraints are derived as the complementary relations associated to the dynamical equations of motion (Crandall, Karnopp, Kurz, and Pridmore-Brown,1968; Arthurs,1970). Let $F = (F_1, ..., F_n)$ be cotangent vectors denoting admissible forces and let $v = \dot{x}$. The nonenergic condition is given by

$$\langle F, \dot{x} \rangle = \sum_{i=1}^{n} F_i \, \dot{x}^i = 0.$$

Note that $F_1, ..., F_n$ are not independent because there exist the following dynamical constraints:

$$\sum_{i=1}^{n} B_{i\,\alpha} \, F_i = 0, \quad \alpha = m+1, ..., n, \quad \text{that is,} \quad B^T(x) \, F = o_{n-m}.$$

Since we have

$$F_i = \sum_{r=1}^{m} N_{ri} f_r, \quad i = 1, ..., n, \quad \text{that is,} \quad F = N^T f_1,$$

the following relation is satisfied:

$$\sum_{r=1}^{m} \left(\sum_{i=1}^{n} N_{ri} \dot{x}^i \right) f_r = 0.$$

In matrix representation,

$$\langle N \, \dot{x}, f_1 \rangle = 0.$$

Because the above relation is satisfied for all $f_1 = (f_1, ..., f_m)$, the kinematical constraints are derived by

$$\sum_{i=1}^{n} N_{ri} \dot{x}^i = 0, \quad r = 1, ..., m,$$

that is, in matrix,

$$N \, \dot{x} = o_m. \tag{5.3}$$

Note that equation (5.3) may be regarded as complementary kinematical relations associated to the dynamical equations of motion in equation (5.2). Hence, we obtain equations (5.2) and (5.3) as a set of n differential algebraic equations for constrained mechanical systems such that

$$\sum_{i=1}^{n} B_{i\,\alpha}\left(\frac{d}{dt}\frac{\partial L}{\partial \dot{x}^i} - \frac{\partial L}{\partial x^i}\right) = 0, \qquad \alpha = m+1, ..., n,$$

$$\sum_{i=1}^{n} N_{ri}\dot{x}^i = 0, \qquad r = 1, ..., m.$$

In matrix, we have

$$\begin{aligned} B^T\left(\frac{d}{dt}\frac{\partial L}{\partial \dot{x}} - \frac{\partial L}{\partial x}\right) &= o_{n-m}, \\ N\,\dot{x} &= o_m. \end{aligned} \tag{5.4}$$

The first $n-m$ ordinary differential equations indicate the (primal) dynamical equations of motion and the second m equations imply the (dual) kinematical equations, which are complementary with each other throughout the nonenergic condition. Obviously, equation (5.4) is a set of n differential algebraic equations obtained by eliminating the Lagrange multipliers from the original $m+n$ differential algebraic equations in equation (2.2).

6. Summary

We have studied the complementarity and duality appearing in classical mechanics, specifically focusing upon mechanical systems with ideal constraints. We have developed a conceptual idea of structural relations of constrained mechanical systems using the nonenergic condition and also have demonstrated that the structural relations are represented by dual connection matrices N and B^T. Then, we have illustrated that the duality principle plays an essential role in structurally understanding kinematical constraints, induced dynamical constraints, and the nonenergic condition. We have finally shown that the original kinematical constraints can be regarded as complementary and dual relations associated to the dynamical equations of motion in the context of the variational principle of Hamilton.

Acknowledgments

The research is partially supported by Waseda University Grant for Special Research Projects No.2002A-554.

References

R. Abraham and J. E. Marsden [1978], *Foundations of Mechanics*, Second Edition, Addison-Wesley.

V. I. Arnold [1984], *Mathematical Methods of Classical Mechanics*, Springer Verlag.

M. G. Arsove [1953], A note on the network postulates, *J. Math. and Phys.*, 32, pp.203-206.

A. M. Arthurs [1970], *Complementary Variational Principles*, Oxford University Press.

G. D. Birkhoff [1927], *Dynamical Systems*, AMS.

A. M. Bloch, P. S. Krishnaprasad, J. E. Marsden, and R. M. Murray [1996], Nonholonomic mechanical systems with symmetry, *Arch. Rational Mech. Anal.*, Vol.136, pp.21-99.

S. H. Crandall, D. C. Karnopp, E. F. Kurz, Jr., and D. C. Pridmore-Brown [1968],*Dynamics of Mechanical and Electromechanical Systems*, MacGraw-Hill.

C. A. Desoer and E. S. Kuh [1969], *Basic Circuit Theory*, McGraw-Hill.

D. Y. Gao [1999], *Duality Principles in Nonconvex Systems : Theory, Methods and Applications*, Kluwer Academic Publishers.

M. Iri, I. Shirakawa, Y. Kajitani and S. Shinoda [1983], *Graph Theory with Exercises* (In Japanese), Corona.

D. L. Jones and F. J. Evans [1973], Variational analysis of electrical networks, *Journal of the Franklin Institute*, Vol.295, No.1, pp.9-23.

T. R. Kane [1965], Dynamics of nonholonomic systems, *Trans. ASME, Journal of Applied Mechanics*, 28, pp.574-578.

S. Kobayashi and K. Nomizu [1963], *Foundations of Differential Geometry, Vol. I and II*, Interscience Tracts in Pure and Applied Mathematics, No.15.

J. Koiller [1992], Reduction of some classical non-holonomic systems with symmetry, *Arch. Rational Mech. Anal.*, 118, pp.113-148.

K. Kondo [1962], *RAAG Memoirs of the Unifying Study of Basic Problems in Engineering and Physical Sciences By Means of Geometry, Vol. III*, Edited by Kazuo Kondo. Published for the Research Association of Applied Geometry, Gakujutsu Bunken Fukyu-kai,Tokyo.

G. Kron [1934], Non-Riemannian dynamics of rotating electrical machinery, *Journal of Mathematics and Physics*, 13, pp.103-194.

G. Kron [1963], *Diakoptics*, MacDonald.

C. Lanczos [1962], *The Variational Principles of Mechanics*, University of Toronto Press, Toronto.

J. E. Marsden and T. S. Ratiu [1999], *Introduction to Mechanics and Symmetry*, Second Edition, Springer-Verlag.

L. A. Pars [1965], *A Treatise on Analytical Dynamics*, OX BOW Press.

H. M. Paynter [1961], *Analysis and Design of Engineering Systems*, Class Note, MIT Press.

R. E. Roberson and J. Wittenburg [1966], A dynamical formalism for an arbitrary number of interconnected rigid bodies with reference to the problem of satellite attitude control, *Proc. of the 3rd Congr. Int. Fed. Automatic Control*, 1: 46D.1-46D.8.

W. Schiehlen [1997], Multibody system dynamics: roots and perspectives, *Multibody System Dynamics*, 1, No.2, pp.149-188.

T. Takahashi [1974], *Seminars on Basic Engineering* (in Japanese). Gendai Sugaku-sha.

van der Schaft, A. J. and B. M. Maschke [1995], The Hamiltonian formulation of energy conserving physical systems with external ports, *Archiv für Elektronik und Übertragungstechnik* **49**, 362–371.

A. M. Vershik and L. D. Faddeev [1981], Lagrangian mechanics and invariant form, *Sel. Math. Sov.*, Vol.1, No.4, pp.339-350.

R. W. Weber [1986], Hamiltonian systems with constraints and their meaning in mechanics, *Arch. Rat. Mech. Anal.*, 91, pp.309-335.

H. Yoshimura [1995], *A Study on Flexible Multibody Dynamics* (in Japanese). Doctoral Dissertation, Waseda University.

H. Yoshimura and T. Kawase [2001], A duality principle in nonholonomic mechanical systems, *Nonsmooth/nonconvex mechanics* (Blackburg,VA, 1999), pp.447-471, Nonconvex Optim. Appl., 50, Kluwer Acad. Publ., Dordrecht.

Chapter 22

MIXED ENERGY METHOD FOR SOLUTION OF QUADRATIC PROGRAMMING PROBLEMS

Zhong Wanxie and Zhang Hongwu
State Key Laboratory of Structural Analysis for Industrial Equipment
Department of Engineering Mechanics,
Dalian University of Technology, Dahlia, China, 116024.
zhanghw@dlut.edu.cn

Abstract A new algorithm for the solution of quadratic programming problems is put forward in terms of the mixed energy theory. The algorithm is used to get the incremental solution of the elastic-plastic truss structures. Different with the traditional approach where the unknown variables are selected only in one class such as displacement or stress, the present method selects the variables in the mixed form with both of displacement and stress. Due to that the method is established in the hybrid space, so the information found in the previous incremental step can be used for the solution of the present step, and makes the algorithm highly efficient in the solution process of the quadratic programming problems. Results obtained in the examples of elastic-plastic solution of truss structures demonstrate what have been predicted in the theoretical analysis.

1. Introduction

It is well known that the quadratic programming method has been used widely in mathematics and operational research, particularly in computational mechanics such as the structural optimization [1, 2], elastic-plastic stability and contact mechanics [3, 4]. However, when the quadratic programming method is used to solve the problems with multi-loading cases in structural mechanics, the equations need to be solved one by one. This solution method expenses a great deal of computer time, particularly when multi-incremental loading steps are needed. This is one of the key disadvantages of the mathematical programming approach. Therefore, it is worthwhile to study the new solution technique for the problem, especially to consider the use of genetic information generated in the previous solution step. With the use of information obtained in the last load step, it is possible to decrease the computational expense for solution of the present load step. This will bring us a totally new solution technique for the quadratic programming problems.

2. Mixed Energy of the Quadratic Programming Problem

Considering a simple quadratic programming problem, for a given positive definite $n \times n$ matrix $\mathbf{K}$ and a vector $\mathbf{r}$ with dimension n, to find the n dimensional vector $\boldsymbol{\lambda}$, makes the objective function of potential "energy" expression $U(\boldsymbol{\lambda})$ to be minimized, i.e.

$$U(\boldsymbol{\lambda}) = \left[\boldsymbol{\lambda}^T\mathbf{K}\boldsymbol{\lambda}/2 - \boldsymbol{\lambda}^T\mathbf{r}\right], \quad \min_{\boldsymbol{\lambda}|\lambda_i \geq 0} U(\boldsymbol{\lambda}) \tag{2.1}$$

The corresponding governing equations can be derived from Kuhn-Tucker conditions

$$\boldsymbol{\lambda}^T\boldsymbol{\nu} = 0, \quad \boldsymbol{\lambda} \geq \mathbf{0}, \quad \boldsymbol{\nu} \geq \mathbf{0}; \quad \text{where} \quad \boldsymbol{\nu} = \partial U/\partial\boldsymbol{\lambda} \tag{2.2}$$

We define respectively here that $\boldsymbol{\lambda}$, $\boldsymbol{\nu}$ and U are the generalized "displacement", generalized "force" vectors and generalized "energy".

The well known Lemke algorithm can be used for the solution of equation (2.1) and (2.2). However, it needs to be noticed that in the Lemek algorithm, the iteration process will always satisfy the condition $\boldsymbol{\lambda} \geq \mathbf{0}$. Therefore it is called the "displacement" method for the solution of "displacement" is always in the feasible domain. In the final solution, all the components of "displacement" $\lambda_i > 0$, $i = i_1, i_2, \cdots, i_a$ are called a collection, while the remaining components are called b collection, i.e., $\nu_j > 0$. These two kinds of variables, $\boldsymbol{\lambda}$ and $\boldsymbol{\nu}$, are dual to each other.

For a given matrix $\mathbf{K}$ and a series of force vectors $\mathbf{r}_i$, $i = 1, 2, \cdots, e$ in which any two consecutive vectors are not far apart to each other, our problem now is to solve the equations described by equation (2.1) with the force vector $\mathbf{r}$ be substituted consecutively with $\mathbf{r}_i$, $i = 1, 2, \cdots, e$. It is easily understood that the consecutive solutions are not greatly different, because the consecutive force vectors are also in the similar forms. This implies that the changes of collections of a and b are not too much.

However, when the traditional quadratic programming method is used to solve the problem, the algorithm will always start from the full a collection and null b collection. The solution information obtained in the previous load step is not applied for the solution of the present load step. In other means, the basis pivot should be carried out from the very beginning, which leads quite a big computational expense.

The purpose of this paper now is to find a new methodology for the serial solutions of the problem with consecutive load vectors. The basic idea in the new algorithm is the use of information obtained in last loading case, so that in solving the present load step the initial approximate collections a and b can be set up by using the collections in the last loading step.

To make the description simpler, without lose of generality, it can be arranged that the first m components being in collection a and the other n-m components being in collection b, which can be realized through basic matrix transformation. The displacement vector is then described as

$$\boldsymbol{\lambda} = \left\{ \begin{array}{c} \boldsymbol{\lambda}_a \\ \boldsymbol{\lambda}_b \end{array} \right\}, \quad \boldsymbol{\lambda}_a = \{x_1, \ldots, x_m\}^T, \quad \boldsymbol{\lambda}_a = \{x_{m+1}, \ldots, x_n\}^T \tag{2.3}$$

The objective function of "energy" expression $U(\boldsymbol{\lambda}_a, \boldsymbol{\lambda}_b)$ thus can be written in the following form

$$U(\boldsymbol{\lambda}_a, \boldsymbol{\lambda}_b) = \frac{1}{2} \left\{ \begin{array}{c} \boldsymbol{\lambda}_a \\ \boldsymbol{\lambda}_b \end{array} \right\}^T \left[\begin{array}{cc} \mathbf{K}_{aa} & \mathbf{K}_{ab} \\ \mathbf{K}_{ba} & \mathbf{K}_{bb} \end{array} \right] \left\{ \begin{array}{c} \boldsymbol{\lambda}_a \\ \boldsymbol{\lambda}_b \end{array} \right\} - \left\{ \begin{array}{c} \boldsymbol{\lambda}_a \\ \boldsymbol{\lambda}_b \end{array} \right\}^T \left\{ \begin{array}{c} \boldsymbol{\lambda}_a \\ \boldsymbol{\lambda}_b \end{array} \right\}$$

Thus, we have the minimization problem:

$$\mathcal{D} \min_{\boldsymbol{\lambda}_a | \boldsymbol{\lambda}_a \geq \mathbf{0}} \mathcal{D} \min_{\boldsymbol{\lambda}_b | \boldsymbol{\lambda}_b \geq \mathbf{0}} U(\boldsymbol{\lambda}_a, \boldsymbol{\lambda}_b). \tag{2.4}$$

Introducing the dual vector $\boldsymbol{\nu}_b$ of collection b as

$$\boldsymbol{\nu}_b = \partial U / \partial \boldsymbol{\lambda}_b = \mathbf{K}_{ba} \boldsymbol{\lambda}_a + \mathbf{K}_{bb} \boldsymbol{\lambda}_b - \mathbf{r}_b \tag{2.5}$$

Solving with respect to λ_b gives

$$\boldsymbol{\lambda}_b = \mathbf{K}_{bb}^{-1} \left[\boldsymbol{\nu}_b - \mathbf{K}_{ba} \boldsymbol{\lambda}_a + \mathbf{r}_b \right] \tag{2.6}$$

The mixed energy $V(\boldsymbol{\lambda}_a, \boldsymbol{\nu}_b)$ can be obtained by Legendre transformation, which derives the following form

$$V(\boldsymbol{\lambda}_a, \boldsymbol{\nu}_b) = \boldsymbol{\nu}_b^T \boldsymbol{\lambda}_b - U(\boldsymbol{\lambda}_a, \boldsymbol{\nu}_b) \tag{2.7}$$

Substituting (2.6) into (2.7) and through a series of derivation gives

$$V(\boldsymbol{\lambda}_a, \boldsymbol{\nu}_b) = \frac{\boldsymbol{\nu}_b^T \mathbf{G} \boldsymbol{\nu}_b}{2} + \boldsymbol{\nu}_b^T \mathbf{F} \boldsymbol{\lambda}_a - \frac{\boldsymbol{\lambda}_a^T \mathbf{Q} \boldsymbol{\lambda}_a}{2} + \mathbf{h}_a^T \boldsymbol{\lambda}_a + \mathbf{h}_b^T \boldsymbol{\nu}_b + \frac{\mathbf{r}_b^T \mathbf{M}_{bb}^{-1} \mathbf{r}_b}{2} \tag{2.8}$$

where

$$\mathbf{Q} = \mathbf{K}_{aa} - \mathbf{K}_{ab}\mathbf{K}_{bb}^{-1}\mathbf{K}_{ba} \quad \mathbf{G} = \mathbf{K}_{bb}^{-1} \quad \mathbf{F} = -\mathbf{K}_{bb}^{-1}\mathbf{K}_{ba} \tag{2.9}$$

$$\mathbf{h}_b = \mathbf{G}\mathbf{r}_b, \quad \mathbf{h}_a = \mathbf{r}_a + \mathbf{F}^T \mathbf{r}_b \tag{2.10}$$

The last term in (2.8) is a constant and is useless. Introducing the dual vector $\boldsymbol{\nu}_a$ of $\boldsymbol{\lambda}_a$

$$\boldsymbol{\nu}_a = \partial U / \partial \boldsymbol{\lambda}_a = \mathbf{K}_{aa}\boldsymbol{\lambda}_a + \mathbf{K}_{ab}\boldsymbol{\lambda}_b - \mathbf{r}_a \tag{2.11}$$

If $\mathbf{K}$ is a positive definite matrix, then it can be proven that $\mathbf{Q}$, $\mathbf{G}$ should also be positive definite. On the other hand, if $\mathbf{Q}$ and $\mathbf{G}$ are positive definite matrices, $\mathbf{K}$ must be positive definite.

Due to the introduction of mixed energy, the "displacement" vector is divided into two collections of a and b. The corresponding equation of (2.2) is then updated into

$$\boldsymbol{\lambda}_a^T \boldsymbol{\nu}_a = 0, \quad \boldsymbol{\lambda}_a \geq \mathbf{0}, \quad \boldsymbol{\nu}_a \geq \mathbf{0}; \quad \boldsymbol{\lambda}_b^T \boldsymbol{\nu}_b = 0, \quad \boldsymbol{\lambda}_b \geq \mathbf{0}, \quad \boldsymbol{\nu}_b \geq \mathbf{0} \tag{2.12}$$

and the equations (2.11) and (2.5) can be expressed as

$$\begin{aligned} \boldsymbol{\nu}_a &= \mathbf{Q}\boldsymbol{\lambda}_a - \mathbf{F}^T \boldsymbol{\nu}_b - \mathbf{h}_a \\ \boldsymbol{\lambda}_b &= \mathbf{F}\boldsymbol{\lambda}_a + \mathbf{G}\boldsymbol{\nu}_b + \mathbf{h}_b \end{aligned} \tag{2.13}$$

From the mixed energy, it can be derived

$$\boldsymbol{\nu}_a = -\partial V / \partial \boldsymbol{\lambda}_a, \quad \boldsymbol{\lambda}_b = \partial V / \partial \boldsymbol{\nu}_b \tag{2.14}$$

It is possible to express equations (2.13) in the matrix form, i.e.

$$\begin{bmatrix} \mathbf{Q} & \mathbf{0} \\ \mathbf{F} & -\mathbf{I}_b \end{bmatrix} \begin{Bmatrix} \boldsymbol{\lambda}_a \\ \boldsymbol{\lambda}_b \end{Bmatrix} + \begin{bmatrix} -\mathbf{I}_b & -\mathbf{F}^T \\ \mathbf{0} & \mathbf{G} \end{bmatrix} \begin{Bmatrix} \boldsymbol{\nu}_a \\ \boldsymbol{\nu}_b \end{Bmatrix} + \begin{Bmatrix} -\mathbf{h}_a \\ \mathbf{h}_b \end{Bmatrix} = 0 \tag{2.15}$$

If $\boldsymbol{\lambda}_a = \mathbf{0}$, $\boldsymbol{\nu}_b = \mathbf{0}$, then $\boldsymbol{\nu}_a = -\mathbf{h}_a$, $\boldsymbol{\lambda}_b = \mathbf{h}_b$. If $\boldsymbol{\nu}_a$, $\boldsymbol{\lambda}_b$ are not negative, then the complementary conditions (2.12) are satisfied. This means that the final solution of the problem is reached. Obviously the equality

$U = -V$ holds under the dual conditions.

For the application of the algorithm, it generally needs to express the energy in mixed form and then to find the minimum solution of the problem. To find the variational principle for the mixed energy, the following equation needs to be proven

$$\min_{\boldsymbol{\lambda}_a} \max_{\boldsymbol{\nu}_b} [-V(\boldsymbol{\lambda}_a, \boldsymbol{\nu}_b)] = \min_{\boldsymbol{\lambda}_a|\boldsymbol{\lambda}_a \geq \mathbf{0}} \min_{\boldsymbol{\lambda}_b|\boldsymbol{\lambda}_b \geq \mathbf{0}} U(\boldsymbol{\lambda}_a, \boldsymbol{\lambda}_b) \tag{2.16}$$

where $\boldsymbol{\lambda}_a$ plays the role as a group of parameters, U is the positive definite quadratic form of $\boldsymbol{\lambda}_a$ and $\boldsymbol{\lambda}_b$; $\boldsymbol{\nu}_a$ is the dual vector of $\boldsymbol{\lambda}_b$, V is the complementary energy of U expressed by $\boldsymbol{\nu}_b$, $\boldsymbol{\lambda}_b$. While $-V$ is the maximum quadratic form expressed by vector $\boldsymbol{\nu}_b$.

The key step now is to prove that

$$\mathcal{D} \max_{\boldsymbol{\nu}_b} [-V(\boldsymbol{\lambda}_a, \boldsymbol{\nu}_b)] = \min_{\boldsymbol{\lambda}_b|\boldsymbol{\lambda}_b \geq \mathbf{0}} U(\boldsymbol{\lambda}_a, \boldsymbol{\lambda}_b)$$

holds for arbitrary $\boldsymbol{\lambda}_a$, which is a group of parameters and expressed as

$$V(\boldsymbol{\lambda}_a, \boldsymbol{\nu}_b) = \boldsymbol{\nu}_b^T \boldsymbol{\lambda}_b - U(\boldsymbol{\lambda}_a, \boldsymbol{\lambda}_b) \tag{2.17}$$

so we have

$$\frac{\partial V}{\partial \boldsymbol{\nu}_b} = \boldsymbol{\lambda}_b + \left(\boldsymbol{\nu}_b - \frac{\partial U}{\partial \boldsymbol{\lambda}_b}\right)^T \frac{\partial \boldsymbol{\lambda}_b}{\partial \boldsymbol{\nu}_b} = \boldsymbol{\lambda}_b \tag{2.18}$$

Because

$$\boldsymbol{\nu}_b^T \cdot \frac{\partial V}{\partial \boldsymbol{\nu}_b} = \boldsymbol{\nu}_b^T \cdot \boldsymbol{\lambda}_b = 0 \tag{2.19}$$

holds. So $-V = U$. Furthermore, because for collection b, $\boldsymbol{\lambda}_b > \mathbf{0}$, so $\boldsymbol{\nu}_b = \mathbf{0}$. Due to that $\mathbf{G}$ is a positive definite matrix, so the equation $\boldsymbol{\lambda}_b = \partial V / \partial \boldsymbol{\nu}_b$ is obtained. In this case, the maximum value derived from $-V$ equals U, and U takes the minimum value with respect to the vector $\boldsymbol{\lambda}_b$. So that

$$\max_{\boldsymbol{\nu}_b|\boldsymbol{\nu}_b \geq \mathbf{0}} [-V(\boldsymbol{\lambda}_a, \boldsymbol{\nu}_b)] = \min_{\boldsymbol{\lambda}_b|\boldsymbol{\lambda}_b \geq \mathbf{0}} U(\boldsymbol{\lambda}_b, \boldsymbol{\lambda}_b)$$

holds, and the above equality (2.16) is thus proven.

For the implementation of the algorithm, if the stiffness matrix is constant and only "force" vector $\mathbf{r}$ is changed a little in each load step, then the a, b divisions obtained in the previous load step can be taken as the approximate base of the Lemek algorithm in this load step. This also means that mixed energy matrices $\mathbf{Q}, \mathbf{G}, \mathbf{F}$ are not necessary to be re-calculated. The only computations are $\mathbf{h}_a, \mathbf{h}_b$ which are expressed by equations (2.10).

Because the variables are distinguished into the groups a and b, then $\mathbf{h}_a, \mathbf{h}_b$ are obtained first. It is assumed that $\boldsymbol{\lambda}_a = \mathbf{0}$, $\boldsymbol{\nu}_b = \mathbf{0}$, then $\boldsymbol{\nu}_a = -\mathbf{h}_a$, $\boldsymbol{\lambda}_b = \mathbf{h}_b$ are obtained. If $\boldsymbol{\nu}_b, \boldsymbol{\lambda}_b$ are not negative, then the complementary conditions are satisfied, which means that the solution is obtained.

In the implementation of the algorithm, the stiffness matrix is unchanged in each load step. Only the "force" vector $\mathbf{r}$ is slightly changed in each load step. Then the group a and group b subdivisions obtained in the previous load step can be taken as the approximate basis of the algorithm in the present load step. After $\mathbf{h}_a, \mathbf{h}_b$ are computed, the displacement $\boldsymbol{\lambda}_a$ is then a known vector, generally, which can be selected as a zero vector. In this case, $\boldsymbol{\lambda}_b = \mathbf{h}_b$ will be obtained and may appear some negative entry, so the pivoting to find the feasible solution of displacement method $\boldsymbol{\lambda}_b$, $\boldsymbol{\nu}_b$ is needed. The first stage of solution is to maximize the $n-m$ dimensional vector $\boldsymbol{\nu}_b$. So the complementary energy $V(\boldsymbol{\lambda}_a, \boldsymbol{\nu}_b)$ needs to be used. The usual Lemke algorithm can then be used to find the solution. The feasible global displacement vector $\boldsymbol{\lambda} = \left\{ \begin{array}{cc} \boldsymbol{\lambda}_a^T & \boldsymbol{\lambda}_b^T \end{array} \right\}^T$ is thus easily composed. When vectors $\boldsymbol{\lambda}_b, \boldsymbol{\nu}_b$ are known, the total "displacement" vector is easily obtained. The same iteration progress can be repeated for the solution of next load case.

What have been mentioned above is the basic principle of the mixed energy method for the solution of quadratic programming problems. To apply the method in numerical analysis of structural elastic-plastic problems, we select here the plastic analysis of the truss structures to show the efficiency of the algorithm presented.

3. The Basic Principle for the Programming Solution of Elastic-Plastic Truss Structures

A 3-bar structure is adopted to describe the basic concept of the programming method described above. The structure is shown in Figure 1(a). The strain-stress relationship is shown in Figure 1(b) where Young's tensile and compressive moduli have different values.

The governing equations for the problem shown in Figure 1 can be concluded as:

Balance equations

$$\begin{array}{ll} \mathcal{D}N_1 + N_2 + N_3 = P \quad \text{or} \quad \sum_i N_i - P = 0 \\ N_1 t_1 + N_2 t_2 + N_3 t_3 = 0 \quad \text{or} \quad \mathcal{D}\sum_i N_i t_i = 0 \end{array} \tag{3.1}$$

where $N_i (i = 1, 2, 3)$ are internal force of the bars $t_1 = 0,\ t_2,\ t_3$ indicate the distance between any two bars, the meanings of the other parameters can be found in Figure 1.

Consistent equations

$$\left.\begin{array}{l} \Delta_1 = y - \theta \cdot t_1 \\ \Delta_2 = y - \theta \cdot t_2 \\ \Delta_3 = y - \theta \cdot t_3 \end{array}\right\} \quad \text{or} \quad \Delta_i = y - \theta \cdot t_i \quad (i = 1, 2, 3) \tag{3.2}$$

Constitutive relationships

$$\Delta_i = \left\{ \begin{array}{lll} N_i / K_i^{(+)} & when & N_i > 0 \\ N_i / K_i^{(-)} & when & N_i \leq 0 \end{array} \right. \quad (i = 1, 2, 3) \tag{3.3}$$

If the displacements (y, θ) are taken as unknown variables, the solution procedure can be expressed in the following way:

- Substituting (3.3) into (3.2), then variable Δ_i is cancelled, the relationship between $(N_i \sim y, \theta)$ is obtained.
- Substituting the relationships of $(N_i \sim y, \theta)$ into equations (3.1), then y and θ can be solved. There is also another method to solve y and θ, i.e., to write first the potential energy $\Pi_1(y, \theta)$ of the system in terms of the relationships of $(N_i \sim y, \theta)$, then to derive the two governing equations for the solution of y and θ by means of the variational principle $\delta\Pi_1 = 0$.
- With the use of $(N_i \sim y, \theta)$, the forces N_i are obtained.

When a linear constitutive relationship is used, i.e., $K_i^{(+)} = K_i^{(-)}$, the solution is easily obtained. However, when the constitutive relationship presents nonlinear behaviour $(K_i^{(+)} \neq K_i^{(-)})$, the difficulty is that we do not know the tensile or compressive state of each bar in advance. So the first solution step described above is difficult to deal with. A possible solution technique is the so called "pre-test" algorithm, i.e. the force state of each bar is assumed before the computation. Then the corresponding solution is obtained. To check if the assumption is right or not, if not, then the new force state is put on the truss structure,

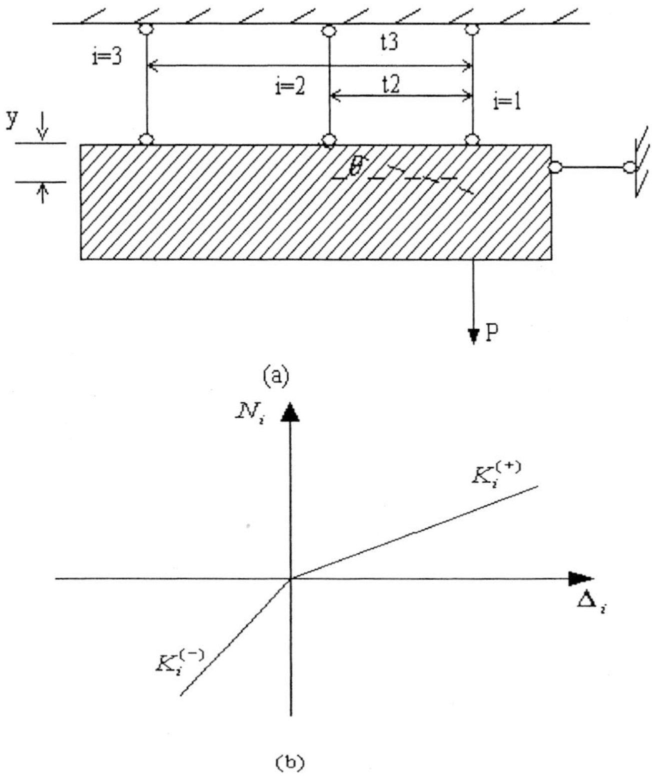

Figure 22.1. (a) A 3-bar structure. (b) Stress-strain relationship

and the new solution is obtained. This iteration progress is repeated until that the obtained solution is consistent with the assumptions. It has been found that the iteration method is effective for some problems. However, this method sometimes will not be effective and will present oscillation phenomenon for some problems. So it is still an important work to find the new effective algorithm.

Assuming that all bars in the structure are in compressive state, the force in each bar equals to $N_i = K_i^{(-)}\Delta_i$, the total potential energy of the system can be written as

$$\Pi_1 = -P \cdot y + \sum_i \frac{1}{2} N_i \Delta_i = -P \cdot y + \sum_i \frac{1}{2} K_i^{(-)} (y - \theta \cdot t_i)^2 \qquad (3.4)$$

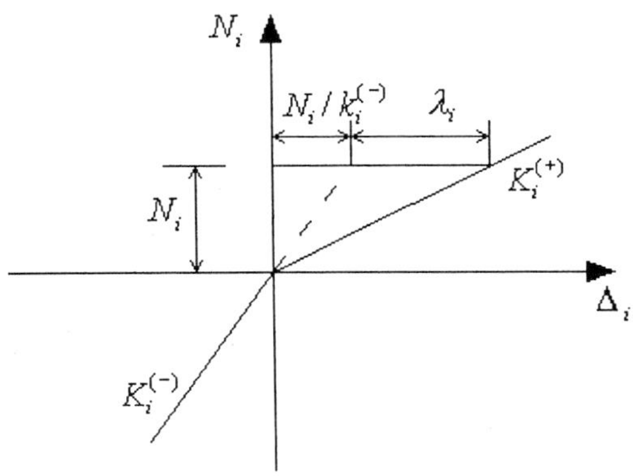

Figure 22.2. The additional tensile value of the plastic bar

With the use of minimum potential energy principle, i.e. $\frac{\partial \Pi_1}{\partial y} = 0$, $\frac{\partial \Pi_1}{\partial \theta} = 0$, the two balance equations (3.1) for the solution of y and θ are obtained. However, the bar (i) may be in tensile state, we have (see Figure 2)

$$\Delta_i = N_i/K_i^{(-)} + \lambda_i \tag{3.5}$$

where λ_i is called additional tensile value of the bar. Its value is

$$\lambda_i = N_i/K^{(+)} - N_i/K^{(-)} \tag{3.6}$$

Due to

$$N_{io} = K^{(-)}(\Delta_i - \lambda_i) = K_i^{(-)}(y - \theta t_i - \lambda_i) \tag{3.7}$$

Substituting (3.7) into (3.6), we have

$$f_i(y, \theta, \lambda_i) = (y - \theta t_i)\left(K_i^{(-)} - K_i^{(+)}\right) - \lambda_i K_i^{(-)} = 0 \tag{3.8}$$

If $f_i = 0$, the bar i will be in tension and $\lambda_i \geq 0$; on the other hand, if $f_i < 0$, then the bar i will be in compression and $\lambda_i = 0$. So the following relationships hold

$$\lambda_i \begin{cases} \geq 0, & f_i = 0 \\ = 0, & f_i < 0 \end{cases} \tag{3.9}$$

With the introduction of the slack variable ν_i in to (3.8), the constitutive relationship is then rewritten as

$$f_i(y, \theta, \lambda_i) + \nu_i = 0, \qquad \nu_i \cdot \lambda_i = 0, \nu_i, \lambda_i \geq 0 \tag{3.10}$$

which is the constitutive law of the bar i for both of the tension and compression cases. Equation $f_i + \nu_i = 0$ is called constitutive equation, $\nu_i \cdot \lambda_i = 0$ is the complementary condition, $\nu_i \cdot \lambda_i \geq 0$ is the positive condition. Complementary condition means that at least on of λ_i and ν_i will be zero, or both of them are equal to zero. It is obviously noticed that when $\lambda_i = 0, \nu_i > 0 \rightarrow f_i < 0$, the bar i is under compression; when $\lambda_i \geq 0, \nu_i = 0 \rightarrow f_i = 0$, bar i under tension. So equation (3.10) is the same as (3.9).

4. The Finite Element Equation

Now the parametric minimum potential energy principle is given. For any y and θ which satisfy the consistent conditions, the total potential energy of the system can be written as

$$\begin{gathered} \mathcal{D}\Pi_2(y,\theta,\lambda_i) = -P\cdot y + \\ \sum_i \left[\tfrac{1}{2}K_i^{(-)}(y-\theta t_i)^2 - \lambda_i K_i^{(-)}(y-\theta t_i) + \tfrac{1}{2}K_i^{(-)}\lambda_i^2\right] \end{gathered} \tag{4.1}$$

where the strain energy of the bar i can be defined as

$$\begin{gathered} \mathcal{D}\tfrac{1}{2}N_i\Delta_i = \tfrac{1}{2}K_i^{(-)}(y-\theta t_i-\lambda_i)^2 = \\ \tfrac{1}{2}K_i^{(-)}(y-\theta t_i)^2 - \lambda_i K_i^{(-)}(y-\theta t_i) + \tfrac{1}{2}K_i^{(-)}\lambda_i^2 \end{gathered} \tag{4.2}$$

The basic idea used in the parametric variational principle is to get the two balance equations through the variation of Π_2 with respect to θ and y. The importance is that the control variables λ_i do not join the variation process. The variables λ_i in the balance equations will be determined from the so called state equation (3.10). Due to that λ_i do not join the variation process, so the item $\frac{1}{2}K_i^{(-)}\lambda_i^2$ in equation (4.1) can be cancelled. The energy Π_2 reduces into

$$\Pi_3(\lambda_i(\cdot)) = \sum_i \left[\frac{1}{2}K_i^{(-)}(y-\theta t_i)^2 - \lambda_i K_i^{(-)}(y-\theta t_i)\right] - P\cdot y \tag{4.3}$$

Expression $\Pi_3[\lambda_i(\cdot)]$, which is a function in positiveness, means that λ_i dose not join the variation of Π_3. The governing equations for the solution of the problem now change into the following quadratic programming problem

$$\min.\ \Pi_3[\lambda_i(\cdot)] \tag{4.4}$$

$$\text{s.t. } f_i(y,\theta,\lambda_i) + \nu_i = 0, \qquad \nu_i\cdot\lambda_i = 0; \qquad \nu_i,\lambda_i \geq 0, (i = 1,2,3) \tag{4.5}$$

where state variables y and θ joint the variation process, while λ_i do not, but are controlled by the constitutive relationship. This means that λ_i will also control the variational process.

For a general elastic-plastic problem, with the use of the discretized method described in [4], we can obtain the general finite element equations

$$\Pi_4 = \frac{1}{2}\boldsymbol{\delta}^T \mathbf{K}\boldsymbol{\delta} - (\boldsymbol{\phi}\boldsymbol{\lambda} + \mathbf{t})^T \boldsymbol{\delta} \tag{4.6}$$

$$\mathbf{C}\boldsymbol{\delta} - \mathbf{M}\boldsymbol{\lambda} - \mathbf{d} + \boldsymbol{\nu} = \mathbf{0}, \quad \boldsymbol{\nu}^T \cdot \boldsymbol{\lambda} = 0,\ \boldsymbol{\nu},\ \boldsymbol{\lambda} \geq \mathbf{0} \tag{4.7}$$

where

$$\mathbf{K} = \mathcal{D}\sum_e \int_{\Omega^c} \mathbf{B}^T \mathbf{D}\mathbf{B} d\Omega;\ \ \mathbf{t} = \sum_e \left(\int_{\Omega^c} \mathbf{N}^T \mathbf{b} d\Omega + \int_{S_p^c} \mathbf{N}^T \overline{\mathbf{p}} dS \right);$$
$$\boldsymbol{\phi} = \mathcal{D}\sum_e \int_{\Omega^e} (\mathbf{D}\mathbf{B})^T d\Omega$$

$$\begin{aligned} \mathbf{C} &= \mathcal{D}\sum_e \int_{\Omega^e} \frac{\partial \mathbf{f}}{\partial \boldsymbol{\delta}} d\Omega = \sum_e \int_{\Omega^e} (\mathbf{D}\mathbf{B})\, d\Omega = \boldsymbol{\phi}^T; \\ \mathbf{M} &= \mathcal{D}\sum_{e=1} \int_{\Omega^e} \frac{\partial \mathbf{f}}{\partial \boldsymbol{\lambda}} d\Omega; \quad \mathbf{d} = -\mathcal{D}\sum_e \int_{\Omega^e} \mathbf{f}_{oe} d\Omega \end{aligned} \tag{4.8}$$

where $\boldsymbol{\delta}$ is displacement vector, $\mathbf{K}$ and $\mathbf{t}$ are elastic stiffness matrix and incremental load vector, matrix $\mathbf{M}$ indicates the hardening property of the materials and called hardening matrix, $\mathbf{M}$ has the sub-block diagonal form, matrices $\boldsymbol{\phi}$ and $\mathbf{C}$ indicate respectively the plastic potential and yield constraints of the system, and call plastic potential matrix and constraint matrix. Matrices $\mathbf{K}$, $\mathbf{C}$ and $\mathbf{M}$ are constant and do not change for the different load cases. The variables which have not been illustrated are the same as usual.

The real solution of the elastic-plastic problem is to take the first order of variation of Π_4 with respect to $\boldsymbol{\delta}$, i.e.

$$\frac{\partial \Pi_4}{\partial \boldsymbol{\delta}} = \mathbf{K}\boldsymbol{\delta} - (\boldsymbol{\phi}\boldsymbol{\lambda} + \mathbf{t}) = 0 \tag{4.9}$$

Because $\mathbf{K}$ is a matrix in positiveness, we have

$$\boldsymbol{\delta} = \mathbf{K}^{-1} (\boldsymbol{\phi}\boldsymbol{\lambda} + \mathbf{t}) \tag{4.10}$$

With combination of (4.9) and (4.10), we derive

$$\boldsymbol{\nu} - \left(\mathbf{M} - \mathbf{C}\mathbf{K}^{-1}\boldsymbol{\phi}\right)\boldsymbol{\lambda} = -\mathbf{C}\mathbf{K}^{-1}\mathbf{t} + \mathbf{d}, \qquad \boldsymbol{\lambda}^T \cdot \boldsymbol{\nu} = 0,\ \boldsymbol{\lambda},\ \boldsymbol{\nu} \geq \mathbf{0} \tag{4.11}$$

It is easily to check that $\left(\mathbf{M} - \mathbf{C}\mathbf{K}^{-1}\boldsymbol{\phi}\right)$ is positiveness matrix, so the algorithm described in section II can be used directly for the solution of the problem (4.11).

5. Numerical Examples

We consider here only the efficiency of the algorithm developed above. For the validity of the quadratic programming method, the demonstration examples can be found in reference [4]. Because the results obtained by the new algorithm are exactly the same as those obtained by the original algorithm. We show in what follows only the comparison of the efficiency of the different algorithms. Two structures are computed, shown in Figures 5 and 22.4 respectively.

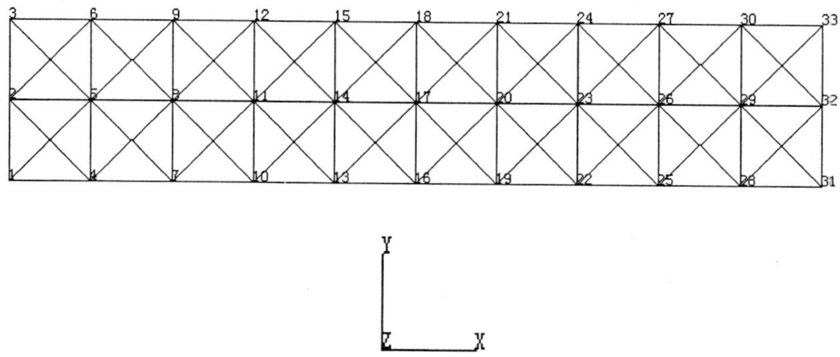

Figure 22.3. Truss structure

Material parameters of the two structures are: Young's modulus $E = K^{(-)} = 210$GPa, cross-section A=0.01m^2, yield stress 300Mpa, hardening parameter $H = \frac{K^{(-)}K^{(+)}}{K^{(-)}-K^{(+)}} = 210$GPa. The boundary and load conditions are: For structure 1, nodes 1, 2, 3, 31, 32 and 33 are fixed in both x and y directions, a load value P=30MN is acted at node 18. For structure 2, nodes 1, 18, 19 and 20 are fixed in x, y and z directions, nodes 21, 22, 23 and 24 are fixed in x and y directions and free in z direction, nodes 2, 3, 4 and 5 are loaded respectively by the P=10MN.

Ten load steps are computed for all the three structures. The incremental load step (ratio) is 0.1. The efficiency of the Lemek algorithm depends on mainly the number of base exchanges. Table 1 gives the number of base exchanges required in the computations of the two structures. The efficiency ratio between the old and the new algorithms are 3.743 and 4.736 respectively. It can be found the efficiency of the new algorithm is highly increased. It demonstrates the efficiency of the new algorithm developed.

Table 1 Efficiency comparison between the old and the new algorithms (base exchange numbers)

Stru.	*Algo.*	*Step1*	*2*	*3*	*4*	*5*	*6*	*7*	*8*	*9*	*10*	*Sum*
Stru1	Old	0	0	5	7	9	19	23	25	25	33	146
Stru1	New	0	0	5	3	3	11	5	3	0	9	39
Stru2	Old	2	5	13	14	17	21	25	26	28	29	180
Stru2	New	2	4	9	2	4	5	5	2	3	2	38

6. Conclusions

Based on the mixed energy theory, a new algorithm for the solution of quadratic programming problem is developed in the paper. The formulations of the algorithm are described in detail, and the equations for the parametric quadratic programming solution of elastic-plastic truss structures are derived and applied in the structural analysis. Numerical examples are presented. The results demonstrate the efficiency of the new algorithm.

The basic concept in the development of the new algorithm is the use of the information obtained in last load step in the solution of the present load step. So the efficiency of the algorithm is increased. The algorithm developed and idea used in this paper can be extended also in the other structural mechanics problems which can be solved by the mathematical programming method.

Acknowledgments

The project was supported by the National Key Basic Research Special Foundation (G1999032805), the National Natural Science Foundation and the Foundation for University Key Teacher by the Ministry of Education of China

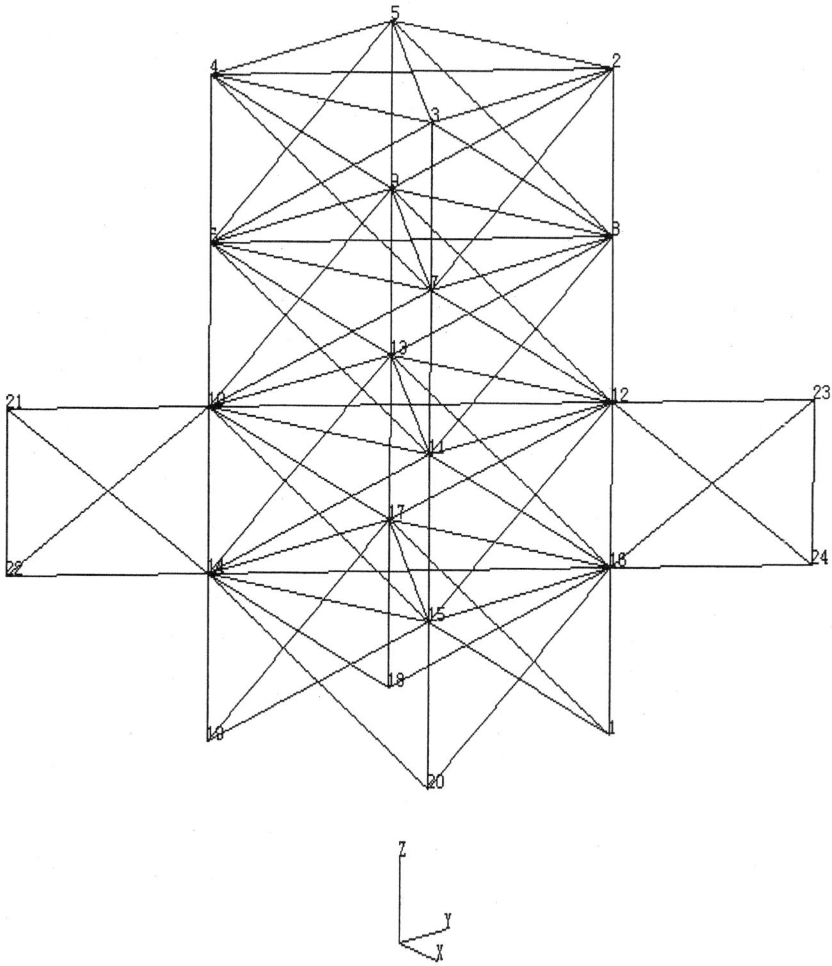

Figure 22.4. 82 bar truss structure

References

[1] L. X. Qian. *Engineering structural optimization design.* Beijing: Hydraulic-Electric Press, 1983.

[2] G.D. Cheng. *Fundamental of engineering structural optimization design.* Beijing: Hydraulic-Electric Press, 1984.

[3] V. F. Demyanov, G. E. Stavroulakis, and L. N. Polyakova. *Panagiotopoulos. Quasidifferentiability and nonsmooth modelling in me-*

chanics, engineering and economics. Dordrecht: Kluwer Academic Publishers, 1996.

[4] W. X. Zhong, H. W. Zhang, and C. W. Wu. *Parametric variational principle and its applications in engineering.* Beijing: Science Press, 1997.